CW00810054

Understanding FACTS

Technical Reviewers

Seth Hutchinson, *University of Illinois, Urbana-Champaign*
Dagmar Niebur, *Drexel University*
Yahia Baghzouz, *University of Nevada, Las Vegas*

Understanding FACTS

Concepts and Technology of Flexible AC Transmission Systems

Narain G. Hingorani
Hingorani Power Electronics
Los Altos Hills, CA

Laszlo Gyugyi
Siemens Power Transmission & Distribution
Orlando, FL

Mohamed E. El-Hawary, *Consulting Editor*

IEEE Power Engineering Society, *Sponsor*

The Institute of Electrical and Electronics Engineers, Inc., New York

This book and other books may be purchased at a discount
from the publisher when ordered in bulk quantities. Contact:

IEEE Press Marketing
Attn: Special Sales
445 Hoes Lane
P.O. Box 1331
Piscataway, NJ 08855-1331
Fax: +1 732 981 9334

For more information about IEEE Press products,
visit the IEEE Press Home Page: http://www.ieee.org/press

Printed in the United States of America

10 9 8 7 6 5 4 3 2 1

ISBN 0-7803-3455-8
IEEE Order No. PC5713

Library of Congress Cataloging-in-Publication Data

Hingorani, Narain G.
Understanding FACTS : concepts and technology of flexible AC
transmission systems / Narain G. Hingorani, Laszlo Gyugyi.
p. cm.
Includes bibliographical references and index.
ISBN 0-7803-3455-8
1. Flexible AC transmission systems. I. Gyugyi, Laszlo.
II. Title.
TK3148.H54 1999
621.319′13—dc21 99-29340
CIP

This book is dedicated to all the engineers
who have participated in the pioneering development
of FACTS technology.

CONTENTS

Preface

Both authors of this book, Hingorani and Gyugyi, have been deeply involved in pioneering work in this new technology of Flexible AC Transmission System (FACTS). Hingorani pioneered the concept and managed a large R&D effort from EPRI, and Gyugyi invented and pioneered several key FACTS Controllers while leading a development team at Westinghouse. In fact, both have been involved in pioneering advances in many other applications of power electronics.

FACTS is one aspect of the power electronics revolution that is taking place in all areas of electric energy. A variety of powerful semiconductor devices not only offer the advantage of high speed and reliability of switching but, more importantly, the opportunity offered by a variety of innovative circuit concepts based on these power devices enhance the value of electric energy. This introduction is partly devoted to briefly conveying this perspective before discussing various specifics of Flexible AC Transmission, the subject matter of this book. After all, technologies from the transistor to microelectronics have revolutionized many aspects of our lives; there is no reason why power devices shouldn't have a significant impact on our lives as well, at least where energy is concerned. The power electronics revolution is happening, and applications of power electronics will continue to expand.

In the generation area, the potential application of power electronics is largely in renewable generation. Photo voltaic generation and fuel cells require conversion of dc to ac. Generation with variable speed is necessary for the economic viability of wind and small hydrogenerators. Variable-speed wind generators and small hydrogenerators require conversion of variable frequency ac to power system frequency. These applications of power electronics in the renewable generation area generally require converter sizes in the range of a few kilowatts to a few megawatts. Continuing breakthroughs will determine if these technologies will make a significant impact on electric power generation. In any case, they serve the vital needs of small, isolated loads where taking utility wires would be more expensive. In thermal power plants, considerable energy could be saved with the use of variable speed drives for pumps and compressors.

In the coming decades, electrical energy storage is expected to be widely used in power systems as capacitor, battery, and superconducting magnet technologies move forward. Batteries are widely used already for emergency power supplies. These require ac/dc/ac converters in the range of a few kilowatts to a few tens of megawatts. On the other hand, variable speed hydrostorage requires converters of up to a few hundred megawatts.

In the distribution area, an exciting opportunity called *Custom Power* enables at-the-fence solutions for delivery to industrial and commercial customers, value-added reliable electric service (which is free from significant voltage reductions) distortions, and over-voltages. It is now well known that voltage reductions of greater than 15 or 20% and of duration greater than a few cycles (resulting from lightning faults and switching events on the transmission and distribution system) lead to significant losses for the increasingly automated processing and manufacturing industry. The Custom Power concept incorporates power electronics Controllers and switching equipment, one or more of which can be used to provide a value-added service to the customers. In general, these Custom Power applications represent power electronics in the range of a few tens of kilowatts to a few tens of megawatts of conversion or switching equipment between the utility supply and the customer.

In the transmission area, application of power electronics consists of High-Voltage Direct Current (HVDC) power transmission and FACTS. HVDC, a well-established technology, is often an economical way to interconnect certain power systems, which are situated in different regions separated by long distances (over 50 km submarine or 1000 km overhead line), or those which have different frequencies or incompatible frequency control. HVDC involves conversion of ac to dc at one end and conversion of dc to ac at the other end. In general, HVDC represents conversion equipment sizes in the range of a hundred megawatts to a few thousand megawatts. Worldwide, more than 50 projects have been completed for a total transmission capacity of about 50,000 MW (100,000 MW conversion capacity) at voltages up to ±600 kV. For remote, modest loads of a few to ten MW, breakeven distance for HVDC may be as low as 100 km.

In general, FACTS—the subject matter of this book and a relatively new technology—has the principal role to enhance controllability and power transfer capability in ac systems. FACTS involves conversion and/or switching power electronics in the range of a few tens to a few hundred megawatts.

On the end-use side, power electronics conversion and switching technology has been a fast-growing area for over two decades for a wide range of needs. The fact is that electricity is an incredible form of energy, which can be converted to many different forms to bring about new and enabling technologies of high value. Conversion to pulses and electromagnetic waves has given us computers and communications. Conversion to microwave has led to microwave ovens, industrial processes, and radar. In arc form, electricity serves its high value in arc furnaces, welding, and so on. Efficient lighting, lasers, visuals, sound, robots, medical tools, and of course, variable speed drives and the expanding need for dc power supplies are among the many other examples. Complementing the Custom Power technology is the whole area of power conditioning technology used by customers, under the term *Power Quality*. Uninterruptible power supplies (UPS) and voltage regulators represent a major growth area in power electronics. In end use, the converter sizes range from a few watts to tens of megawatts.

Considering the opportunities in power electronics through reduction in cost, size and losses, we are in an early stage of the power electronic revolution, and there is a bright future ahead for those who are involved. Potentially, there is a significant commonality and synergism between the different areas of applications in generation, transmission, distribution, and end use. FACTS technology, being new, has a lot to borrow from the power electronics conversion, switching, and control ideas in other areas. Also there is considerable overlap in the megawatt size, and hence there is potential use of standard components and subassemblies among many applications noted above and new ones in the future. Therefore, it is suggested that those individuals involved in power electronics not confine their interest to one narrow application area.

In this book the term "FACTS Controller" or just "Controller" with capital C, is used to generally characterize the various power electronic circuit topologies or equipment that perform a certain function such as current control, power control, and so on. In many papers and articles, the term "FACTS device" is used. Since the Power Semiconductor device is also referred to as a "device", the authors have chosen to use the term "Controller". The reason for using capital C is to distinguish Controllers from the controllers used for industrial controls. Besides, the word "device" sounds like a component, and the authors request the readers to use the word Controller for FACTS Controllers.

The authors' intent in writing this book on FACTS is to provide useful information for the application engineers rather than for a detailed post-graduate college course. Therefore, there is an emphasis on physical explanations of the principles involved, and not on the mathematically supported theory of the many design aspects of the equipment. Nevertheless, post-graduate students will also greatly benefit from this book before they launch into the theoretical aspect of their research. This book will help post-graduate students acquire a broad understanding of the subject and a practical perspective enabling them to use their talents on real problems that need solutions.

The book does not go into the details of transmission design and system analysis, on which there are already several good published books.

Chapter 1: "FACTS Concept and General Considerations" explains all about FACTS to those involved in corporate planning and management. Engineers who wish to acquire sufficient knowledge to sort out various options, participate in equipment specifications, and become involved with detailed engineering and design will find significant value in reading the entire book in preparation for more lifelong learning in this area.

Chapter 2: "Power Semiconductor Devices" is a complex subject, and the subject matter of many books. In this book, sufficient material is provided for the FACTS application engineer for knowing those options.

Those familiar with the subject of HVDC know that practically all the HVDC projects are based on use of thyristors with no gate turn-off capability, assembled into 12-pulse converters, which can be controlled to function as a voltage-controlled rectifier (ac to dc) or as inverter (dc to ac). The voltage can be controlled from maximum positive to maximum negative, with the current flowing in the same direction; that is, power flow reverses with reversal of voltage and unidirectional current. Such converters, known as *current-sourced converters,* are clearly more economical for large HVDC projects, but are also useful in FACTS technology. Current-sourced converters based on thyristors with no gate turn-off capability only consume but cannot supply reactive power, whereas the *voltage-sourced converters* with gate turn-off thyristors can supply reactive power.

The most dominant converters needed in FACTS Controllers are the voltage-sourced converters. Such converters are based on devices with gate turn-off capability. In such unidirectional-voltage converters, the power reversal involves reversal of current and not the voltage. The voltage-sourced converters are described in Chapter 3 and the current-sourced converters in Chapter 4.

Chapters 5 to 9 are specific chapters on the main FACTS Controllers. There are a wide variety of FACTS Controllers, and they have overlapping and competing attributes in enhancing the controllability and transfer capability of transmission. The best choice of a Controller for a given need is the function of the benefit-to-cost consideration.

Chapter 5 describes various shunt Controllers, essentially for injecting reactive power in the transmission system, the Static VAR Compensator (SVC) based on conventional thyristors, and the Static Compensator (STATCOM) based on gate turn-off (GTO) thyristors.

Chapter 6 describes various series Controllers essentially for control of the transmission line current, mainly the Thyristor-Controlled Series Capacitor (TCSC) and the Static Synchronous Series Compensator (SSSC).

Chapter 7 describes various static voltage and phase angle regulators, which are a form of series Controllers, mainly the Thyristor-Controlled Voltage Regulator (TCVR) and Thyristor-Controlled Phase Angle Regulator (TCPAR).

Chapter 8 describes the combined series and shunt controllers, which are in a way the ultimate controllers that can control the voltage, the active power flow, and the reactive power flow. These include the Unified Power Flow Controller (UPFC) and the Interline Power Flow Controller (IPFC).

Chapter 9 describes two of the special-purpose Controllers, the NGH SSR Damping Controller and the Thyristor-Controlled Braking Resistor (TCBR).

Chapter 10 describes some FACTS applications in operation in the United States, including WAPA Kayenta TCSC, BPA Slatt TCSC, TVA STATCOM, and AEP Inez UPFC.

There already is a large volume of published literature. At the end of each chapter, authors have listed those references that represent the basis for the material in that chapter, as well as a few other references that are directly relevant to that chapter.

Narain G. Hingorani
Hingorani Power Electronics
Los Altos Hills, CA

Laszlo Gyugyi
Siemens Power Transmission & Distribution
Orlando, FL

Acknowledgments

First and foremost, both authors acknowledge EPRI and its members, for providing an organization and support that enabled Dr. Hingorani to conceptualize, fund, and manage an R&D strategy with the necessary resources. This support also was vital for Dr. Gyugyi to conceptualize the converter-based approach and direct the development of several key FACTS Controllers at Westinghouse under the sponsorship of EPRI and its members.

Dr. Hingorani acknowledges several members of existing and past EPRI staff, including Stig Nilsson, Dr. Neal Balu, Ben Damsky, the late Dr. Gil Addis, Dr. Ram Adapa, Dr. Aty Edris, Dr. Harshad Mehta, and Dominic Maratukulam for competent management of many projects funded with various companies and universities. The authors further acknowledge the continued support of FACTS technology advancement by the current management of EPRI's Power Delivery Group, including Dr. Karl Stahlkopf, Mark Wilhelm, and Dr. Robert Schainker, with special thanks to Dr. Aty Edris who has been an active participant in the more recent FACTS projects.

Dr. Gyugyi would like to acknowledge his past and present colleagues at the Westinghouse Science & Technology Center, who significantly contributed to the FACTS and related power electronics technology development. In the 1960s and 1970s, John Rosa, Brian R. Pelly, and the late Peter Wood were part of the early development efforts on circuit concepts, along with Eric Stacey who joined in these development efforts. Subsequently, Michael Brennen, Frank Cibulka, Mark Gernhardt, Ronald Pape, and Miklos Sarkozi were the core members of the technical team that developed the Static Var Compensator. Special acknowledgment is due to the present team that developed the converter-based FACTS Controller technology and whose work provided the basis for part of this book. In particular, Dr. Colin Schauder, whose conceptual and lead-design work were instrumental in the practical realization of the high-power converter-based Controllers, and whose publications provided important contributions to this book; Eric Stacey, whose participation generated many novel ideas and practical designs for power converter circuits; Gary L. Rieger, who effectively coordinated several joint development programs, and also read the manuscript of Chapters 5 through 8 and made suggestions to improve the text; Thomas

Lemak and Leonard Kovalsky, who competently managed the major FACTS projects; George Bettencourt, Don Carrera, Yu-Fu Lin, Mack Lund, Ronald Pape, Donald Ramey, Dr. Kalyan Sen, Matthew Weaver, and others who contributed to the details of these projects. A sincere, personal gratitude is expressed to Miklos Sarkozi, who constructed many of the illustrations used in this book. Special thanks are also due to Dr. Kalyan Sen, who performed some computations and simulations for the book.

Dr. Gyugyi wishes also to express his thanks to the executives of the Westinghouse Electric Corporation, who supported and funded the FACTS technology development, and to Siemens Power Transmission and Distribution who, having acquired Westinghouse FACTS and Custom Power business, continue to embrace the technology and pursue its application to utility systems. A particular debt of gratitude is extended to John P. Kessinger, General Manager, Siemens FACTS & Power Quality Division for his encouragement and support of this book.

Dr. Gyugyi acknowledges the Institution of Electrical Engineers whose kind permission allowed some of the material he earlier provided primarily on the Unified Power Flow Controller for the planned IEE book, *Flexible AC Transmission Systems Technology: Power Electronics Applications in Power Systems* be utilized herein.

Special credit is due to the pioneering utilities who are at the forefront of exploiting advanced technologies and maintaining high-level technical and management expertise to undertake first-of-a-kind projects. Principals among these are Jacob Sabath and Bharat Bhargava of Southern California Edison in collaboration with EPRI and Siemens for the NGH SSR Damping Scheme; Charles Clarke, Mark Reynolds, Bill Mittelstadt, and Wayne Litzenberger of Bonneville Power Administration in collaboration with EPRI and GE for the 500 kV modular Thyristor-Controlled Series Capacitor; William Clagett, Thomas Weaver, and Duane Torgerson of Western Area Power Administration in collaboration with Siemens for the Advanced Series Compensator; Terry Boston, Dale Bradshaw, T. W. Cease, Loring Rogers, and others of Tennessee Valley Authority in collaboration with EPRI and Westinghouse (now Siemens) for the Static Compensator; Bruce Renz, Ray Maliszewski, Ben Mehraban, Manny Rahman, Albert Keri, and others of American Electric Power in collaboration with EPRI and Westinghouse (now Siemens) for the Unified Power Flow Controller; and Philip Pellegrino, Shalom Zelingher, Bruce Fardanesh, Ben Shperling, Michael Henderson, and others of New York Power Authority in collaboration with EPRI and Siemens for the Convertible Static Compensator/Interline Power Flow Controller.

The authors recognize the pioneering role of General Electric in introduction of Static VAR Compensation in the utility industry. Recognition is also due to Donald McGillis of Hydro-Québec, Canada, and Dr. Arslan Erinmez, of the National Grid Company, England, and their colleagues, and for their pioneering work in the large scale application of Static Var Compensators. They played a significant role in making those two utilities the largest users of SVCs, each with over a dozen installations.

The authors appreciate the professional organizations—IEEE Power Engineering Society, IEE, and CIGRE—as the source of invaluable learning and information made available through publications, special reports, Working Groups, and paper and panel sessions. Indeed IEEE Power Engineering Society executives, including Ted Hissey and Wally Behnki, were quick to recognize new opportunities of FACTS technology and took initiatives to jump start Power Engineering Society activities through special sessions and Working Group activities through the T&D Committees. Under the chairmanship of Dr. Hingorani, and now Dr. Dusan Povh, FACTS became a thriving activity alongside HVDC in Study Committee 14 of CIGRE. In the Power Engineering

Society, the authors are thankful to FACTS and HVDC Subcommittee and Power Electronics Equipment Subcommittee under the Chairmanships of Dr. John Reeve, Stig Nilson, Einar Larsen, Richard Piwko, Wayne Litzenberger, Kara Clark, Peter Lips, Gerhard Juette, and others for spearheading FACTS agenda in their meetings. In addition, the authors acknowledge Dr. Arslan Erinmez, Dr. Pierre-Guy Therond, Dr. Adel Hammad, Dennis Woodford, and Michael Baker for generating important source material to the author's knowledge base. Special acknowledgment is due to Professor Willis Long for orchestrating material and a diverse faculty for an excellent course on FACTS for professional development at the University of Wisconsin. Dr. Hingorani further acknowledges Dr. Vic Temple for reviewing the chapter on power semiconductor devices, and his son Naren for his editorial help. It would make a long list for the authors to acknowledge their professional colleagues working worldwide, who are among the leading innovators and contributors to the transmission and power electronics technologies.

Narain G. Hingorani
Hingorani Power Electronics
Los Altos Hills, CA

Laszlo Gyugyi
Siemens Power Transmission & Distribution
Orlando, FL

1

FACTS Concept and General System Considerations

1.1 TRANSMISSION INTERCONNECTIONS

Most if not all of the world's electric power supply systems are widely interconnected, involving connections inside utilities' own territories which extend to inter-utility interconnections and then to inter-regional and international connections. This is done for economic reasons, to reduce the cost of electricity and to improve reliability of power supply.

1.1.1 Why We Need Transmission Interconnections

We need these interconnections because, apart from delivery, the purpose of the transmission network is to pool power plants and load centers in order to minimize the total power generation capacity and fuel cost. Transmission interconnections enable taking advantage of diversity of loads, availability of sources, and fuel price in order to supply electricity to the loads at minimum cost with a required reliability. In general, if a power delivery system was made up of radial lines from individual local generators without being part of a grid system, many more generation resources would be needed to serve the load with the same reliability, and the cost of electricity would be much higher. With that perspective, transmission is often an alternative to a new generation resource. Less transmission capability means that more generation resources would be required regardless of whether the system is made up of large or small power plants. In fact small distributed generation becomes more economically viable if there is a backbone of a transmission grid. One cannot be really sure about what the optimum balance is between generation and transmission unless the system planners use advanced methods of analysis which integrate transmission planning into an integrated value-based transmission/generation planning scenario. The cost of transmission lines and losses, as well as difficulties encountered in building new transmission lines, would often limit the available transmission capacity. It seems that there are many cases where economic energy or reserve sharing is constrained by transmission capacity, and the situation is not getting any better. In a deregulated electric service environment, an effective electric grid is vital to the competitive environment of reliable electric service.

On the other hand, as power transfers grow, the power system becomes increasingly more complex to operate and the system can become less secure for riding through the major outages. It may lead to large power flows with inadequate control, excessive reactive power in various parts of the system, large dynamic swings between different parts of the system and bottlenecks, and thus the full potential of transmission interconnections cannot be utilized.

The power systems of today, by and large, are mechanically controlled. There is a widespread use of microelectronics, computers and high-speed communications for control and protection of present transmission systems; however, when operating signals are sent to the power circuits, where the final power control action is taken, the switching devices are mechanical and there is little high-speed control. Another problem with mechanical devices is that control cannot be initiated frequently, because these mechanical devices tend to wear out very quickly compared to static devices. In effect, from the point of view of both dynamic and steady-state operation, the system is really uncontrolled. Power system planners, operators, and engineers have learned to live with this limitation by using a variety of ingenious techniques to make the system work effectively, but at a price of providing greater operating margins and redundancies. These represent an asset that can be effectively utilized with prudent use of FACTS technology on a selective, as needed basis.

In recent years, greater demands have been placed on the transmission network, and these demands will continue to increase because of the increasing number of nonutility generators and heightened competition among utilities themselves. Added to this is the problem that it is very difficult to acquire new rights of way. Increased demands on transmission, absence of long-term planning, and the need to provide open access to generating companies and customers, all together have created tendencies toward less security and reduced quality of supply. The FACTS technology is essential to alleviate some but not all of these difficulties by enabling utilities to get the most service from their transmission facilities and enhance grid reliability. It must be stressed, however, that for many of the capacity expansion needs, building of new lines or upgrading current and voltage capability of existing lines and corridors will be necessary.

1.1.2 Opportunities for FACTS

What is most interesting for transmission planners is that FACTS technology opens up new opportunities for controlling power and enhancing the usable capacity of present, as well as new and upgraded, lines. The possibility that current through a line can be controlled at a reasonable cost enables a large potential of increasing the capacity of existing lines with larger conductors, and use of one of the FACTS Controllers to enable corresponding power to flow through such lines under normal and contingency conditions.

These opportunities arise through the ability of FACTS Controllers to control the interrelated parameters that govern the operation of transmission systems including series impedance, shunt impedance, current, voltage, phase angle, and the damping of oscillations at various frequencies below the rated frequency. These constraints cannot be overcome, while maintaining the required system reliability, by mechanical means without lowering the useable transmission capacity. By providing added flexibility, FACTS Controllers can enable a line to carry power closer to its thermal rating. Mechanical switching needs to be supplemented by rapid-response power electronics. It must be emphasized that FACTS is an enabling technology, and not a one-on-one substitute for mechanical switches.

The FACTS technology is not a single high-power Controller, but rather a collection of Controllers, which can be applied individually or in coordination with others to control one or more of the interrelated system parameters mentioned above. A well-chosen FACTS Controller can overcome the specific limitations of a designated transmission line or a corridor. Because all FACTS Controllers represent applications of the same basic technology, their production can eventually take advantage of technologies of scale. Just as the transistor is the basic element for a whole variety of microelectronic chips and circuits, the thyristor or high-power transistor is the basic element for a variety of high-power electronic Controllers.

FACTS technology also lends itself to extending usable transmission limits in a step-by-step manner with incremental investment as and when required. A planner could foresee a progressive scenario of mechanical switching means and enabling FACTS Controllers such that the transmission lines will involve a combination of mechanical and FACTS Controllers to achieve the objective in an appropriate, staged investment scenario.

Some of the Power Electronics Controllers, now folded into the FACTS concept predate the introduction of the FACTS concept by co-author Hingorani to the technical community. Notable among these is the shunt-connected Static VAR Compensator (SVC) for voltage control which was first demonstrated in Nebraska and commercialized by GE in 1974 and by Westinghouse in Minnesota in 1975. The first series-connected Controller, NGH-SSR Damping Scheme, invented by co-author Hingorani, a low power series capacitor impedance control scheme, was demonstrated in California by Siemens in 1984. It showed that with an active Controller there is no limit to series capacitor compensation. Even prior to SVCs, there were two versions of static saturable reactors for limiting overvoltages and also powerful gapless metal oxide arresters for limiting dynamic overvoltages. Research had also been undertaken on solid-state tap changers and phase shifters. However, the unique aspect of FACTS technology is that this umbrella concept revealed the large potential opportunity for power electronics technology to greatly enhance the value of power systems, and thereby unleashed an array of new and advanced ideas to make it a reality. Co-author Gyugyi has been at the forefront of such advanced ideas. FACTS technology has also provided an impetus and excitement perceived by the younger generation of engineers, who will rethink and re-engineer the future power systems throughout the world.

It is also worth pointing out that, in the implementation of FACTS technology, we are dealing with a base technology, proven through HVDC and high-power industrial drives. Nevertheless, as power semiconductor devices continue to improve, particularly the devices with turn-off capability, and as FACTS Controller concepts advance, the cost of FACTS Controllers will continue to decrease. Large-scale use of FACTS technology is an assured scenario.

1.2 FLOW OF POWER IN AN AC SYSTEM

At present, many transmission facilities confront one or more limiting network parameters plus the inability to direct power flow at will.

In ac power systems, given the insignificant electrical storage, the electrical generation and load must balance at all times. To some extent, the electrical system is self-regulating. If generation is less than load, the voltage and frequency drop, and thereby the load, goes down to equal the generation minus the transmission losses. However, there is only a few percent margin for such a self-regulation. If voltage is propped up with reactive power support, then the load will go up, and consequently frequency will keep dropping, and the system will collapse. Alternately, if there is inadequate reactive power, the system can have voltage collapse.

When adequate generation is available, active power flows from the surplus generation areas to the deficit areas, and it flows through all parallel paths available which frequently involves extra high-voltage and medium-voltage lines. Often, long distances are involved with loads and generators along the way. An often cited example is that much of the power scheduled from Ontario Hydro Canada to the North East United States flows via the PJM system over a long loop, because of the presence of a large number of powerful low impedance lines along that loop. There are in fact some major and a large number of minor loop flows and uneven power flows in any power transmission system.

1.2.1 Power Flow in Parallel Paths

Consider a very simple case of power flow [Figure 1.1(a)], through two parallel paths (possibly corridors of several lines) from a surplus generation area, shown as an equivalent generator on the left, to a deficit generation area on the right. Without any control, power flow is based on the inverse of the various transmission line impedances. Apart from ownership and contractual issues over which lines carry how much power, it is likely that the lower impedance line may become overloaded and thereby limit the loading on both paths even though the higher impedance path is not fully loaded. There would not be an incentive to upgrade current capacity of the overloaded path, because this would further decrease the impedance and the investment would be self-defeating particularly if the higher impedance path already has enough capacity.

Figure 1.1(b) shows the same two paths, but one of these has HVDC transmission. With HVDC, power flows as ordered by the operator, because with HVDC power electronics converters power is electronically controlled. Also, because power is electronically controlled, the HVDC line can be used to its full thermal capacity if adequate converter capacity is provided. Furthermore, an HVDC line, because of its high-speed control, can also help the parallel ac transmission line to maintain stability. However, HVDC is expensive for general use, and is usually considered when long distances are involved, such as the Pacific DC Intertie on which power flows as ordered by the operator.

As alternative FACTS Controllers, Figures 1.1(c) and 1.1(d) show one of the transmission lines with different types of series type FACTS Controllers. By means of controlling impedance [Figure 1.1(c)] or phase angle [Figure 1.1(d)], or series injection of appropriate voltage (not shown) a FACTS Controller can control the power flow as required. Maximum power flow can in fact be limited to its rated limit under contingency conditions when this line is expected to carry more power due to the loss of a parallel line.

1.2.2 Power Flow in a Meshed System

To further understand the free flow of power, consider a very simplified case in which generators at two different sites are sending power to a load center through a network consisting of three lines in a meshed connection (Figure 1.2). Suppose the lines AB, BC, and AC have continuous ratings of 1000 MW, 1250 MW, and 2000 MW, respectively, and have emergency ratings of twice those numbers for a sufficient

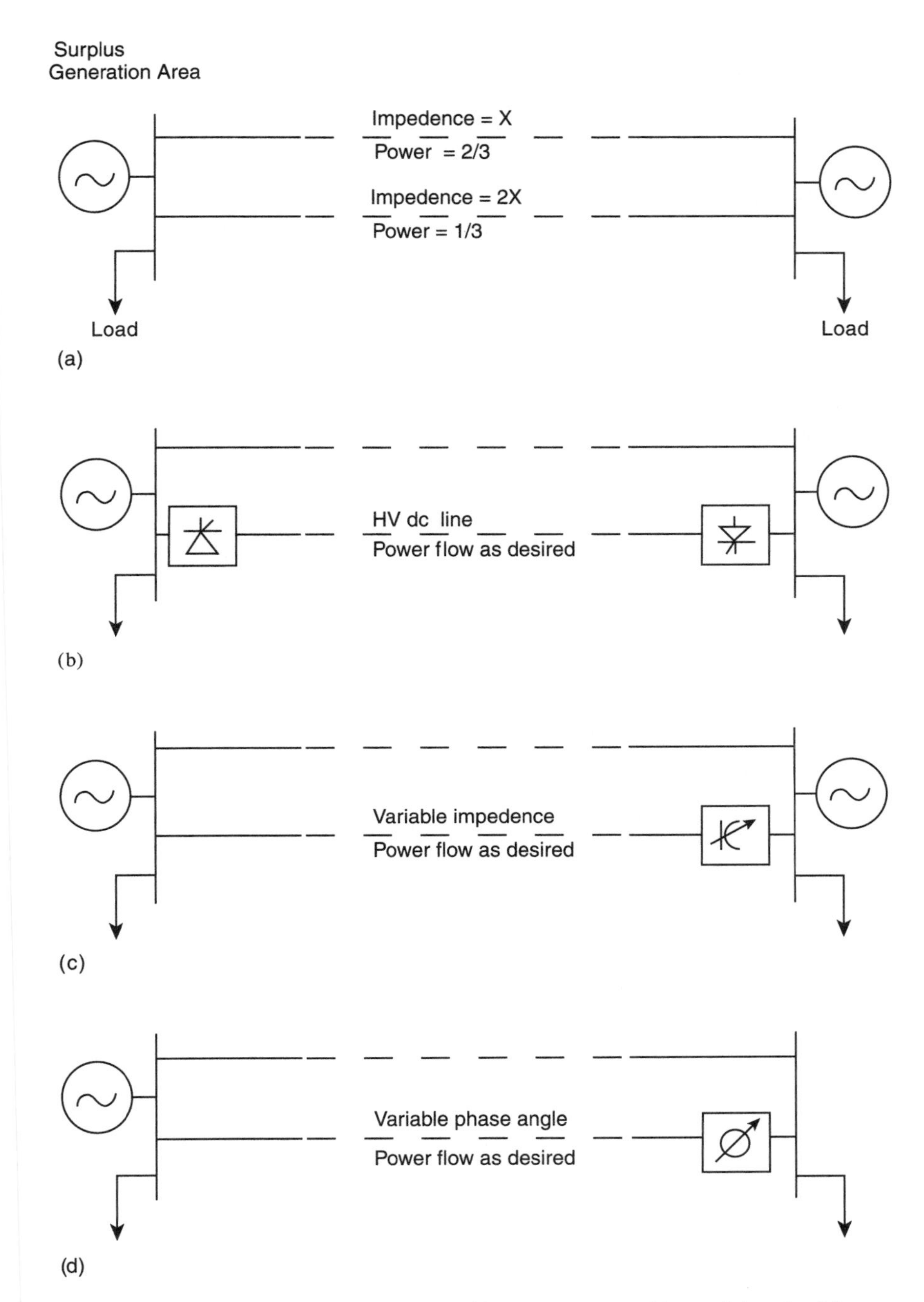

Figure 1.1 Power flow in parallel paths: (a) ac power flow with parallel paths; (b) power flow control with HVDC; (c) power flow control with variable impedance; (d) power flow control with variable phase angle.

length of time to allow rescheduling of power in case of loss of one of these lines. If one of the generators is generating 2000 MW and the other 1000 MW, a total of 3000 MW would be delivered to the load center. For the impedances shown, the three lines would carry 600, 1600, and 1400 MW, respectively, as shown in Figure 1.2(a). Such a situation would overload line BC (loaded at 1600 MW for its continuous rating of 1250 MW), and therefore generation would have to be decreased at B, and increased at A, in order to meet the load without overloading line BC.

Power, in short, flows in accordance with transmission line series impedances (which are 90% inductive) that bear no direct relationship to transmission ownership, contracts, thermal limits, or transmission losses.

If, however, a capacitor whose reactance is -5 ohms (Ω) at the synchronous frequency is inserted in one line [Figure 1.2(b)], it reduces the line's impedance from 10 Ω to 5 Ω, so that power flow through the lines AB, BC, and AC will be 250, 1250, and 1750 MW, respectively. It is clear that if the series capacitor is adjustable, then other power-flow levels may be realized in accordance with the ownership, contract, thermal limitations, transmission losses, and a wide range of load and generation schedules. Although this capacitor could be modular and mechanically switched, the number of operations would be severely limited by wear on the mechanical components because the line loads vary continuously with load conditions, generation schedules, and line outages.

Other complications may arise if the series capacitor is mechanically controlled. A series capacitor in a line may lead to subsynchronous resonance (typically at 10–50 Hz for a 60 Hz system). This resonance occurs when one of the mechanical resonance frequencies of the shaft of a multiple-turbine generator unit coincides with 60 Hz

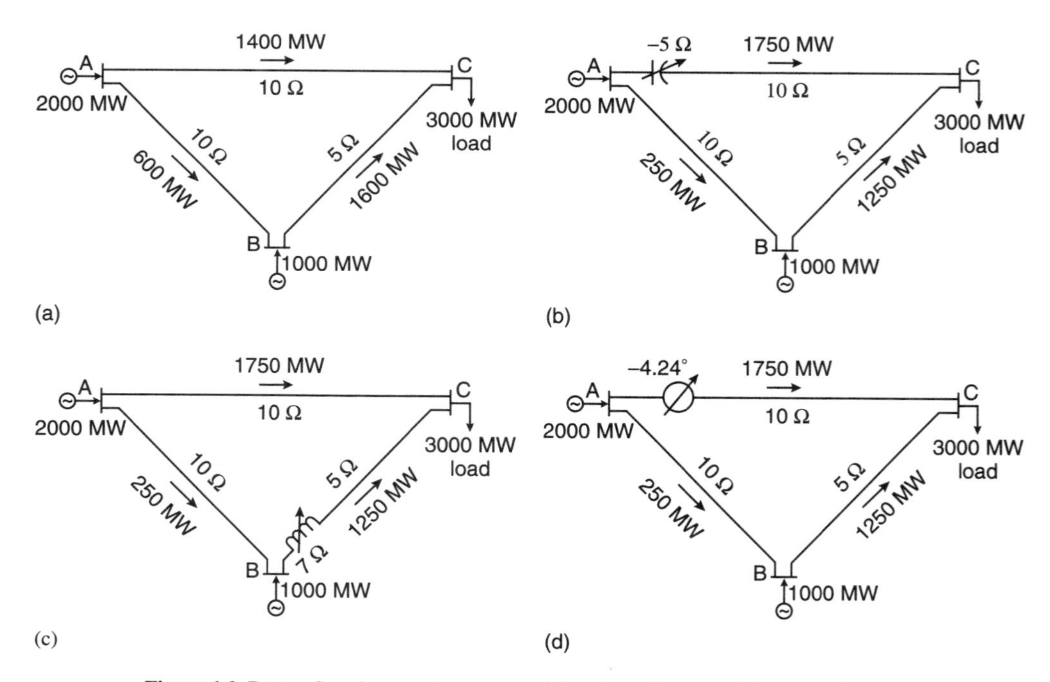

Figure 1.2 Power flow in a mesh network: (a) system diagram; (b) system diagram with Thyristor-Controlled Series Capacitor in line AC; (c) system diagram with Thyristor-Controlled Series Reactor in line BC; (d) system diagram with Thyristor-Controlled Phase Angle Regulator in line AC.

minus the electrical resonance frequency of the capacitor with the inductive impedance of the line. If such resonance persists, it will soon damage the shaft. Also while the outage of one line forces other lines to operate at their emergency ratings and carry higher loads, power flow oscillations at low frequency (typically 0.3–3 Hz) may cause generators to lose synchronism, perhaps prompting the system's collapse.

If all or a part of the series capacitor is thyristor-controlled, however, it can be varied as often as required. It can be modulated to rapidly damp any subsynchronous resonance conditions, as well as damp low frequency oscillations in the power flow. This would allow the transmission system to go from one steady-state condition to another without the risk of damage to a generator shaft and also help reduce the risk of system collapse. In other words, a thyristor-controlled series capacitor can greatly enhance the stability of the network. More often than not though, it is practical for part of the series compensation to be mechanically controlled and part thyristor controlled, so as to counter the system constraints at the least cost.

Similar results may be obtained by increasing the impedance of one of the lines in the same meshed configuration by inserting a 7 Ω reactor (inductor) in series with line AB [Figure 1.2(c)]. Again, a series inductor that is partly mechanically and partly thyristor-controlled, it could serve to adjust the steady-state power flows as well as damp unwanted oscillations.

As another option, a thyristor-controlled phase-angle regulator could be installed instead of a series capacitor or a series reactor in any of the three lines to serve the same purpose. In Figure 1.2(d), the regulator is installed in the third line to reduce the total phase-angle difference along the line from 8.5 degrees to 4.26 degrees. As before, a combination of mechanical and thyristor control of the phase-angle regulator may minimize cost.

The same results could also be achieved by injecting a variable voltage in one of the lines. Note that balancing of power flow in the above case did not require more than one FACTS Controller, and indeed there are options of different controllers and in different lines.

If there is only one owner of the transmission grid, then a decision can be made on consideration of overall economics alone. On the other hand, if multiple owners are involved, then a decision mechanism is necessary on the investment and ownership.

1.3 WHAT LIMITS THE LOADING CAPABILITY?

Assuming that ownership is not an issue, and the objective is to make the best use of the transmission asset, and to maximize the loading capability (taking into account contingency conditions), what limits the loading capability, and what can be done about it?

Basically, there are three kinds of limitations:

- Thermal
- Dielectric
- Stability

Thermal Thermal capability of an overhead line is a function of the ambient temperature, wind conditions, condition of the conductor, and ground clearance. It varies perhaps by a factor of 2 to 1 due to the variable environment and the loading history. The nominal rating of a line is generally decided on a conservative basis, envisioning a statistically worst ambient environment case scenario. Yet this scenario occurs but rarely which means that in reality, most of the time, there is a lot more real time capacity than assumed. Some utilities assign winter and summer ratings, yet this still leaves a considerable margin to play with. There are also off-line computer programs that can calculate a line's loading capability based on available ambient environment and recent loading history. Then there are the on-line monitoring devices that provide a basis for on-line real-time loading capability. These methods have evolved over a period of many years, and, given the age of automation (typified by GPS systems and low-cost sophisticated communication services), it surely makes sense to consider reasonable, day to day, hour to hour, or even real-time capability information. Sometimes, the ambient conditions can actually be worse than assumed, and having the means to determine actual rating of the line could be useful.

During planning/design stages, normal loading of the lines is frequently decided on a loss evaluation basis under assumptions which may have changed for a variety of reasons; however losses can be taken into account on the real-time value basis of extra loading capability.

Of course, increasing the rating of a transmission circuit involves consideration of the real-time ratings of the transformers and other equipment as well, some of which may also have to be changed in order to increase the loading on the lines. Real-time loading capability of transformers is also a function of ambient temperature, aging of the transformer and recent loading history. Off-line and on-line loading capability monitors can also be used to obtain real time loading capability of transformers. Also, the transformer also lends itself to enhanced cooling.

Then there is the possibility of upgrading a line by changing the conductor to that of a higher current rating, which may in turn require structural upgrading. Finally, there is the possibility of converting a single-circuit to a double-circuit line. Once the higher current capability is available, then the question arises of how it should be used. Will the extra power actually flow and be controllable? Will the voltage conditions be acceptable with sudden load dropping, etc.? The FACTS technology can help in making an effective use of this newfound capacity.

Dielectric From an insulation point of view, many lines are designed very conservatively. For a given nominal voltage rating, it is often possible to increase normal operation by +10% voltage (i.e., 500 kV–550 kV) or even higher. Care is then needed to ensure that dynamic and transient overvoltages are within limits. Modern gapless arresters, or line insulators with internal gapless arresters, or powerful thyristor-controlled overvoltage suppressors at the substations can enable significant increase in the line and substation voltage capability. The FACTS technology could be used to ensure acceptable over-voltage and power flow conditions.

Stability There are a number of stability issues that limit the transmission capability. These include:

- Transient stability
- Dynamic stability
- Steady-state stability
- Frequency collapse
- Voltage collapse
- Subsynchronous resonance

Excellent books are available on this subject. Therefore, discussion on these topics in this book will be brief, and limited to what is really essential to the explanation of FACTS Controllers. The FACTS technology can certainly be used to overcome any of the stability limits, in which case the ultimate limits would be thermal and dielectric.

1.4 POWER FLOW AND DYNAMIC STABILITY CONSIDERATIONS OF A TRANSMISSION INTERCONNECTION

Figure 1.3(a) shows a simplified case of power flow on a transmission line. Locations 1 and 2 could be any transmission substations connected by a transmission line. Substations may have loads, generation, or may be interconnecting points on the system and for simplicity they are assumed to be stiff busses. E_1 and E_2 are the magnitudes of the bus voltages with an angle δ between the two. The line is assumed to have inductive impedance X, and the line resistance and capacitance are ignored.

As shown in the phasor diagram [Figure 1.3(b)] the driving voltage drop in the line is the phasor difference $\boldsymbol{E}_L$ between the two line voltage phasors, $\boldsymbol{E}_1$ and $\boldsymbol{E}_2$. The line current magnitude is given by:

$$I = E_L/X, \text{ and lags } E_L \text{ by } 90°$$

It is important to appreciate that for a typical line, angle δ and corresponding driving voltage, or voltage drop along the line, is small compared to the line voltages. Given that a transmission line may have a voltage drop at full load of perhaps 1%/10 km, and assuming that a line between two stiff busbars (substations) is 200 km long, the voltage drop along this line would be 20% at full load, and the angle δ would be small. If we were to assume, for example, that with equal magnitudes of E_1 and E_2, and X of 0.2 per unit magnitude, the angle δ would be only 0.2 radians or 11.5 degrees.

The current flow on the line can be controlled by controlling E_L or X or δ. In order to achieve a high degree of control on the current in this line, the equipment required in series with the line would not have a very high power rating. For example, a 500 kV (approximately 300 kV phase-ground), 2000 A line has a three-phase throughput power of 1800 MVA, and, for a 200 km length, it would have a voltage drop of about 60 kV. For variable series compensation of say, 25%, the series equipment required would have a nominal rating of 0.25×60 kV $\times$ 2000 A = 30 MVA per phase, or 90 MVA for three phases, which is only 5% of the throughput line rating of 1800 MVA. Voltage across the series equipment would only be 15 kV at full load, although it would require high-voltage insulation to ground (the latter is not a significant cost factor). However, any series-connected equipment has to be designed to carry contingency overloads so that the equipment may have to be rated to 100% overload capability.

Nevertheless the point of this very simple example is that generally speaking the rating of series FACTS Controllers would be a fraction of the throughput rating of a line.

Figure 1.3(b) shows that the current flow phasor is perpendicular to the driving voltage (90° phase lag). If the angle between the two bus voltages is small, the current flow largely represents the active power. Increasing or decreasing the inductive impedance of a line will greatly affect the active power flow. Thus impedance control, which in reality provides current control, can be the most cost-effective means of controlling the power flow. With appropriate control loops, it can be used for power flow control and/or angle control for stability.

Figure 1.3(c), corresponding to Figure 1.3(b), shows a phasor diagram of the

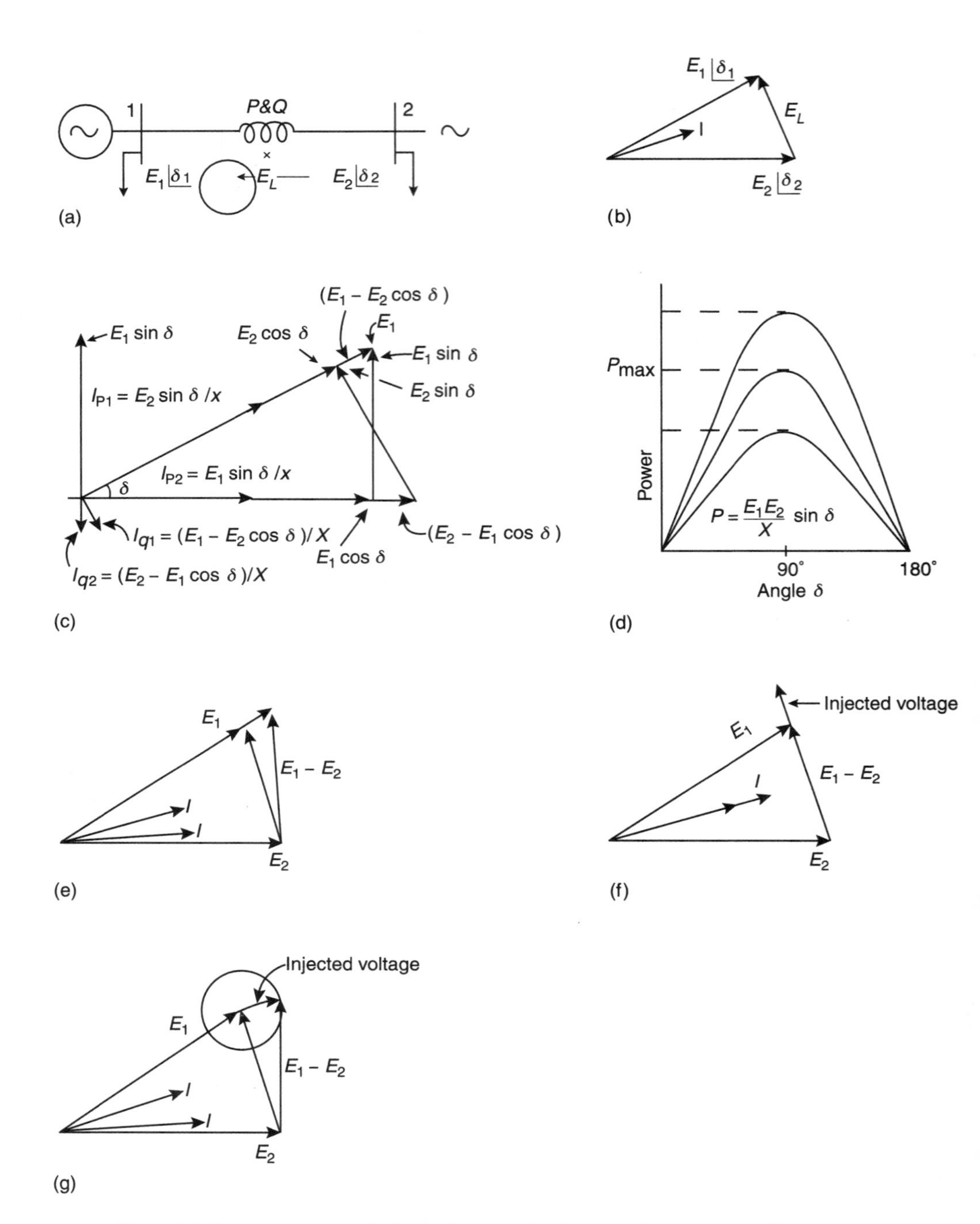

Figure 1.3 Ac power flow control of a transmission line: (a) simple two-machine system; (b) current flow perpendicular to the driving voltage; (c) active and reactive power flow phasor diagram; (d) power angle curves for different values of X; (e) regulating voltage magnitude mostly changes reactive power; (f) injecting voltage perpendicular to the line current mostly changes active power; (f) injecting voltage phasor in series with the line. (Note that for clarity the phasors are identified by their magnitudes in this figure.)

relationship between the active and reactive currents with reference to the voltages at the two ends.

Active component of the current flow at E_1 is:

$$I_{p1} = (E_2 \sin \delta)/X$$

Reactive component of the current flow at E_1 is:

$$I_{q1} = (E_1 - E_2 \cos \delta)/X$$

Thus, active power at the E_1 end:

$$P_1 = E_1 (E_2 \sin \delta)/X$$

Reactive power at the E_1 end:

$$Q_1 = E_1 (E_1 - E_2 \cos \delta)/X \tag{1.1}$$

Similarly, active component of the current flow at E_2 is:

$$I_{p2} = (E_1 \sin \delta)/X$$

Reactive component of the current flow at E_2 is:

$$I_{q2} = (E_2 - E_1 \cos \delta)/X$$

Thus, active power at the E_2 end:

$$P_2 = E_2 (E_1 \sin \delta)/X$$

Reactive power at the E_2 end:

$$Q_2 = E_2 (E_2 - E_1 \cos \delta)/X \tag{1.2}$$

Naturally P_1 and P_2 are the same:

$$P = E_1 (E_2 \sin \delta)/X \tag{1.3}$$

because it is assumed that there are no active power losses in the line. Thus, varying the value of X will vary P, Q_1, and Q_2 in accordance with (1.1), (1.2), and (1.3), respectively.

Assuming that E_1 and E_2 are the magnitudes of the internal voltages of the two equivalent machines representing the two systems, and the impedance X includes the internal impedance of the two equivalent machines, Figure 1.3(d) shows the half sine-wave curve of active power increasing to a peak with an increase in δ to 90 degrees. Power then falls with further increase in angle, and finally to zero at $\delta = 180°$. It is easy to appreciate that without high-speed control of any of the parameters E_1, E_2, $E_1 - E_2$, X and δ, the transmission line can be utilized only to a level well below that corresponding to 90 degrees. This is necessary, in order to maintain an adequate margin needed for transient and dynamic stability and to ensure that the system does not collapse following the outage of the largest generator and/or a line.

Increase and decrease of the value of X will increase and decrease the height of the curves, respectively, as shown in Figure 1.3(d). For a given power flow, varying of X will correspondingly vary the angle between the two ends.

Power/current flow can also be controlled by regulating the magnitude of voltage phasor $\boldsymbol{E}_1$ or voltage phasor $\boldsymbol{E}_2$. However, it is seen from Figure 1.3(e) that with change in the magnitude of $\boldsymbol{E}_1$, the magnitude of the driving voltage phasor $\boldsymbol{E}_1 - \boldsymbol{E}_2$ does not change by much, but its phase angle does. This also means that regulation of the magnitude of voltage phasor $\boldsymbol{E}_1$ and/or $\boldsymbol{E}_2$ has much more influence over the reactive power flow than the active power flow, as seen from the two current phasors corresponding to the two driving voltage phasors $\boldsymbol{E}_1 - \boldsymbol{E}_2$ shown in Figure 1.3(e).

Current flow and hence power flow can also be changed by injecting voltage in series with the line. It is seen from Figure 1.3(f) that when the injected voltage is in phase quadrature with the current (which is approximately in phase with the driving voltage, Figure 1.3(f), it directly influences the magnitude of the current flow, and with small angle influences substantially the active power flow.

Alternatively, the voltage injected in series can be a phasor with variable magnitude and phase relationship with the line voltage [Figure 1.3(g)]. It is seen that varying the amplitude and phase angle of the voltage injected in series, both the active and reactive current flow can be influenced. Voltage injection methods form the most important portfolio of the FACTS Controllers and will be discussed in detail in subsequent chapters.

1.5 RELATIVE IMPORTANCE OF CONTROLLABLE PARAMETERS

With reference to the above discussion and Figure 1.3, it is worth noting a few basic points regarding the possibilities of power flow control:

- Control of the line impedance X (e.g., with a thyristor-controlled series capacitor) can provide a powerful means of current control.
- When the angle is not large, which is often the case, control of X or the angle substantially provides the control of active power.
- Control of angle (with a Phase Angle Regulator, for example), which in turn controls the driving voltage, provides a powerful means of controlling the current flow and hence active power flow when the angle is not large.
- Injecting a voltage in series with the line, and perpendicular to the current flow, can increase or decrease the magnitude of current flow. Since the current flow lags the driving voltage by 90 degrees, this means injection of reactive power in series, (e.g., with static synchronous series compensation) can provide a powerful means of controlling the line current, and hence the active power when the angle is not large.
- Injecting voltage in series with the line and with any phase angle with respect to the driving voltage can control the magnitude and the phase of the line current. This means that injecting a voltage phasor with variable phase angle can provide a powerful means of precisely controlling the active and reactive power flow. This requires injection of both active and reactive power in series.
- Because the per unit line impedance is usually a small fraction of the line voltage, the MVA rating of a series Controller will often be a small fraction of the throughput line MVA.
- When the angle is not large, controlling the magnitude of one or the other line voltages (e.g., with a thyristor-controlled voltage regulator) can be a very cost-effective means for the control of reactive power flow through the interconnection.
- Combination of the line impedance control with a series Controller and voltage

regulation with a shunt Controller can also provide a cost-effective means to control both the active and reactive power flow between the two systems.

1.6 BASIC TYPES OF FACTS CONTROLLERS

In general, FACTS Controllers can be divided into four categories:

- Series Controllers
- Shunt Controllers
- Combined series-series Controllers
- Combined series-shunt Controllers

Figure 1.4(a) shows the general symbol for a FACTS Controller: a thyristor arrow inside a box.

Series Controllers: [Figure 1.4(b)] The series Controller could be a variable impedance, such as capacitor, reactor, etc., or a power electronics based variable source of main frequency, subsynchronous and harmonic frequencies (or a combination) to serve the desired need. In principle, all series Controllers inject voltage in series with the line. Even a variable impedance multiplied by the current flow through it, represents an injected series voltage in the line. As long as the voltage is in phase quadrature with the line current, the series Controller only supplies or consumes variable reactive power. Any other phase relationship will involve handling of real power as well.

Shunt Controllers: [Figure 1.4(c)] As in the case of series Controllers, the shunt Controllers may be variable impedance, variable source, or a combination of these. In principle, all shunt Controllers inject current into the system at the point of connection. Even a variable shunt impedance connected to the line voltage causes a variable current flow and hence represents injection of current into the line. As long as the injected current is in phase quadrature with the line voltage, the shunt Controller only supplies or consumes variable reactive power. Any other phase relationship will involve handling of real power as well.

Combined series-series Controllers: [Figure 1.4(d)] This could be a combination of separate series controllers, which are controlled in a coordinated manner, in a multiline transmission system. Or it could be a unified Controller, Figure 1.4(d), in which series Controllers provide independent series reactive compensation for each line but also transfer real power among the lines via the power link. The real power transfer capability of the unified series-series Controller, referred to as *Interline Power Flow Controller,* makes it possible to balance both the real and reactive power flow in the lines and thereby maximize the utilization of the transmission system. Note that the term "unified" here means that the dc terminals of all Controller converters are all connected together for real power transfer.

Combined series-shunt Controllers: [Figures 1.4(e) and 1.4(f)] This could be a combination of separate shunt and series Controllers, which are controlled in a coordinated manner [Figure 1.4(e)], or a *Unified Power Flow Controller* with series and shunt elements [Figure 1.4(f)]. In principle, combined shunt and series Controllers inject current into the system with the shunt part of the Controller and voltage in series in the line with the series part of the Controller. However, when the shunt and series Controllers are unified, there can be a real power exchange between the series and shunt Controllers via the power link.

1.6.1 Relative Importance of Different Types of Controllers

It is important to appreciate that the series-connected Controller impacts the driving voltage and hence the current and power flow directly. Therefore, if the purpose of the application is to control the current/power flow and damp oscillations, the series Controller for a given MVA size is several times more powerful than the shunt Controller.

As mentioned, the shunt Controller, on the other hand, is like a current source, which draws from or injects current into the line. The shunt Controller is therefore a good way to control voltage at and around the point of connection through injection of reactive current (leading or lagging), alone or a combination of active and reactive current for a more effective voltage control and damping of voltage oscillations.

This is not to say that the series Controller cannot be used to keep the line voltage within the specified range. After all, the voltage fluctuations are largely a consequence of the voltage drop in series impedances of lines, transformers, and generators. Therefore, adding or subtracting the FACTS Controller voltage in series (main frequency, subsynchronous or harmonic voltage and combination thereof) can be the most cost-effective way of improving the voltage profile. Nevertheless, a shunt controller is much more effective in maintaining a required voltage profile at a substation bus. One important advantage of the shunt Controller is that it serves the bus node independently of the individual lines connected to the bus.

Series Controller solution may require, but not necessarily, a separate series Controller for several lines connected to the substation, particularly if the application calls for contingency outage of any one line. However, this should not be a decisive reason for choosing a shunt-connected Controller, because the required MVA size of the series Controller is small compared to the shunt Controller, and, in any case, the shunt Controller does not provide control over the power flow in the lines.

On the other hand, series-connected Controllers have to be designed to ride through contingency and dynamic overloads, and ride through or bypass short circuit currents. They can be protected by metal-oxide arresters or temporarily bypassed by solid-state devices when the fault current is too high, but they have to be rated to handle dynamic and contingency overload.

The above arguments suggest that a combination of the series and shunt Controllers [Figures 1.4(e) and 1.4(f)] can provide the best of both, i.e., an effective power/current flow and line voltage control.

For the combination of series and shunt Controllers, the shunt Controller can be a single unit serving in coordination with individual line Controllers [Figure 1.4(g)]. This arrangement can provide additional benefits (reactive power flow control) with unified Controllers.

FACTS Controllers may be based on thyristor devices with no gate turn-off (only with gate turn-on), or with power devices with gate turn-off capability. Also, in general, as will be discussed in other chapters, the principal Controllers with gate turn-off devices are based on the dc to ac converters, which can exchange active and/or reactive power with the ac system. When the exchange involves reactive power only, they are provided with a minimal storage on the dc side. However, if the generated ac voltage or current is required to deviate from 90 degrees with respect to the line current or voltage, respectively, the converter dc storage can be augmented beyond the minimum required for the converter operation as a source of reactive power only. This can be done at the converter level to cater to short-term (a few tens of main

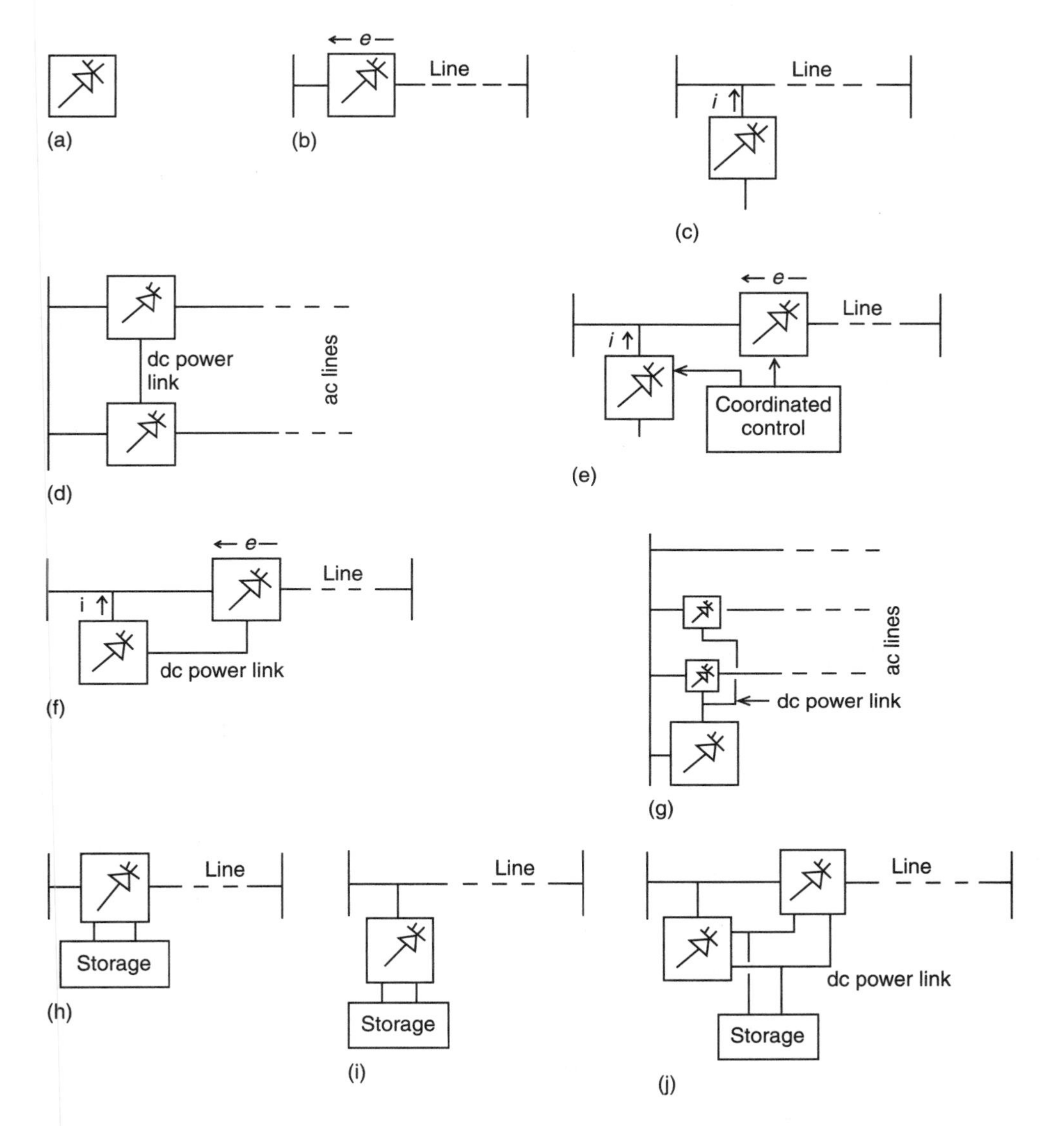

Figure 1.4 Basic types of FACTS Controllers: (a) general symbol for FACTS Controller; (b) series Controller; (c) shunt Controller; (d) unified series-series Controller; (e) coordinated series and shunt Controller; (f) unified series-shunt Controller; (g) unified Controller for multiple lines; (h) series Controller with storage; (i) shunt Controller with storage; (j) unified series-shunt Controller with storage.

frequency cycles) storage needs. In addition, another storage source such as a battery, superconducting magnet, or any other source of energy can be added in parallel through an electronic interface to replenish the converter's dc storage. Any of the converter-based, series, shunt, or combined shunt-series Controllers can generally accommodate storage, such as capacitors, batteries, and superconducting magnets, which bring an added dimension to FACTS technology [Figures 1.4(h), 1.4(i), and 1.4(j)].

The benefit of an added storage system (such as large dc capacitors, storage batteries, or superconducting magnets) to the Controller is significant. A Controller with storage is much more effective for controlling the system dynamics than the corresponding Controller without the storage. This has to do with dynamic pumping of real power in or out of the system as against only influencing the transfer of real power within the system as in the case with Controllers lacking storage. Here also, engineers have to rethink the role of storage, particularly the one that can deliver or absorb large amounts of real power in short bursts.

A converter-based Controller can also be designed with so-called high pulse order or with pulse width modulation to reduce the low order harmonic generation to a very low level. A converter can in fact be designed to generate the correct waveform in order to act as an active filter. It can also be controlled and operated in a way that it balances the unbalance voltages, involving transfer of energy between phases. It can do all of these beneficial things simultaneously if the converter is so designed.

Given the overlap of benefits and attributes, it can be said that for a given problem one needs to have an open mind during preliminary evaluation of series versus shunt and combination Controllers and storage versus no storage.

1.7 BRIEF DESCRIPTION AND DEFINITIONS OF FACTS CONTROLLERS

The purpose of this section is to briefly describe and define various shunt, series, and combined Controllers, leaving the detailed description of Controllers to their own specific chapters.

Before going into a very brief description of a variety of specific FACTS Controllers, it is worth mentioning here that for the converter-based Controllers there are two principal types of converters with gate turn-off devices. These are the so-called voltage-sourced converters and the current-sourced converters. As shown in the left-hand side of Figure 1.5(a), the voltage-sourced converter is represented in symbolic form by a box with a gate turn-off device paralleled by a reverse diode, and a dc capacitor as its voltage source. As shown in the right-hand side of Figure 1.5(a), the current-sourced converter is represented by a box with a gate turn-off device with a diode in series, and a dc reactor as its current source.

Details of a variety of voltage-sourced converters suitable for high power applications are discussed in Chapter 3, and current-sourced converters in Chapter 4. It would suffice to say for now that for the voltage-sourced converter, unidirectional dc voltage of a dc capacitor is presented to the ac side as ac voltage through sequential switching of devices. Through appropriate converter topology, it is possible to vary the ac output voltage in magnitude and also in any phase relationship to the ac system voltage. The power reversal involves reversal of current, not the voltage. When the storage capacity of the dc capacitor is small, and there is no other power source connected to it, the converter cannot supply or absorb real power for much more than a cycle. The ac output voltage is maintained at 90 degrees with reference to the ac current, leading or lagging, and the converter is used to absorb or supply reactive power only.

For the current-sourced converter, the dc current is presented to the ac side through the sequential switching of devices, as ac current, variable in amplitude and also in any phase relationship to the ac system voltage. The power reversal involves reversal of voltage and not current. The current-sourced converter is represented symbolically by a box with a power device, and a dc inductor as its current source.

From overall cost point of view, the voltage-sourced converters seem to be preferred, and will be the basis for presentations of most converter-based FACTS Controllers.

One of the facts of life is that those involved with FACTS will have to get used to a large number of new acronyms. There are and will be more acronyms designated by the manufacturers for their specific products, and by the authors of various papers

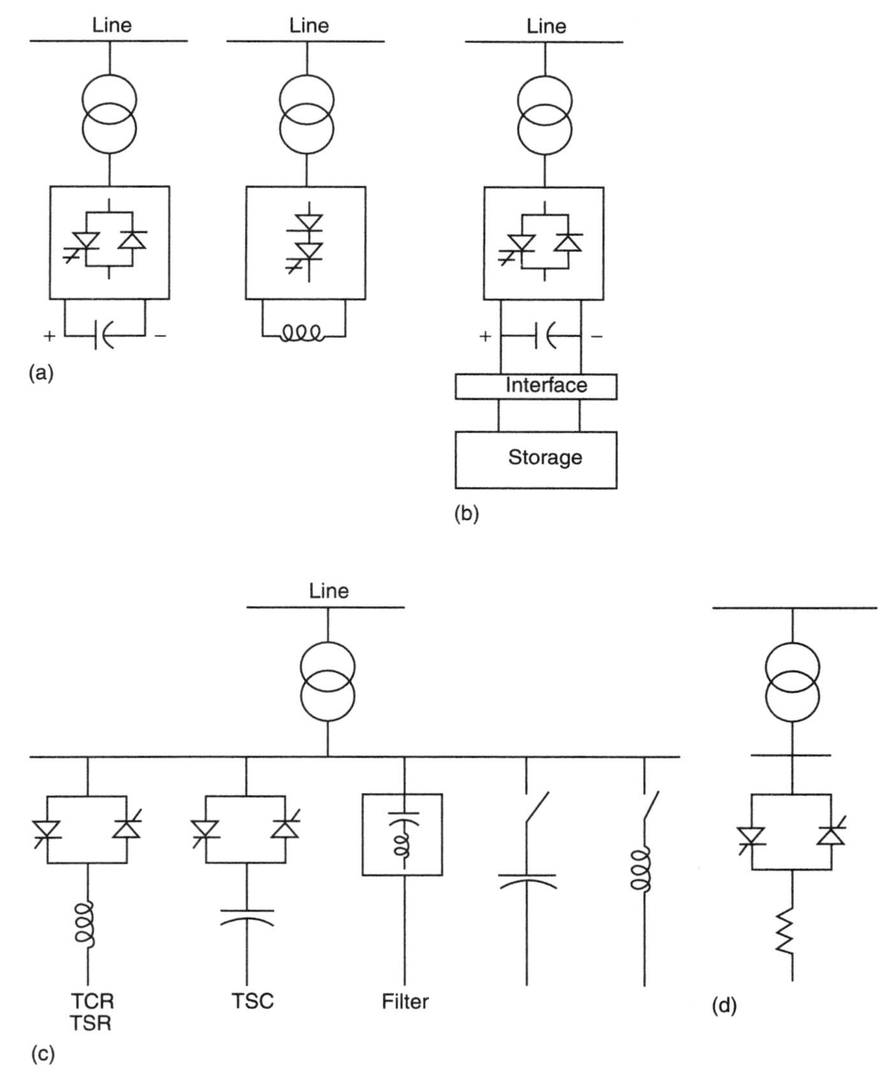

Figure 1.5 Shunt-connected Controllers: (a) Static Synchronous Compensator (STATCOM) based on voltage-sourced and current-sourced converters; (b) STATCOM with storage, i.e., Battery Energy Storage System (BESS) Superconducting Magnet Energy Storage and large dc capacitor; (c) Static VAR Compensator(SVC), Static VAR Generator (SVG), Static VAR System (SVS), Thyristor-Controlled Reactor (TCR), Thyristor-Switched Capacitor (TSC), and Thyristor-Switched Reactor (TSR); (d) Thyristor-Controlled Braking Resistor.

on new Controllers or variations of known Controllers. The IEEE PES Task Force of the FACTS Working Group defined Terms and Definitions for FACTS and FACTS Controllers. Along with a brief description of FACTS Controllers, appropriate IEEE Terms and Definitions are also presented in this section in italic for reference. Generally, this book will use the IEEE terms and definitions.

Flexibility of Electric Power Transmission. *The ability to accommodate changes in the electric transmission system or operating conditions while maintaining sufficient steady-state and transient margins.*

Flexible AC Transmission System (FACTS). *Alternating current transmission systems incorporating power electronic-based and other static controllers to enhance controllability and increase power transfer capability.*

It is worthwhile to note the words "other static Controllers" in this definition of FACTS implying that there can be other static Controllers which are not based on power electronics.

FACTS Controller. *A power electronic-based system and other static equipment that provide control of one or more AC transmission system parameters.*

1.7.1. Shunt Connected Controllers

Static Synchronous Compensator (STATCOM): *A Static synchronous generator operated as a shunt-connected static var compensator whose capacitive or inductive output current can be controlled independent of the ac system voltage.*

STATCOM is one of the key FACTS Controllers. It can be based on a voltage-sourced or current-sourced converter. Figure 1.5(a) shows a simple one-line diagram of STATCOM based on a voltage-sourced converter and a current-sourced converter. As mentioned before, from an overall cost point of view, the voltage-sourced converters seem to be preferred, and will be the basis for presentations of most converter-based FACTS Controllers.

For the voltage-sourced converter, its ac output voltage is controlled such that it is just right for the required reactive current flow for any ac bus voltage dc capacitor voltage is automatically adjusted as required to serve as a voltage source for the converter. STATCOM can be designed to also act as an active filter to absorb system harmonics.

STATCOM as defined above by IEEE is a subset of the broad based shunt connected Controller which includes the possibility of an active power source or storage on the dc side so that the injected current may include active power. Such a Controller is defined as:

Static Synchronous Generator (SSG): *A static self-commutated switching power converter supplied from an appropriate electric energy source and operated to produce a set of adjustable multiphase output voltages, which may be coupled to an ac power system for the purpose of exchanging independently controllable real and reactive power.*

Clearly SSG is a combination of STATCOM and any energy source to supply or absorb power. The term, *SSG*, generalizes connecting any source of energy including a battery, flywheel, superconducting magnet, large dc storage capacitor, another rectifier/inverter, etc. An electronic interface known as a "chopper" is generally needed between the energy source and the converter. For a voltage-sourced converter, the energy source serves to appropriately compensate the capacitor charge through the electronic interface and maintain the required capacitor voltage.

Within the definition of SSG is also the Battery Energy Storage System (BESS), defined by IEEE as:

> ***Battery Energy Storage System (BESS):*** *A chemical-based energy storage system using shunt connected, voltage-source converters capable of rapidly adjusting the amount of energy which is supplied to or absorbed from an ac system.*

Figure 1.5(b) shows a simple one-line diagram in which storage means is connected to a STATCOM. For transmission applications, BESS storage unit sizes would tend to be small (a few tens of MWHs), and if the short-time converter rating was large enough, it could deliver MWs with a high MW/MWH ratio for transient stability. The converter can also simultaneously absorb or deliver reactive power within the converter's MVA capacity. When not supplying active power to the system, the converter is used to charge the battery at an acceptable rate.

Yet another subset of SSG, suitable for transmission applications, is the Superconducting Magnetic Energy Storage (SMES), which is defined by IEEE as:

> ***Superconducting Magnetic Energy Storage (SMES):*** *A Superconducting electromagnetic energy storage device containing electronic converters that rapidly injects and/or absorbs real and/or reactive power or dynamically controls power flow in an ac system.*

Since the dc current in the magnet does not change rapidly, the power input or output of the magnet is changed by controlling the voltage across the magnet with a suitable electronics interface for connection to a STATCOM.

> ***Static Var Compensator (SVC):*** *A shunt-connected static var generator or absorber whose output is adjusted to exchange capacitive or inductive current so as to maintain or control specific parameters of the electrical power system (typically bus voltage).*

This is a general term for a thyristor-controlled or thyristor-switched reactor, and/or thyristor-switched capacitor or combination [Figure 1.5(c)]. SVC is based on thyristors without the gate turn-off capability. It includes separate equipment for leading and lagging vars; the thyristor-controlled or thyristor-switched reactor for absorbing reactive power and thyristor-switched capacitor for supplying the reactive power. SVC is considered by some as a lower cost alternative to STATCOM, although this may not be the case if the comparison is made based on the required performance and not just the MVA size.

> ***Thyristor Controlled Reactor (TCR):*** *A shunt-connected, thyristor-controlled inductor whose effective reactance is varied in a continuous manner by partial-conduction control of the thyristor valve.*

TCR is a subset of SVC in which conduction time and hence, current in a shunt reactor is controlled by a thyristor-based ac switch with firing angle control [Figure 1.5(c)].

> ***Thyristor Switched Reactor (TSR):*** *A shunt-connected, thyristor-switched inductor whose effective reactance is varied in a stepwise manner by full- or zero-conduction operation of the thyristor valve.*

TSR [Figure 1.5(c)] is another a subset of SVC. TSR is made up of several shunt-connected inductors which are switched in and out by thyristor switches without any firing angle controls in order to achieve the required step changes in the reactive power consumed from the system. Use of thyristor switches without firing angle control results in lower cost and losses, but without a continuous control.

Thyristor Switched Capacitor (TSC): *A shunt-connected, thyristor-switched capacitor whose effective reactance is varied in a stepwise manner by full- or zero-conduction operation of the thyristor valve.*

TSC [Figure 1.5(c)] is also a subset of SVC in which thyristor based ac switches are used to switch in and out (without firing angle control) shunt capacitors units, in order to achieve the required step change in the reactive power supplied to the system. Unlike shunt reactors, shunt capacitors cannot be switched continuously with variable firing angle control.

Other broadbased definitions of series Controllers by IEEE include:

Static Var Generator or Absorber (SVG): *A static electrical device, equipment, or system that is capable of drawing controlled capacitive and/or inductive current from an electrical power system and thereby generating or absorbing reactive power. Generally considered to consist of shunt-connected, thyristor-controlled reactor(s) and/or thyristor-switched capacitors.*

The SVG, as broadly defined by IEEE, is simply a reactive power (var) source that, with appropriate controls, can be converted into any specific- or multipurpose reactive shunt compensator. Thus, both the SVC and the STATCOM are *static var generators* equipped with appropriate control loops to vary the var output so as to meet specific compensation objectives.

Static Var System (SVS): *A combination of different static and mechanically-switched var compensators whose outputs are coordinated.*

Thyristor Controlled Braking Resistor (TCBR): *A shunt-connected thyristor-switched resistor, which is controlled to aid stabilization of a power system or to minimize power acceleration of a generating unit during a disturbance.*

TCBR involves cycle-by-cycle switching of a resistor (usually a linear resistor) with a thyristor-based ac switch with firing angle control [Figure 1.5(d)]. For lower cost, TCBR may be thyristor switched, i.e., without firing angle control. However, with firing control, half-cycle by half-cycle firing control can be utilized to selectively damp low-frequency oscillations.

1.7.2 Series Connected Controllers

Static Synchronous Series Compensator (SSSC): *A static synchronous generator operated without an external electric energy source as a series compensator whose output voltage is in quadrature with, and controllable independently of, the line current for the purpose of increasing or decreasing the overall reactive voltage drop across the line and thereby controlling the transmitted electric power. The SSSC may include transiently rated energy storage or energy absorbing devices to enhance the dynamic behavior of the power system by additional temporary real power compensation, to increase or decrease momentarily, the overall real (resistive) voltage drop across the line.*

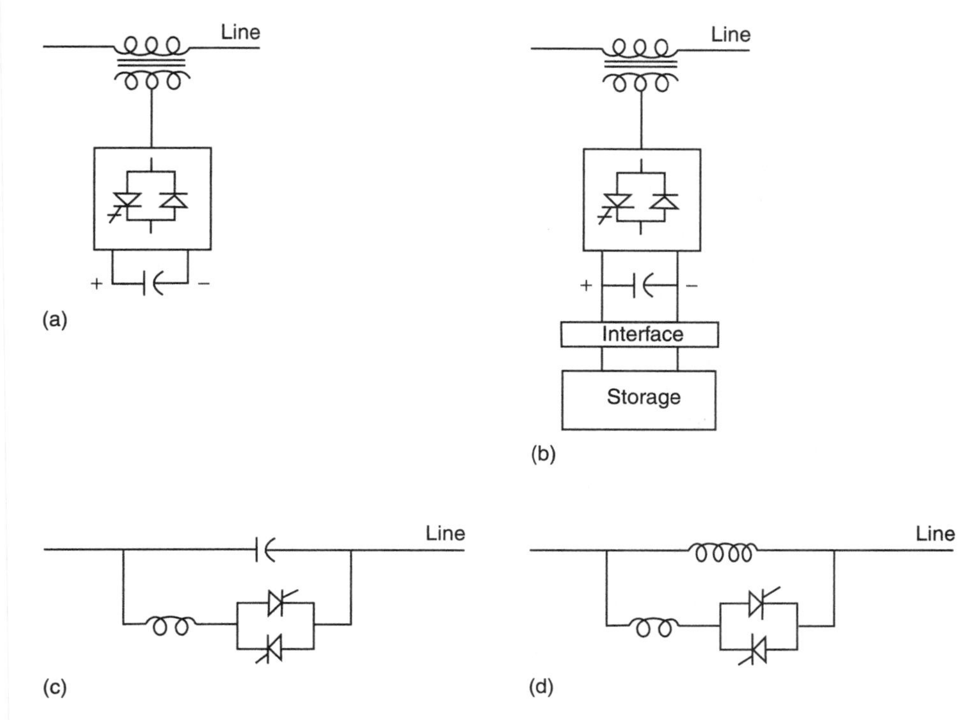

Figure 1.6 (a) Static Synchronous Series Compensator (SSSC); (b) SSSC with storage; (c) Thyristor-Controlled Series Capacitor (TCSC) and Thyristor-Switched Series Capacitor (TSSC); (d) Thyristor-Controlled Series Reactor (TCSR) and Thyristor-Switched Series Reactor (TSSR).

SSSC is one the most important FACTS Controllers. It is like a STATCOM, except that the output ac voltage is in series with the line. It can be based on a voltage-sourced converter [Figure 1.6(a)] or current-sourced converter. Usually the injected voltage in series would be quite small compared to the line voltage, and the insulation to ground would be quite high. With an appropriate insulation between the primary and the secondary of the transformer, the converter equipment is located at the ground potential unless the entire converter equipment is located on a platform duly insulated from ground. The transformer ratio is tailored to the most economical converter design. Without an extra energy source, SSSC can only inject a variable voltage, which is 90 degrees leading or lagging the current. The primary of the transformer and hence the secondary as well as the converter has to carry full line current including the fault current unless the converter is temporarily bypassed during severe line faults.

Battery-storage or superconducting magnetic storage can also be connected to a series Controller [Figure 1.6(b)] to inject a voltage vector of variable angle in series with the line.

Interline Power Flow Controller (IPFC): The IPFC is a recently introduced Controller and thus has no IEEE definition yet. A possible definition is: *The combination of two or more Static Synchronous Series Compensators which are coupled via a common dc link to facilitate bi-directional flow of real power between the ac terminals of the SSSCs, and are controlled to provide independent reactive compensation for the adjustment of real power flow in each line and maintain the desired distribution of reactive power flow among* *the lines. The IPFC structure may also include a STATCOM, coupled to the IFFC's common dc link, to provide shunt reactive compensation and supply or absorb the overall real power deficit of the combined SSSCs.*

Thyristor Controlled Series Capacitor (TCSC): *A capacitive reactance compensator which consists of a series capacitor bank shunted by a thyristor-controlled reactor in order to provide a smoothly variable series capacitive reactance.*

The TCSC [Figure 1.6(c)], is based on thyristors without the gate turn-off capability. It is an alternative to SSSC above and like an SSSC, it is a very important FACTS Controller. A variable reactor such as a Thyristor-Controlled Reactor (TCR) is connected across a series capacitor. When the TCR firing angle is 180 degrees, the reactor becomes nonconducting and the series capacitor has its normal impedance. As the firing angle is advanced from 180 degrees to less than 180 degrees, the capacitive impedance increases. At the other end, when the TCR firing angle is 90 degrees, the reactor becomes fully conducting, and the total impedance becomes inductive, because the reactor impedance is designed to be much lower than the series capacitor impedance. With 90 degrees firing angle, the TCSC helps in limiting fault current. The TCSC may be a single, large unit, or may consist of several equal or different-sized smaller capacitors in order to achieve a superior performance.

Thyristor-Switched Series Capacitor (TSSC): *A capacitive reactance compensator which consists of a series capacitor bank shunted by a thyristor-switched reactor to provide a stepwise control of series capacitive reactance.*

Instead of continuous control of capacitive impedance, this approach of switching inductors at firing angle of 90 degrees or 180 degrees but without firing angle control, could reduce cost and losses of the Controller [Figure 1.6(c)]. It is reasonable to arrange one of the modules to have thyristor control, while others could be thyristor switched.

Thyristor-Controlled Series Reactor (TCSR): *An inductive reactance compensator which consists of a series reactor shunted by a thyristor controlled reactor in order to provide a smoothly variable series inductive reactance.*

When the firing angle of the thyristor controlled reactor is 180 degrees, it stops conducting, and the uncontrolled reactor acts as a fault current limiter [Figure 1.6(d)]. As the angle decreases below 180 degrees, the net inductance decreases until firing angle of 90 degrees, when the net inductance is the parallel combination of the two reactors. As for the TCSC, the TCSR may be a single large unit or several smaller series units.

Thyristor-Switched Series Reactor (TSSR): *An inductive reactance compensator which consists of a series reactor shunted by a thyristor-controlled switched reactor in order to provide a stepwise control of series inductive reactance.*

This is a complement of TCSR, but with thyristor switches fully on or off (without firing angle control) to achieve a combination of stepped series inductance [Figure 1.6(d)].

1.7.3 Combined Shunt and Series Connected Controllers

Unified Power Flow Controller (UPFC): *A combination of static synchronous compensator (STATCOM) and a static series compensator (SSSC) which are coupled via a common dc link, to allow bidirectional flow of real power between the series output terminals of the SSSC and the shunt output terminals of the STATCOM, and are controlled to provide concurrent real and reactive series line compensation without an external electric energy source. The UPFC, by means of angularly unconstrained series voltage injection, is able to control, concurrently or selectively, the transmission line voltage, impedance, and angle or, alternatively, the real and reactive power flow in the line. The UPFC may also provide independently controllable shunt reactive compensation.*

In UPFC [Figure 1.7(b)], which combines a STATCOM [Figure 1.5(a)] and an SSSC [Figure 1.6(a)], the active power for the series unit (SSSC) is obtained from the line itself via the shunt unit STATCOM; the latter is also used for voltage control with control of its reactive power. This is a complete Controller for controlling active and reactive power control through the line, as well as line voltage control.

Additional storage such as a superconducting magnet connected to the dc link via an electronic interface would provide the means of further enhancing the effectiveness of the UPFC. As mentioned before, the controlled exchange of real power with an external source, such as storage, is much more effective in control of system dynamics than modulation of the power transfer within a system.

Thyristor-Controlled Phase Shifting Transformer (TCPST): *A phase-shifting transformer adjusted by thyristor switches to provide a rapidly variable phase angle.*

In general, phase shifting is obtained by adding a perpendicular voltage vector in series with a phase. This vector is derived from the other two phases via shunt connected transformers [Figure 1.7(a)]. The perpendicular series voltage is made variable with a variety of power electronics topologies. A circuit concept that can handle voltage reversal can provide phase shift in either direction. This Controller is also referred to as *Thyristor-Controlled Phase Angle Regulator* (TCPAR).

Interphase Power Controller (IPC): *A series-connected controller of active and reactive power consisting, in each phase, of inductive and capacitive branches subjected to separately*

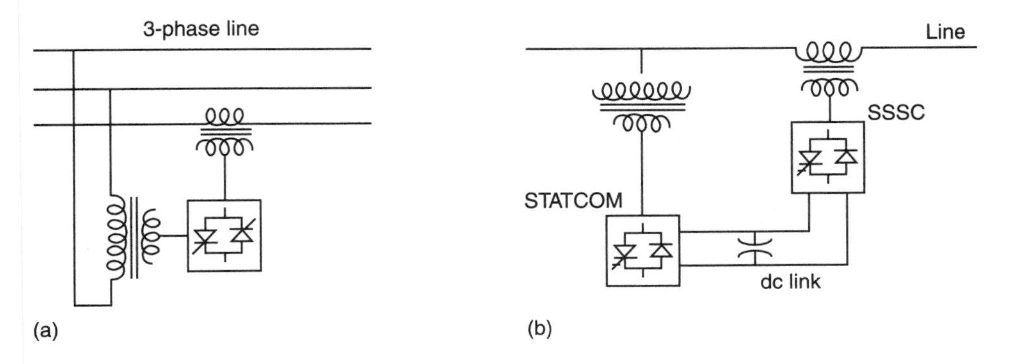

Figure 1.7 (a) Thyristor-Controlled Phase-Shifting Transformer (TCPST) or Thyristor-Controlled Phase Angle Regulator (TCPR); (b) Unified Power Flow Controller UPFC).

phase-shifted voltages. The active and reactive power can be set independently by adjusting the phase shifts and/or the branch impedances, using mechanical or electronic switches. In the particular case where the inductive and capacitive impedance form a conjugate pair, each terminal of the IPC is a passive current source dependent on the voltage at the other terminal.

This is a broadbased concept of series Controller, which can be designed to provide control of active and reactive power.

1.7.4 Other Controllers

Thyristor-Controlled Voltage Limiter (TCVL): *A thyristor-switched metal-oxide varistor (MOV) used to limit the voltage across its terminals during transient conditions.*

The thyristor switch can be connected in series with a gapless arrester, or [as shown in Figure 1.8(a)] part of the gapless arrester (10–20%) can be bypassed by a

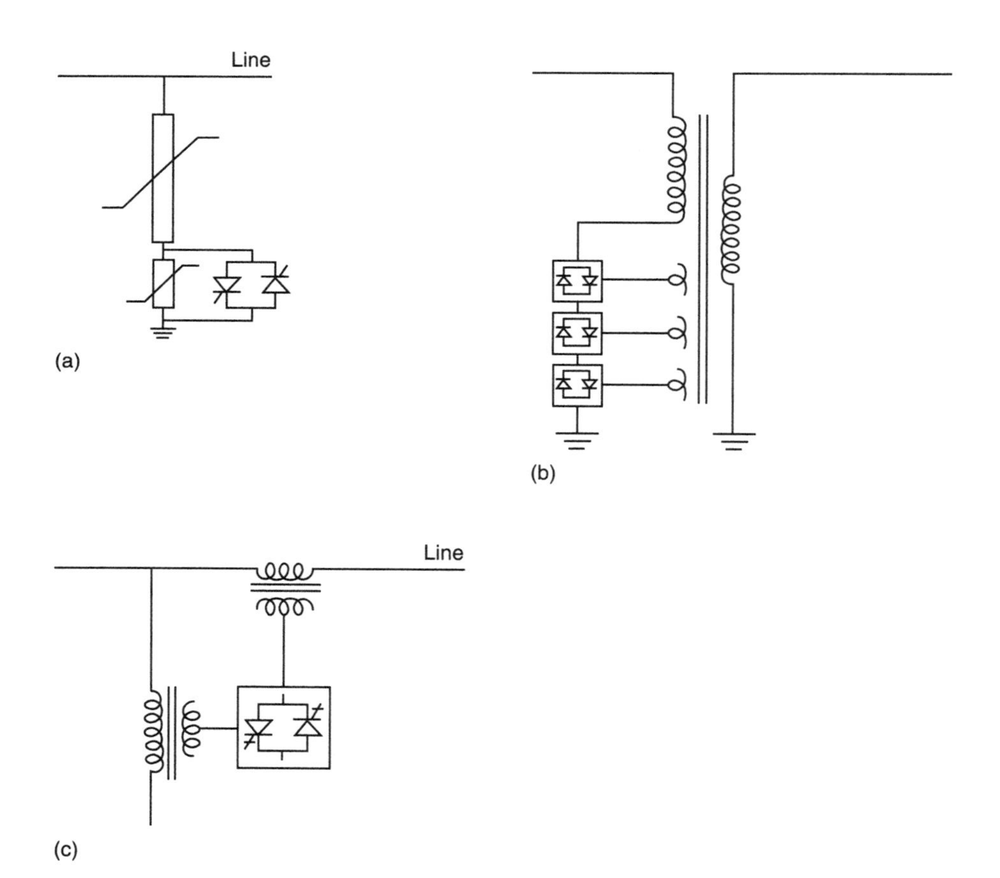

Figure 1.8 Various other Controllers: (a) Thyristor-Controlled Voltage Limiter (TCVL); (b) Thyristor-Controlled Voltage Regulator (TCVR) based on tap changing; (c) Thyristor-Controlled Voltage Regulator (TCVR) based on voltage injection.

thyristor switch in order to dynamically lower the voltage limiting level. In general, the MOV would have to be significantly more powerful than the normal gapless arrester, in order that TCVL can suppress dynamic overvoltages, which can otherwise last for tens of cycles.

Thyristor-Controlled Voltage Regulator (TCVR): *A thyristor-controlled transformer which can provide variable in-phase voltage with continuous control.*

For practical purposes, this may be a regular transformer with a thyristor-controlled tap changer [Figure 1.8(b)] or with a thyristor-controlled ac to ac voltage converter for injection of variable ac voltage of the same phase in series with the line [Figure 1.8(c)]. Such a relatively low cost Controller can be very effective in controlling the flow of reactive power between two ac systems.

Another family of Controllers, not defined here, but mentioned in Section 1.6 are the ones that link two or more transmission lines, in order to balance the real and reactive power flows. These are discussed in Chapter 8.

1.8 CHECKLIST OF POSSIBLE BENEFITS FROM FACTS TECHNOLOGY

Details of various FACTS Controllers will be discussed throughout the book. It is appropriate to state here that within the basic system security guidelines these Controllers enable the transmission owners to obtain, on a case-by-case basis, one or more of the following benefits:

- Control of power flow as ordered. The use of control of the power flow may be to follow a contract, meet the utilities' own needs, ensure optimum power flow, ride through emergency conditions, or a combination thereof.
- Increase the loading capability of lines to their thermal capabilities, including short term and seasonal. This can be accomplished by overcoming other limitations, and sharing of power among lines according to their capability. It is also important to note that thermal capability of a line varies by a very large margin based on the environmental conditions and loading history.
- Increase the system security through raising the transient stability limit, limiting short-circuit currents and overloads, managing cascading blackouts and damping electromechanical oscillations of power systems and machines.
- Provide secure tie line connections to neighboring utilities and regions thereby decreasing overall generation reserve requirements on both sides.
- Provide greater flexibility in siting new generation.
- Upgrade of lines.
- Reduce reactive power flows, thus allowing the lines to carry more active power.
- Reduce loop flows.
- Increase utilization of lowest cost generation. One of the principal reasons for transmission interconnections is to utilize lowest cost generation. When this cannot be done, it follows that there is not enough cost-effective transmission capacity. Cost-effective enhancement of capacity will therefore allow increased use of lowest cost generation.

There is a natural overlap among the above-cited benefits, and in reality, any one or

TABLE 1.1 Control Attributes for Various Controllers

FACTS Controller	Control Attributes	Figure
Static Synchronous Compensator (STATCOM without storage)	Voltage control, VAR compensation, damping oscillations, voltage stability	1.5(a)
Static Synchronous Compensator (STATCOM with storage, BESS, SMES, large dc capacitor)	Voltage control, VAR compensation, damping oscillations, transient and dynamic stability, voltage stability, AGC	1.5(b)
Static VAR Compensator (SVC, TCR, TCS, TRS)	Voltage control, VAR compensation, damping oscillations, transient and dynamic stability, voltage stability	1.5(c)
Thyristor-Controlled Braking Resistor (TCBR)	Damping oscillations, transient and dynamic stability	1.5(d)
Static Synchronous Series Compensator (SSSC without storage)	Current control, damping oscillations, transient and dynamic stability, voltage stability, fault current limiting	1.5(a)
Static Synchronous Series Compensator (SSSC with storage)	Current control, damping oscillations, transient and dynamic stability, voltage stability	1.6(b)
Thyristor-Controlled Series Capacitor (TCSC, TSSC)	Current control, damping oscillations, transient and dynamic stability, voltage stability, fault current limiting	1.6(c)
Thyristor-Controlled Series Reactor (TCSR, TSSR)	Current control, damping oscillations, transient and dynamic stability, voltage stability, fault current limiting	1.6(d)
Thyristor-Controlled Phase-Shifting Transformer (TCPST or TCPR)	Active power control, damping oscillations, transient and dynamic stability, voltage stability	1.7(a)
Unified Power Flow Controller (UPFC)	Active and reactive power control, voltage control, VAR compensation, damping oscillations, transient and dynamic stability, voltage stability, fault current limiting	1.7(b)
Thyristor-Controlled Voltage Limiter (TCVL)	Transient and dynamic voltage limit	1.8(a)
Thyristor-Controlled Voltage Regulator (TCVR)	Reactive power control, voltage control, damping oscillations, transient and dynamic stability, voltage stability	1.8(b) 1.8(c)
Interline Power Flow Controller (IPFC)	Reactive power control, voltage control, damping oscillations, transient and dynamic stability, voltage stability	1.4(d)

two of these benefits would be a principal justification for the choice of a FACTS Controller. It is, however, important to check the list of these benefits on a value-added basis.

Because the voltage, current, impedance, real power, and reactive power are interrelated, each Controller has multiple attributes of what they can do in terms of controlling the voltage, power flow, stability and so on. These controllers can have multiple open loop and closed loop controls to accomplish multiple benefits. Table 1.1 gives a checklist of control attributes for various Controllers. The list is not exhaustive, in particular it does not include the Controllers that link multiple lines, discussed in Chapter 8. Of course the degree of controllability, in particular the control of stability and power flow will vary from Controller to Controller. For understanding the value of these Controllers, one needs to read detail in various chapters of this book.

1.9 IN PERSPECTIVE: HVDC OR FACTS

It is important to recognize that generally HVDC and FACTS are complementary technologies. HVDC is not a grid network in the way that an ac system is, nor is it expected to be. The role of HVDC, for economic reasons, is to interconnect ac systems where a reliable ac interconnection would be too expensive.

There are now over 50 HVDC projects in the world. These can be divided into four categories:

1. **Submarine cables.** Cables have a large capacitance, and hence ac cables require a large charging current (reactive power) an order of magnitude larger than that of overhead lines. As a result, for over a 30 km or so stretch of ac submarine cable, the charging current supplied from the shore will fully load the cable and leave no room for transmitting real power. The charging current flowing in the cables can only be reduced by connecting shunt inductors to the cable at intervals of 15–20 km, thus requiring appropriate land location. With HVDC cable on the other hand, distance is not a technical barrier. Also, the cost of dc cable transmission is much lower than that of ac which works to HVDC's advantage to cover new markets for long distance submarine transmission. In this area, FACTS technology (e.g., the UPFC) can provide an improvement by controlling the magnitude of one of the end (e.g., the receiving-end) voltages so as to keep it identical to that of the other one. In this way, the effective length of the cable from the standpoint of the charging current can be halved. This approach may provide an economical solution for moderate submarine distances, up to about 100 km, but for long distance transmission HVDC will remain unchallenged.
2. **Long distance overhead transmission.** If the overhead transmission is long enough, say 1000 km, the saving in capital costs and losses with a dc transmission line may be enough to pay for two converters (note that HVDC represents total power electronics rating of 200% of the rated transmission capacity). This distance is known as the break-even distance. This *break-even distance* is very subject to many factors including the cost of the line, right-of-way, any need to tap the line along the way, and often most important, the politics of obtaining permission to build the line. Nevertheless, it is important to recognize that while FACTS can play an important role in an effective use of ac transmission, it probably does not have too much influence on the break-even distance. Thus, the principal role of FACTS is in the vast ac transmission market where HVDC is generally not economically viable.
3. **Underground transmission.** Because of the high cost of underground cables, the break even distance for HVDC is more like 100 km as against 1000 km for overhead lines. Again, FACTS technology probably does not have much influence in this break-even distance. In any case, to date there have been no long distance underground projects, either ac or dc, because, in an open landscape, overhead transmission costs so much less than underground transmission (about 25% of the costs of underground transmission). Cable transmission, on the other hand, has a significant potential of cost reduction, both in the cost of cables and construction cost.
4. **Connecting ac systems of different or incompatible frequencies.** For historical reasons, the oceans in effect separate the globe's electric systems into 50 Hz and 60 Hz groups. The 60 Hz normal frequency pervades all the countries of the Americas, excepting Argentina and Paraguay. Those two countries and all the rest of the world have a 50 Hz frequency except Japan, which is partly 50 Hz and partly 60 Hz. In general, the oceans are too huge and deep to justify interconnections of 50 and 60 Hz systems. Thus there is a limited market for HVDC for connecting 50 and 60 Hz systems.

The related, but larger, HVDC market consists of tying together ac systems that have incompatible frequency control such that the phase angle between the two systems frequently straddles the full circle or very large part of a circle. Even once or twice per day excursion of phase angle in excess of 60 degrees would be often enough to rule out any contrived mechanical phase shifter. HVDC can solve this problem. To date there are many HVDC projects, most of them back-to-back ties, which serve this need, and some include some length of dc line as well. Experts agree that some ties of this kind will drop out of use once the neighbors agree upon and implement a common frequency control strategy. This is in fact what has happened in Europe. Not so very long ago, members of the former Comecon (East Europe and the European part of the former USSR) were interconnected as one ac system, but could not be connected to the West European system, known as UCPTE, because they employed different kinds of frequency control and available reserve margins. Then the desire to exchange power with East Europe prompted the installation of three back-to-back ties: the Dürnrohr 550 MW tie connecting Austria to Poland, the Etzenricht 600 MW tie connecting Germany to the Czech Republic, and the Vienna East-West 600 MW tie connecting Austria to Hungary. However, with the subsequent unification of Germany and other political developments, East Germany and the East European countries of Poland, the Slovak and Czech Republics, and Hungary formed themselves into the Centrel Group and installed and field-calibrated their controls at their power plants and control centers. Then in September and October of 1995, their systems were connected to the West European system by ac. Other members of the former Comecon are either operating in isolation or connected to the Integrated Power System with Russia. One side effect is that the three back-to-back HVDC ties, the Dürnrohr and Sud-Ost in Austria and the Etzenricht in Germany, were bypassed and no longer used. This will also happen in India where electricity regions have been connected with back-to-back ties.

There are also many locations in the world where frequency control strategies are the same, but, because of the remoteness of centers of electrical gravity from the desired points of connection, phase angle drifts and varies over a wide angle with daily variations in load and even go full circle. The Eastern and Western United States systems are an example of this, and therefore they are connected with several back-to-back HVDC ties. Conceptually, FACTS technology can also solve this problem of wide variations of phase angle. It requires, for example, a phase-angle Controller for one side to chase or lead the other as required through full circle. This is in fact technically feasible with known concepts of phase angle Controller. However, the economics are perhaps in favor of HVDC. After all, HVDC is independent of phase angle and therefore, in a way, equivalent to a 360 degree phase angle Controller, and a true 360 degree phase angle Controller may be more expensive.

Large market potential for FACTS is within the ac system on a value-added basis, where:

- the existing steady-state phase angle between bus nodes is reasonable,
- the cost of a FACTS solution is lower than the HVDC cost, and
- the required FACTS Controller capacity is less than, say, 100% of the transmission throughput rating.

Figure 1.9 conveys the basic attributes of HVDC and FACTS, and also general cost comparison between HVDC and FACTS where FACTS solutions are viable and could have scope of applications.

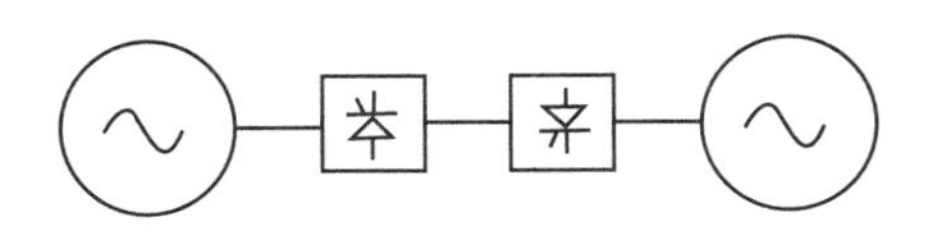

HVDC:

- Independent frequency and control
- Lower line costs
- Power control, voltage control, stability control

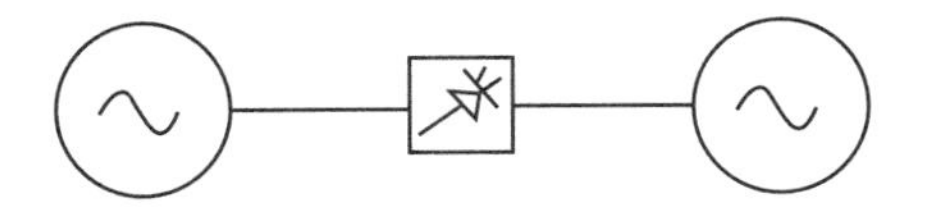

FACTS:

- Power control, voltage control, stability control

Installed Costs (millions of dollars)

Throughput MW	HVDC 2 Terminals	FACTS
200 MW	$ 40–50 M	$ 5–10 M
500 MW	75–100	10–20
1000 MW	120–170	20–30
2000 MW	200–300	30–50

Figure 1.9 HVDC and FACTS: Complementary solutions.

REFERENCES

Adapa, R., "Summary of EPRI's FACTS System Studies," *CIGRE SC 14 International Colloquium on HVDC & FACTS,* Montreal, September 1995.

Ainsworth, J. D., Davies, M., Fitz, P. J., Owen, K. E., and Trainer, D. R., "A Static VAR Compensator Based on Single-Phase Chain Circuit Converters," *IEE Proc. Gen. Trans. and Dist.,* vol. 145, no. 4, July 1998.

Arabi, S., Kundur, P., and Adapa, R., "Innovative Techniques in Modeling UPFC for Power System Analysis," *Paper PE-231-PWRS-0-10-1998 IEEE PES Winter Meeting,* New York, January–February 1999.

Baker, M. H., "An Assessment of FACTS Controllers for Transmission System Enhancement," *CIGRE SC 14 International Colloquium on HVDC & FACTS,* Montreal, September 1995.

Baker, M. H., Povh, D., and Larsen, E. V., "FACTS Equipment," *Special Report CIGRE Symposium on Power Electronics in Electric Power Systems,* Tokyo, May 1995.

Baker, M. H., Thanawala, H. L., Young, D. J., and Erinmez L. A. "Static VAR Compensators Enhance a Meshed Transmission System," *CIGRE Special Report: Technology and Benefits of Flexible AC Transmission Systems, Joint Session of SC14/37/38,* Paris, 1992.

Breuer, G., "Flexible AC Transmission Scoping Studies," *EPRI Proceedings,* FACTS Conference 1, Cincinnati, OH, November 1990.

Brochu, J., Pelletier, P., Beauregard, F., and Morin, G., "The Interphase Power Controller, A new Concept for Managing Power Flow Within AC Networks," *IEEE Transactions on Power Delivery,* vol. 9, no. 2, pp. 833–841, April 1994.

Canizares, C. A., "Analysis of SVC and TCSC in Voltage Collapse," *IEEE PES 1998 Summer Meeting,* San Diego, CA, July 1998.

Cannon, D. Jr., Stevenson, C., Hanson, D., Lyons, P. F., and McCafferty, M., "ASCR Conductor Performance Under High Thermal Loading" *EPRI Conference on the Future of Power Delivery,* Washington, DC, April 1996.

Casazza, J. A., and Lekang, D. J., "New FACTS Technology: Its Potential Impact on Transmission System Utilization," *EPRI Conference on Flexible AC Transmission System,* Cincinnati, OH, November 1990.

Chamia, M., "Power Flow Control in Highly Integrated Transmission Network," *EPRI Conference on Flexible AC Transmission System,* Cincinnati, OH, November 1990.

Davriu, A., Douard, G., Mallet, P., and Therond, P. G., "Taking Account of FACTS Investment Selection Studies," *CIGRE Symposium on Power Electronics in Electric Power Systems,* Tokyo, May 1995.

Davriu, A., Mallet, P., and Pramayon, P., "Evaluation of FACTS Functionalities Within the Planning of the Very High Voltage Transmission Network: The EDF Approach," *Proceedings of the IEEE/Royal Institute of Technology Stockholm Power Tech: Power Electronics,* Stockholm, June 1995.

DeMarco, C. L., "Security Measures To Evaluate the Impact of FACTS Devices in Preventing Voltage Collapse," *Proceedings of EPRI Conference on Flexible AC Transmission Systems (FACTS),* Boston, MA, May 1992.

Deuse, J., Stubbe, M., Meyer, B., and Panciatici, P., "Modeling of FACTS for Power System Analysis," *CIGRE Symposium on Power Electronics in Electric Power Systems,* Tokyo, May 1995.

Douglass, D. A., Edris, A. A., and Pritchard, G. A., "Field Application of A Dynamic Thermal Circuit Rating Method," *IEEE PES 1995 Winter Meeting,* New York, January–February 1995.

Edris A., Adapa, R., Baker, M. H., Bohmann, L., Clark, K., Habashi, K., Gyugui, L., Lemay, J., Mehraban, A. S., Meyers, A. K., Reeve, J., Sener, F., Torgerson, D. R., and Wood, R. R., "Proposed Terms and Definitions for Flexible AC Transmission System (FACTS)," *IEEE Trans on Power Delivery,* vol. 12, no. 4, October 1997, pp. 1848–1853.

Edris, E., "Enhancement of First-Swing Stability Using a High Speed Phase Shifter," *IEEE Trans. Power Systems,* vol. 6, no. 3, pp. 1113–1118, 1991.

Erche, M., Lerch, E., Povh, D., and Mihalic, R., "Improvement of Power System Performance Using Power Electronic Equipment," *CIGRE Special Report: Technology and Benefits of Flexible AC Transmission Systems, Joint Session of SC14/37/38,* Paris, 1992.

Ewart, D. N., Koessler, R. J., Mountford, J. D., and Maratukulam, D., "FACTS Options Permit the Utilization of the Full Thermal Capacity of AC Transmission," *IEE Conference Publication Series 5, International Conference on AC and DC Power Transmission,* London, *IEE Publication No. 345,* pp. 1–7, September 1991.

Eunson, E. M., and Meslier, F., "Planning of the Power System Taking Into Account the Application of Power Electronics," Special Report for Session 2, *CIGRE 1995 Symposium on Power Electronics in Electric Power Systems,* Tokyo, May 1995.

Eunson, E. M., Meslier, Daniel, D., Le Du, A., Poumarede, C., Therond, P. G., Langlet, B., Taisne, J. P., and Collet-Billon, V., "Power Electronics: An Effective Tool for Network Development? An Electricite de France Answer Based on the Development of a Prototype Unified Power Flow Controller," *CIGRE Symposium on Power Electronics in Electric Power Systems,* Tokyo, May 1995.

Falk, Christensen, J., "More Effective Networks," CIGRE Special Report P1-00, Paris, Session, 1996.

Feak, S. D., "Superconducting Magnetic Energy Storage (SMES): Utility Application Studies," *IEEE PES 1996 Summer Meeting,* Denver, Paper 96 SM 472-1-PWRS, July–August 1996.

Ferguson, J., Woolery, B., and Lyons, P. F., "Innovative Upgrades of Transmission Lines," *EPRI Conference on the Future of Power Delivery,* Washington, DC, April 1996.

Galiana, F. D., Almeida, K., Toussaint, M., Griffen, J., Ooi, B. T., and McGillis, D. T., "Assessment and Control of the Impact of FACTS Devices on Power System Performance," *IEEE PES 1996 Winter Meeting,* Baltimore, Paper 96 WM 256-8-PWRS, January 1996.

Gama, C. A., Ellery, E. H., Azvedo, D. C. B., and Ponte, J. R. R., "Static VAR Compensators (SVC) Versus Synchronous Condesners (SC) for Inverter Stations Compensation—Technical and Economical Aspects in Electronorte Studies," *CIGRE Group 14 Paper 14-103,* 1992.

Gjengedal, T., Gjerde, J. O., and Flolo, R., "Prospective Use of HVDC and FACTS—

Components for Enhancement of Power System Stability in the Norwegian Main Grid," *EPRI Conference on the Future of Power Delivery,* Washington, DC, April 1996.

Gyugyi, L., "Solid State Control of AC Power Transmission," *EPRI Conference on Flexible AC Transmission System, Proceedings,* Cincinnati, OH, November 1990.

Hammad, A. E., "Comparing the Voltage Control Capabilities of Present and Future VAR Compensating Techniques in Transmission Systems," *IEEE Transactions on Power Delivery,* vol. 11, no. 1, January 1995.

Hammad, A. E., "FACTS Control Concepts: Case Study for Voltage Stability," *EPRI Conference on the Future of Power Delivery,* Washington, DC, April 1996.

Hammad, A. E., "FACTS Specification: Criteria and Associated Studies," *EPRI Conference on the Future of Power Delivery,* Washington, DC, April 1996.

Henderson, M., and Lemay, J., "Design and Implementation of the Interphase Power Controller Between NYPA and VELCO Systems," *IEEE PES 98 Summer Meeting,* San Diego, CA. July 1998.

Henderson, M. I., "Operating Issues for FACTS Devices: An Operations Planning Perspective," *EPRI Conference on Flexible AC Transmission System,* Baltimore, MD, October 1994.

Hingorani, N. G., "Power Electronics in AC Transmission Systems," *CIGRE Special Report P1-02,* Paris Session, 1996.

Hingorani, N. G., "Future Opportunities for Electric Power Systems," IEEE PES Summer Power Meeting Luncheon Speech, San Francisco, CA, July 1987.

Hingorani, N. G., "Power Electronics in Electric Utilities: Role of Power Electronics in Future Power Systems," *Proceedings of the IEEE Special Issue,* vol. 76, no. 4, April 1988.

Hingorani, N. G., "High Power Electronics and Flexible AC Transmission System," *IEEE Power Engineering Review,* vol. 8, no. 7, July 1988. Reprint of Joint American Power Conference/IEEE Luncheon Speech, Chicago, IL, April 1988.

Hingorani, N. G., "Flexible AC Transmission Systems (FACTS)—Overview," Paper presented at the Panel Session of FACTS, *IEEE PES 1990 Winter Meeting,* Atlanta, February 1990.

Hingorani, N. G., "FACTS: Flexible AC Transmission Systems," *EPRI Conference on Flexible AC Transmission System,* Cincinnati, OH, November 1990.

Hingorani, N. G., "FACTS—Flexible AC Transmission System," Proceedings of the Fifth International Conference on AC and DC Power Transmission, London, *IEE Publication No. 345,* pp. 1–7, September 1991.

Hingorani, N. G., "Flexible AC Transmission," *IEEE Spectrum,* vol. 30, no. 4, April 1993.

Hingorani, N. G., and Stahlkopf, K. E., "High Power Electronics," *Scientific American,* vol. 269, no. 5, pp. 77–85, November 1993.

Hingorani, N. G., "Power Electronics," *IEEE Power Engineering Review,* vol. 15, no. 10, October 1995.

IEEE Power Electronics Modeling Task Force, "Guidelines for Modeling Power Electronics in Electric Power Engineering Applications," *IEEE 1996 PES Summer Meeting,* Denver, Paper 96 SM 439-0-PWRD, July–August 1996.

Kapoor, S. C., Tripathy, A. K., Kumar, M., Senthil, J., Adhikari, T., and Chaukiyal, G. P., "A Proposed Rapid Network Impedance Control Scheme for an Existing Power Corridor in India," *Proceedings of CIGRE Group 14, Paper 14-106,* August–September 1994.

Klein, M., Le, L. X., Rogers, G. J., Farrokhpay, S., and Balu, N. J., "H-Infinity Damping Controller Design in Large Power Systems," *IEEE Transactions on Power Systems,* vol. 10, no. 1, pp. 158–166, February 1995.

Koessler, R., "Investigation of FACTS Options to Utilize the Full Thermal Capacity of AC Transmission," *EPRI Conference on Flexible AC Transmission System,* Cincinnati, OH, November 1990.

Kurita, A., Okubo, H., Miller, N. W., and Sanchez-Gasca, J. J., "Application of Fast Controllable

Electronic Devices for Performance Enhancement in Tightly Interconnected Systems," *CIGRE 1995 Symposium on Power Electronics in Electric Power Systems,* Tokyo, May 1995.

Lachs, W. R., Sutanto, D., and Logothetis, D. N., "Power System Control in the Next Century," *IEEE Transactions on Power Systems,* vol. 11, no. 1, pp. 11–18, February 1996.

Lacoste, J., Cholley, P., Trotignon, M., Daniel, D., and Nativel, G., "FACTS Equipment and Power System Dynamics," *CIGRE 1995 Symposium on Power Electronics in Electric Power Systems,* Tokyo, May 1995.

Larsen, E., Bowler, C., Damsky, B., and Nilsson, S., "Benefits of Thyristor Controlled Series Compensation," *CIGRE Special Report: Technology and Benefits of Flexible AC Transmission Systems, Joint Session of SC14/37/38,* Paris, 1992.

Larsen, E., and Weaver, T., "FACTS Overview," *IEEE/CIGRE FACTS Working Group, IEEE Technical Publication 95 TP108,* 84 pages, 1995.

Larsen, E., and Sener, F., "Current Activity in Flexible AC Transmission System (FACTS)," *IEEE Publication No. 92TH0465-5-PWR,* April 1992.

Le Du, A., Tontini, G., and Winfield, M., "Which FACTS Equipment for Which Need? Identification of the Technology Developments to Meet the Needs of Electricite de France (EDF), Ente Nazionale per l'Energia Eletrica (ENEL) and National Grid Company (NGC)," *CIGRE Special Report: Technology and Benefits of Flexible AC Transmission Systems, Joint Session of SC14/37/38,* Paris, 1992.

Lei, X., Lerch, E., Povh, D., Wang, X., and Haubrich, H. J., "Global Settings of FACTS Controllers in Power Systems," *CIGRE Session Group 14,* Paris, Paper 14-305, 1996.

Lemay, J., "A Planner's View of FACTS: Meeting the Needs of Modern Networks Power," *Technology International,* United Kingdom, pp. 109–111, 1996.

Lerch, E., and Povh, D., "Performance of AC Systems Using FACTS Equipment," *Proceedings of EPRI Conference on Flexible AC Transmission Systems (FACTS),* Boston, MA, May 1992.

Lu, C. F., Liu, C. C., and Wu, C. J., "Dynamic Modeling of Battery Energy Storage System and Application to Power System Stability," *IEE Proceedings, Generation, Transmission and Distribution,* vol. 142, no. 4, pp. 429–435, July 1995.

Maguire, T. L., and Gole, A. M., "Digital Simulation of Flexible Topology Power Electronic Apparatus in Power Systems," *IEEE SM 414-3 PWRD,* 1991.

Maliszewski, R. M., Eunson, E. M., Meslier, F., Balazs, P., Schwarz, J., Takahashi, K., and Wallace, P., "Interaction of System Planning in the Development of the Transmission Networks of the 21st Century," *CIGRE 1995 Symposium on Power Electronics in Electric Power Systems,* Tokyo, May 1995.

Maliszewski, R. M., and Meslier, F., "More Effective Network: An Answer to Question 10 of the Special Report," *CIGRE Special Report P1-06,* Session 1996.

Maliszewski, R. M., Pasternack, B. M., Scherer, Jr., H. N., Chamia, M., Frank, H., and Paulsson, L., "Power Flow in a Highly Interconnected Transmission Network," *CIGRE Paper 37-303,* Paris, 1990 Session, 1990.

Manzoni, G., Salvaderi, Eunson, E. M., Schwarz, J., and Casazza, J. A., "Power System Planning with Changing Institutional Arrangements," *CIGRE Panel Session Paper No. P2-01,* 1996.

Miller, N. W., Mukerji, R., and Clayton, R. E., "The Role of Power Electronics in Open Access Markets," *EPRI Conference on the Future of Power Delivery,* Washington, DC, April 1996.

Mitani, Y., Uranaka, T., and Tsuji, K., "Power System Stabilization by Superconducting Magnetic Energy Storage with Solid State Phase Shifter," *IEEE Transactions on Power Systems,* vol. 10, no. 3, August 1995.

Mittelstadt, W. A., "Considerations in Planning Use of FACTS Devices on a Utility System," *EPRI Conference on Flexible AC Transmission System, Proceedings,* Cincinnati, OH, November 1990.

Nilsson, S. L., "Security Aspects of Flexible AC Transmission System Controller Applications,"

International Journal of Electrical Power and Energy Systems (United Kingdom), vol. 17, no. 3, pp. 173–179, June 1995.

Norris, W. T., and James, D. A., "Series Compensation Using Power Electronic Switches: The SERCOM," *Proceedings of the IEE Sixth International Conference on AC and DC Transmission,* pp. 405–410, April–May 1996.

Oh, T. K., Choi, K. H., Byon, S. B., Kim, Y. H., Hwang, J. Y., and Lee, K. J., "Power Transfer Capability Increase and Transient Stability Enhancement by FACTS Devices Application," *EPRI Conference on the Future of Power Delivery,* Washington, DC, April 1996.

Okamoto, H., Kurita, A., and Sekine, Y., "A Method for Identification of Effective Locations of Variable Impedance Apparatus on Enhancement of Steady-State Stability in Large Scale Power Systems," *IEEE Transactions on Power Systems,* vol. 10, no. 3, August 1995.

Ooi, B. T., Kazerani, M., Marceau, R., Wolanski, Z., Galiana, F. D., McGillis, D., and Joos, G., "Mid-Point Siting of FACTS Devices in Transmission Lines," *Paper No. PE-292-PWRD-0-01-1997,* IEEE PES, 1997.

Padiyar, K. R., Geetha, M. K., and Rao, K. U., "A Novel Power Flow Controller for Controlled Series Compensation," *Proceedings, of the IEE Sixth International Conference on AC and DC Transmission,* pp. 3290–3334, April–May 1996.

Padiyar, K. R., and Kulkarni, A. M., "Application of Static Condenser for Enhancing Power Transfer in Long AC Lines," *CIGRE 1995 Symposium on Power Electronics in Electric Power Systems,* Tokyo, May 1995.

Padiyar, K. R., and Rao, K. U., "Discrete Control of TCSC for Stability Improvement in Power Systems," Proceedings, *IEEE Fourth Conference on Control Applications,* Albany, NY, pp. 246–251, September 1995.

Paserba, J. J., Condordia, C., Lerch, E., Lysheim, D. P., Ostojic, D., Thorvaldsson, B. H., Dagle, J. E., Trudnowski, D. J., Hauer, J. F., and Janssens, N., "Opportunities for Damping Oscillations by Applying Power Electronics in Electric Power Systems," *CIGRE 1995 Symposium on Power Electronics in Electric Power Systems,* Tokyo, May 1995.

Pastos, D. A., Vovos, N. A., and Giannakopoulos, G. B., "Comparison of Compensators Capability to Modulate the Real Power Flow in Integrated AC/DC Systems," *IEEE PES 97 Summer Meeting,* Berlin, Germany, July 1997.

Pilotto, L. A. S., Ping, W. W., Carvalho, A. R., Prado, S., Wey, A., Long, W. F., and Edris, A. A., "First Results on the Analysis of Control Interactions on FACTS Assisted Power Systems," EPRI Conference on Flexible AC Transmission System, Baltimore, MD, October 1994.

Pilotto, L. A. S., Ping, W. W., Carvalho, A. R., Wey, A., Long, W. F., Alvarado, F. L., DeMarco, C. L., and Edris, A. A., "Determination of Needed FACTS Controllers That Increase Asset Utilization of Power Systems," *IEEE PES 1996 Winter Meeting,* Baltimore, Paper 96 WM 090-1-PWRD, January 1996.

Povh, D., "Advantages of Power Electronics Equipment in AC Systems," *CIGRE SC14 International Colloquium on HVDC and FACTS Systems,* Wellington, New Zealand, September 1993.

Povh, D., and Mihalic, R., "Enhancement of Transient Stability on AC Transmission by Means of Controlled Series and Parallel Compensation," *IEE Conference Publication Series 5, International Conference on AC and DC Power Transmission,* London, *IEE Publication No. 345,* pp. 8–12, September 1991.

Povh, D., Mihalic, R., and Papic, I., "FACTS Equipment for Load Flow Control In High Voltage Systems," *CIGRE 1995 Symposium on Power Electronics in Electric Power Systems,* Tokyo, May 1995.

Pramayon, P., Nonnon, P., Mallet, P., Trotignon, M., and Lauzanne, B., "Methodology for Technical and Economical Evaluation of FACTS Impact and Efficiency on EHV Transmission System," *EPRI Conference on the Future of Power Delivery,* Washington DC, April 1996.

Rajaraman, R., "Oscillations in Power Systems with Thyristor Switching Devices," *IEEE PES 1996 Winter Meeting,* Baltimore, Paper 96 WM 255-0-PWRS, January 1996.

Rajaraman, R., and Dobson, I., "Damping Estimates of Subsynchronous and Power Swing Oscillations in Power Systems with Thyristor Switching Devices," *IEEE PES Winter Meeting,* Baltimore, January 1996.

Rajaraman, R., Maniaci, A., and Camfield, R., "Determination of Location and Amount of Series Compensation to Increase Over Transfer Capability," *IEEE PES 97 Summer Meeting,* Berlin, Germany, July 1997.

Ray, C., "Meeting Security and Quality Requirements In an Open Market," *EPRI Conference on the Future of Power Delivery,* Washington, DC, April 1996.

Rehtanz, C., "Systemic Use of Multifunctional SMES in Electric Power Systems," *Paper PE-224-PWRS-0-10-1998, IEEE PES Winter Meeting,* New York, January–February 1999.

Sanchez, J. J., "Coordinated Control of Two FACTS Devices for Damping Inter-Area Oscillations," *IEEE PES 97 Summer Meeting,* Berlin, Germany, July 1997.

Sekine, Y., Hayashi, T., Abe, K., Inoue, Y., and Horiuchi, S., "Application of Power Electronics Technologies to Future Interconnected Power System in Japan," *CIGRE 1995 Symposium on Power Electronics in Electric Power Systems,* Tokyo, May 1995.

Sekine, Y., Takahashi, K., and Hayashi, T., "Application of Power Electronics Technologies to the 21st Century's Bulk Power Transmission in Japan," *International Journal of Electrical Power and Energy Systems* (United Kingdom), vol. 17, no. 3, pp. 181–193, June 1995.

Simo, J. B., and Kamwa, I., "Exploratory Assessment of the Dynamic Behavior of Multi-Machine System Stabilized by a SMES Unit," *IEEE Transactions on Power Systems,* vol. 10, no. 3, August 1995.

Singh, B., Al-Haddad, K., and Chandra, A., "A New Control Approach to Three Phase Active Power for Harmonics and Reactive Power Compensation," *IEEE PES 97 Summer Meeting,* Berlin, Germany, July 1997.

Smith, P., Thanawala, H. L., and Corcoran, J. W. C., "Potential Application of Power Electronic Equipment in the Irish Power Transmission System," *CIGRE 1995 Symposium on Power Electronics in Electric Power Systems,* Tokyo, May 1995.

Strbac, G., and Jenkins, N., "FACTS Devices in Uplift Control," *Proceedings of the IEE Sixth International Conference on AC and DC Transmission,* pp. 214–219, April–May 1996.

Sugimoto, S., Kida, J., Arita, H., Fukui, C., and Yamagiwa, T., "Principle and Characteristics of a Fault Current Limiter With Series Compensation," *IEEE Transactions on Power Delivery,* vol. 11, no. 2, April 1996.

Sultan, M., Reeve, J. R., and Adapa, R., "Combined Transient and Dynamic Analysis of HVDC and FACTS Systems," *Paper No. PE-83-PWRD-0-1-1998, IEEE PES,* 1998

Sybille, G., Haj-Maharsi, Y., Morin, G., Beauregard, F., Brochu, J., Lemay, J., and Pelletier, P., "Simulator Demonstration of the Interphase Power Controller Technology" *IEEE PES 1996 Winter Meeting,* Baltimore, January 1996.

Tang, Y., and Meliopoulos, A. P. S., "Power System Small Signal Stability Analysis with FACTS Elements," *IEEE PES 1996 Summer Meeting,* Denver, Paper 96 SM 467-1-PWRD, July–August 1996.

Taranto, G. N., Chow, J. H., and Othman, H. A., "Robust Redesign of Power System Damping Controllers," *IEEE Transactions on Control Systems Technology,* vol. 3, no. 3, pp. 290–298, September 1995.

Taranto, G. N., Pinto, L. M. V. G., and Pereira, M. V. F., "Representation of FACTS Devices in Power System Economic Dispatch," *IEEE Trans Power Systems,* vol. 7, no. 2, pp. 572–576, May 1992.

Tenorio, A. R. M., Ekanayake, J. B., and Jenkins, N. "Modeling of FACTS Devices," Proceedings of the *IEE Sixth International Conference on AC and DC Transmission,* pp. 340–345, April–May 1996.

Tripathy, S. C., and Juengst, K. P., "Sampled Data Automatic generation Control with Superconducting Magnetic Energy Storage in Power System," *IEEE PES 1996 Winter Meeting,* Baltimore, Paper 96 WM 046-3-EC, January 1996.

Vieira, X., Szechtman, M., Salgado, E., Praca, J. C. G., Ping, W. W., and Bianco, A., "Prospective Applications of FACTS in the Brazilian Interconnected Power Systems," *CIGRE 1995 Symposium on Power Electronics in Electric Power Systems,* Tokyo, May 1995.

Wang, Y., Mohler, R. R., Spee, R., and Mittelstadt, S. W., "Variable Structure FACTS Controllers for Power System Transient Stability," *IEEE Trans Power Systems,* vol. 7, no. 1, pp. 307–313, February 1992.

Wang, H. F., and Swift, F. J., "A Unified Model for the Analysis of FACTS Devices in Damping Power System Oscillations. Part 1: Single-Machine Infinite-Bus Power Systems," *IEEE PES 1996 Summer Meeting,* Denver, Paper 96 SM 464-8-PWRD, July–August 1996.

Yokoyama, A., Okamoto, H., and Sekine, Y., "Simulation Study on Steady State and Transient Stability Enhancement of Multi-Machine Power System by Dynamic Control of Variable Series Capacitor," *CIGRE 1995 Symposium on Power Electronics in Electric Power Systems,* Tokyo, May 1995.

2

Power Semiconductor Devices

2.1 PERSPECTIVE ON POWER DEVICES

The intent of this section is to give only general information about power semiconductor devices suitable for FACTS Controllers. Sufficient information is provided for power systems engineers to understand the options and their relevance to FACTS applications. There are books available, which provide in-depth information for those interested in device details.

Generally, FACTS applications represent a three-phase power rating from tens to hundreds of megawatts. Basically, FACTS Controllers are based on an assembly of ac/dc and/or dc/ac converters and/or high power ac switches. A converter is an assembly of valves (with other equipment), and each valve in turn is an assembly of power devices along with snubber circuits (damping circuits) as needed, and turn-on/turn-off gate drive circuits. Similarly, each ac switch is an assembly of back-to-back connected power devices along with their snubber circuits and turn-on/turn-off gate-drive circuits. Nominal rating of large power devices is in the range of 1–5 kA and 5–10 kV per device, and their useable circuit rating may only be 25 to 50% of their nominal rating. This conveys that the converters and ac switches would be an assembly of a large number of power devices. The converters, ac switches, and devices are connected in series and/or parallel, to achieve the desired FACTS Controller rating and performance, and a Controller in some cases may also be separated into single-phase assemblies. These considerations provide an interesting possibility and indeed a necessity for a supplier to adapt modularity for an effective use of the power devices. Modularity if properly utilized, cannot only reduce the cost through standardization of modules and submodules, but it can also be an asset from the user perspective in terms of reliability, redundancy, and staged investment.

The device ratings and characteristics and their exploitation have a significant leverage on the cost, performance, size, weight, and losses of FACTS Controllers, or for that matter, all power device applications. The leverage includes the cost of all that surrounds the devices including snubber circuits, gate-drive circuits, transformers and other magnetic equipment, filters, cooling equipment, losses, operating performance, and maintenance requirements. For example, faster switching capability leads to fewer snubber components, lower snubber losses and adaptation of concepts that produce less harmonics and faster FACTS Controller response. They are also important for successful implementation of particular concepts of FACTS Controllers, such as active filters.

There are many advanced circuit concepts used in low power industrial applications, mostly driven by first cost, the economic application of which at a high power level is largely a function of advances in devices. These concepts include *pulse width modulation* (*PWM*), soft switching, resonant converters, choppers, and others. Therefore the design of FACTS Controller equipment would usually be based on the devices with best available characteristics, even at higher prices. Although the cost of devices is a factor, it would be correct to say that availability of devices with better characteristics provides an important leverage for the FACTS options and a competitive edge for a supplier of FACTS technology to meet a certain specified performance at the lowest evaluated cost. Thus the cost, performance, and market success of FACTS Controllers is very much tied to progress in power semiconductor devices and their packaging. In fact the designer of the FACTS Controller has much to gain from negotiating the device characteristics, packaging, and subassemblies, with the device supplier and not assume the characteristics stated in the device catalog as the basis for the Controller design. It is therefore important for a user of FACTS technology to have a general idea of the power semiconductor options, state of the device technology and future trends, and also circuit concepts used in utility and industrial applications.

In general terms, high-power electronic devices are fast switches based on high-purity single-crystal silicon wafers, designed for a variety of switching characteristics. In their forward-conducting direction, the devices may have control to turn on and turn off the current flow when ordered to do so by means of gate control. Some power devices are designed without the capability to block in reverse direction, in which case they are provided with another reverse blocking device (diode) in series or they are bypassed in the reverse direction by another parallel device (diode).

Basically, power semiconductor devices consist of a variety of diodes, transistors, and thyristors. Principle devices in these categories are symbolically shown in Figure 2.1. Often these symbols are drawn differently than shown in Figure 2.1. In the following paragraphs these three categories are described briefly and then specific devices are described in somewhat greater detail.

Diodes. The diodes are a family of two-layer devices with unidirectional conduction. A diode conducts in a forward (conducting) direction from anode to cathode, when its anode is positive with respect to the cathode. It does not have a gate to control conduction in its forward direction. The diode blocks conduction in the reverse direction, when its cathode is made positive with respect to its anode. The diode is a key component for several FACTS Controllers.

Transistors. The transistors are a family of three-layer devices. A transistor conducts in its forward direction when one of its electrodes, called a collector, is positive with respect to its other electrode, called an emitter, and when a turn-on voltage or current signal is applied to the third electrode, called the base. When the base voltage or current is less than what is needed for full turn-on, it will conduct while still holding partial anode to cathode voltage. Transistors are widely used in low- and medium-power applications. One type of transistor known as the Insulated Gate Bipolar Transistor (IGBT, Section 2.11) has progressed to become a choice in

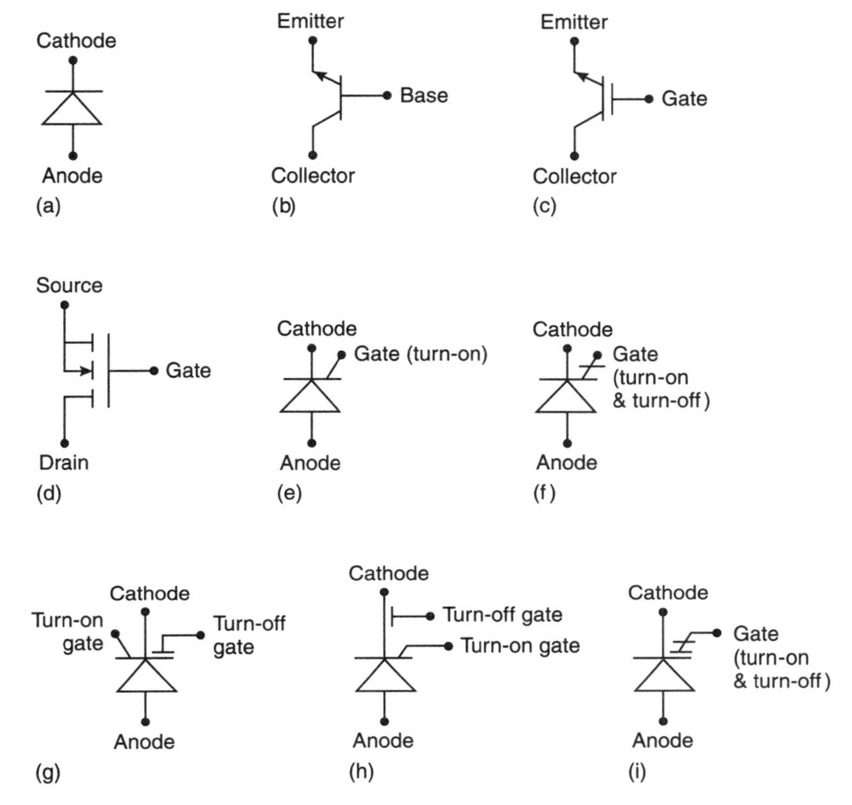

Figure 2.1 Power semiconductor devices: (a) Diode, (b) Transistor, (c) Integrated Gate Bipolar Transistor (IGBT), (d) MOS Field Effect Transistor (MOSFET), (e) Thyristor, (f) Gate Turn-Off (GTO) Thyristor- and Gate-Controlled Thyristor (GCT), (g) MOS Turn-Off Thyristor (MTO), (h) Emitter Turn-Off (ETO) Thyristor, and (i) MOS-Controlled Thyristor (MTO).

a wide range of low and medium power applications going up to several megawatts and even a few tens of megawatts. Thus IGBT is of some importance to FACTS Controllers. The MOS Field Effect Transistor (MOSFET), another type of transistor is only suitable for low voltages but with very fast turn-on and turn-off capability and is often used as a pilot gate device for thyristors.

Thyristors. The thyristors (Section 2.6) are a family of four-layer devices. A thyristor latches into full conduction in its forward direction when one of its electrodes (anode) is positive with respect to its other electrode (cathode) and a turn-on voltage or current signal (pulse) is applied to its third electrode (gate). Latched conduction is a key to low on-state conduction losses, as will be explained later in Section 2.6. Some thyristors are designed without gate-controlled turn-off capability, in which case the thyristor recovers from its latched conducting state to a nonconducting state only when the current is brought to zero by other means. Other thyristors are designed to have both gate-controlled turn-on and turn-off capability. The thyristor may be designed to block in both the forward and reverse direction (referred to as a symmetrical device) or it may be designed to block only in the forward direction (referred to as an asymmetrical device). Thyristors are the most important devices for FACTS Controllers.

Compared to thyristors, transistors generally have superior switching performance, in terms of faster switching and lower switching losses. On the other hand thyristors have lower on-state conduction losses and higher power handling capability than transistors. Advances are continuously being made to achieve devices with the best of both, i.e., low on-state and switching losses, while increasing their power handling capability.

2.1.1 Types of High-Power Thyristor Devices

Technically, the terms "thyristor" and the "Silicon Controlled Rectifier" apply to a basic family of four-layer controlled semiconductor devices, in which turn-on and turn-off depends on pnpn regenerative feedback (further described in Section 2.6). The name Silicon Controlled Rectifier (SCR) was given by the inventors and commercially pioneered by GE. In the context of a device which has turn-on but no turn-off capability, the term SCR was later changed by others to thyristor. With the emergence of a device with both turn-on and turn-off capability, named Gate Turn-Off Thyristor referred to as a GTO, the device with just the turn-on capability began to be referred to as "conventional thyristor" or just "thyristor." Other members of the thyristor or SCR family have acquired other names based on acronyms. In this book, use of the term thyristor is generally meant to be the conventional thyristor.

The thyristor starts conduction in a forward direction when a trigger current pulse is passed from gate to cathode, and rapidly latches on into full conduction with a low forward voltage drop (1.5 to 3 V depending on the type of thyristor and the current). As mentioned, the conventional thyristor cannot force its current back to zero; instead, it relies on the circuit itself for the current to come to zero. When the circuit current comes to zero, the thyristor recovers in a few tens of microseconds of reverse blocking voltage, following which it can block the forward voltage until the next turn-on pulse is applied.

Because of their low cost, high efficiency, ruggedness, and high voltage and current capability, conventional thyristors are extensively used when circuit configuration and cost-effective applications do not call for a turn-off capability. Often the turn-off capability does not offer sufficient benefits to justify higher cost and losses of the devices. The conventional thyristor has been the device of choice for almost all HVDC projects, some FACTS Controllers, and a large percentage of industrial applications. It is often referred to as the workhorse of the power electronics business.

There are several versions of thyristors with turn-off capability; key among these and relevant to the FACTS technology are:

- *Gate Turn-Off Thyristor* (Section 2.7), invented at GE, is now referred to as a GTO thyristor or simply a GTO. Like a conventional thyristor, it turns on in a fully conducting mode (latched mode) with a low forward voltage drop, when a turn-on current pulse is applied to its gate with respect to its cathode. Like a conventional thyristor, the GTO will turn off when the current naturally comes to zero, but the GTO also has turn-off capability when a turn-off pulse is applied to the gate in reverse direction. With an adequate turn-off pulse, the GTO rapidly turns off and recovers to withstand the forward voltage and be ready for the next turn-on pulse. The GTO is a widely used device for FACTS Controllers; however, because of its bulky gate drivers, and slow turn-off and costly snubbers, it is likely to be replaced in the coming years by more advanced GTOs and thyristors. These advanced GTOs, which in turn are a part of the thyristor family, have their own acronyms mentioned below and explained in a little more detail later in this chapter.

- *MOS Turn-Off Thyristor* (MTO, Section 2.8), invented at Silicon Power Corporation (SPCO) by Harshad Mehta, uses transistors to assist in turn-off and achieves fast turn-off capability with low turn-off switching losses. This device has been commercially introduced only recently, and has a good potential for use in the medium- to high-power industrial and FACTS Controllers.
- *Emitter Turn-Off Thyristor* (ETO, Section 2.9), developed at Virginia Power Electronics Center in collaboration with SPCO, is another variation on the GTO, and incorporates low voltage transistors in series with a high voltage GTO to achieve fast turn-off and low turn-off switching losses.
- *Integrated Gate-Commutated Thyristor* (GCT and IGCT, Section 2.10), developed by Mitsubishi and ABB, is basically a GTO with hard turn-off, which in combination with other advances in the packaging, achieves a fast turn-off and low turn-off switching losses. These devices have also been commercially introduced recently and have a good potential for application in industrial and FACTS Controllers.
- *MOS-Controlled Thyristor* (MCT, Section 2.12), invented by Victor Temple at GE, is the near ultimate in thyristor device, which includes integrated MOS structure for both fast turn-on and turn-off. Along with very low switching losses it also has low conduction losses. These devices have been commercially introduced for low power applications and have a good potential for use in FACTS Controllers.

Given the importance in terms of FACTS Controllers, the devices briefly discussed in this chapter include the diode, transistor, MOSFET, thyristor, GTO, MTO, ETO, IGCT, IGBT, and the MCT.

2.2 PRINCIPAL HIGH-POWER DEVICE CHARACTERISTICS AND REQUIREMENTS

2.2.1 Voltage and Current Rating

Device cells for high power are usually single crystal silicon wafers, 75–125 mm in diameter, and pushing toward 150 mm in diameter. The same diameter device can be made for higher voltage with lower current and vice versa.

Potentially, silicon crystal has a very high voltage breakdown strength of 200 kV/cm and a resistivity somewhere in between metals and insulators. Doping with impurities can alter its conduction characteristics. With doping, the number of carriers is increased, and as a result, its withstand voltage decreases and its current capability increases. Lower doping means higher voltage capability, but also higher forward voltage drop and lower current capability. To some extent the current and voltage capabilities are interchangeable, as mentioned above. A larger diameter naturally means higher current capability. A 125 mm device may have a current-carrying capability of 3000–4000 amperes and a voltage-withstand capability in the range of 6000–10,000 volts. It is not appropriate for this book to go into a detailed explanation of various trade-offs, but it is worth noting the important parameters of different devices.

With higher device ratings, the total number of devices as well as the cost of all the surrounding components decreases. The highest blocking capability along with other desirable characteristics is somewhere in the range of 8–10 kV for thyristors, 5–8 kV for GTOs, and 3–5 kV for IGBTs. In a circuit, after making various allowances for overvoltages and redundancy, the useable device voltage will be about half the blocking voltage capability. More often than not, it will be necessary to connect devices in series for high-voltage valves. Ensuring equal sharing of voltage during turn-on, turn-off, and dynamic voltage changes becomes a major exercise for a valve designer in considering trade-off among various means to do so and deciding on the best mix. One of these means is the matching of devices, especially the device-switching characteristics.

Large power devices can be designed to handle several thousand amperes of load current, which generally makes it unnecessary to connect devices in parallel. However, it is often the short-circuit current duty that determines the required current capacity, in which case connecting two matched devices directly in parallel on the same heat sink is a good solution. Devices are usually required to ride through to a blocked state after one cycle of offset fault current in the application circuit. While it is a common practice to use fuses in industrial power electronics, use of fuses is undesirable in high-voltage applications such as FACTS Controllers. The device selection must therefore consider all possible fault and protection scenarios to decide on the current and also voltage margins and redundancy. The thyristor family of devices can carry a large overload current for short periods and a very large single-cycle fault current without failures. The thyristor and diode family of devices fail in a short circuit with low-voltage drop, so the circuit may continue to operate if the remaining devices in the circuit can perform the needed function.

As dictated by the market needs of converters (discussed in Chapter 3: *Voltage-Sourced Converters*), most of the devices made with turn-off capability, are made with no reverse blocking capability. They are therefore referred to as asymmetric turn-off devices, often just turn-off devices. Without the reverse voltage capability requirement the device can be thinner, have lower forward conduction and lower switching losses. Conversely higher forward withstand voltage can be achieved with asymmetric devices. As it turns out the voltage-sourced converters also require a reverse diode in parallel with each main device. These diodes are usually special diodes with lowest possible reverse leakage current because of their impact on the turn-on requirements of the main devices.

However, converters discussed in Chapter 4, *Current-Sourced Converters,* need devices with reverse voltage withstand capability. Nevertheless, because of the high-volume of asymmetric power devices, it is not uncommon for many industrial applications with a focus on the first cost, to consider use of a diode in series with the asymmetric main device to obtain reverse blocking capability.

2.2.2 Losses and Speed of Switching

Apart from the voltage withstand and current-carrying capabilities, there are many characteristics that are important to the devices. The most important among these are:

- Forward-voltage drop and consequent losses during full conducting state (on-state losses). Losses have to be rapidly removed from the wafer through the package and ultimately to the cooling medium and removing that heat represents a high cost.

- Speed of switching. Transition from a fully conducting to a fully nonconducting state (turn-off) with corresponding high dv/dt just after turn-off, and from a fully nonconducting to a fully conducting state (turn-on) with corresponding high di/dt during the turn-off are very important parameters. They dictate the size, cost, and losses of snubber circuits needed to soften high dv/dt and di/dt, ease of series connection of devices, and the useable device current and voltage rating.
- Switching losses. During the turn-on, the forward current rises; before the forward voltage falls and during turn-off of the turn-off devices, the forward voltage rises before the current falls. Simultaneous existence of high voltage and current in the device represents power losses. Being repetitive, they represent a significant part of the losses, and often exceed the on-state conduction losses. In a power semiconductor design, there is a trade-off between switching losses and forward voltage drop (on-state losses), which also means that the optimization of device design is a function of the application circuit topology. Even though normal system frequency is 50 or 60 Hz, as will be seen later in Chapters 3 and 4, a type of converters called "pulse-width modulation (PWM)" converters have high internal frequency of hundreds of Hz, to even a few kilo-Hz for high-power applications. With many times more switching events, the switching losses can become a dominant part of the total losses in PWM converters.
- The gate-driver power and the energy requirement are a very important part of the losses and total equipment cost. With large and long current pulse requirements, for turn-on and turn-off, not only can these losses be important in relation to the total losses, the cost of the driver circuit and power supply can be higher than the device itself. The size of all components that accompany a power device increases the stray inductance and capacitance, which in turn impacts the stresses on the devices, switching time and snubber losses. Given the high importance of coordination of the device and the driver design and packaging, the future trend is to purchase the device and the driver as a single package from the device supplier.

Serious attention to losses is important for two reasons:

1. For the obvious reason that losses are a cost liability for the user. Losses are invariably evaluated by utilities and often by the industrial customers on a lifetime present worth basis, and this value can be from $1000 to $5000 per kilowatt losses for evaluation of purchase price. If a FACTS Converter costs $100 per kilowatt and its losses are 2% (0.02 kW loss per kW rating), the value of losses for an evaluated value of $2000 per kilowatt will be $40 per kilowatt, equal to 40% of the purchase price of the converter itself. Therefore the efficiency of a complete FACTS Controller of several hundred MW rating needs to be better than 98% and the converter valve losses have to be less than 1%.
2. The device losses have to be efficiently removed from inside the wafer to outside the sealed, high-voltage, insulating package and on to the external cooling medium. For this reason, packaging and cooling of the devices is a formidable challenge to ensure that its wafer temperature does not exceed

the safe operating level, about 100°C, with safe switching characteristics and adequate margin for the overload and short circuit currents. More often than not, fault current determines the normal useable rating of the devices. Higher losses mean higher cost of packaging, further losses and cost in disposing the thermal losses to water or air, as well as the size and weight of the complete equipment.

2.2.3 Parameter Trade-Off of Devices

The cost of devices is also related to production yield of good devices, which are then graded into various ratings. This therefore calls for good quality control all the way from the starting material to the finished product and including the quality of the electric power supply in the production plant. All power devices for high-power Controllers are individually tested, as is the practice with HVDC converters, and their record kept for future replacement service.

Apart from the trade-off between voltage and current capability, other trade-off parameters include:

- power requirements for the gate
- di/dt capability
- dv/dt capability
- turn-on time and turn-off time
- turn-on and turn-off capability (so-called Safe Operating Area [SOA])
- uniformity of characteristics
- quality of starting silicon wafers
- class of clean environment for manufacturing of devices, etc.

Advanced design and processing methods have been developed and continue to be developed. Given the large number of variables, a device manufacturer divides up the market into various types of devices tailored in accordance with the application and market size. It is also common for device manufacturers to tailor make the devices for individual large customers and even for individual large project orders, such as HVDC and FACTS projects.

The switching speed, switching losses, size, and cost of snubber circuits and associated losses, usually attributed to the power semiconductor devices, largely result from the fact that the devices are sold separately from the gate-drive circuits and from the snubber circuits. It should be clear from the discussion later in this chapter that the device performance is intertwined with the gate driver, snubber, and circuit-bus design for connecting the device-modules into a converter, in that order of priority. Major improvements can be made in the application cost if the device, the gate driver, the snubber, and the associated bus-work very close to the devices, were assembled and sold by the device supplier as a building block. In fact the electrical–mechanical integration of the device wafer and its gate-drive circuit provides major benefits downstream all the way to the application. For the low- and medium-power industrial applications, there has been an increasing practice to sell assemblies of several devices in a mold or a package, representing a circuit or a part of a circuit. This practice, while reducing the packaging cost, does not really serve the need for integration, which should start with the device wafer and its gate driver. It is with this intent that the U.S. Office of Naval Research (ONR), undertook a power electronics program named Power Electronics Building Block (PEBB), addressing all aspects of integration including the device, gate driver, packaging, and bus-work, which have leverage on reduction of overall conversion cost, losses, weight, and size. This program has set in motion and enabled significant advances, and as a matter of fact this trend has begun and the potential benefits are now being recognized. Devices are being offered along with prepackaged gate driver and snubber circuits under a variety of trade names, not necessarily with the name PEBB.

2.3 POWER DEVICE MATERIAL

Power semiconductor devices are based on high-purity, single-crystal silicon. Single crystals several meters long and with the required diameter (up to 150 mm) are grown in the so-called *Float Zone* furnaces. Then this huge crystal is sliced into thin wafers to be turned into power devices through numerous process steps.

Pure silicon atoms have four electron bonds per atom with neighboring atoms in the lattice. It has high resistivity and very high dielectric strength (over 200 kV/cm). Its resistivity and charge carriers available for conduction can be changed, shaped in layers, and graded by implantation of specific impurities (doping). With different impurities, levels and shapes of doping, along with the high technology of photolithography, laser cutting, etching, insulation, and packaging, large finished devices are produced.

There are two types of impurities for implantation in silicon wafers: *donor* and *acceptor*. Phosphorus is a donor because it has five electrons as against four for silicon. So when a phosphorus atom is implanted in a silicon lattice, it becomes a fixed atom site with one extra electron. This extra electron can be dislodged easily with an electric field. When an electron leaves the phosphorus atom site, it results in a positively charged site (called a hole) waiting to be filled by another electron from another site, which in turn acquires a hole. Thus in a directed conduction, with an applied electric field, there are electrons and holes available for the conduction process. Phosphorus doping is known as *n* doping for adding negative particles (electrons) for the conduction process. When the silicon is lightly doped with phosphorus, the doping is denoted as n^- doping and when it is heavily doped, it is denoted as n^+ doping.

Another doping agent, boron, is the opposite of phosphorus. It has three electrons per atom, and when a boron atom is implanted in the silicon lattice, it becomes a site of one empty location (a hole which can be filled by a passing electron). When a hole is filled at a boron atom site, it results in a negatively charged site waiting to be neutralized by a hole from another site which then acquires a negative charge, thereby creating a specter of moving holes. Boron doping is known as p doping for adding positive holes for the conduction process. P doped silicon can be lightly doped p^-, or heavily doped p^+.

Thus in a directed conduction with an applied electric field, there are electrons available in a n doped silicon, and holes available in a p doped silicon, for the conduction process.

The holes in p doped silicon are called the majority carriers and any electrons in a p doped silicon are called the minority carriers. In the n doped layer, electrons are called the majority carriers, and holes as the minority carriers.

In addition to the carriers introduced by doping in a power device, there are also the so-called intrinsic carriers of equal number of electrons and holes, which are generated by thermal excitation. These carriers are continuously generated and recombine in accordance to their lifetime, and achieve an equilibrium density of carriers from about 10^{10}–$10^{13}/cm^3$ over a range of about 0°C to 100°C.

Achieving high withstand voltage requires low doping (fewer carriers), which results in intrinsic carriers becoming significant in numbers for the conduction process. Being temperature dependent, the intrinsic carriers become significant and even dominant carriers during high current levels.

As a starting material for high-voltage, high-power semiconductor devices, silicon slices are irradiated with neutrons in a reactor. Depending on the radiation, the right amount of silicon atoms are converted to phosphorous atoms, thus producing an n doped silicon, but with a low and uniform doping concentration in the range of $5 \times 10^{12}/cm^3$, comparable to the intrinsic carrier concentration. With diffusion in high-temperature furnaces and other processes, this thin slice with very low n doping is modified to have multiple doping profiles in layers, channels, etc. required for specific devices. Doping processes are not discussed in this book.

2.4 DIODE (Pn Junction)

The diode is symbolically shown in Figure 2.2(a), and the cross-sectional structure of its wafer junction is conceptually shown in Figure 2.2(b). The importance of diodes for FACTS Controllers stems from the possibility that:

1. A diode converter can be used as a simple low cost and efficient converter, to supply active power in a FACTS Controller.
2. A diode is connected across each turn-off thyristor in voltage-sourced converters, and also connections of intermediate levels in multilevel voltage-sourced converters (discussed in Chapter 3).
3. A diode may be connected in series with each turn-off thyristor for reverse blocking of voltage (discussed in Chapter 4).
4. Diodes are used in snubber and gate-drive circuits.

As a matter of fact, on an overall basis almost half of the devices used in FACTS may well be diodes.

A diode is a single junction device of p and n layers in a silicon wafer [Figure 2.2(b)].

Layer p is deficient in electrons (has holes as majority carriers), and similarly layer n has surplus electrons as majority carriers. As mentioned earlier, these p and n layers are obtained by doping impurities in a silicon slice. To explain the one-way

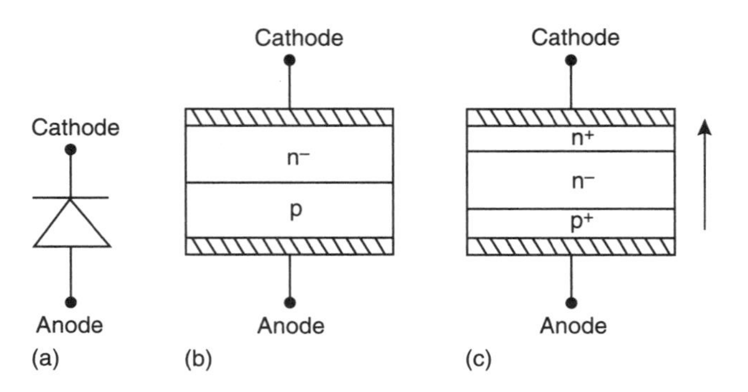

Figure 2.2 Diode: (a) diode symbol, (b) diode structure, and (c) diode structure.

conduction process across a p-n junction (a diode), the carriers are generated and participate in conduction, with application of voltage that makes p side positive and n side negative. This external force causes holes from p side to go over the junction to the n side and electrons from the n side to go over the junction to the p side. However, if the voltage is reversed, the holes and electrons move away from the junction creating an internal counter field, which stops the current from flowing. An extended explanation of a diode is necessary in order that the concepts of multiple junctions for other devices are better understood.

These electrons and holes can move under two physical mechanisms:

1. By diffusion caused by differences in carrier density
2. By drift in a direction dictated by an externally applied voltage

Without any external voltage, the pn junction acquires a very small electric field (less than 1 V). This occurs due to diffusion of some holes from the p side into the n side and diffusion of some electrons from the n into the p side. The boundary of this space charge is formed on the two sides of the junction and is tightly constrained around the junction by a counter force (opposite electric field) created by the sites vacated by the diffused holes and electrons. This small electric field is positive on the p side and negative on the n side. When the anode is made positive with respect to the cathode, electrons are drawn from the n side to the p side, and holes from the p side to the n side. Once the small electric field barrier caused by diffusion is overcome with a small voltage of less than 1 V, a large current can readily flow by drift with the positive driving voltage. The voltage drop will rise further with current on account of the resistance of the silicon, to about 1.5–3.0 V at full-rated current.

When the cathode is made positive with respect to the anode, electrons are pulled away from its side of the junction in the n layer, and holes are pulled away from its side of the junction in the p layer. This creates a large electric field near the junction, positive on the cathode side and negative on the anode side to counter the external applied voltage and therefore no conduction takes place (in an ideal diode).

This electric field region formed at the junction is known as the depletion region. For higher doping, the field is more intense and therefore the depletion region is thinner and vice versa. Within the depletion region, the field is highest at the junction. As the reverse voltage increases, the depletion layer expands, essentially on the n^- side and the diode will break down if the reverse voltage is made high enough to expand the depletion layer to full width of the n^- region.

When during the conduction-state, the holes cross the boundary and go into the n layer, they become the minority carriers. Similarly, as the electrons leave the n region and go into the p region, they also become the minority carriers. Thus the device is termed a minority carrier device, as long as carriers from the doping sites dominate the conduction.

In n-type power diodes [Figure 2.2(c)], the p side is heavily doped (p^+), which results in a very narrow depletion region on the p^+ side, and the n side is lightly doped (n^-) near the junction, which results in a wide depletion region on the n^- side. When a reverse voltage is applied (cathode made positive with respect to anode), the n^- side expands much more than the p side. Therefore the n^- side is made thick and supports most (almost all) of the reverse voltage. The n^- layer may in fact be so lightly doped that the intrinsic carriers become a significant part of the carriers on the n^- side. Thickness of the device is increased for higher reverse voltage capability of the device to accommodate the expanded depletion layer. Increase in thickness in turn will increase the resistance and hence the on-state losses. The n^- side is known as the drift region because past the actual depletion layer, corresponding to the applied reverse voltage, the conduction takes place by diffusion of a small number of thermal carriers across the remaining thickness of the n^- layer. Almost all silicon devices are designed for the widest region of the device to be n type.

In power diodes, the n layer is also heavily doped (n^+) but well away from the junction and near the end where the cathode plate is connected [Figure 2.2(c)]. The function of both the p^+ and n^+ regions at the ends, both with very large numbers of doping-based carriers, is to avoid the depletion region during reverse voltage from reaching the metal itself. Another important function of the n^+ layer is that once the depletion layer reaches the n^+ boundary, the stress across the n^- layer will start to even out and thereby take higher voltage. This is known as "punch-through" operation and for the same reverse voltage rating, it allows reduction of the thickness of the n^- layer and consequent reduction of on-state losses. The on-state losses are also further reduced because of the availability of carriers from the n^+ layer during forward conduction. This heavy doping on the n side next to the anode plate is also done in several other devices as will be discussed later.

For high-power diodes as well as other high-power silicon power devices, the edge is contoured (physically and with doping profile) and insulated to prevent edge flashover. This is necessary because the breakdown level of the atmosphere around the edge of a device is much lower (about 1/10th) than the voltage stress across the device thickness. The transition from the silicon wafer edge to the atmosphere outside the edge (passivation) is a very complex matter in itself, and not discussed in this book. The device is packaged to provide sealed and rugged wafer containment with adequate external insulation between anode and cathode and good thermal contact between the wafer and outside for efficient heat removal from inside to the outside. The packaging of devices, which effectively addresses a combination of electrical, thermal and mechanical stresses, is a major challenge for all power electronics devices.

Usually, in application circuits, when the current comes to zero after conduction, voltage across the diode jumps to some negative value. This causes some reverse current to flow for a short time (microseconds or tens of microseconds), to draw out the excess internal charges and restore the depletion layer corresponding to the applied reverse voltage. This reverse current flow in diodes results in an increase in the current that the turn-off devices have to turn on in voltage-sourced converters (Section 2.7.1), which in turn increases the turn-on losses of these devices. Therefore the diodes used in the voltage-sourced converters in parallel with the turn-off devices must have very fast turn-off and low stored charge. Advanced diodes have been developed in terms of doping profiles in order to achieve higher speed and reduced stored charge, but they are not discussed here. Improved diodes in terms of low reverse turn-off current will have a significant leverage on the voltage-sourced converter cost.

2.5 TRANSISTOR

The transistor is a family of three layer (with two junctions) devices. Some basics of the transistor are covered in this section so that the concepts of high-power devices are better understood.

The transistor is equivalent to two p-n diode junctions stacked in opposite directions to one another. There are two types of transistors:

1. The pnp transistor [Figure 2.3(a,b)], which corresponds to a stack of pn (diode) over np (reversed diode), giving a device which has two p layers with an n layer in between. The anode (emitter) side p layer is made wide, n (base) is narrow, and the cathode (collector) side p layer is narrow and heavily doped.
2. npn transistor [Figure 2.3(c,d)] which corresponds to a stack of np (reverse diode) over a pn (diode) giving a device with a p layer sandwiched between two n layers.

Considering just the npn transistor, preferred for power transistors, one of the outer n layers is designed with heavy doping (n^+) called the *emitter,* the other n layer is called the *collector,* and the middle p layer becomes the base. When the collector is made positive with respect to the emitter by an external driving voltage, no current flows because it is blocked by the depletion layer formed at the n-p junction on the collector side. This junction is made to withstand high voltage with low doping of the p layer. Now if another small external voltage is applied to the gate, with the gate positive with respect to the emitter, electrons flow from the n^+ emitter to the p base (current from the gate to the emitter). As the electrons flow from the n^+ emitter to the base, electrons are also accelerated by the electric field of the depletion layer to the collector, i.e., current flows through the device as shown by the arrow in Figure 2.3(d).

Because the injected electron from n^+ layer is a function of the base current,

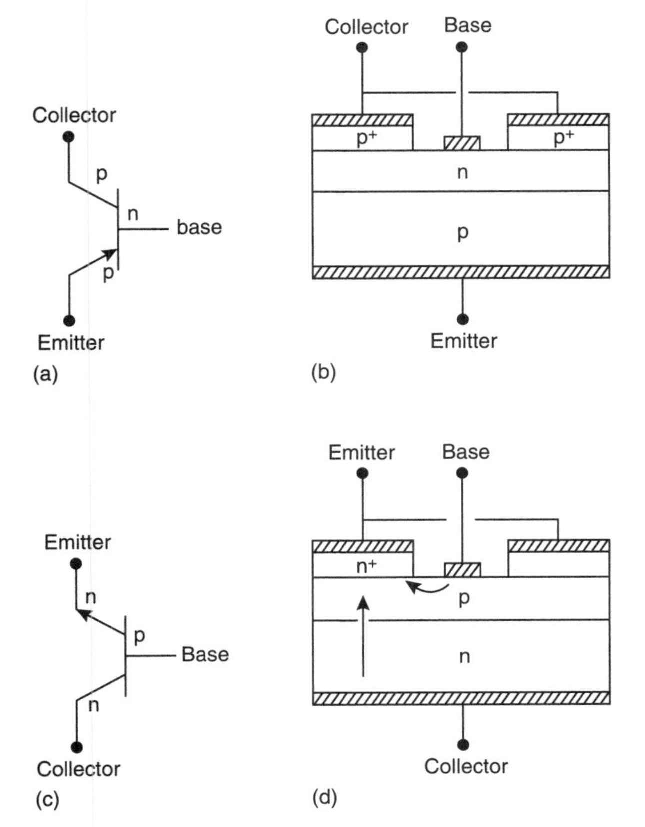

Figure 2.3 Transistor: (a) transistor symbol (pnp), (b) pnp structure, (c) transistor symbol (npn), and (d) npn structure.

the current flow saturates, and is limited by the depletion layer voltage. Figure 2.4 shows the forward-conducting characteristics of device current versus device voltage for different values of base currents. The base current determines the device saturation current. In normal operation, with high base current, the device current and the forward voltage drop in a power device will be limited along the steep line on the left side of the curves, and the voltage drop and hence the losses will be low. But if the base current is limited, the device itself will hold some of the voltage, and the device current will be limited at the saturation line for the corresponding base current. As a matter of fact this feature is used in low power converters for current limiting during an external fault, and then the devices are rapidly turned off in a safe manner.

It should be noted that for a power device, the wafer is fabricated in a way that a large number of gate connection lines are brought out through the top layer, and effectively a power transistor may have a large number of small devices in parallel.

Because of the relatively low gain (base-to-device current ratio), the devices are made with amplifying stage(s) as shown in Figure 2.5. Such transistors are known as *darlington* transistors.

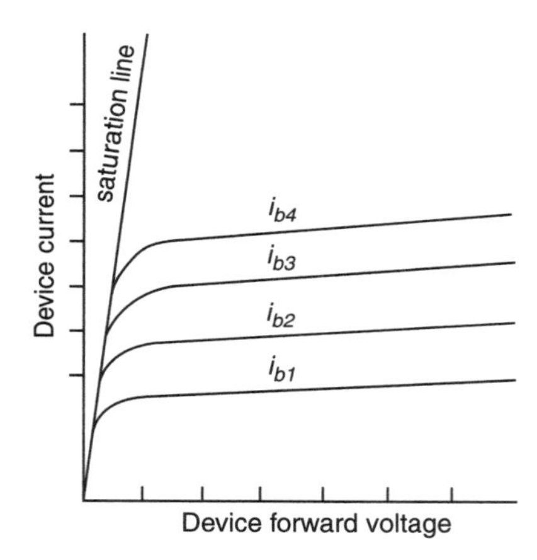

Figure 2.4 Transistor voltage-current characteristic for different values of base current.

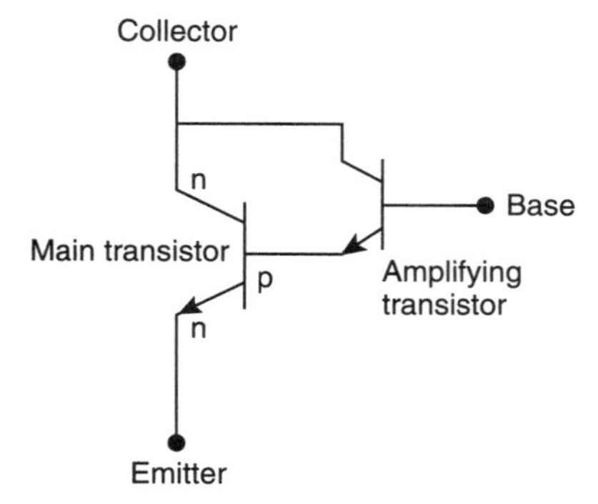

Figure 2.5 Transistor with amplifying stage.

2.5.1 MOSFET

There are many types of transistors. A type of transistor capable of fast-switching speed and low-switching losses is the so-called Metal-Oxide Semiconductor Field Effect Transistor (MOSFET) which is gate controlled by electric field (voltage) rather than current. This is achieved by capacitive coupling of the gate to the device. Figure 2.6 shows its structure and equivalent circuit. MOSFET is extensively used for low power (kilowatts) applications and is unsuitable for high power. However, they are useful devices when in conjunction with advanced GTOs, as explained later under Sections 2.8 on MTO and 2.9 on ETO; for this reason, MOSFET is briefly described here.

MOSFET can be a pnp or npn device—only npn structure is shown in Figure 2.6. There is a silicon-oxide (SiO) dielectric layer between the gate metal and the n^+ and p junction. The principal advantage of MOS gate is that voltage, instead of current, is applied to the gate with respect to the source to fully or partially block the device by creating space charge around the tiny gate areas. When the gate is given a sufficiently positive voltage with respect to the emitter, the effect of its electric field pulls electrons from the n^+ layer into the p layer. This opens a channel closest to the gate, which in turn allows the current to flow from the drain (collector) to the source (emitter).

MOSFET is heavily doped on the drain side to create an n^+ buffer below the n^- drift layer. As discussed in Section 2.4 on diodes this buffer prevents the depletion layer from reaching the metal, evens out the voltage stress across the n^- layer, and also reduces the forward voltage drop during conduction. Provision of the buffer layer also makes it an asymmetrical device with rather low reverse voltage capability.

MOSFETs require low gate energy, and have very fast switching speed and low switching losses. Unfortunately MOSFETs have high forward on-state resistance, and hence high on-state losses, which makes them unsuitable for power devices, but they are excellent as gate amplifying devices.

MOSFETs have similar forward current-voltage characteristics, as that for a transistor shown in Figure 2.4; however, the base current is replaced by the gate voltage.

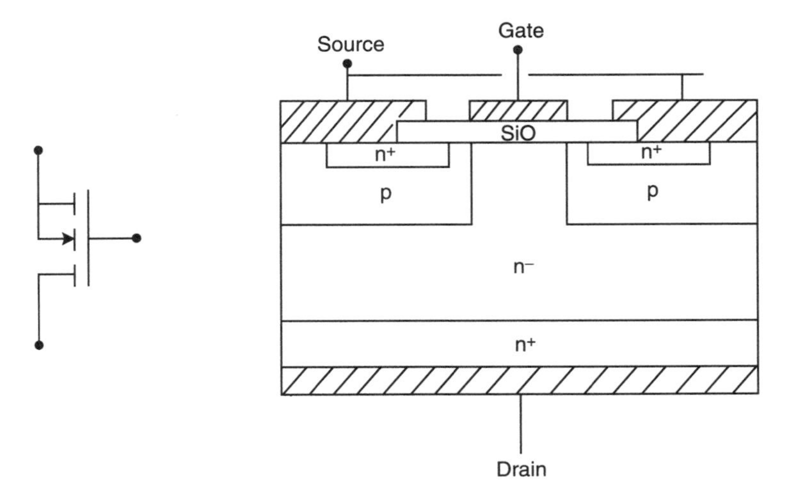

Figure 2.6 Power MOSFET: (a) MOSFET symbol, (b) MOSFET structure.

2.6 THYRISTOR (without Turn-Off Capability)

The thyristor, Figure 2.7, is a three-junction, four-layer device shown by its symbol in Figure 2.7(a) and structure in Figure 2.7(b). The thyristor is a unidirectional switch, which once turned on by a trigger pulse, latches into conduction with the lowest forward voltage drop of 1.5 to 3 V at its continuous rated current. It does not have capability to turn off its current, so that it recovers its turned-off state only when the external circuit causes the current to come to zero.

The thyristor is referred to as a workhorse of power electronics. In a large number of applications, turn-off capability is not necessary. Without turn-off capability, the resulting device can have higher voltage and/or rating, cost less than one-half, require a simple control circuit, has lower losses, etc. compared to a device with turn-off capability. Therefore the choice in favor of a more expensive and higher loss device with turn-off capability will occur when there is a decisive application advantage, which is often the case for FACTS Controllers, as would be evident in other chapters.

As shown in Figure 2.7(c) and (d), the thyristor is equivalent to the integration of two transistors, pnp and npn. When a positive gate trigger is applied to the p gate of the upper npn transistor with respect to the n^+ emitter [cathode Figure 2.7(d)] it starts to conduct. Current through the npn transistor becomes the gate current of the pnp transistor as shown by the arrows, causing it to conduct as well. The current through this pnp transistor in turn becomes the gate current of the npn transistor giving a regenerative effect to latched conduction with low forward voltage drop with the current flow essentially limited by the external circuit. What is important is that due to the internal regenerative action into saturation, once the thyristor is turned on, the internal p and n layers become saturated with electrons and holes and act like a short circuit in the forward direction. The whole device behaves like a single pn junction device (a diode). Thus its forward on-state voltage drop corresponds to only one junction (even though it has three junctions) compared to two junctions in transistor like devices such as the MOSFET and IGBT.

It is obvious from the diagram that the n base of the lower transistor can also be used for turn-on, however, the n base requires more current, and therefore the p base is used as the gate for turn-on in thyristors.

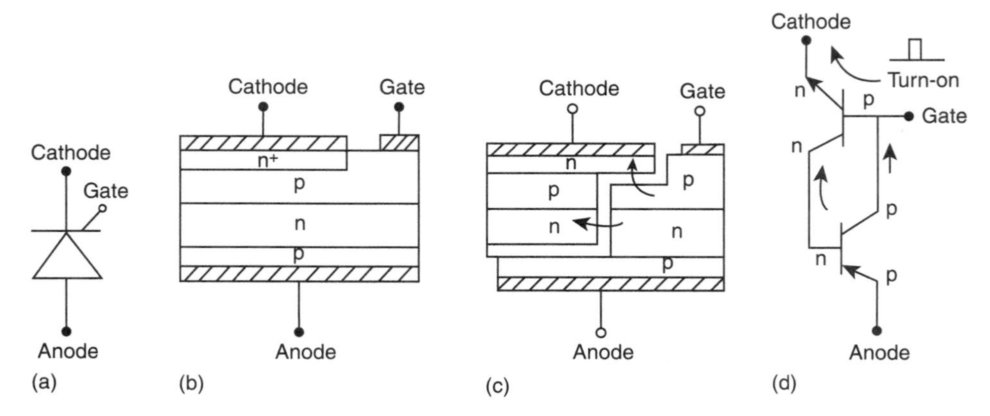

Figure 2.7 Thyristor: (a) thyristor symbol, (b) thyristor structure, (c) two-transistor structure, and (d) thyristor equivalent circuit.

When current comes to zero (due to external circuit), the thyristor is still full of electron and hole carriers in the center pn region, which must be removed or recombined for the device to recover and be ready for blocking the voltage when it becomes positive again. Fortunately in practical thyristor-based circuits, this stored charge is removed with the application of negative voltage across the device immediately after the current zero, in addition to the slower process of recombination of charge carriers to the point of thermal equilibrium. Thus the turn-off time, which can be a few to a few tens of microseconds, depends on the reverse voltage after current zero, and has to be carefully considered for specific applications. This turn-off time must elapse before any positive voltage can be safely applied.

In a large thyristor wafer, the gate structure is brought out through the cathode side at the top, as shown in the wafer photograph in Figure 2.8. Furthermore, several amplifying stages are provided in concentric circles at the center, in order to decrease the required external gate pulse current. It is essential to rapidly spread out the turn-on current over the whole device. This is done through a variety of gate structures; one such structure is shown in Figure 2.8. The required structure is fabricated through the use of masks, photolithography, etching, oxide insulation, etc., starting with a slice of n^- doped single-crystal material.

Spreading out the turn on rapidly across the whole wafer is a very important feature, particularly to ensure that the device can handle high fault current for short periods as well as to minimize the turn on losses.

It is also appropriate to consider adding another high-voltage, very low current external pilot thyristor in order to increase the gain and reduce the gate turn-on power at the thyristor level. Such a device would be inexpensive because of its very low current rating.

A thyristor can also be turned on by hitting the gate region with light of appropriate bandwidth. The direct light triggered thyristor allows triggering of the thyristor directly from the control circuits via an optical fiber. As an alternative, the external

Figure 2.8 A 125 mm diameter thyristor wafer standing on top of the complete thyristor package, rated for 5 kV and 5000 A. The gate structure is of involute type with an amplifying pilot gate at the center. The package itself is of a unique lightweight silicon sandwich (LSS) with bonded inert silicon slice. The two terminals are for application of the gate pulse between the gate and cathode. (Courtesy Silicon Power Corporation SPCO.)

pilot thyristor (mentioned above) may be a light triggered thyristor with the main thyristor as an electrically triggered thyristor.

Application of positive anode to cathode voltage with high rate of rise (dv/dt) can also turn on the device. This happens because the capacitive coupling of the cathode to the gate and high dv/dt causes just enough current to turn the device on. This is not a safe way to turn on a thyristor because such a turn-on can occur at a weak spot, does not spread rapidly, and could damage the device. Unsafe turn-on will also occur if the forward voltage is too high, thus creating charge carriers in a weak spot through acceleration of internal charge carriers. This also suggests that a device can be made with a deliberately designed weak spot from where safe turn-on can be designed into the device. Such devices with self-protection and optional triggering have been introduced in recent HVDC projects.

Another important aspect is that when a turn-on pulse is applied, there has to be enough anode-to-cathode forward voltage, or rate of rise of voltage, to cause a rapid turn-on. Insufficient voltage can lead to soft turn-on with device voltage falling slowly while the current is rising. This can lead to a high turn-on loss in certain areas of the device and possible damage. Depending on the application, the device has to be designed for the specified minimum turn-on voltage and the turn-on pulse is blocked if the forward voltage is inadequate.

At high temperatures, the thyristor has a negative temperature coefficient. Thus it has to be designed to ensure a uniform internal turn-on and turn-off. Being a high-voltage device, it includes doping-based carriers as well as a large number of intrinsic carriers. With higher temperatures the number of thermal carriers and hence total carriers increase and this leads to lower forward voltage drop.

Once a thyristor is turned on there is a need to sustain a minimum anode-cathode current for the device to stay turned on. This minimum current is usually a few percent of the rated current. The gate drive is usually arranged to send another turn-on pulse as needed.

Generally thyristors have a large overload capability. They may have two times normal over-current capability for several seconds, ten times for several cycles, and 50 times fully offset short-circuit current for one cycle.

2.7 GATE TURN-OFF THYRISTOR (GTO)

Basically, the gate turn-off (GTO) thyristor is similar to the conventional thyristor and essentially most aspects discussed above in Section 2.6 apply to GTOs as well. The GTO (Figure 2.9) like the thyristor, is a latch-on device, but it is also a latch-off device. Discussion of GTO in this section refers to the conventional GTO without the recent advances made in devices made under different acronyms, which are discussed in later sections.

Consider the equivalent circuit, Figure 2.9(c), which is the same as the circuit of Figure 2.7(c) for the thyristor, except that the turn-off has been added between the gate and the cathode in parallel with the gate turn-on (shown only by arrows in the equivalent circuit). If a large pulse current is passed from the cathode to the gate to take away sufficient charge carriers from the cathode, i.e., from the emitter of the upper pnp transistor, the npn transistor will be drawn out of the regenerative action. As the upper transistor turns off, the lower transistor is left with an open gate, and the device returns to a nonconducting state. However, the required gate current for turn-off is quite large. Whereas the gate current pulse required for turn-on may be 3–5%, i.e., 30 A, for only 10 μsec for a 1000 A device, the gate current required for

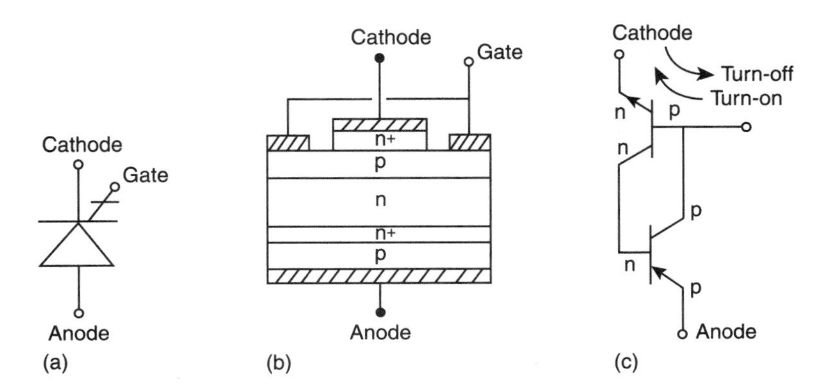

Figure 2.9 Gate turn-off (GTO) thyristor: (a) GTO symbol, (b) GTO structure, and (c) GTO equivalent circuit.

turn-off would be more like 30–50%, i.e., 300 A or larger for 20–50 μsec. The voltage required to drive the high-current pulse is low (about 10–20 V), and being a pulse of 20–50 μsec duration, the energy required for turn-off is not very large. Yet the losses are large enough to be a significant economic liability in terms of losses and cooling requirements, when considering the number of valves and turn-off events in a converter. Turn-off energy required is 10 to 20 times that required for GTO turn-on, and the GTO turn-on energy required is 10 to 20 times that for a thyristor. The cost and size of turn-off circuits for GTO are comparable to the device cost itself.

Another consideration is that the turn-off has to be uniformly effective over the entire device. Whereas in a thyristor, there is one cathode with a single gate structure spread out across the device, successful GTO turn-off requires dividing up the cathode into several thousand islands with a common gate-line, which surrounds each and every cathode island (Figure 2.10). Thus a GTO consists of a large number of thyristor cathodes with a common gate, drift region, and anode. Given the complex structure, state-of-the-art GTOs do not have built-in amplifying gates. Consequently, the total available area on the device for a cathode decreases to about 50% compared to a thyristor. Therefore a GTO's forward voltage drop is about 50% higher than for a thyristor but still 50% lower than that of a transistor (IGBT) of the same rating.

The general process of making GTOs is about the same as that for thyristors, although due to complications of the cathode and grid distribution, its process requires a cleaner room, yield may be less, and cost perhaps twice that of a thyristor for the same converter ratings. As for a thyristor, there are trade-offs between voltage, current, *di*/*dt*, *dv*/*dt*, switching times, forward losses, switching losses, etc., for a GTO design.

Most of the market for GTOs is for voltage-sourced converters in which a fast recovery diode is connected in reverse across each GTO, which means that GTOs do not need reverse voltage capability. This also provides beneficial tradeoffs for other parameters, particularly the voltage drop and higher voltage and current ratings. This is achieved by a so-called buffer layer, a heavily doped n^+ layer at the end of the n^- layer. Such GTOs are known as *asymmetric* GTOs.

Like a thyristor, the continuous operating junction temperature limit is about 100°C, after making allowances for the fault current requirements. Like a thyristor, a GTO is capable of surviving a high, short-time over-current (10 times for one offset cycle) as long as it is not required to turn off that current. Failure mechanisms are

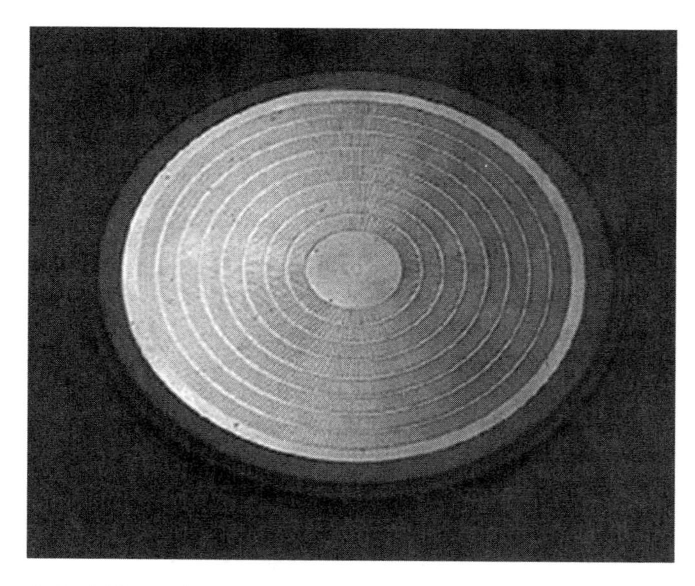

Figure 2.10 A 77 mm diameter Gate Turn-Off (GTO) thyristor wafer rated 4.5 kV and 2000 A. The cathode structure consists of a large number of finger-like islands arranged in ring formation. The remaining surface is the gate. [Courtesy Silicon Power Corporation (SPCO).]

also similar, and the edge requires appropriate contouring to reduce voltage stress and passivation to avoid a flashover around the edge.

In a thyristor, since the current zero is brought about by the external system, the voltage across the device automatically becomes negative immediately after the current zero. In a GTO on the other hand, the GTO is turned off while the circuit is driving in the forward direction. Therefore for a successful turn-off it is necessary to reduce the rate-of-rise of forward voltage with the help of a damping circuit.

In a GTO, the anode side pn^- junction is lightly doped and designed to support almost all of the blocking voltage, essentially on the n^- side. On the other hand, the cathode side pn junction is heavily doped on both sides and its breakdown voltage may be about 20 V.

2.7.1 Turn-On and Turn-Off Process

Apart from the gate-drive power, GTOs also have high switching losses and it is important to appreciate the turn-on and turn-off process with associated device stresses and losses. Figure 2.11 shows simplified waveforms for turn-on and turn-off process.

For turn-on, a 10 μsec current pulse of >5% of the load current and with a fast rise time limited largely by the gate circuit inductance is applied from the gate to cathode. However, there is a delay of a few microseconds, before the anode-cathode current begins to rise and voltage begins to fall. The current rises at the rate limited by the circuit as required for safe turn-on of the device such that all cathode islands turn on evenly. Also, given the circuit topology of voltage-sourced converters (Chapter 3), the GTO turn-on is accompanied by turning off of a reverse-conducting diode in another valve of the same phase. Therefore the GTO has to turn-on the main circuit

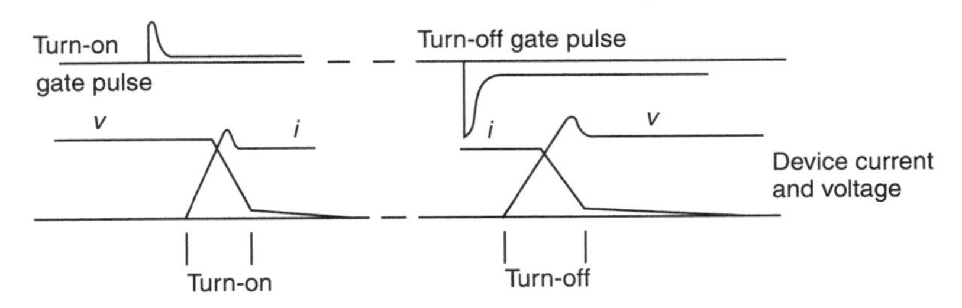

Figure 2.11 GTO turn-on and turn-off process: (a) turn-on and (b) turn-off.

current plus a large reverse leakage current of that diode. During this process of rising current, the anode-cathode voltage falls slowly in accordance with the plasma spreading time, ultimately to its on-state low-voltage level.

Following the full turn-on, it is then necessary to maintain some gate current of about 0.5% to ensure that the gate does not unlatch; this current is known as *back-porch*. GTO turn-on losses result from simultaneous existence of voltage and current, made more difficult by the current overshoot corresponding to the reverse current of a diode, mentioned above.

The turn-off process requires a much larger reverse gate current pulse of magnitude greater than 30% of the device current, which removes part of the current from the cathode to the gate. With the application of the turn-off pulse, there is a significant time delay (known as storage time), in the cathode region, before the current begins to fall and voltage begins to rise. This delay results in a significant energy requirement for the gate driver. The anode-cathode current then falls rapidly to a low level and then continues to decay slowly, until the charge carriers recombine in the pn region of the anode side of the device. This tail is responsible for a significant portion of the turn-off losses.

During turn-off, the rate of voltage rise has to be limited in order to ensure safe turn-off of all the cathode islands.

The GTO's principal handicap compared to IGBT discussed later has been its large gate turn-off drive requirements. This in turn results in long turn-off time, lower di/dt and dv/dt capability, and therefore costly turn-on and turn-off snubber circuits adding to the cost and losses. Because of its slow turn-off, it can be operated in PWM converters at a relatively low frequency (up to a few hundred Hz), which is, however, sufficient for high-power converters. On the other hand, it has lower forward voltage drop and is available in larger ratings than IGBT. GTO has been utilized in FACTS Controllers of several hundred MWs.

It would be a big advantage if the devices had low on-state voltage drop (as for a thyristor), as well as low gate-drive requirements and fast turn-off (as for IGBT). In fact, there are a number of such devices coming into the marketplace, and in time they could replace conventional GTOs. They are in fact GTOs with advancements, consistent with the concept of Power Electronics Building Block (PEBB) to integrate and reduce the gate-drive requirements and achieve fast switching. The key is to achieve fast turn-off, which essentially means fast transfer of current out of the cathode to the gate of the upper transistor. This has been accomplished in a variety of ways in the emerging advanced GTOs. These include *MOS Turn-Off Thyristor* (*MTO*), *Emitter Turn-Off Thyristor* (*ETO*), and *Integrated Gate-Commutated Thyristor* (*IGCT*); these are briefly described below.

2.8 MOS TURN-OFF THYRISTOR (MTO)

SPCO has developed the MTO thyristor, which is a combination of a GTO and MOSFETs, which together overcome the limitations of the GTO regarding its gate-drive power, snubber circuits, and dv/dt limitations. Unlike IGBT (Section 2.12), the MOS structure is not implanted on the entire device surface, but instead the MOSFETs are located on the silicon, all around the GTO to eliminate need for high-current GTO turn-off pulses. The GTO structure is essentially retained for its advantages of high voltage (up to 10 kV), high current (up to 4000 A), and lower forward conduction losses than IGBTs. With the help of these MOSFETs, and tight packaging to minimize the stray inductance in the gate-cathode loop, the MTO becomes significantly more efficient than conventional GTO, requiring drastically smaller gate drive while reducing the charge storage time on turn-off, providing improved performance and reduction of system costs. As before, the GTO is still provided with double-sided cooling, and lends itself to thin packaging technology for even more efficient removal of heat from the GTO.

Figure 2.12 shows the symbol, structure, and equivalent circuit; Figure 2.13 shows a photograph of an MTO. The structure shown is that of a thyristor with four layers,

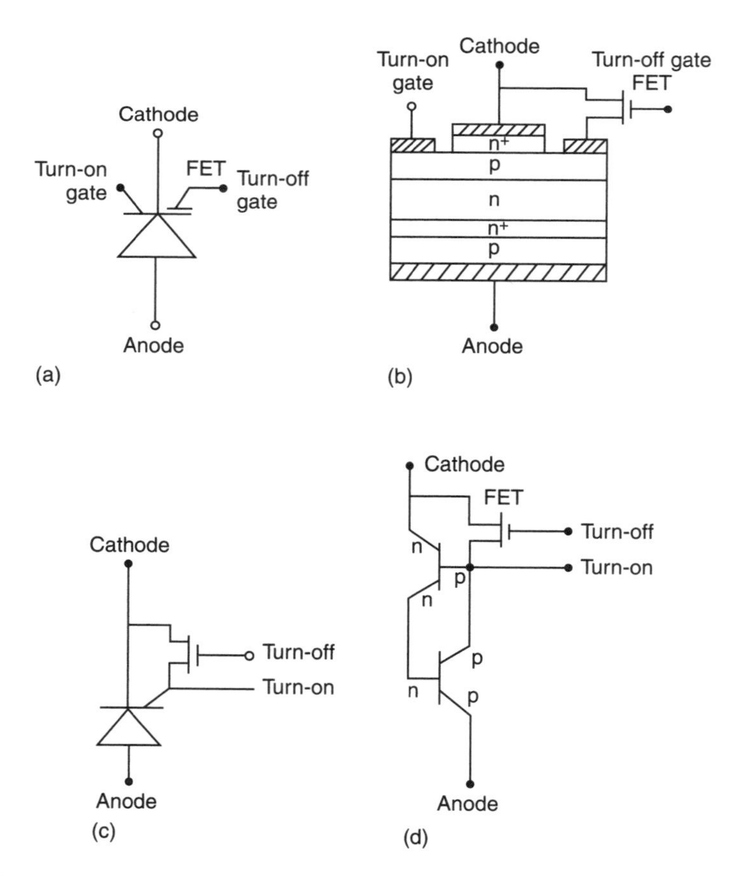

Figure 2.12 MOS Turn-Off (MTO) thyristor: (a) MTO symbol, (b) MTO structure, (c) MTO equivalent circuit, and (d) MTO more detailed equivalent circuit.

Figure 2.13 A 50 mm diameter MOS Turn-Off thyristor (MTO™), rated for 4.5 kV and 500 A. Complete package is shown with the lid off. Inside the device package is a GTO thyristor, just like the one in Figure 2.10, which is surrounded by a ring of very low voltage MOSFETs for gate turn-off. The ring is partially cut away to show the MOSFETs. The turn-on gate is the same as for the GTO. [Courtesy Silicon Power Corporation (SPCO). MTO is SPCO Trademark.]

and has two gate structures, one for turn-on and the other for turn-off. For both gates, the metal is directly bonded on the p-layer.

Just as in GTO, turn-on is achieved by a turn-on current pulse of about one-tenth of the main current for 5–10 μsec followed by a small back-porch current. Turn-on pulse turns on the upper npn transistor, which in turn turns on the lower pnp transistor leading to a latched turn-on.

Turn-off is carried out by application of just a voltage pulse of about 15 V to the MOSFET gates, thereby turning the MOSFETs on, which shorts the emitter and base of the npn transistor, shunting off the latching process. In contrast, as discussed before, the conventional GTO turn-off is carried out by sweeping enough current out of the emitter base of the upper npn transistor with a large negative pulse to stop the regenerative latching action. What is equally important with the new approach is that the turn-off can be much faster, (1–2 μsec as against 20–30 μsec) and the losses corresponding to the storage time are almost eliminated. This also means high dv/dt, and much smaller snubber capacitors and elimination of the snubber resistor.

Small turn-off time also means that MTOs can be connected in series without matching of devices because virtually all devices turn off simultaneously resulting in all devices taking their share of current. Since MOSFETs are essentially turned on in parallel to the GTO's gate cathode, the rapid turn-off requires MOSFETs with a very low forward voltage drop. MOSFETs are small, inexpensive, and commercially produced in large quantities. Fast turn-off of MTO and other advanced GTOs can essentially overcome the disadvantage of GTOs compared to the IGBTs with regard to the over-current protection as will be discussed in the Section 2.11 on IGBTs.

It must be mentioned that the long turn-off tail shown at the end of the turn-off in Figure 2.11 is still present and the next turn-on must wait until the residual charge on the anode side is dissipated through recombination process. This also applies to the other advanced thyristor devices discussed below, except the MCT. It would be advantageous if there was another gate on the anode side to accelerate charge dissipation in the anode region. Such a device would represent another leap in the advancement of high-power devices. SPCO has suggested such an approach and has also suggested a monolithic design in which MOSFETs are embedded onto the p-layers of a GTO.

2.9 EMITTER TURN-OFF THYRISTOR (ETO)

Like MTO, ETO is another variation on exploiting the virtues of both the thyristor and the transistor, i.e., GTO and MOSFET. ETO was invented at Virginia Power Electronics Center in collaboration with SPCO. The ETO symbol and its equivalent circuit are shown in Figure 2.14. As shown, a MOSFET T1 is connected in series with the GTO and a second MOSFET T2 is connected across this series MOSFET and the GTO gate. Actually T1 consists of several N-MOSFETs and T2 consists of several P-MOSFETs packaged around the GTO in order to minimize inductance between the MOSFETs and the gate cathode of the GTO. N- and P-MOSFETs and GTOs are commercially available devices made in large quantities.

ETO has two gates: one is the GTO's own gate used for turn-on, and the other is the series MOSFET gate used for turn-off. When the turn-off voltage signal is applied to the N-MOSFET, it turns off and transfers all the current away from the cathode (n emitter of the upper npn transistor of the GTO) into the base via MOSFET T2, thus stopping the regenerative latched state and a fast turn-off. It is important to note that neither MOSFETs see high voltage, no matter how high the ETO voltage. T2 is connected with its gate shorted to its drain, and hence voltage across it is clamped at a value slightly higher than its threshold voltage and the maximum voltage across T1 cannot exceed that of T2.

The advantage of series MOSFET is that the transfer of current from the cathode is complete and rapid, giving a uniform simultaneous turn-off of all the individual cathodes. The disadvantage of series MOSFETs is that they have to carry the main GTO current, thus increasing the total voltage drop and corresponding losses. However, because these MOSFETs are low-voltage devices, the added voltage drop is low (about 0.3–0.5V), although not insignificant.

Thus ETO is essentially a GTO, which, with the help of auxiliary MOSFETs, enhances the GTO with fast switching and therefore much lower driver losses, and

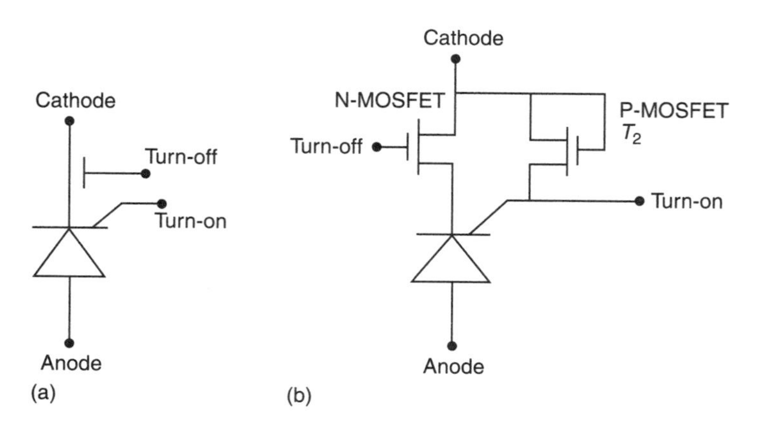

Figure 2.14 Emitter Turn-Off (ETO) thyristor: (a) ETO symbol and (b) ETO equivalent circuit.

greatly reduces the costly gate drive and snubber circuits while enhancing the GTO's high power capability.

The long turn-off tail shown at the end of the turn-off in Figure 2.11 is still present and the next turn-on must wait until the residual charge on the anode side is dissipated through the recombination process.

2.10 INTEGRATED GATE-COMMUTATED THYRISTOR (GCT AND IGCT)

The gate-commutated thyristor (GCT) is a hard-switched GTO involving very fast and large current pulse, as large as the full rated current, that draws out all the current from the cathode into the gate in 1 μsec to ensure a fast turn-off. Its structure and equivalent circuit is the same as that of a GTO shown in Figure 2.9. IGCT is a device with added value on GCT, including a multilayered printed circuit board gate drive supplied with the main device, and may also include a reverse diode, as shown in a structure diagram in Figure 2.15 and a photograph in Figure 2.16.

In order to apply a fast-rising and high-gate current, GCT (IGCT) incorporates a special effort to reduce the inductance of the gate circuit (gate-driver-gate-cathode loop) to lowest possible, as required also for MTO and ETO to the extent possible. Essentially the key to GCT (IGCT) is to achieve a very fast gate drive and this is achieved by coaxial cathode-gate feed through and a multilayered gate-driver circuit boards, which enable gate current to rise at 4 kA/μsec with gate-cathode voltage of 20 V. In 1 μsec the GTO's upper transistor is totally turned off and the lower pnp transistor is effectively left with an open base turn-off. Being a very short duration pulse, the gate-drive energy is greatly reduced. Also by avoiding the gate overdrive, the gate-drive energy consumption is minimized. The gate-drive power requirement is decreased by a factor of 5 compared to the conventional GTO. As in a conventional GTO and applicable to MTO and ETO as well, buffer layer is provided on the anode side of the n-layer which decreases the on-state conduction losses and makes the device asymmetrical.

The anode p-layer is made thin and lightly doped to allow faster removal of charges from the anode-side during turn-off. An IGCT may also have an integrated

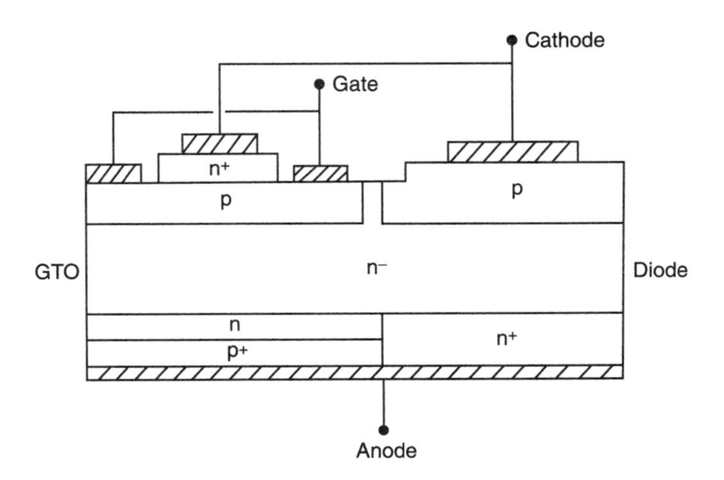

Figure 2.15 IGCT structure with a Gate-Commutated Thyristor and a reverse diode.

Figure 2.16 Photo shows an Integrated Gate-Commutated Thyristor (IGCT) which consists of a Gate-Commutated Thyristor (GCT) device and a very low inductance gate driver package. Two wafers of different design are also shown on the side. The lower wafer is a GTO of advanced design, and the upper wafer is a GTO along with a reverse diode as part of the device. (Photo courtesy ABB Semiconductor Inc., USA.)

reverse diode, as shown by the n^+n^-p junction on the right side of the structural diagram in Figure 2.15. As mentioned before, a reverse diode is needed in voltage-sourced converters. Inclusion of the n buffer layer evens out the voltage stress across the n^- layer, the n^- layer thickness is reduced by 40%, which enables inclusion of diode with comparable forward on-state voltage drop as that of a separate diode. Naturally integration of diode means allocation of appropriate silicon active surface to the diode which in turn reduces the area for the GTO on a given wafer.

It is seen from the description of the MTO, ETO, and GCT that getting the most out of the GTO's capability is all about getting the current out of the cathode into the base of the upper transistor as fast as possible. Reducing the inductance of the gate drive and cathode loop is common to all advanced GTOs discussed above and applies to the conventional GTO as well. All of them lead to high dv/dt, uniform and fast current turn-off, thus increasing the current turn-off capability to maximum possible. This in turn results in much smaller snubber capacitor with no resistor, much easier series connection of GTOs, and the turn-on which is already a low energy drive remains the same as for conventional GTO. As was discussed in Section 2.2, these devices and MCT discussed below represent the essence of the Power Electronics Building Block (PEBB) concept. Through integration of gate drive, major advantage has been delivered and these advanced GTOs should replace conventional GTOs, at least in applications in which device characteristics are highly leveraged as in FACTS Controllers.

Thus advanced GTO concepts represent a major advance based on the recognition of the PEBB concept that stray inductances and capacitances of the gate drive and bus connections have a major impact on the overall device losses, snubber circuits, and all that surrounds the devices in any specific application.

2.11 INSULATED GATE BIPOLAR TRANSISTOR (IGBT)

A modern power transistor is the Insulated Gate Bipolar Transistor (IGBT). It operates as a transistor with high-voltage and high-current capability and a moderate forward voltage drop during conduction.

The IGBT is a device that is part way to being a thyristor, but is designed to not latch into full conduction equivalent to a voltage drop of one junction, but instead IGBT part way to latching stays as a transistor. In addition it also has an integrated MOS structure with insulated gate, like a MOSFET. Its structural cross section and equivalent circuit are shown in Figure 2.17. Like the thyristor and the GTO, it has a two-transistor structure. But the turn-on and turn-off are carried out by a MOSFET structure across its npn transistor instead of the np gate emitter of the upper npn transistor. With turn-on, there is a current flow through the base and the emitter of the npn transistor as in a thyristor, but not enough for the device to avalanche into a latched conduction. As shown in Figure 2.17, the base emitter junction is shunted by a resistance, which is built into the device structure. This resistance bypasses part of the cathode current rather than all of it.

In the structural cross section shown, the upper n^+ is the MOS source of n carriers, p is the base, n^- layer is the drift region, lower p^+ is the buffer layer, and finally p^+ is the substrate. Like a MOSFET, when the gate is made positive with respect to the emitter for turn-on, n carriers are drawn into the p channel near the gate region, which forward biases the base of the npn transistor, which thereby turns on. The IGBT is turned on by just applying a positive base voltage to open the channel for n carriers and is turned off by removing the base voltage to close the channel, which results in a very simple driver circuit. Basically this can be achieved in MTOs and ETOs if MOSFETs were also added for the turn-on.

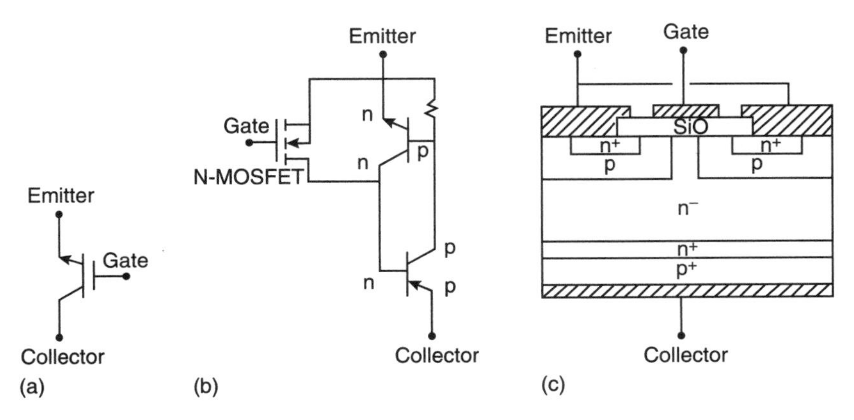

Figure 2.17 Insulated Gate Bipolar Transistor (IGBT): (a) IGBT symbol, (b) IGBT equivalent circuit, and (c) IGBT structure.

Given the complex MOS technology on the entire surface of the device, IGBTs are made in sizes of about 1 cm^2. To make high power-devices, several IGBTs are connected in parallel, wire bonded, and potted inside a larger package to resemble a single device.

The advantage of the IGBT is its fast turn-on and turn-off because it is more like a majority carrier (electrons) device. It can be therefore used in Pulse Width Modulation (PWM) Converters operating at high frequency. On the other hand being a transistor device, it has higher forward voltage drop compared to thyristor type devices such as GTOs. Nevertheless the IGBT has become a workhorse for industrial applications, and has reached sizes capable of applications in the range of 10 MW or more.

The transistor devices, such as MOSFETs and IGBTs, potentially have current-limiting capability by controlling the gate voltage. During this current-limiting mode, the device losses are very high, and in high-power applications, current-limiting action can only be used for very short periods of a few microseconds. Yet this time can be enough to allow other protective actions to be taken for safe turn-off of the devices. This feature is extremely valuable in voltage-sourced converters, in which fault current can rise to high levels very rapidly due to the presence of a large dc capacitor across the converter. On the other hand, with fast sensing, combined with the fast turn-off of the advanced GTOs, an effective turn-off can be achieved within 2–3 microseconds. This method will also spare the devices from high-power dissipation and sacrifice of their useable capacity. The turn-off time of the conventional GTOs is too long for high-speed protective turn-off.

IGBT, coming from a low-power end, has been pushing out the conventional GTO, as viable packaged parallel-IGBTs ratings go up. This is because the conventional GTOs have serious disadvantages of large gate-drive requirements, slow-switching and high-switching losses. Evolution of the GTO into MTO, ETO and IGCT/GCT has shown that these are the result of previously unattended issue of gate-drive packaging and stray inductance of the gate cathode loop.

IGBT has its own generic limitations, including: higher forward voltage drop, complexities with providing double-sided cooling, nature of the repetitive MOS on the chip limits what can be achieved in increased blocking voltage, and IGBT production needs much cleaner production facility.

A major advantage for IGBT for high-power applications is its low-switching losses, fast switching, and current-limiting capability. However, with the advanced GTOs and MCTs (discussed in the next section), there is a prospect for major advances for devices suitable for wide range of FACTS Controllers.

On the other hand, future outcome often depends on the market forces of volume production, and this is in favor of the IGBTs continuing to push its applications to higher power levels.

2.12 MOS-CONTROLLED THYRISTOR (MCT)

An MOS Controlled Thyristor (MCT) incorporates a MOSFET-like structure in the device for both the turn-on and turn-off.

Figure 2.18 shows an n-type MCT. Equivalent circuit for the n-MCT shows that for turn-on there is an n-type MOSFET (shown as n-FET) connected across the cathode side npn transistor, like that for an IGBT. Another p-type MOSFET (shown as p-FET) is connected across the gate cathode of the cathode side npn transistor for turn-off, like that for an MTO.

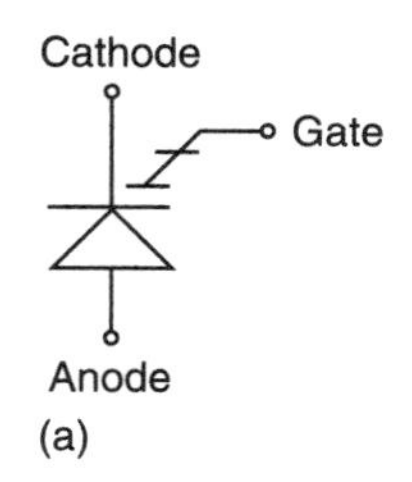

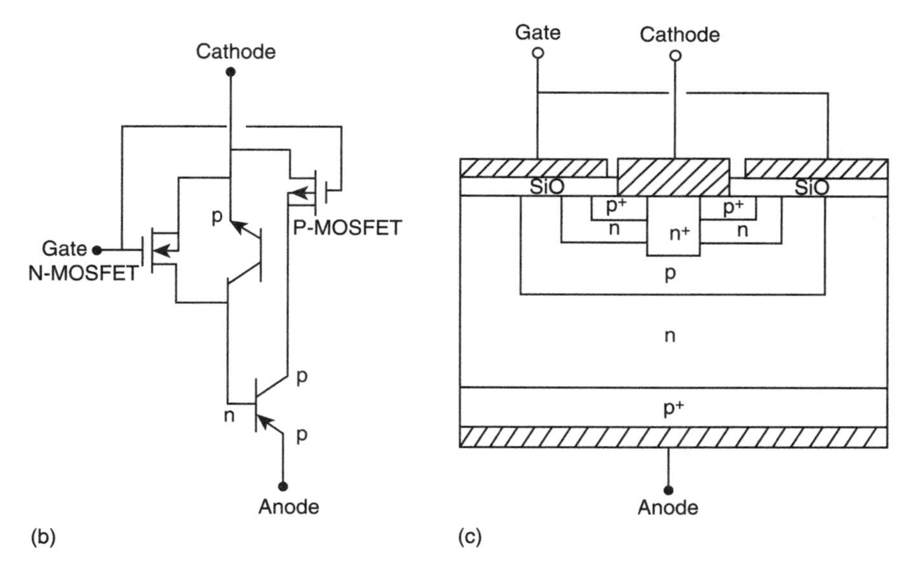

Figure 2.18 MOS Controlled Thyristor (MCT): (a) MCT symbol, (b) MCT equivalent circuit, and (c) MCT structure.

As n-FET is turned on with the application of positive voltage to the gate with respect to the cathode, current flows from the anode to the base of the lower npn transistor, which turns on and leads to the latched turn-on of the thyristor. As shown, the same gate voltage is also applied to the base of the p-FET, which ensures that the p-FET stays off.

When the gate voltage is made negative, it turns off the n-FET and turns on the p-FET. The p-FET thereby bypasses the gate cathode thus unlatching the thyristor.

The MOS structure is spread across the entire surface of the device giving a fast turn-on and turn-off with low-switching losses. The power/energy required for the turn-on and turn-off is very small, and so is the delay time (storage time). Furthermore, being a latching device, it has a low on-state voltage drop as for a thyristor. Its processing technology is essentially the same as that of IGBT.

The key advantage for MCT compared to other turn-off thyristors is that it brings distributed MOS gates for both turn-on and turn-off, very close to the distributed cathodes, resulting in fast-switching and low switching losses for a thyristor device. Therefore MCT represents the near-ultimate turn-off thyristor with low on-state and switching losses, and fast-switching device needed for high-power advanced converters with active filtering capability.

REFERENCES

Ajit, J. S., and Kinzer, D. M., "New MOS-Gate Controlled Thyristor (MGCT)," *Proceedings of The 7th International Symposium on Power Semiconductor Devices and ICs,* Yokohama, Japan, May 1995.

Baliga, B. J., "Power Semiconductor Devices for Variable Frequency Drives," Chapter 1, *Power Electronics and Variable Frequency Drives—Technology and Applications,* edited by B. K. Bose, ed., IEEE Press, Piscataway, NJ, 1996.

Ballad, J. P., Bassett, R. J., and Davidson, C. C., "Power Electronic Devices and Their Impact for Power Transmission," *Proceedings, IEE Sixth International Conference on AC and DC Transmission,* April–May 1996.

Bhalla, A., and Paul Chow, T., "ESTD: An Emitter Switched Thyristor with a Diverter," *IEEE Electronic Device Letters,* vol. 16, no. 2, February 1995.

Bose, B. K., "Evaluation of Modern Power Semiconductor Devices and Future Trends of Converters," *IEEE Transactions on Industry Applications,* vol. 28, no. 2, pp. 403–413, March–April 1992.

Bose, B. K., "Power Electronics: A Technology Review," *Proceedings of the IEEE,* vol. 80, no. 8, August 1992.

De Doncker, R., "High Power Semiconductor Device Developments for FACTS and Custom Power Applications," *EPRI Conference on the Future of Power Delivery,* Washington DC, April 1996.

Gruning, H., and Zuckerberger, A., "Hard Drive of High Power GTOs: Better Switching Capability Obtained through Improved Gate Units, A New High Power Bipolar Thyristor," *IEEE IAS Annual Meeting,* pp. 1474–1480, New Orleans, October 1996.

Hingorani, N. G., Mehta, H., Levy, S., Temple, V., and Glascock, H. H., "Research Coordination for Power Semiconductor Technology," *Proceedings of the IEEE,* vol. 77, no. 9, September 1989.

Iwamuro, M., Hoshi, Y., Iwaana, T., Ueno, K., Seki, Y., Otsuki, M., and Sakurai K., "Experimental Demonstration of Dual Gate MOS Thyristor," *Proceedings of the 7th International Symposium on Power Semiconductor Devices and ICs,* Yokohama, Japan, May 1995.

Li, Y., and Huang, Q., "The Emitter Turn-Off Thyristor—A New MOS Bipolar High Power Device" *VPEC Seminar Proceedings,* pp. 179–183, September 1997.

Lips, P., et al., "Semiconductor Power Devices for Use in HVDC and FACTS Controllers," *CIGRE Report of CIGRE Working Group 14.17,* Paris, 1997.

Mohan, N., Undeland, T. M., and Robbins W. T., "Power Electronics," *Converters, Applications and Design.* Book copublished IEEE Press, Piscataway, NJ, and John Wiley & Sons, New York, 1989.

Piccone, D. E., De Doncker, R. W., Barrow, J. A., and Tobin W. H., "The MTO Thyristor—A New High Power Bipolar MOS Thyristor," *IEEE IAS Annual Meeting,* pp. 1472–1473, New Orleans, October 1996.

Rodrigues, R., Piccone, P., Huang, A., and DeDoncker, R., "MTO Thyristor Power Switches," *Power Systems World Conference Records,* Baltimore, MD, pp. 3.53–3.64.

Steimer, P. K., Gruning, H. E., Werninger, J., Carroll, E., Klaka, S., and Linder, S., "IGCT—A New Emerging Technology for High Power, Low Cost Inverters," *IEEE IAS Annual Meeting,* pp. 1592–1599, October 1997.

Temple, V. A. K., "MOS-Controlled Thyristors—a New Class of Power Devices," *IEEE Transaction Electron Devices,* vol. 33, no. 10, pp. 1609–1618, October 1986.

Temple, V. A. K., Arthur, S., Watrous, D., De Doncker, R., and Mehta, H. "Megawatt MOS Controlled Thyristor for High Voltage Power Circuits," *IEEE Power Electronics Specialists Conference,* pp. 1018–1025, June 1992.

User Guides for MCT and MTO, available from Silicon Power Corporation, Malvern, PA.

3

Voltage-Sourced Converters

3.1 BASIC CONCEPT OF VOLTAGE-SOURCED CONVERTER

Discussion of FACTS Controller concepts in Chapter 1 conveyed that the voltage-sourced converter is the building block of STATCOM, SSSC, UPFC, IPFC, and some other Controllers. Therefore, this converter is generically discussed in this chapter.

It was explained in Chapter 2 that the so-called conventional thyristor device has only the turn-on control; its turn-off depends on the current coming to zero as per circuit and system conditions. Devices such as the Gate Turn-Off Thyristor (GTO), Integrated Gate Bipolar Transistor (IGBT), MOS Turn-off Thyristor (MTO), and Integrated Gate-Commutated Thyristors (IGCT), and similar devices have turn-on and turn-off capability. These devices (referred to as turn-off devices) are more expensive and have higher losses than the thyristors without turn-off capability; however, turn-off devices enable converter concepts that can have significant overall system cost and performance advantages. These advantages in principle result from the converters, which are self-commutating as against the line-commutating converters. Compared to the self-commutating converter, the line-commutating converter must have an ac source connected to the converter, it consumes reactive power, and suffers from occasional commutation failures in the inverter mode of operation. Therefore, unless a converter is required to function in the two lagging-current quadrants only (consuming reactive power while converting active power), converters applicable to FACTS Controllers would be of the self-commutating type. There are two basic categories of self-commutating converters:

1. Current-sourced converters in which direct current always has one polarity, and the power reversal takes place through reversal of dc voltage polarity.
2. Voltage-sourced converters in which the dc voltage always has one polarity, and the power reversal takes place through reversal of dc current polarity.

Conventional thyristor-based converters, being without turn-off capability, can only be current-sourced converters, whereas turn-off device-based converters can be of either type.

For reasons of economics and performance, voltage-sourced converters are often preferred over current-sourced converters for FACTS applications, and in this chapter various self-commutating voltage-sourced converter concepts, which form the basis of several FACTS Controllers, will be discussed.

Since the direct current in a voltage-sourced converter flows in either direction, the converter valves have to be bidirectional, and also, since the dc voltage does not reverse, the turn-off devices need not have reverse voltage capability; such turn-off devices are known as asymmetric turn-off devices. Thus, a voltage-sourced converter valve is made up of an asymmetric turn-off device such as a GTO [as shown in Figure 3.1(a)] with a parallel diode connected in reverse. Some turn-off devices, such as the IGBTs and IGCTs, may have a parallel reverse diode built in as part of a complete integrated device suitable for voltage-sourced converters. However, for high power converters, provision of separate diodes is advantageous. In reality, there would be several turn-off device-diode units in series for high-voltage applications. In general,

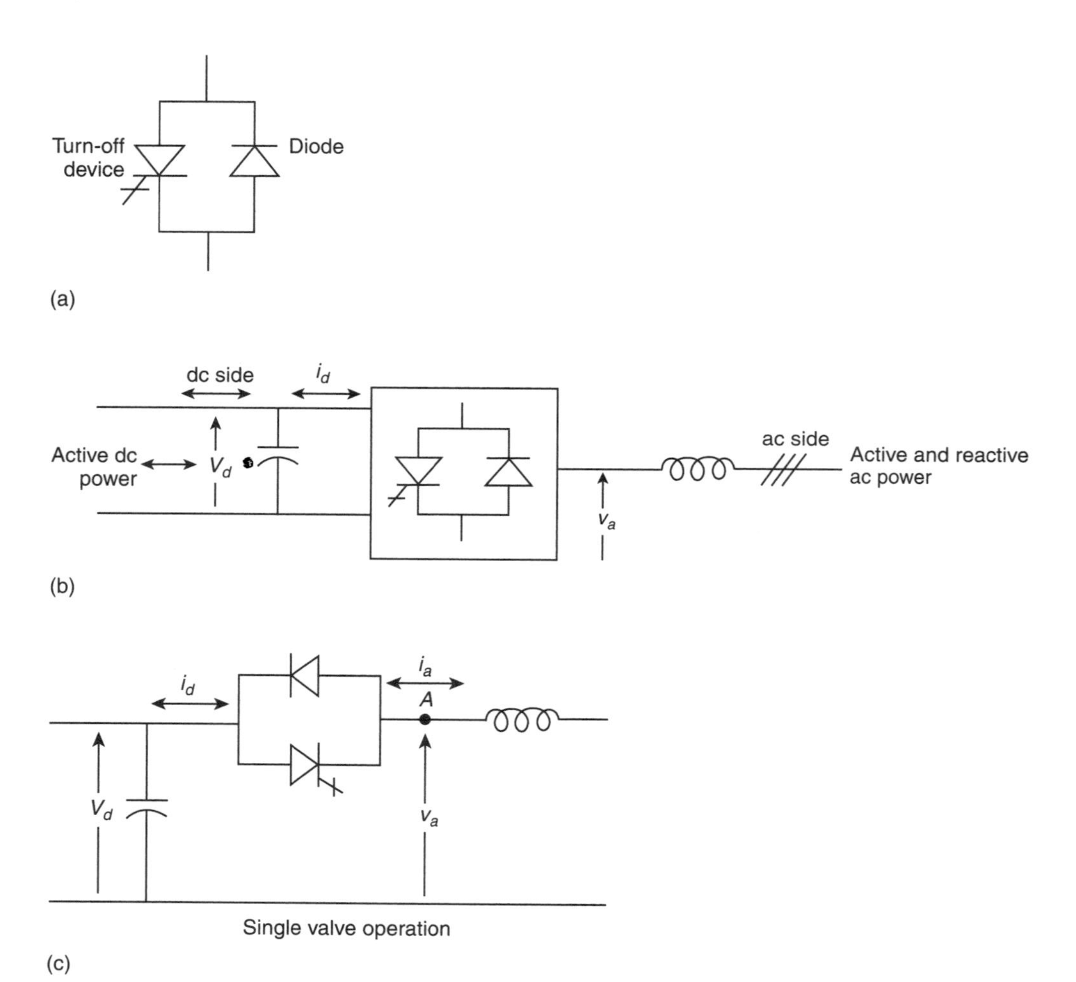

Figure 3.1 Basic principles of voltage-sourced converters: (a) Valve for a voltage-sourced converter; (b) Voltage-sourced converter concept; (c) Single-valve operation.

the symbol of one turn-off device with one parallel diode, as shown in Figure 3.1(a), will represent a valve of appropriate voltage and current rating required for the converter.

Within the category of voltage-sourced converters, there are also a wide variety of converter concepts. The ones relevant to FACTS Controllers are described in this chapter. There are some converter topologies that are suitable for supplying and consuming reactive power only and not for converting active power; they are not discussed in this chapter.

Figure 3.1(b) shows the basic functioning of a voltage-sourced converter. The internal topology of converter valves is represented as a box with a valve symbol inside. On the dc side, voltage is unipolar and is supported by a capacitor. This capacitor is large enough to at least handle a sustained charge/discharge current that accompanies the switching sequence of the converter valves and shifts in phase angle of the switching valves without significant change in the dc voltage. For the purposes of discussion in this chapter, the dc capacitor voltage will be assumed constant. It is also shown on the dc side that the dc current can flow in either direction and that it can exchange dc power with the connected dc system in either direction. Shown on the ac side is the generated ac voltage connected to the ac system via an inductor. Being an ac voltage source with low internal impedance, a series inductive interface with the ac system (usually through a series inductor and/or a transformer) is essential to ensure that the dc capacitor is not short-circuited and discharged rapidly into a capacitive load such as a transmission line. Also an ac filter may be necessary (not shown) following the series inductive interface to limit the consequent current harmonics entering the system side.

Basically a voltage-sourced converter generates ac voltage from a dc voltage. It is, for historical reasons, often referred to as an inverter, even though it has the capability to transfer power in either direction. With a voltage-sourced converter, the magnitude, the phase angle and the frequency of the output voltage can be controlled.

In order to further explain the principles, Figure 3.1(c) shows a diagram of a single-valve operation. DC voltage, V_d, is assumed to be constant, supported by a large capacitor, with the positive polarity side connected to the anode side of the turn-off device. When turn-off device 1 is turned on, the positive dc terminal is connected to the ac terminal, A, and the ac voltage would jump to $+V_d$. If the current happens to flow from $+V_d$ to A (through device 1), the power would flow from the dc side to ac side (inverter action). However, if the current happens to flow from A to $+V_d$ it will flow through diode 1′ even if the device 1 is so called turned on, and the power would flow from the ac side to the dc side (rectifier action). Thus, a valve with a combination of turn-off device and diode can handle power flow in either direction, with the turn-off device handling inverter action, and the diode handling rectifier action. This valve combination and its capability to act as a rectifier or as an inverter with the instantaneous current flow in positive (ac to dc side) or negative direction, respectively, is basic to voltage-sourced converter concepts.

3.2 SINGLE-PHASE FULL-WAVE BRIDGE CONVERTER OPERATION

Although FACTS Controllers will generally utilize three-phase converters, a single-phase, full-wave bridge converter may also be used in some designs. In any case, it is important to first understand the operation of a single-phase bridge converter and operation of a phase-leg to further understand the principles of voltage-sourced converters.

Figure 3.2(a) shows a single-phase full-wave bridge converter consisting of four valves, (1–1′) to (4–4′), a dc capacitor to provide stiff dc voltage, and two ac connection points, a and b. The designated valve numbers represent their sequence of turn-on and turn-off. The dc voltage is converted to ac voltage with the appropriate valve turn-on, turn-off sequence as explained below.

As shown by the first waveform of Figure 3.2(b), with turn-off devices 1 and 2 turned on, voltage v_{ab} becomes $+V_d$ for one half-cycle, and, with 3 and 4 turned on and devices 1 and 2 turned off, v_{ab} becomes $-V_d$ for the other half-cycle. This voltage waveform occurs regardless of the phase angle, magnitude and waveform of the ac current flow. The ac current is the result of interaction of the converter generated ac voltage with the ac system voltage and impedance. For example, suppose that the current flow from the ac system, as shown by the second waveform, is a sinusoidal waveform i_{ab}, angle θ, leading with respect to the square-wave voltage waveform. Starting from instant t_1, it is seen from the circuit and the waveform that:

1. From instant t_1 to t_2, with turn-off devices 1 and 2 on and 3 and 4 off, v_{ab} is positive and i_{ab} is negative. The current flows through device 1 into ac phase a, and then out of ac phase b through device 2, with power flow from dc to ac (inverter action).
2. From instant t_2 to t_3, the current reverses, i.e., becomes positive, and flows through diodes 1′ and 2′ with power flow from ac to dc (rectifier action). Note that during this interval, although devices 1 and 2 are still on and voltage v_{ab} is $+V_d$, devices 1 and 2 cannot conduct in a reverse direction. In reality, devices 1 and 2 are ready to turn on by turn-on pulses when required by the direction of actual current flow.
3. From instant t_3 to t_4, devices 1 and 2 are turned off and devices 3 and 4 are turned on, thereby v_{ab} becomes negative while i_{ab} is still positive. The current now flows through devices 3 and 4 with power flow from dc to ac (inverter action).
4. From instant t_4 to t_5, with devices 3 and 4 still on, and 1 and 2 off, and v_{ab} negative, current i_{ab} reverses and flows through diodes 3′ and 4′ with power flow from ac to dc (rectifier action).

From instant t_5, the cycle starts again as from t_1 with devices 1 and 2 turned on and 3 and 4 turned off. Table 3.1 summarizes the four operating modes in a cycle.

Figure 3.2(b) also shows the waveform of current flow i_d in the dc bus with the positive side flowing from ac to dc (rectifier action), and the negative side flowing from dc to ac (inverter action). Clearly the average dc current is negative. The current I_d contains the dc current and the harmonics. The dc current must flow into the dc system and for a large dc capacitor, virtually all of the harmonic current will flow through the capacitor. Being a single-phase, full-wave bridge, the dc harmonics have an order of $2k$, where k is an integer, i.e., 2nd, 4th, 6th, . . ., all of the even harmonics.

Voltage across valve 1-1′ is shown as the last waveform in Figure 3.2(b).

The relationship between ac voltage and current phasors is shown in Figure 3.2(c), showing power flow from ac to dc with a lagging power factor.

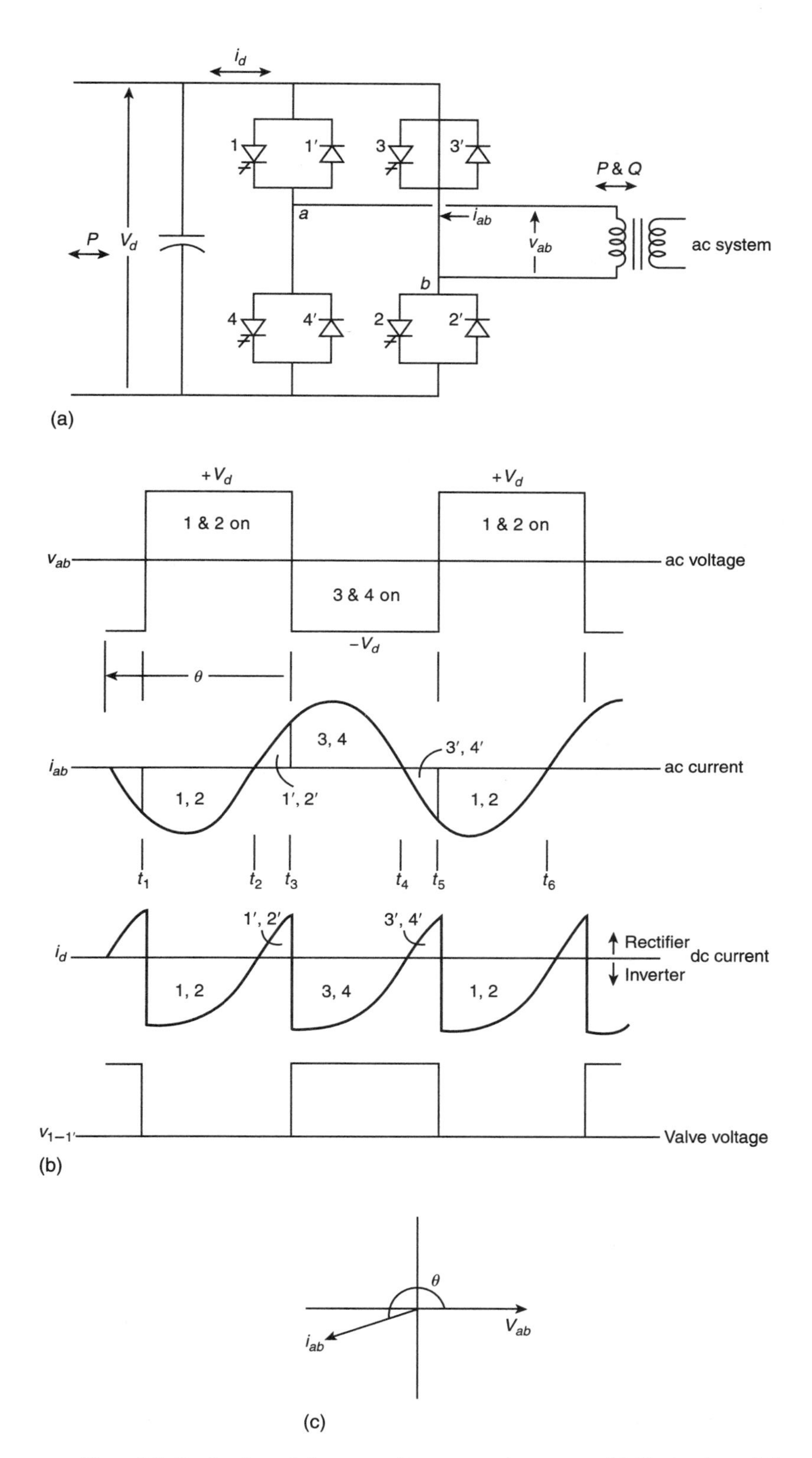

Figure 3.2 Single-phase, full-wave, voltage-sourced converter: (a) Single-phase, full-wave circuit; (b) Operation waveform; (c) Phase relationship between current and voltage.

TABLE 3.1 Four Operating Modes in One Cycle of a Single-Phase Converter

Devices	V_{ab}	Current Flow	Conducting Devices	Conversion
1 & 2 on, 3 & 4 off	Positive	Negative	1 & 2	Inverter
1 & 2 on, 3 & 4 off	Positive	Positive	1′ & 2′	Rectifier
1 & 2 off, 3 & 4 on	Negative	Positive	3 & 4	Inverter
1 & 2 off, 3 & 4 on	Negative	Negative	3′ & 4′	Rectifier

3.3 SINGLE PHASE-LEG (POLE) OPERATION

Now consider operation of just one-leg (single-pole) circuit shown in Figure 3.3, in which the capacitor is split into two series-connected halves with the neutral point of the ac side connected to the midpoint N of the dc capacitor. With the two turn-off devices alternately closing/opening, the ac voltage waveform is a square wave with peak voltage of $V_d/2$. Note that when two phase-legs are operated in a full-wave bridge mode, Figure 3.2(b), the ac square wave is the sum of the two halves of Figure 3.3(b), giving a peak voltage of V_d. In a full-wave circuit, the neutral connection is no longer needed, because the current has a return path through the other phase-leg.

It should now be obvious that:

1. AC current and voltage can have any phase relationship, that is, the converter phase angle between voltage and current can cover all four quadrants, i.e., act as a rectifier or an inverter with leading or lagging reactive power. This

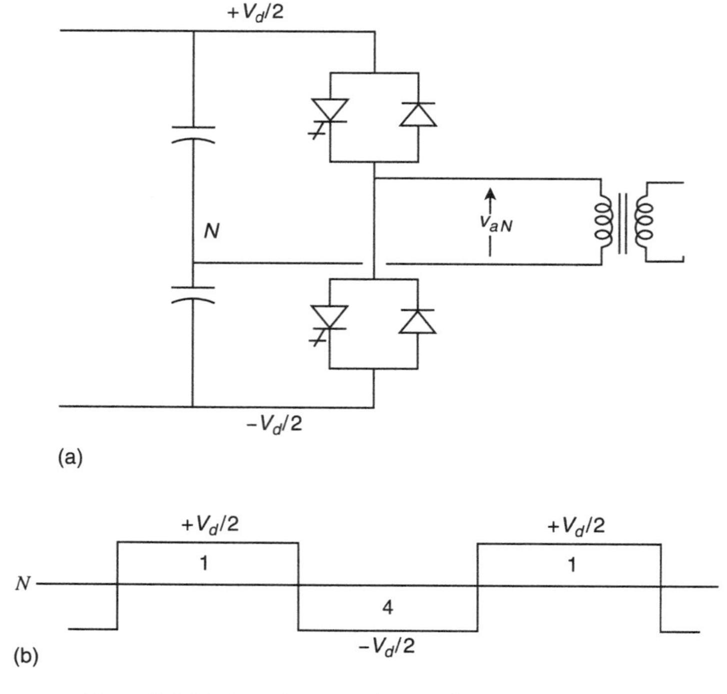

Figure 3.3 (a) One-phase-leg circuit; (b) Output ac voltage.

assumes that there is a dc and an ac system connected on the two sides of the converter, as in Figure 3.1(b), to exchange real power. If the converter is used for reactive power only then there is no need for the dc system and the converter will terminate at the dc capacitor.

2. The active and reactive power can be independently controlled with control of magnitude and angle of the converter generated ac voltage with respect to the ac current.
3. Diodes carry out instantaneous rectifier function, and turn-off devices carry out instantaneous inverter function. Of course, each ac cycle is made up of periods of rectifier and inverter actions in accordance with the phase angle, and the average current determines the net power flow and hence the net rectifier or inverter operation. When the converter operates as a rectifier with unity power factor, only diodes are involved with conduction, and when it operates as an inverter with unity power factor, only turn-off devices are involved in conduction.
4. When any turn-off device turns off, the ac bus current is not actually interrupted at all, but is transferred from a turn-off device to a diode when the power factor is not unity, and to another turn-off device when power factor is unity.
5. Turn-off devices 1 and 4 (or turn-off devices 2 and 3) in the same phase-leg are not turned on simultaneously. Otherwise this would cause a "shoot-through" (short circuit) of the dc side and a very fast discharge of the dc capacitor through the shorted phase-leg, which will destroy the devices in that phase-leg. In a phase-leg, when one turn-off device is on, the other is off. The gate control is arranged to ensure that only one of the two devices in a phase-leg receives a turn-on pulse, and that the current in the other device was indeed zero. Regardless, sensing and protection means are provided, usually to ensure safe shutdown of the converter.
6. Each phase-leg is independently capable of operating at any frequency or timing with the two valves in a leg alternately switching.
7. In principle, any number of phase-legs can be connected in parallel and each operated independently although being connected to the ac system, there is a need to have appropriate sequence and system interface through transformers in order to achieve the desired converter performance.
8. It is important to note that turn-on and turn-off of the turn-off devices establish the voltage waveform of the ac bus voltage in relation to the dc voltage, and do not necessarily conduct current if the direction of current flow results in a corresponding diode to carry the current.

Operation of each phase leg is further discussed in Section 3.6.

3.4 SQUARE-WAVE VOLTAGE HARMONICS FOR A SINGLE-PHASE BRIDGE

The square wave, shown in Figure 3.2(b) as the ac voltage v_{ab}, has substantial harmonics in addition to the fundamental. These harmonics are of the order $2n \pm 1$ where n is an integer, i.e., 3rd, 5th, 7th . . . The magnitude of the 3rd is 1/3rd of the fundamental, the 5th is 1/5th of fundamental, and so on.

As mentioned earlier, an inductive interface with the ac system (usually through an inductor and/or transformer) is essential to ensure that the dc capacitor does not discharge rapidly into a capacitive load such as a transmission line but it is also essential to reduce the consequent harmonic current flow. Generally, an ac filter would be necessary following the inductive interface to limit the consequent current harmonics on the system side although the filters will only increase the harmonic current in the converter itself. It would therefore be preferable if the converter generated less harmonics so that it does not require ac filters in the first place.

Integration of the waveform in Figure 3.2(b) gives the rms value of the square-wave ac voltage with a peak voltage of V_d:

$$V_{ab} = \sqrt{\frac{1}{\pi}\int_{-\pi/2}^{+\pi/2} V_d^2\, d\omega t} = V_d$$

which includes the fundamental and the harmonics. The fundamental and individual harmonics are given by

$$v = \frac{4}{\pi}(V_d)\left[\cos\omega t - \frac{1}{3}\cos 3\omega t + \frac{1}{5}\cos 5\omega t - \frac{1}{7}\ldots\right]$$

which gives

$$v_n = \frac{4}{\pi}(V_d)\left[\frac{1}{n}\cos n\omega t\right]$$

for $n = 1, 3, 5, 7. \ldots$

and its rms value given by

$$V_n = \frac{1}{n}\frac{2\sqrt{2}}{\pi}V_d$$

Thus, the rms fundamental component of a square wave ac voltage v_{ab} is

$$V_1 = \frac{2\sqrt{2}}{\pi}V_d = 0.9V_d$$

and the magnitude of each voltage harmonic is $1/n$th of the fundamental. These voltage harmonics will cause current harmonics to flow into the system, the magnitude of each determined by the system impedance. It is therefore essential to provide an inductive interface followed by shunt capacitive filters if necessary. Since for the nth harmonic the voltage is $1/n$th the fundamental voltage and the inductive impedance is n times the fundamental frequency impedance, it follows that lower frequency harmonics are the biggest concern.

3.5 THREE-PHASE FULL-WAVE BRIDGE CONVERTER

3.5.1 Converter Operation

Figure 3.4(a) shows a three-phase, full-wave converter with six valves, (1–1′) to (6–6′). The designated order 1 to 6 represents the sequence of valve operation in time. It consists of three phase-legs, which operate in concert, 120 degrees apart. The three phase-legs operate in a square wave mode, in accordance with the square wave mode described in the section above and with reference to Figure 3.3. Each valve alternately closes for 180 degrees as shown by the waveforms v_a, v_b, and v_c in Figure 3.4(b). These three square-wave waveforms are the voltages of ac buses a, b, and c with respect to a hypothetical dc-capacitor midpoint, N, with peak voltages of $+V_d/2$ and $-V_d/2$. The three phase legs have their timing 120 degrees apart with respect to each other in what amounts to a 6-pulse converter operation. Phase-leg 3–6 switches 120 degrees after phase-leg 1–4, and phase-leg 5–2 switches 120 degrees after phase-leg 3–6, thus completing the cycle as shown by the valve close-open sequence.

Figure 3.4(b) also shows the three phase-to-phase voltages, v_{ab}, v_{bc}, and v_{bc}, where $v_{ab} = v_a - v_b$, $v_{bc} = v_b - v_c$, and $v_{ca} = v_c - v_a$. It is interesting to note that these phase-to-phase voltages have 120 degrees pulse-width with peak voltage magnitude of V_d. The periods of 60 degrees, when the phase-to-phase voltages are zero, represent the condition when two valves on the same side of the dc bus are closed on their dc bus.

For example, the waveform for v_{ab} shows voltage V_d when turn-off device 1 connects ac bus a to the dc bus $+V_d/2$, and turn-off device 6 connects ac bus b to the dc bus $-V_d/2$, giving a total voltage $v_{ab} = v_a - v_b = V_d$. It is seen that 120 degrees later, when turn-off device 6 is turned off and turn-off device 3 is turned on, both ac buses, a and b, become connected to the same dc bus $+V_d/2$, giving zero voltage between buses a and b. Another 60 degrees later, as turn-off device 1 turns off, and turn-off device 4 connects bus a to $-V_d/2$, v_{ab} becomes $-V_d$. Another 120 degrees later, turn-off device 3 turns off, and turn-off device 6 connects bus b to $-V_d/2$, giving $v_{ab} = 0$. The cycle is completed when, after another 60 degrees turn-off device 4 turns off, and turn-off device 1 turns on. The other two voltages v_{bc} and v_{ca} have the same sequence 120 degrees apart.

As mentioned earlier, the turn-on and turn-off of the devices establish the waveforms of the ac bus voltages in relation to the dc voltage, the current flow itself is the result of the interaction of the ac voltage with the ac system. Also as mentioned earlier, each converter phase-leg can handle resultant current flow in either direction. Figure 3.4(b) shows an assumed ac current i_a in phase a, with positive current representing current from the ac to the dc side. For simplicity, the current is assumed to have fundamental frequency only. From point t_1 to t_2, for example, phase a current is negative and has to flow through either valve 1–1′ or valve 4–4′. It is seen, when comparing the phase a voltage (top curve) with the waveform of the phase a current, that when turn-off device 4 is on and turn-off device 1 is off and the current is negative, the

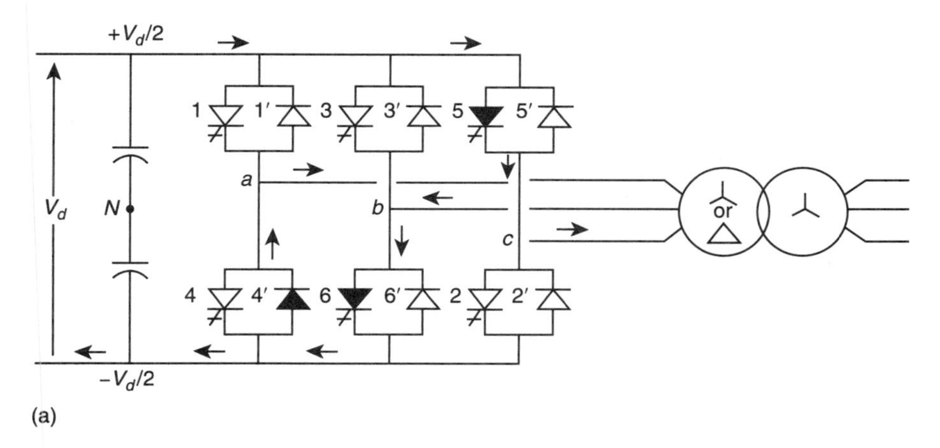

Figure 3.4 Operation of a three-phase, full-wave voltage-sourced converter: (a) Three-phase, full-wave converter; (b) AC waveforms of a three-phase, full-wave converter; (c) DC current waveforms of a three-phase, full-wave voltage-sourced converter.

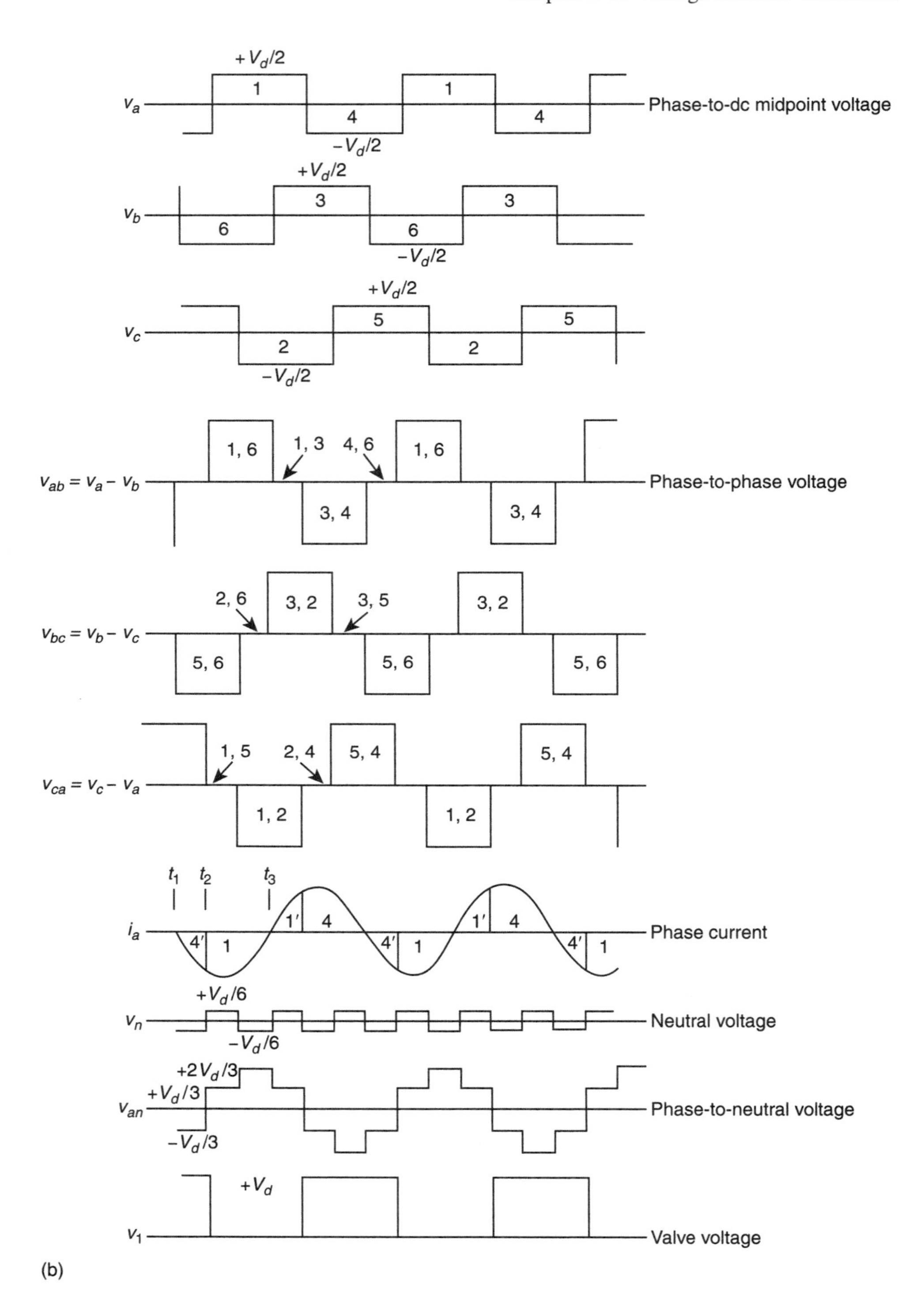

Figure 3.4 Continued

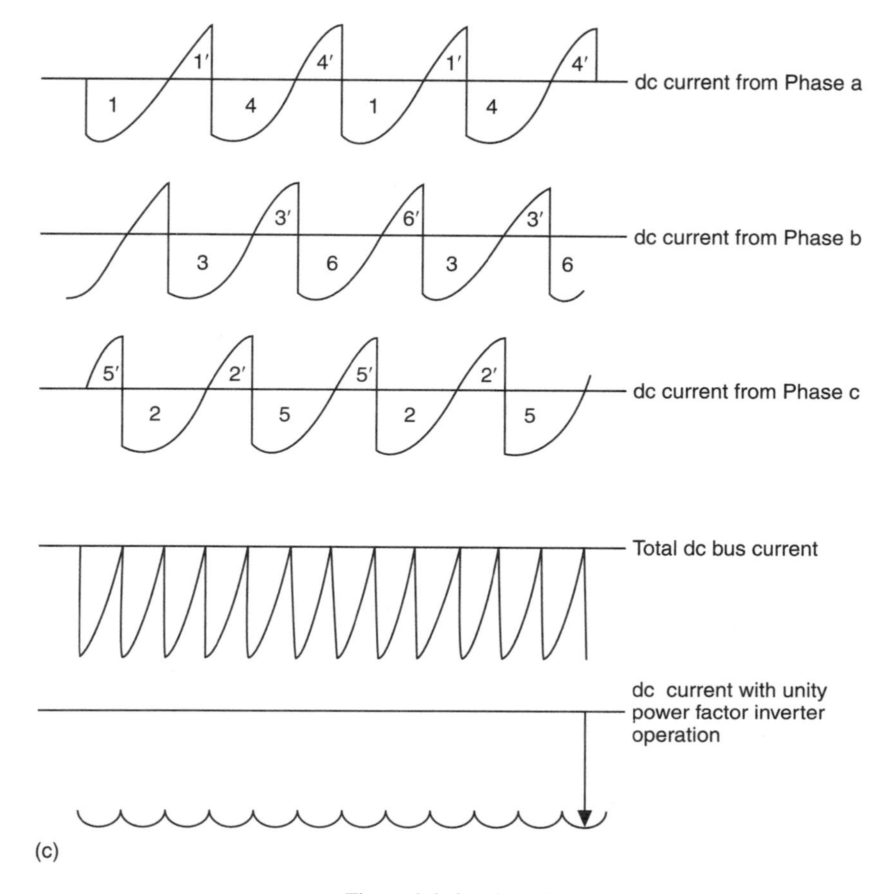

Figure 3.4 Continued

current would actually flow through diode 4′. But later, say from point t_2 to t_3, when device 4 is turned off and device 1 is turned on, the negative current flows through device 1, the current having transferred from diode 4′ to device 1. Figure 3.4(a) shows the current flow path during t_1–t_2; the current coming out of phase b flows through device 6, but then part of this current returns back through diode 4′ into phase a, and part goes into the dc bus. The dc current returns via turn-off device 5 into phase c. At any time, three valves are conducting in a three-phase converter system. In fact only the active power part of the ac current and part of the harmonics flows into the dc side as illustrated in Figure 3.4(c) and further explained later in this section.

3.5.2 Fundamental and Harmonics for a Three-Phase Bridge Converter

It should be noted that the square wave waveforms of v_a, v_b, and v_c are the phase terminal voltages with respect to the hypothetical midpoint N of the dc voltage and not the neutral point of the ac side. These voltages would be phase-to-neutral ac voltages only if the ac neutral is physically connected to the midpoint of the dc voltage, in which case the converter would in effect become a series connection of two three-pulse three-phase half-wave converters and not a six-pulse three-phase full-wave converter.

For a square wave with amplitude of $V_d/2$, the instantaneous values of v_a, v_b, and v_c based on Fourier analysis are given by

$$v_a = \frac{4}{\pi}\left(\frac{V_d}{2}\right)\left[\cos \omega t - \frac{1}{3}\cos 3\omega t + \frac{1}{5}\cos 5\omega t - \frac{1}{7}\cos \omega t + \ldots\right]$$

ν_b is obtained by replacing ωt by

$$\left(\omega t - \frac{2\pi}{3}\right)$$

and ν_c is obtained by replacing ωt by

$$\left(\omega t + \frac{2\pi}{3}\right)$$

For all the triplen harmonics (i.e., 3rd, 9th . . .), the multipliers 3, 9 . . . in the terms

$$\cos 3\left(\omega t \pm \frac{2\pi}{3}\right),$$

$$\cos 9\left(\omega t \pm \frac{2\pi}{3}\right), \text{etc.}$$

reduce these terms to cos 3 ωt, which means that all the triplen harmonics of all three phases are in phase.

Since the ac neutral in a bridge converter is floating, it is necessary to work out phase-to-neutral voltages, which appear across the transformer secondaries. If it is assumed that the three phases are connected to a wye transformer secondary with floating neutral, then the floating neutral will acquire a potential with respect to the dc midpoint which is one-third of the sum of all three voltages of phase terminals a, b, and c. Figure 3.4 shows that v_n is a square-wave of magnitude $V_d/6$ with three times the frequency, i.e., it has all the 3n harmonics of the terminal voltages.

Subtracting v_n from the phase terminal voltages with respect to dc neutral give the phase voltages across the wye-connected transformer secondaries, as shown only for v_{an}, phase a to transformer neutral voltage in Figure 3.4(b). It consists of steps of $V_d/3$, a six-pulse waveform free from $3n$ harmonics. It now has harmonics of only the order of $6n \pm 1$, i.e., 5th, 7th, 11th, 13th, etc. Waveforms v_{bn} and v_{cn} would be the same except phase shifted by 120 degrees and 240 degrees, respectively, from v_{an}. Note that the ac phase-to-neutral voltages are still in phase with the ac phase to dc neutral voltages, i.e., v_{aN} and v_{an} are in phase. The only difference between v_{aN} and v_{an} is that v_{an} is with the triplen harmonics of v_{an}.

$$v_{an} = \frac{4}{\pi}\left(\frac{V_d}{2}\right)\left[\cos \omega t + \frac{1}{5}\cos 5\omega t - \frac{1}{7}\cos \omega t - \frac{1}{11}\cos 11\omega t + \frac{1}{13}\cos 13\omega t + \ldots\right]$$

The phase-to-phase waveform v_{ab} shown in Figure 3.4(b) is also a six-pulse waveform, but of different waveshape than v_{an}. Apart from the observation that v_{ab} is a two-level voltage with 0 or V_d and v_{an} is a three-level voltage with levels of 0, 1/3 V_d, and 2/3 V_d, comparison of waveforms v_{ab} and v_{an} shows that the two are phase shifted and v_{ab} is larger than V_{an}. Being phase-to-phase voltage, fundamental component of v_{ab} is phase shifted by 30°, and its amplitude is $\sqrt{3}$ times the v_{an}.

The rms value of the phase-to-phase voltage (120 degrees, square wave with an amplitude of V_d) is given by

$$V = \sqrt{\frac{1}{\pi}\int_{-\pi/3}^{+\pi/3} V_d^2 \, d\omega t} = \sqrt{2} * V_d/\sqrt{3} = 0.816 V_d$$

The value of the fundamental and harmonic components of the phase-to-phase voltage is given by

$$v_{ab} = \frac{2\sqrt{3}}{\pi} V_d \left[\cos \omega t - \frac{1}{5}\cos 5\omega t + \frac{1}{7}\cos 7\omega t - \frac{1}{11}\cos 11\omega t + \frac{1}{13}\cos 13\omega t - \ldots\right]$$

The rms value of the fundamental is given by

$$V_1 = \frac{\sqrt{6}}{\pi} V_d = 0.78 V_d$$

compared to the total rms value (including harmonics) of 0.816 V_d.

Individual harmonic voltage is given by

$$V_n = V_1/n$$

The value of fundamental and harmonic component of the phase-to-neutral voltage is given by

$$v_{an} = \frac{2}{\pi} V_d \left[\cos \omega t + \frac{1}{5}\cos 5\omega t - \frac{1}{7}\cos 7\omega t - \frac{1}{11}\cos 11\omega t + \frac{1}{13}\cos 13\omega t + \ldots\right]$$

Note that both v_{ab} and v_{an} are defined above with their own zero reference and in fact the two are 30 degrees out of phase.

Figure 3.4(c) shows the dc side current waveforms. Consider first the waveform i_a for current in phase a, shown in Figure 3.4(b), this current flows through the phase-leg with valves 1–1′ and 4–4′. Reversing the waveform sections of the valve 4–4′ gives the dc current contribution of the phase-leg a, to the total current in the dc bus on converter side of the dc capacitors, as shown by the top waveform in Figure 3.4(c). Contribution of current from the other two phase-legs is shown by the next two waveforms in Figure 3.4(c). Adding up the three currents gives the total dc current i_d in the dc bus, as shown by the third waveform of Figure 3.4(c). It consists of direct current component and harmonics of the order of $n = 6k$, i.e., 6th, 12th, 18th. . . . The direct current component of this current is given by:

$$I_d = (3\sqrt{2}/\pi) I \cos\theta = 1.35\, I \cos\theta$$

where I is the rms ac phase current and θ is the power factor angle. The current is maximum at 1.35 I, when power factor is unity and changes from +1.35 I to −1.35 I, and vice versa, as the angle changes from full rectification to inversion of power.

*N*th-harmonic current is at its minimum when power factor is unity and corresponds to

$$I_n/I_{dm} = \sqrt{2}/(n^2 - 1), \text{ where } I_{dm} \text{ is the peak value of the dc bus current}$$

and increases to maximum when power factor is zero and corresponds to

$$I_n/I_{dm} = \sqrt{2}n/(n^2 - 1)$$

The higher the n, the lower is the harmonic amplitude, and clearly the three-phase, full-wave converter has much lower harmonics than the single-phase, full-wave converter because of the elimination of low-order and other harmonics, particularly the second harmonic. However, even in a six-pulse operation, the second harmonic will reappear during ac voltage unbalances, and the system needs to be designed to suppress and/or ride through low frequency harmonics during ac system faults and other reasons for system unbalance.

3.6 SEQUENCE OF VALVE CONDUCTION PROCESS IN EACH PHASE-LEG

It is necessary to discuss the operation of each phase-leg in greater detail, in order to relate to a variety of converter topologies.

It is clear from the discussion in the sections above, that each phase-leg operates independently, and involves alternate turn-on and turn-off of the devices. For instantaneous current (power) flow from ac to dc, the current flows through the diodes, and, for instantaneous current (power) flow from dc to ac, the current flows through the turn-off devices. Figure 3.5 shows an ac voltage waveform of one phase-leg (with respect to the dc midpoint), with a varying phase angle with respect to an assumed sinusoidal current flow. It is not important to be concerned about where the current is going on the ac or dc sides, because in a complete circuit there will be other phase-legs and ac and dc system connections for a complete current loop.

The ac voltage waveform represents at the start an inverter with a unity power factor for a one-cycle segment. It then makes a phase delay of 60 degrees to show the operation for the next one cycle of inverter operation with phase delay of 60 degrees from the unity power factor. This is then followed by delayed steps of 30 degrees, 60 degrees, 30 degrees, 60 degrees, 30 degrees, and 30 degrees to illustrate one full cycle of operation at each of the angles in all four quadrants. The specific device carrying the current is noted inside the current waveform. For each one-cycle segment, angle between the phase current and the phase voltage is shown by a phasor diagram just below the one cycle waveform. Number 1 on top and 4 below the square wave is the number of the device turned on during each half-cycle.

Starting from the beginning of the waveform, during the first one-cycle segment of inverter operation with unity power factor, device 1 is turned on with the current flow from the $+V_d/2$ bus into the ac phase for the first full half-cycle. This is followed by turn-off of device 1 and turn-on of device 4, which results in the current flow from the ac phase into the $-V_d/2$ bus via device 4. During this full cycle, the phase-leg works as an inverter with a unity power factor. Note that no diodes are involved in conduction. It is also worth noting that the current transfer is at the natural current zero, i.e., device 1 turns off and device 4 turns on (and vice versa) when the current is zero. With zero-current switching, so-called "soft-switching," turn-on and turn-off events involve much lower device stresses and switching losses, compared to the switching when current is at its highest operating level.

For the next half-cycle, the turn-off of device 1 and turn-on of device 4 is delayed by 60 degrees in order to change the phase angle for the following one cycle [segment 2 in Figure 3.5(b)] by 60 degrees. It is seen that, as the current polarity reverses, the current transfers from device 1 to diode 1′. For the one-cycle segment 2, the converter operates as an inverter with current lagging the voltage by 120 degrees, i.e., with inductive reactive power. In this one-cycle segment, turn-off device 4 conducts for 120 degrees feeding power from dc to ac (inverter action), and then diode 4′ conducts

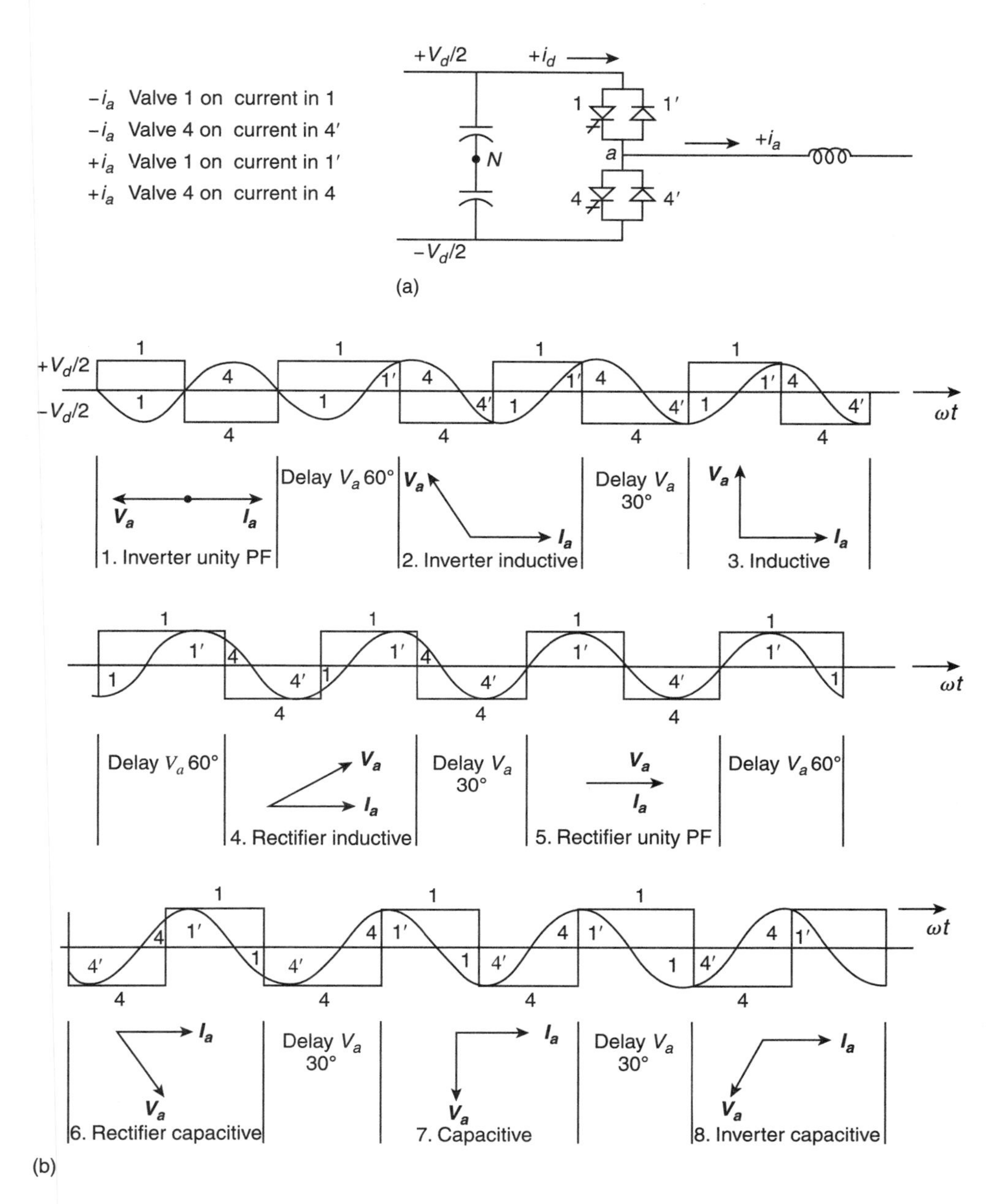

Figure 3.5 Operation of a phase-leg through four quadrants: (a) Phase-leg; (b) Waveforms and phasor diagrams through all four quadrants.

for 60 degrees feeding power back from ac to dc (rectifier action). This is then followed by 120 degrees conduction of device 1 feeding power from dc to ac and diode 1′ feeding power from ac to dc for the final 60 degrees. Note now that the transfer from turn-off device 4 to turn-off device 1 takes place via diode 4′ and from 1 to 4 via diode 1′. Also turn-off of devices 1 and 4 is at natural current zero (soft turn-off). However, turn-on of devices 1 and 4 occurs when current is large and voltage across the valves is V_d. This is known as hard turn-on with significant associated device losses. However, the device capability is often limited by the hard turn-off current requirements so that, notwithstanding the switching losses, the turn-off duty is the most onerous for device capability.

The sequence is repeated by delaying transfer from 4′ to 1 by another 30 degrees. Operation in segment 3 now corresponds to current lagging the voltage by 90 degrees, the converter acting as a pure inductor. In this operation, device 1 conducts for 90 degrees feeding power from dc to ac, then current is transferred to diode 1′ which conducts for 90 degrees feeding power back from ac to dc. For the next half cycle, current is transferred from diode 1′ to device 4 feeding power from dc to ac for 90 degrees, and then from device 4 to diode 4′ feeding power back from ac to dc for 90 degrees. Note again that turn-off of devices 1 and 4 occurs at natural current zero when current transfers to diodes 1′ and 4′, respectively, but the turn-on is hard.

With further delay of 60 degrees, the converter now operates (one-cycle waveform in segment 4) as a rectifier in inductive mode with current lagging voltage by 30 degrees. This is followed by further delay of 30 degrees and one cycle of segment 5 when the converter operates as a rectifier with a unity power factor. Note that in this mode, only the diodes are involved in conduction. Current transfers naturally from 1′ to 4′ and vice versa during current polarity reversal.

With another delay of 60 degrees, segment 6 shows operation as a rectifier, now in capacitive mode with current leading the voltage by 60 degrees. Note that in capacitive mode, turn-off devices have to turn off high current (transfer from 1 to 4′ and 4 to 1′) with a corresponding positive voltage jump of V_d. As mentioned earlier for turn-off devices, the maximum current they have to turn off, accompanied by the magnitude of forward voltage jump during turn-off, is the most important limiting parameter of the devices' useable capability, because turn-off is much harsher than turn-on for such devices. In capacitive mode, the turn-on is soft, but the turn-off is hard, which is the reverse of the operation in an inductive mode.

With another delay of 30 degrees, segment 7 shows pure capacitive operation. It is now seen that the sequence of transfer is 1′–1–4′–4–1′, each conducting for 90 degrees, 1 and 4 feeding power from dc to ac, and 1′ and 4′ feeding power from ac to dc. Note that this transfer sequence is the reverse of the sequence for pure inductive mode of 1–1′–4–4′–1 accomplished with 180 degrees phase delay from segment 3. Note also, that in segment 7, the current transfer from 1 to 4′ and from 4 to 1′ involves turn-off of devices 1 and 4 at maximum current, their most severe duty.

From Figure 3.5, it is seen that for the sequential change of phase angle described above, the conduction and transfer of current takes place as follows:

1-4-1-1′-4-4′-1-1′-4-4′-1-1′-4-4′-1-1′-4-4′-1-1′-4-4′-1′-
4′-1′-1-4′-4-1′-1-4′-4-1′-1-4′-4-1’-1-4′-4-1′

It should be noted that in inductive operation, all turn-off devices turn off at current zero during current reversal, and therefore turn-offs are soft, i.e., the current is zero when the voltage across the turn-off device rises to V_d. Thus, the turn-off stresses and losses are minimal. Also, in inductive mode, events of the current transfer are from the turn-off device to its own parallel diode, i.e., 1 to 1′ or 4 to 4′. In capacitive mode, turn-off is hard, i.e., at finite current, and turn-offs are transfers to the opposite diode, i.e., 1 to 4′ or 4 to 1′.

When the transfer is from one turn-off device to another turn-off device, i.e., 1 to 4 or 4 to 1 during inverter operation with unity power factor, it is essential to delay the turn-on for at least several tens of microseconds following the turn-off of the complementary device is complete. This is to ensure that there is no chance of simultaneous conduction of devices 1 and 4, which represents a direct short circuit across the dc bus capacitor. It is important to note from the sequence of transfers shown above with reference to Figure 3.5, that except for unity power-factor inverter operation, all the current transfers are from a device to a diode, or from a diode to a device. Thus there is very little risk of shoot-through in a topology that involves one conduction pulse per half-cycle.

Also, since power devices and transformers have losses, these losses have to be supplied from the dc side or the ac side during inverter or rectifier operation respectively. However, during full inductive or capacitive operation, losses can be supplied from either side by operating very slightly in rectifier or inverter mode.

It should now be clear that by having two phase-legs on the same dc bus, with 180 degrees phase shifted in their pulse sequence, would give a single-phase full-wave converter. A total of three such phase legs with a pulse sequence 120 degrees apart would give a three-phase, full-wave converter.

One important point to note is that, in the converter described above, the ac voltage output is strictly a function of dc voltage. For an effective interaction with the ac system, it is often necessary to vary the converter ac output voltage, which means that dc voltage is variable accordingly. This can be done by charging/discharging the dc capacitor from another source/absorber of power or from the ac side of the converter itself. The speed with which the dc voltage can be changed would determine the response time of the converter. For a number of applications, the response time for changing the dc bus voltage would more than adequate. However, in some applications, the dc source assigned to control the dc bus voltage may have other functional priorities. An important point to make here is that an alternate approach is to have a voltage-sourced converter, which has a stiff dc bus, yet be able to vary the ac voltage of the converter. Indeed, there are such converters, the so-called multistep converters and pulse-width modulation converters. These are discussed later in this chapter.

3.7 TRANSFORMER CONNECTIONS FOR 12-PULSE OPERATION

In Section 3.5, harmonic content of the phase-to-phase voltage and phase-to-neutral voltage was discussed, and it was mentioned that the two voltages were 30 degrees out of phase. If this phase shift is corrected, then for the phase to neutral voltage, i.e., v_{an}, the harmonics, other than those of the order of $12n \pm 1$, would be in phase opposition to those of the phase-to-phase voltage v_{ab} and with $1/\sqrt{3}$ times the amplitude. It follows then, as shown in Figure 3.6(a), that if the phase-to-phase voltages of a second converter were connected to a delta-connected secondary of a second transformer, with $\sqrt{3}$ times the turns compared to the wye-connected secondary, and the pulse train of one converter was shifted by 30 degrees with respect to the other (in order to bring v_{ab} and v_{an} to be in phase), the combined output voltage would have a 12-pulse waveform, with harmonics of the order of $12n \pm 1$, i.e., 11th, 13th, 23rd, 25th . . ., and with amplitudes of 1/11th, 1/13th, 1/23rd, 1/25th . . ., respectively, compared to the fundamental. Figure 3.6(b) shows the two waveforms v_{an} and v_{ab}, adjusted for the transformer ratio and one of them phase displaced by 30 degrees. These two waveforms are then added to give the third waveform, which is seen to be a 12-pulse waveform, closer to being a sine wave than each of the six-pulse waveform.

In the arrangement of Figure 3.6(a), the two six-pulse converters, involving a total of six phase-legs are connected in parallel on the same dc bus, and work together as a 12-pulse converter. It is necessary to have two separate transformers, otherwise

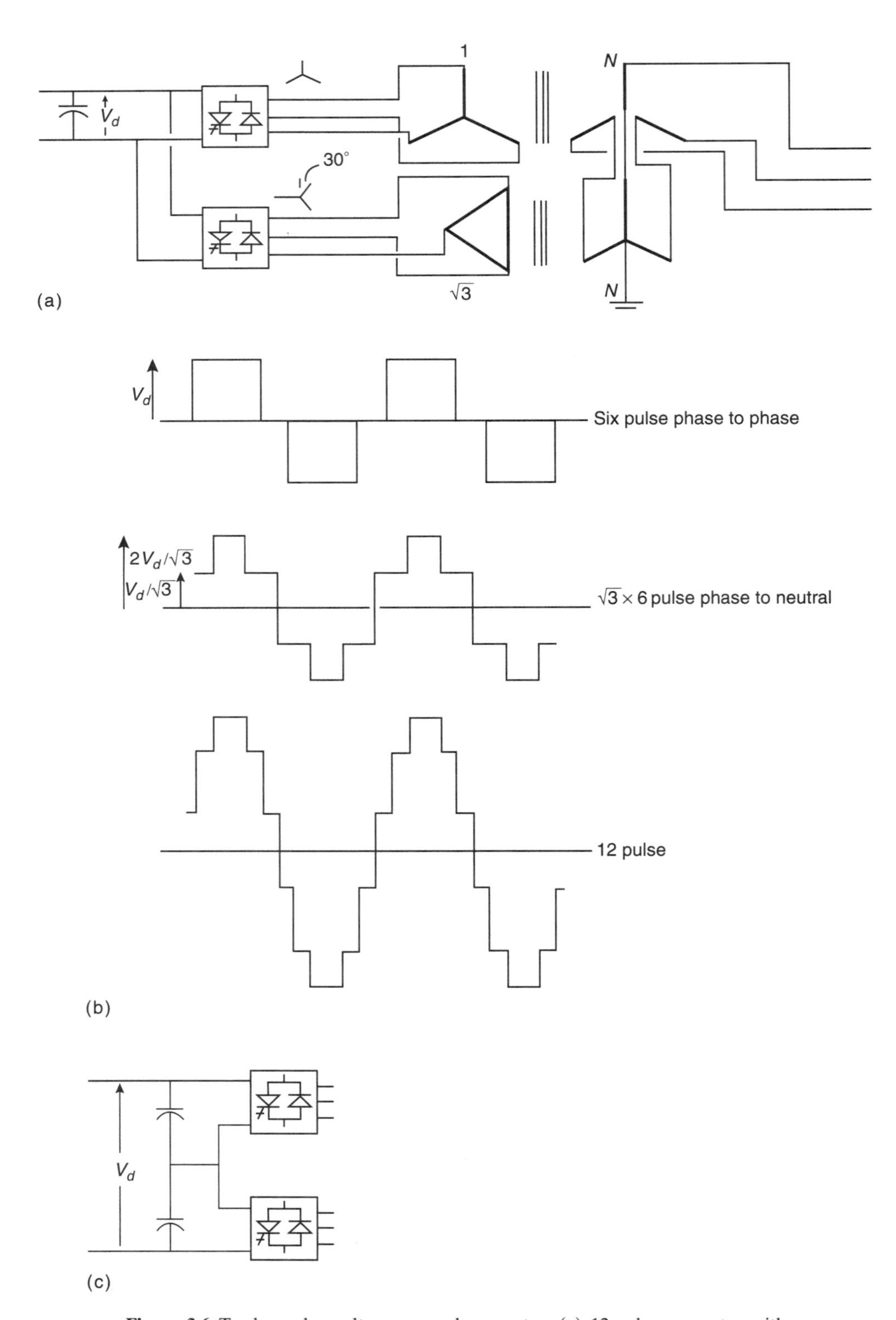

Figure 3.6 Twelve-pulse voltage-sourced converter: (a) 12-pulse converter with wye and delta secondaries; (b) 12-pulse waveform from two six-pulse waveforms; (c) 12-pulse converter with two series connected six-pulse converters.

phase shift in the non-12-pulse harmonics, i.e., 5th, 7th, 17th, 19th . . . in the secondaries will result in a large circulating current due to common core flux. To the non-12-pulse voltage harmonics, common core flux will represent a near short circuit. Also for the same reason, the two primary side windings should not be directly connected in parallel to the same three-phase ac busbars on the primary side. Again this is because the non-12-pulse voltage harmonics, i.e., 5th, 7th, 17th, 19th . . ., while they cancel out looking into the ac system, would be in phase for the closed loop. Consequently, a large current corresponding to these harmonics will also flow in this loop, limited only by the impedance of the loop, which is essentially the leakage inductance of the transformers.

The circulating current of each non–12-pulse harmonic is given by:

$$I_h/I_1 = 100/(X_T * n^2) \text{ percent}$$

where, I_1 is the nominal fundamental current, n is the relevant harmonic number, and X_T is the per unit transformer impedance of each transformer at the fundamental frequency. For example, if X_T is 0.15 per unit at fundamental frequency, then the circulating current for the fifth harmonic will be 26.6%, seventh, 14.9%, eleventh, 5.5%, thirteenth, 3.9% of the rated fundamental current, and so on. Clearly this is not acceptable for practical voltage sourced converters. Therefore, it is necessary to connect the transformer primaries of two separate transformers in series, and connect the combination to the ac bus as shown in Figure 3.6(a). With the arrangement shown in Figure 3.6(a), the 5th, 7th, 17th, 19th . . . harmonic voltages cancel out, and the two fundamental voltages add up, as shown in Figure 3.6(b), and the combined unit becomes a true 12-pulse converter.

The two converters can also be connected in series on the dc side for a 12-pulse converter of twice the dc voltage, Figure 3.6(c). In such a case, it is important to provide a control to ensure that the two dc buses (capacitors) have equal voltages. The dc voltage of either converter can be increased or decreased by shifting the operation in the rectifier or inverter direction, by a dc voltage balancing control.

There are other means of paralleling voltage source converters on the ac side, which involve transformers with special windings. However, it is generally recognized that special transformers would cost more than the means described above.

Increase in pulse number also decreases the dc side current harmonics, which are cancelled out among the phase legs and do not even enter the dc bus. For 12-pulse converters, the harmonics of the order 6th, 18th, . . . are cancelled out, and only the 12-pulse harmonics 12th, 24th . . ., enter the dc bus.

3.8 24- AND 48-PULSE OPERATION

Two 12-pulse converters, phase shifted by 15 degrees from each other, can provide a 24-pulse converter, obviously with much lower harmonics on ac and dc side. Its ac output voltage would have $24n \pm 1$ order harmonics, i.e., 23rd, 25th, 47th, 49th . . . harmonics, with magnitudes of 1/23rd, 1/25th, 1/47th, 1/49th . . ., respectively, of the fundamental ac voltage. The question now is how to arrange this 15 degrees phase shift.

One approach is to provide 15 degrees phase-shift windings on the two transformers of one of the two 12-pulse converters. Another approach is to provide phase-shift windings for +7.5 degrees phase shift on the two transformers of one 12-pulse converter and −7.5 degrees on the two transformers of the other 12-pulse converter, as shown in Figure 3.7(a). The latter is preferred because it requires transformers of the same

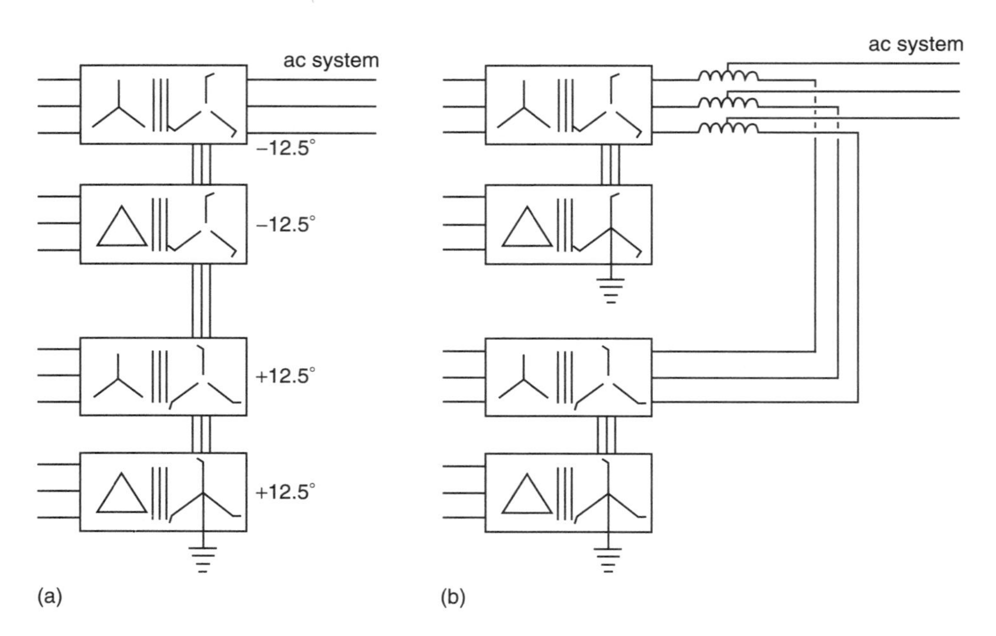

Figure 3.7 Various means to obtain 24-pulse converter operation: (a) 24-pulse converter transformer connections with two 12-pulse converters in series on the ac side; (b) 24-pulse converter transformer connections with two 12-pulse converters in parallel on the ac side.

design and leakage inductances. It is also necessary to shift the firing pulses of one 12-pulse converter by 15 degrees with respect to the other.

All four six-pulse converters can be connected on the dc side in parallel, i.e., 12 phase-legs in parallel. Alternately all four six-pulse converters can be connected in series for high voltage, or two pair of 12-pulse series converters may then be connected in parallel. Each six-pulse converter will have a separate transformer, two with wye-connected secondaries, and the other two with delta-connected secondaries. Primaries of all four transformers can be connected in series as shown in Figure 3.7(b) in order to avoid harmonic circulation current corresponding the 12-pulse order, i.e., 11th, 13th, 23rd, 24th.

It may be worthwhile to consider two 12-pulse converters connected in parallel on the ac system busbars, with interphase reactors as shown in Figure 3.7(b), for a penalty of small harmonic circulation inside the converter loop. While this may be manageable from the point of view of converter rating, care has to be exercised in the design of the converter controls, particularly during light load when the harmonic currents could become the significant part of the ac current flowing through the converter. An increase in the transformer impedance to, say, 0.2 per unit may be appropriate when connecting two 12-pulse transformers to the ac bus directly and less than that when connected through interphase reactors.

For high-power FACTS Controllers, from the point of view of the ac system, even a 24-pulse converter without ac filters could have voltage harmonics, which are higher than the acceptable level. In this case, a single high-pass filter tuned to the 23rd–25th harmonics located on the system side of the converter transformers should be adequate. The alternative, of course, is go to 48-pulse operation with eight six-pulse groups, with one set of transformers of one 24-pulse converter phase-shifted from the other by 7.5 degrees, or one set shifted by +3.75 degrees and the other by −3.75 degrees. Logically, all eight transformer primaries may be connected in series, but because of the small phase shift (i.e., 7.5 degrees), the primaries of the two 24-pulse converters (each with four primaries in series) may be connected in parallel if the consequent circulating current is acceptable. This should not be much of a problem because the higher the order of a harmonic, the lower would be the circulating current. For 0.1 per unit transformer impedance and the 23rd harmonic, the circulating current would be 1.9% only. The circulating current can be further limited by higher transformer inductance or by inter-phase reactors at the point of parallel connection of the two 24-pulse converters. With 48-pulse operation, ac filters should not be necessary.

3.9 THREE-LEVEL VOLTAGE-SOURCED CONVERTER

3.9.1 Operation of Three-Level Converter

It was mentioned earlier in this chapter that it would be desirable to vary the magnitude of ac output voltage without having to change the magnitude of the dc voltage. The three-level converter is one concept that can accomplish that to some extent. One phase-leg of a three-level converter is shown in Figure 3.8(a). The other two phase-legs (not shown) would be connected across the same dc busbars and the clamping diodes connected to the same midpoint N of the dc capacitor. It is seen tha each half of the phase leg is split into two series connected valves, i.e., 1–1′ is sp into 1–1′ and 1A–1′A. The midpoint of the split valves is connected by diodes and D_4 to the midpoint N as shown. On the face of it, this may seem like doubling the number of valves from two to four per phase-leg in addition to providing two extra diode valves. However, doubling the number of valves with the same voltage

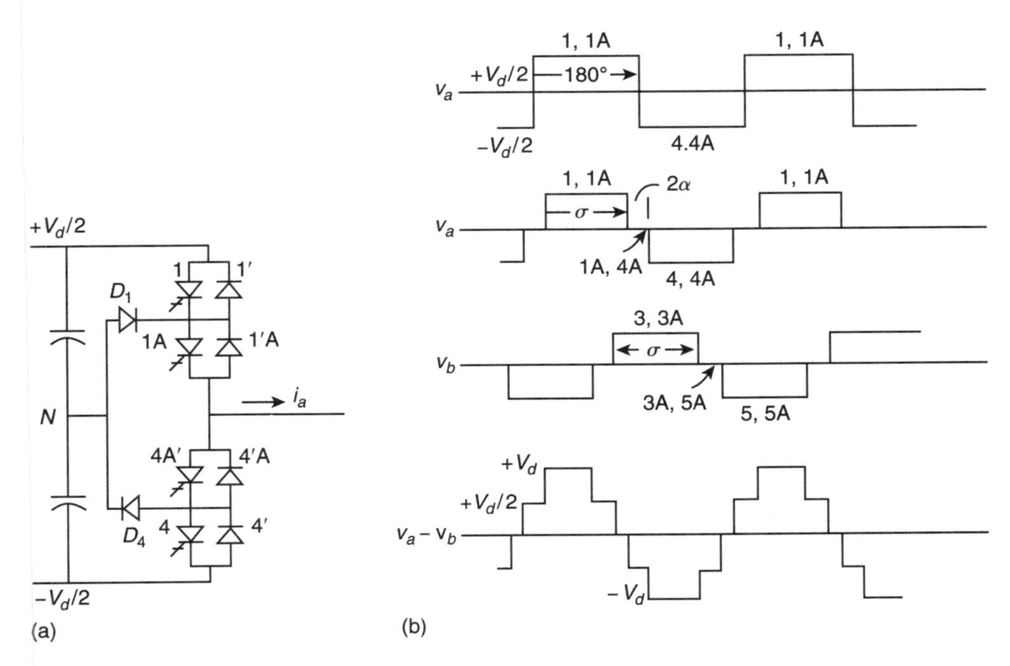

Figure 3.8 Operation of a three-level converter: (a) One phase-leg of a three-level converter; (b) Output ac voltage.

rating would double the dc voltage and hence the power capacity of the converter. Thus only the addition of the diode clamping valves, D_1 and D_2, per phase-leg, Figure 3.8(a), adds to the converter cost. If the converter is a high-voltage converter with devices in series, then the number of main devices would be about the same. A diode clamp at the midpoint may also help ensure a more decisive voltage sharing between the two valve-halves. On the other hand, requirement that a converter continue safe operation with one failed device in a string of series connected devices, may require some extra devices.

Figure 3.8(b) shows output voltage corresponding to one three-level phase leg. The first waveform shown is a full 180 degrees square wave obtained by the closing of devices 1 and 1A to give $+V_d/2$ for 180 degrees, and the closing of valves 4 and 4A for 180 degrees to give $-V_d/2$ for 180 degrees. Now consider the second voltage waveform in Figure 3.8(b) in which the upper device 1 is turned off and device 4A is turned on an angle α earlier than they were due in the 180 degrees square wave operation. This leaves only device 1A and 4A on, which in combination with diodes D_1 and D_2, clamp the phase voltage V_a to zero with respect to the dc midpoint N regardless of which way the current is flowing. This continues for a period 2α until device 1A is turned off, and device 4 is turned on, and the voltage jumps to $-V_d/2$ with both the lower devices 4 and 4A turned on and both the upper devices 1 and 1A turned off, and so on. Of course, angle α is variable, and the output voltage V_a is made up of $\sigma = 180° - 2\alpha°$, square waves. This variable period σ per half-cycle, potentially allows the voltage V_a to be independently variable with potentially a fast response. It is seen that devices 1A and 4A are turned on for 180 degrees during each cycle, devices 1 and 4 are turned on for $\sigma = 180° - 2\alpha°$ during each cycle, while diodes D_1 and D_4 conduct for $2\alpha° = 180°\sigma$ each cycle. The converter is referred to as three-level because the dc voltage has three levels, i.e., $-V_d/2$, 0, and $+V_d/2$.

As explained earlier, these phase-legs can handle the current flow in any phase relationship. Turn-off devices handle any instantaneous inverter action and their parallel diodes handle any instantaneous rectifier action. Clamping diodes D_1 or D_4, along with lower devices 1A and 4A, carry the current during clamping periods, with D_1-1A carrying the negative current (current going into the ac bus) and D_4-4A carrying the positive current.

Also as explained before for the two-level phase-legs, these three-level phase-legs can also be connected in different configurations to obtain the required converter. These configurations include a single phase, full-wave converter with two legs, a three-phase bridge converter with three legs and wye-connected secondaries with floating neutral or delta-connected secondaries, etc. Figure 3.7(b) also shows the output voltage of the second phase v_b and the phase-to-phase voltage v_{ab} for a three-phase converter. As discussed previously, the triplen harmonics will not pass through to the primaries with floating neutral, etc. Two six-pulse converters can provide a 12-pulse converter, and so on.

With the three-level converter it should be noted that for the duration of zero output voltage, Figure 3.8(b), the phase-leg current(s) flows into the midpoint of the two capacitors and thus flows through the dc capacitor. This current is mainly of third harmonic and is substantially independent of the converter pulse number.

3.9.2 Fundamental and Harmonic Voltages for a Three-Level Converter

What the three-level converter buys is the flexibility to rapidly vary the ac voltage or to provide a defined zero voltage "notch" to eliminate or reduce some specific harmonics. However, when considering the harmonics, there is a limitation in exploitation of this flexibility, as explained in the following paragraphs.

Taking into account the ac voltage pulse width of duration σ per half-cycle, the magnitudes of fundamental and harmonic components of phase-to-dc-neutral voltage are defined by

$$v = \frac{4}{\pi}\left(\frac{V_d}{2}\right)\left[\sin\frac{\sigma}{2} * \sin\left(\omega t + \frac{\sigma}{2}\right) - \frac{1}{3}\sin\frac{3\sigma}{2} * \sin 3\left(\omega t + \frac{\sigma}{2}\right) + \frac{1}{5}\dots\right]$$

which gives

$$v_n = \frac{4}{\pi}\left(\frac{V_d}{2}\right)\left[\frac{1}{n}\sin\frac{n\sigma}{2} * \sin n\left(\omega t + \frac{\sigma}{2}\right)\right]$$

RMS value is given by

$$V_n = \frac{2\sqrt{2}}{\pi}\left(\frac{V_d}{2}\right)\frac{1}{2}\sin\frac{n\sigma}{2}$$

Fundamental rms voltage is given by

$$V_1 = \frac{2\sqrt{2}}{\pi}\left(\frac{V_d}{2}\right)\sin\frac{\sigma}{2}$$

which starts from a maximum rms value of $(\sqrt{2}/\pi)V_d$ at $\sigma = 180°$, and reaches zero at $\sigma = 0$.

If V_1 is assumed to be $V_{1\max} = (\sqrt{2}/\pi)\ V_d = 1$ pu at $\sigma = 180°$, Figure 3.9 shows the value of the fundamental voltage V_1 as per unit of $V_{1\max}$ and function of the pulse width, σ. As seen, the fundamental is 1.0 p.u. at $\sigma = 180°$, and decreases with decreasing pulse width, decreasing to zero at $\sigma = 0°$.

Figure 3.9 also shows the values of harmonics, as per unit of the actual fundamental voltage V_1, as a function of the pulse width σ. Harmonic values as per unit of $V_{1\max}$ can be more useful, if the purpose is to consider harmonics in terms of the specified distortion levels. Such values can be obtained by multiplying a harmonic level in Figure 3.9 by corresponding per unit fundamental level.

It is interesting to note the variation in per unit harmonics with decrease in σ (increase in $2\alpha = 180° - \sigma$). It is seen that the 5th, 7th ... harmonics are at their maximum, 0.2 p.u. and 0.143 p.u., respectively, at $\sigma = 180°$, and then decrease and increase according to the equation above. A particular harmonic reaches zero, when

$$180° - \sigma = (180°/n), \text{ where } n \text{ is the harmonic number.}$$

For fifth harmonics, this occurs at $\sigma = 144°$ and again at 72°.

The seventh harmonic is zero when $\sigma = 154.3°$, 102.9°, and 77°.

After the first zero, each harmonic rises again, reaching a peak at

$$180° - \sigma = 2x(180°/n).$$

These peaks are higher than the first peaks because the values shown are per unit of the actual declining magnitude of the fundamental voltage V_1. All harmonics eventually approach 1.0 p.u. at $\sigma = 0$ when the fundamental approaches zero.

It is noted that for operation with σ within 154 degrees to 144 degrees, both the fifth and seventh harmonics are very low, and the converter almost behaves like a 12-pulse converter. For operation at $\sigma = 144°$, the fundamental voltage also drops to 0.95 p.u., which in effect represents 5% loss of capacity.

As an alternative, it may be preferred to normally operate a three-level converter

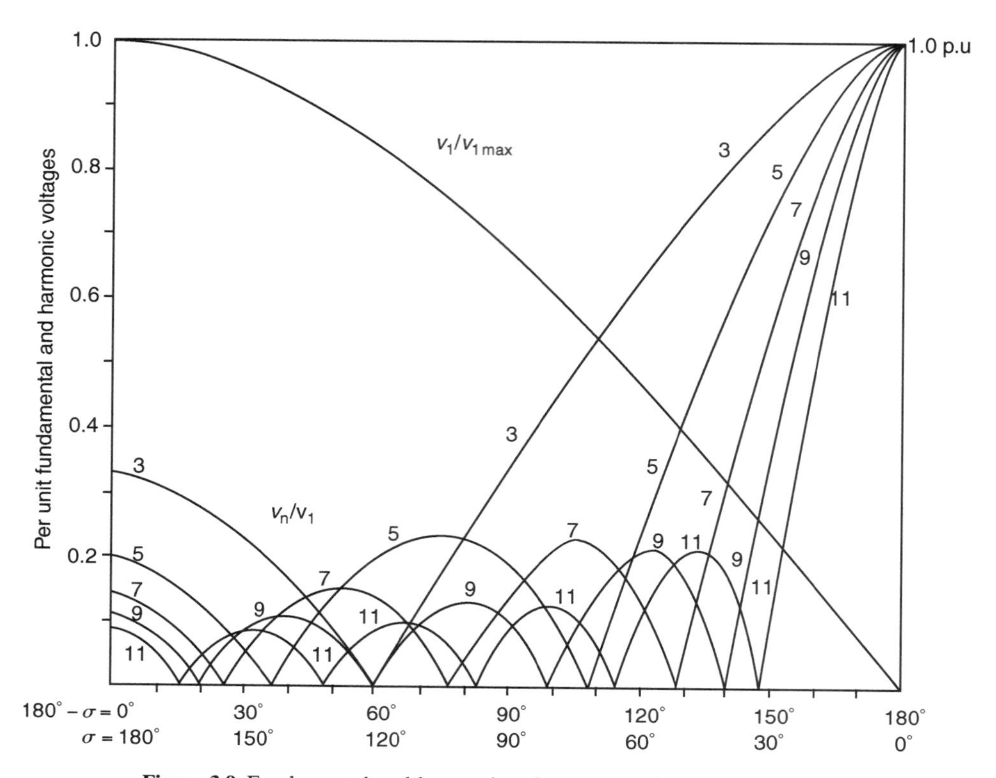

Figure 3.9 Fundamental and harmonic voltages for a three-level converter.

at about 154 degrees, which is where the fourth harmonic is zero, and the fifth harmonic is about half of its maximum value. In any case, it is a matter of compromising flexibility of voltage control, reduction of specific harmonics, and some loss of useable capacity. In many applications, it is not necessary to vary the ac voltage independently of the dc voltage, in which case a three-level converter would be a good way to reduce low order harmonics or to eliminate a specific harmonic. Alternately, one can normally operate the three-level converter with varying σ over a range from $\sigma = 180°$ down to $\sigma = 90°$, giving a variable ac voltage from 100% to 70%. It also points to the consideration of a combination of higher pulse converter, dc bus voltage control and use of three-level phase-legs. Unless the three-step converter is used within a multipulse structure, it is useful for only a limited range of independent ac voltage control. This is because the percentage harmonics increase rapidly with decrease in the fundamental output voltage to below about 70%, as shown by Figure 3.9.

With the split dc capacitor level, it is essential to ensure that the two capacitors are charged to equal voltage because unequal voltages will generate even harmonics. Capacitor voltage balance is achieved by a control, which prolongs or shortens the conduction time of the appropriate devices.

These curves of Figure 3.9 also apply to any other configuration that is based on square wave with pulse width less than 180 degrees. Note that all the triple harmonics are zero at $\sigma = 60°$, which corresponds to the phase-phase six-pulse voltage waveform for a three-phase full wave bridge (discussed in Section 3.5).

A significant disadvantage of a three-level converter is that with increasing σ, an increasing amount of third harmonic flows into the midpoint of the dc capacitor. This current generates a third harmonic voltage across the capacitors. In order to keep this harmonic voltage within acceptable limits (to avoid generation of additional harmonics in the ac output and an increase in the converter current rating), the size of the dc capacitor has to be increased compared to that of the two-level converter.

3.9.3 Three-Level Converter with Parallel Legs

There is yet another method of achieving a three-level converter, which is to connect two phase-legs in parallel per phase as shown in Figure 3.10(a). The two legs are paralleled through an inductor and the ac connection is made at the midpoint of this inductor, and their pulse sets are phase shifted by an angle α each in the opposite directions (a total of 2α). The ac terminal voltage of the two phase-legs, with respect to a hypothetical dc midpoint, are shown in Figure 3.10(b) by the first two waveforms, shifted from each other by an angle, 2α. The net ac terminal voltage, with respect to the dc midpoint, is the average of the two voltages, and is shown by the third waveform. It consists of a half-cycle pulse of variable duration $\sigma = 180 - 2\alpha°$, the same as for the three-level converter with split capacitor discussed above. The harmonic content is the same as shown by Figure 3.9. The inductor voltage is given by the difference of the ac voltage of the two legs, and is shown by the fourth waveform of Figure 3.10(b), which conveys that greater the required range of control, larger would be the inductor size. Its MVA rating would be directly proportional to the integral of the voltage V_L.

It is natural to suppose that one can go to higher levels, i.e., four-level, five-level, and so on. However, detailed consideration of these higher level topologies would reveal a major problem of voltage balancing between the capacitors. It is not reasonable to assume that the current flow though each level is balanced within some tolerable range for a converter to continue to operate in five-level mode on its own. However, if there is a two-converter system, with dc bus in between, it is possible to link the two converters in a way to ensure balanced capacitor voltages of different levels.

3.10 PULSE-WIDTH MODULATION (PWM) CONVERTER

In two-level or multilevel converters, there is only one turn-on, turn-off per device per cycle. With these converters, the ac output voltage can be controlled, by varying the width of the voltage pulses, and/or the amplitude of the dc bus voltage. Another approach is to have multiple pulses per half-cycle, and then vary the width of the pulses to vary the amplitude of the ac voltage. The principal reason for doing so is to be able to vary the ac output voltage and to reduce the low-order harmonics, as will be explained here briefly. It goes without saying that more pulses means more switching losses, so that the gains from the use of PWM have to be sufficient to justify an increase in switching losses. There are also resonant PWM converter topologies that incorporate current-zero or voltage-zero type soft switching, in order to reduce the switching losses. Such converters are being increasingly utilized in some low power applications, but with the known topologies, they have not been justifiable at high power levels due to higher equipment cost.

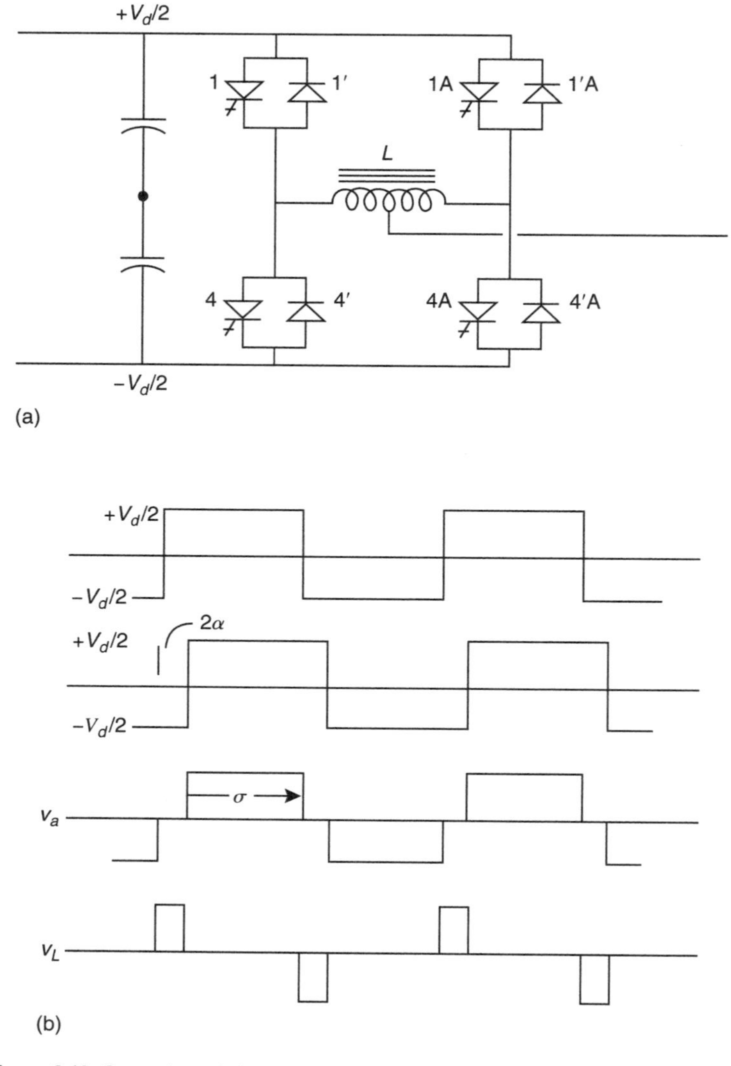

Figure 3.10 Operation of three-level converter with parallel legs: (a) One phase-leg with two parallel legs; (b) Waveforms for the two parallel legs.

PWM converters of low voltage and low power in the range of tens of watts, for, say, printed circuit boards' power supplies, may have internal PWM frequency in the hundreds of kilohertz. Industrial drives in tens of kilowatts may have internal PWM frequency in the tens of kilohertz. For converters in the 1 MW range, such as for Custom Power, the frequency may be in a few kilohertz range. For FACTS technology with high power in the tens of megawatts and converter voltage in kVs and tens of kVs, low frequencies in the few hundred Hertz or maybe the low kilohertz range may seem feasible and worth considering.

Consider again a phase-leg, shown in Figure 3.11(a) [which is the same as Figure 3.5(a)], as part of a three-phase bridge converter. Figure 3.10(b) shows comparison of two types of control signals, three signals of main-frequency sine wave representing three phases, and a sawtooth wave signal of nine times the main frequency (540 Hz

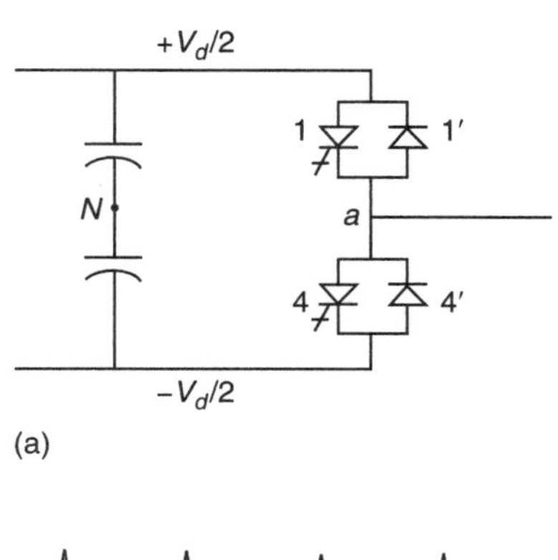

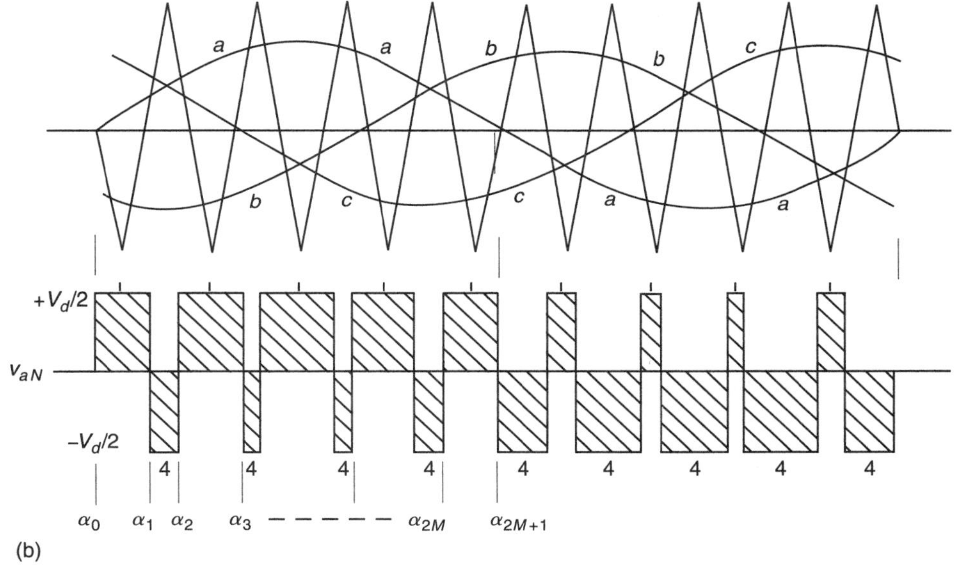

Figure 3.11 Operation of a PWM converter with switching frequency of nine times the fundamental: (a) A phase-leg; (b) PWM waveforms.

for 60 Hz main frequency). Turn-on and turn-off pulses to the devices correspond to the crossing points of the sawtooth wave with the sine wave of corresponding phase. The negative slope of the sawtooth wave crossing the sine wave of phase a, results in a turn-on pulse for device 1 and turn-off pulse for device 4. The positive slope of the sawtooth wave crossing the sine wave of phase a results in a turn-off pulse for device 1 and turn-on pulse for device 4. The resulting voltage of the ac terminal a, with respect to hypothetical midpoint N of the dc capacitor, is shown in hatch in Figure 3.11(b). In comparison to Figure 3.5(b) with two square pulses per cycle, the waveform of Figure 3.11(b) is made up of nine square pulse cycles of varying width per main frequency cycle. The pulses are wider in the middle of each half sine wave compared to the ends of the half sine wave.

The following observations are worth noting with respect to the waveforms of Figure 3.11(b):

1. The output voltage waveform contains a fundamental frequency component and harmonics.

2. The output voltage pulses are symmetrical about the zero crossings of the sine wave, because the sawtooth frequency is an odd integer multiple of the main frequency. Any even multiple will create asymmetry about the zero crossing, which will contain even harmonics. Non-integer multiples are even worse as they create sub- and supersynchronous harmonics as well. When the frequency is high, above a few kilohertz, this asymmetry becomes insignificant, but at the low PWM frequencies synchronization of the control signals is important.
3. With a fixed sawtooth wave, increasing the magnitude of the sine wave will increase the conduction time of device 1, and decrease the conduction time of device 4 for the positive half-cycle and vice versa for the negative half-cycle. This means that the fundamental component of the ac voltage V_{aN} and hence the output ac voltage will increase with an increase in the magnitude of the control sine wave and decrease with a decrease in the magnitude of the control sine wave. For control sine wave peak less than the sawtooth-wave peak, the output ac voltage varies linearly with variation of the control sine wave.
4. As the control sine-wave peak equals the peak of the sawtooth wave, the middle notch in the ac output voltage disappears. If the control sine wave is increased to a higher and higher magnitude, more and more notches will disappear and the output voltage will eventually become a single square wave per half-cycle.
5. It is clear that the ac output voltage can be controlled from zero to maximum.
6. The sine wave itself can be modified with notches, etc. to create other effects on the waveform.

The order of harmonics present in this type of PWM waveform is determined by $k_1 n \pm k_2$, where k_1 is the frequency multiplier (9 for Figure 3.11) of the carrier frequency, and n and k_2 are integers. However, k_2 may only be taken up to 2, after which the magnitude of that harmonic order becomes rather small. Due to half-cycle symmetry, all the even order harmonics also disappear. Furthermore in a three-phase bridge circuit, all of the triple harmonics, i.e., 3rd, 9th, . . ., are eliminated. Also, if the carrier frequency is a multiple of 3, even the harmonics of the order of the carrier frequency are cancelled out in the phase-to-phase and phase to the floating-neutral voltages.

Thus, for the chosen frequency multiplier 9, the order of harmonics is given as: 5th, 7th, 11th, 13th . . . (all the harmonics except the even and triple harmonics), as for the six-pulse three-step converter discussed in Section 3.9. However, in the PWM case shown the fifth harmonic will be very small.

Figure 3.12 shows PWM output voltage waveforms corresponding to a PWM frequency of three times the main frequency. The first waveform shows the control signals, similar to those in Figure 3.11. Second waveform is the phase a to dc neutral voltage v_{aN}. It is seen that it has one notch in the center of each half-cycle and the width of this notch is dynamically controllable every half-cycle. The third waveform is v_{bN}, the output voltage of phase b to dc neutral voltage, which obviously is the same as v_{aN} except the phase delay of 120 degrees. Subtracting v_{bN} from v_{aN} gives the phase–phase voltage v_{ab}, as shown by the fourth waveform. This shows two notches resulting from the crossings of the control signals. The next waveform is that of the v_{nN}, the voltage between the floating neutral n of a wye-connected floating secondary and the dc neutral. This is obtained by adding and averaging the three ac voltages,

v_{aN}, v_{bN}, and v_{cN} (v_{cN} not shown). Subtracting v_{nN} from v_{aN} gives the last waveform shown in Figure 3.12, that of the transformer phase-to-neutral voltage. Because of the half-wave symmetry, all the ac waveforms are free from even harmonics. Waveforms v_{ab} and v_{an} are free from triplen harmonics and the v_{an} lags v_{ab} by 30 degrees. As explained earlier in Section 3.7, combining these two waveforms through separate wye and delta transformers, will result in a 12-pulse converter, which will have adequate flexibility of rapid ac voltage control without having to change the dc voltage level. The control of the dc voltage can then be optimized for other considerations.

It should now be obvious that the ac voltage waveform can be chopped in many different ways with different control waveforms and numerical program. There are a variety of waveforms other than sinusoidal and sawtooth that are used to create ac output voltage with fewer low-frequency harmonics and fewest notches.

While chopping of the waveforms and showing PWM waveforms is easy on paper, it is not a trivial matter when we consider its implications for design of high-power and high-voltage converters. In terms of switching losses, impact of increased higher harmonics, EMI, audible noise, etc. has to be justified by appropriate gains in other areas, particularly if it helps satisfy the harmonic requirements without having to go to a pulse number higher than 12 for modest size converters. Nevertheless, a considered use of PWM or notches at low-frequency level has its merits, particularly for the FACTS Controllers of lower power levels of, say, 10–50 MW, as will be seen from the next section.

3.11 GENERALIZED TECHNIQUE OF HARMONIC ELIMINATION AND VOLTAGE CONTROL

One effective way to have the freedom of controlling the voltage and also eliminating lower order harmonics is a method pioneered by Patel and Hoft in the early 1970s. It involves varying specific notches (also referred to as chops) in the square wave such that specific harmonics are eliminated from the waveform. This is explained below.

Basically, a square wave can be chopped a number of times in a relationship that eliminates a number of harmonics, as well as giving flexibility to vary the fundamental voltage. With M number of chops, there are M degrees of freedom. One of these degrees of freedom can be used to control the fundamental leaving the other M-1 degrees of freedom used to eliminate M-1 selected harmonics.

It is seen from Figure 3.11(b), that a sawtooth wave with nine times fundamental frequency has four notches each in the positive and negative half-cycle waveform. They also have half-wave and quarter-wave symmetry. It follows that with proper firing angles as well as half-wave/quarter-wave symmetry, not only the fundamental frequency component can be controlled, but the three other selected harmonics can be eliminated, i.e., 5th, 7th, and 11th, from the output ac voltage of a three-phase full wave bridge. If there were two three-phase bridges, phase-shifted as previously discussed in this chapter, forming a 12-pulse voltage sourced converter, then the harmonics that would need eliminating would be the three lowest, i.e., 11th, 13th, and 23rd. With a limited number of notches, it is possible to almost achieve the equivalent of 24-pulse operation with two three-phase six-pulse converters. On the other hand, it must be mentioned that with PWM operation, the higher harmonics will have relatively higher magnitudes than with the single-pulse operation, although higher harmonics are easier to filter out. If the low-order harmonics are eliminated, then with a high-pass filter for higher harmonics, a near sinusoidal current flow can be obtained.

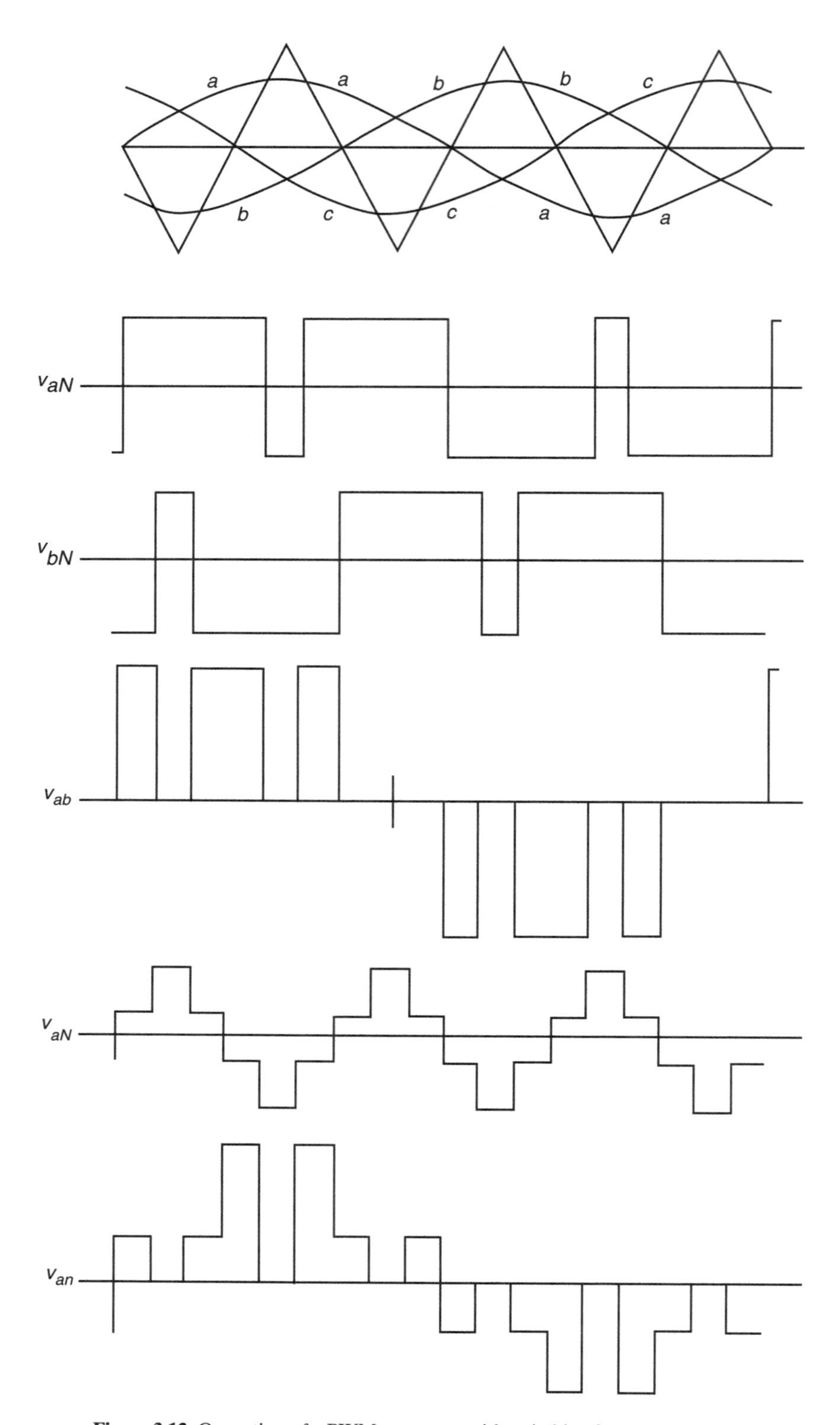

Figure 3.12 Operation of a PWM converter with switching frequency of three times the fundamental frequency.

With three notches per half-cycle, one can eliminate the fifth and seventh harmonics from a six-pulse converter, or eliminate the eleventh and thirteenth from a 12-pulse converter, and so on.

Figure 3.11(b) conveys a generalized half-wave output waveform with M notches defined by angles α_0, α_1, α_2, ... α_{2M+1}, with angles α_1 and α_2 defining the first notch. With $2M = 9$, in the example of Figure 3.11, $M = 4$ representing four notches per half-cycle. The generalized waveform of ac voltage v_{aN} for a phase leg has a voltage amplitude of $+V_d/2$ and $-V_d/2$ with respect to the dc midpoint.

Assuming $V_d/2 = 1$ per unit for generalization, and half-wave/quarter-wave symmetry as in the ac voltage waveforms of Figure 3.11, the waveform can be represented by a Fourier series as

$$f(wt) = \sum_{n=1}^{\alpha} [a_n \sin(n\omega t) + b_n \cos(n\omega t)]$$

where, with half-wave and quarter-wave symmetry, $b_n = 0$, and:

$$a_n = \frac{4}{n\pi}\left[1 + 2\sum_{k=1}^{M} (-1)^k \cos n\alpha_k\right]$$

$M-1$ equations can be assigned for the specific targeted harmonics to be eliminated as $a_n = 0$ for these harmonics. The remaining one equation is used for a specific required value of the fundamental voltage with $a_1 = a$, a finite per unit value. By solving these equations, one can obtain specific values of angles α for a specific value of the fundamental voltage.

Thus for four notches per half-cycle, as for Figure 3.11, there will be four equations to be solved to calculate all values of α for turn-on and turn-off pulses corresponding to each discrete value of the fundamental voltage. These are nonlinear equations, and would require large amounts of real time iterative computation and there may be more than one solution for some values. Thus it is preferred to use lookup tables for, say, each 0.5% change in the magnitude of the required ac voltage. The required shift in angles for the required ac voltage variations per step is quite small, and the number of steps can be reduced and the change from step to step linearized.

Clearly with digital controls, this method is superior to the sawtooth sine wave control or any other wave comparison method. The ac fundamental output voltage can be controlled over a wide range, down to perhaps about 10%, without excessive distortion of the current flow or need for a large harmonic filter.

3.12 CONVERTER RATING—GENERAL COMMENTS

Assuming that the required converter rating is rather low, the least cost and simplest controllable three-phase converter would seem to be a six-valve converter with one turn-off device/diode per valve. In FACTS applications, there will usually be a need for a transformer between the converter valves and the ac system; there is therefore a certain flexibility provided by the transformer turn ratio to match the available device current and voltage ratings. A six-pulse converter based on one device per valve could yield a maximum rating of, say, 5 MVA. However, even at this low rating, a simple six-pulse converter is unlikely to be the best choice for connection to the ac system on account of harmonic distortion requirements. Complex decisions have to be made on the use of large filters versus higher pulse-order and PWM or quasi-PWM converter topologies, etc., apart from the flexibility required for the control.

One area of needed attention for single-device per valve converters is the consequence of failure of a device. Use of fuses would not be desirable. However given the advances in sensing and digital protection technology, it is not unreasonable to sense a fault event, including a shoot-through in a few microseconds, and turn-off other associated converter valves for an effective safe protection strategy.

Most FACTS applications will involve converters with a rating much higher than 5 MVA. The designer now has many options to meet the needs of the higher rating. These options include:

1. Increase pulse-order to 12, 24, or 48, in order to reduce the harmonics to an acceptable level with 2, 4, or 8 six-pulse converters, duly phase-shifted, one can concurrently increase the total converter rating to a maximum of 10, 20, or 40 MVA, still with one turn-off device per valve. Here, one is faced with the choice of series or parallel connections on the dc side, types of transformer arrangements and phase shift among converters, and series/parallel connections on the ac side. These approaches have to be balanced with the other means of reducing harmonics and acquiring flexibility of dynamic and steady state controls, including three-level, PWM, special notches and combination of the above.
2. Adapting a three-level converter topology also doubles the converter voltage and hence the potential maximum single-device per valve converter capacity to, say, 10 MVA per six-pulse, 20 MVA per 12-pulse, and so on. The three-level topology provides the flexibility of a limited range of independent ac voltage control. But again, this has to be balanced against the low frequency PWM or notch-based topologies to achieve independent ac voltage control. Thus it is seen that a modest size FACTS converter rating can be achieved without having devices in series.
3. Connecting devices in series is the most frequently used option for high-power converters. Here the issue is that of ensuring equal voltage distribution among the devices. While the series connection technique is well known, voltage dividers/snubbers have to be provided and some allowance has to be made in the device voltage rating. It is also a practice to include one extra device/diode in series in each valve to ensure continued operation in the event of failure of a device. Note that when a power semiconductor device fails, it must fail into a short circuit and continue to carry current indefinitely without adverse consequences, other than the fact that failure of a second device/diode in the same valve could lead to a catastrophic failure.
4. Double the number of phase-legs and connect them in parallel; this is shown in Figure 3.10, in which the two phase-legs are connected in parallel via a center-tapped inductor. These phase-legs may be of the two-level or three-level variety.
5. Connect converter groups in parallel. In fact a large number of groups can be connected in parallel beyond what may be needed for increasing the pulse number. With parallel connection of converters, it is necessary to have a protection strategy that isolates a faulty converter, with minimal impact on the operation of the other converters. A 48-pulse converter with all six-pulse converters connected on the same dc bus involves 24 phase legs in parallel. With high-speed sensing and fast turn-off capability of devices, parallel connection of a large number of phase legs is quite feasible. However, considerations

of fault currents, high-current bus work would often lead to a combination of series and parallel connections.

6. Use a combination of two or more of the options mentioned above or any other option not mentioned above, in order to arrive at a converter of required rating and performance. Also given the high relative cost of high-voltage isolation with transformers, there is a strong incentive to somehow come up with a platform-based design, and even a transformer-less design, particularly when a small converter is needed in a high-power transmission.

It is obvious that a designer (supplier) has a large number of options to sort out. If two design teams worked separately, the odds are very much in favor of them arriving at different solutions. It is therefore important that the purchaser of FACTS technology pay more attention to the performance specifications rather than to details of the technical design.

REFERENCES

Akagi, H., "New Trends in Active Filters for Power Conditioning," *IEEE Trans. On Ind. Appl.,* vol. 32, no. 6, November–December, 1996.

Bhagwat, P. M., and Stepanovic, V. R., "Generalized Structure of a Multi-Level PWM Inverter," *IEEE Trans. on Industry Applications,* vol. IA-19, no. 6, pp. 1057–69, November–December 1983.

Ekanayake, J. B., and Jenkins, N. "A Three-Level Advanced Static VAR Compensator," *IEEE Transactions on Power Delivery,* vol. 10, no. 2, April 1996.

Ekstrom, A., "Theoretical Analysis and Simulation of Force-Commutated Voltage-Source Converters for FACTS Applications," *EPRI FACTS Conference 1: The Future in High Voltage Transmission,* Cincinnati, November 1990.

Hatziadoniu, C. J., and Chalkiadakis, F. E., "A 12-Pulse Static Synchronous Compensator for the Distribution System Employing the Three- Level GTO Inverter," *IEEE 1997 PES Meeting,* Paper No. PE-542-PWRD-0-01-1997.

Hirakawa, M., Somiya, H., Mino, Y., Bab, K., Murakami, S., and Watanabe, Y., "Application of Self-commutated Inverters to Substation Reactive Power Control," *CIGRE Group 23,* Paper 23–205, 1996.

Horiuchi, S., Watanabe, M., Hasegawa, T., Sampei, M., Hayashi, T., Kawakami, N., Nakamura, T., and Asaeda, T. "Control System for High Performance Self-commutated Power Converter," *CIGRE Session Group 14,* Paper 14–304, 1996.

Ichikawa, F., Suzuki, K. I., Yajima, M., Kawakami, N., Irokawa, S., and Kitahara, T., "Development and Operation of a 50 MVA Self-Commutated SVC at the Shin-Shinano Substation," *CIGRE SC 14 International Colloquium on HVDC & FACTS,* Montreal, September 1995.

Jiang, Y., and Ekstrom, A. "Applying PWM to Control Over-currents at Unbalanced Faults of Force-Commutated VSCs Used as Static VAR Compensators," *Proceedings of the IEEE/ Royal Institute of Technology Stockholm Power Tech.,* Stockholm, June 1995.

Kimura, N., Kishimoto, M., and Matsui, K., "New Digital Control of Forced Commutation HVDC Converter Supplying into Load System Without AC Source," *IEEE Transactions on Power Delivery,* pp. 1425–1431, November 1991.

Kuang, J., and Ooi, B. T., "Series Connected Voltage-Source Converter Modules for Force-Commutated SVC and DC-Transmission," *IEEE Transactions on Power Delivery,* vol. 9, no. 2, pp. 977–983, April 1994.

Lafon, L., Maire, J., Le Du, A., and Coquery, G., "Investigations into the Technology and Specifications of Self-Commutated Valves for Potential Application in High Voltage Power

Systems," *CIGRE 1995 Symposium on Power Electronics in Electric Power Systems,* Tokyo, May 1995.

Lai, J. S., McKeever, J. W., Peng, F. Z., and Stovall, J. P. "Multilevel Converters for Power System Applications," *CIGRE SC 14 International Colloquium on HVDC & FACTS,* Montreal, September 1995.

Larsen, E. V., Miller, N. W., Nilsson, S., and Lindgren, S. R., "Benefits of GTO-Based Compensation System for Electric Utility Application," *IEEE Trans Power Delivery,* vol. 7, no. 4, pp. 2056–2064, October 1992.

Lipphardt, G., "Using A Three-Level GTO Voltage Source Inverter In an HVDC Transmission System Power Electronics in Generation and Transmission," *IEE Conference Publication,* pp. 151–155, 1993.

Menzies, R. W., and Zhuang, Y., "Advanced Static Compensation Using a Multilevel GTO Thyristor Inverter," *IEEE Transactions on Power Delivery,* vol. 10, no. 2, April 1995.

Mohaddes, M., Gole, A. M., and McLarsen, P. G., "A Neural Network Controlled Optimal Pulse Width Modulated STATCOM," *IEEE PES Summer Meeting,* July 1998.

Mori, S., and Matsuno, K., "Development of a Large Static VAR Generator Using Self Commutated Inverters for Improving Power System Stability," *IEEE Trans Power Systems,* vol. 8, no. 1, pp. 371–377, February 1993.

Nakajima, T., Suzuki, H., Izumi, K., and Sugikotio, S., "A Converter Transformer with Series Connected Line Side Windings for a DC Link Using Voltage Source Converters," *IEEE PES Winter Meeting,* New York, January–February 1999.

Ogimoto, K., Saiki, K., Arakawa, F., Nishikawa, H., Shimamura, T., Higa, O., and Momotake, T., "GTO Converters with Snubber Energy Regeneration—Application to Large Current Loads (SMES)," *Proceedings, 5th Annual Conference of Power and Energy Society,* Tokyo, vol. 1, pp. 287–92, 1994.

Ooi, B. T., Dai, S. Z., and Galiana, F. D., "A Solid State PWM Phase-Shifter," *IEEE Trans. Power Delivery,* vol. 8, no. 2, April 1993.

Ooi, B. T., Galiana, F. D., Lee, H. C., Wang, X., Guo, Y., McGillis, D., Dixon, J. W., Nakra, H. L., and Belanger, J., "Research In Pulse Width Modulated HVDC Transmission," *IEE International Conference on AC and DC Power Transmission, Publication Series* 5, pp. 188–193, September 1991.

Ooi, B. T., and Wang, X., "Boost Type PWM HVDC Transmission System," *IEEE Transactions on Power Delivery,* pp. 1557–1563, October 1991.

Patel, H. S., and Hoft, R. G., "Generalized Techniques of Harmonic Elimination and Voltage Control in Thyristor Inverters: Part I—Harmonic Elimination," *IEEE Trans. Ind. Applications,* vol. 1A-9, pp. 310–317, May–June 1973.

Patel, H. S., and Hoft, R. G., "Generalized Techniques of Harmonic Elimination and Voltage Control in Thyristor Inverters: Part II—Voltage Control Techniques," *IEEE Trans. Ind. Appl.,* vol. 1A-10, pp. 666–673, September–October 1974.

Patil, K. V., Mathur, R. M., Jiang, J., and Hosseini, S. H., "A New Binary Multi-Level Voltage Source Inverter," *IEEE PES Summer Meeting,* July 1998.

Peng, F. Z., and Lai, J. S., "A Static VAR Generator Using a Staircase Waveform Multilevel Voltage-Source Converter," *Proceedings of the Seventh International Power Quality Infrastructure Conference,* Power Quality '94 USA, Dallas, TX, September 1994.

Peng, F. A., Lai, J. S., McKeever, J. W., and VanCoevering, J., "Multilevel Voltage-Source Inverter With Separate DC Sources for Static VAR Generation," *IEEE IAS 30th Annual Meeting,* Orlando, vol. 3, pp. 2541–8, 1995.

Peng, F. Z., and Lai J., "Reactive Power and Harmonic Compensation Based on the Generalized Instantaneous Reactive Power Theory for Three Phase Power Systems," *7th International Conference on Harmonics and Quality of Power,* pp. 83–89, Las Vegas, NV, October 1996.

Raju, N. R., and Venkata, S. S., "The Decoupled Converter Topology—A New Approach to

Solid State Compensation of AC Power Systems," *IEEE PES 1996 Summer Meeting,* Denver, July–August 1996.

Scheuer, G., and Stemmler, H. "Analysis of a 3-Level-VSI Neutral Point Control for Fundamental Frequency Modulated SVC Applications," *Proceedings, IEE Sixth International Conference on AC and DC Transmission,* pp. 303–310, April–May 1996.

Skiles, J. J., Kustom, R. L., Vong, F., Wong, V., and Klontz, K., "Performance of a Power Conversion System for Superconducting Magnetic Energy Storage," *IEEE PES 1996 Winter Meeting,* Baltimore, Paper 96 WM 176-8-PWRS, January 1996.

Trainer, D. R., Tennakoon, S. B., and Morrison, R. E. "Analysis of GTO-Based Static VAR Compensators," *IEEE Proceedings Electric Power Applications,* vol. 141, no. 6, November 1994.

Ueda, F., Matsui, M., Asao, M., and Tsuboi, K., "Parallel Connections of Pulse Width Modulated Inverters Using Current Sharing Reactors," *IEEE Trans. On Power Electronics,* vol. 10, no. 6, pp. 673–679, November 1995.

Zhang, Z., Kuang, J., Wang, X., and Ooi, B. T., "Force Commutated HVDC and SVC Based on Phase-Shifted Multi-Converter Modules," *IEEE Trans. Power Delivery,* vol. 8, no. 2, pp. 712–718, April 1993.

Zhou, X., Zhao, H., Zhang, W., Wu, S., and Shaonin, W., "An Experimental and Analytical Research Using Four-Quadrant Converter and UPFC to Improve Power System Stability," *CIGRE 1995 Symposium on Power Electronics in Electric Power Systems,* Tokyo, May 1995.

4

Self- and Line-Commutated Current-Sourced Converters

4.1 BASIC CONCEPT OF CURRENT-SOURCED CONVERTERS

A current-sourced converter is characterized by the fact that the dc current flow is always in one direction and the power flow reverses with the reversal of dc voltage. In this respect, it differs from the voltage-sourced converter in which the dc voltage always has one polarity and the power reversal takes place with reversal of dc current. Figure 4.1 conveys this difference between the current sourced and the voltage-sourced converters.

In Figure 4.1(a), the converter box for the voltage-sourced converter is symbolically shown with a turn-off device with a reverse diode, whereas the converter box for the current-sourced converter, Figure 4.1(b), is shown without a specific type of device. This is because the voltage-sourced converter requires turn-off devices with reverse diodes; the current-sourced converter may be based on diodes, conventional thyristors or the turn-off devices.

Thus there are three principal types of current-sourced converters (Figure 4.2):

1. Diode converter [Figure 4.2(a)], which simply converts ac voltage to dc voltage, and utilizes ac system voltage for commutation of dc current from one valve to another. Obviously the diode-based, line-commutating converter just converts ac power to dc power without any control and also in doing so consumes some reactive power on the ac side.
2. Line-commutated converter, based on conventional thyristors (with gate turn-on but without gate turn-off capability), Figure 4.2(b), utilizes ac system voltage for commutation of current from one valve to another. This converter can convert and control active power in either direction, but in doing so consumes reactive power on the ac side. It can not supply reactive power to the ac system.
3. Self-commutated converter which is based on turn-off devices (GTOs, MTOs, IGCTs, IGBTs, etc.), in which commutation of current from valve to valve takes place with the device turn-off action and provision of ac capacitors, to facilitate transfer of current from valve to valve. Whereas, in a voltage-sourced converter the commutation of current is supported by a stiff dc bus with a dc

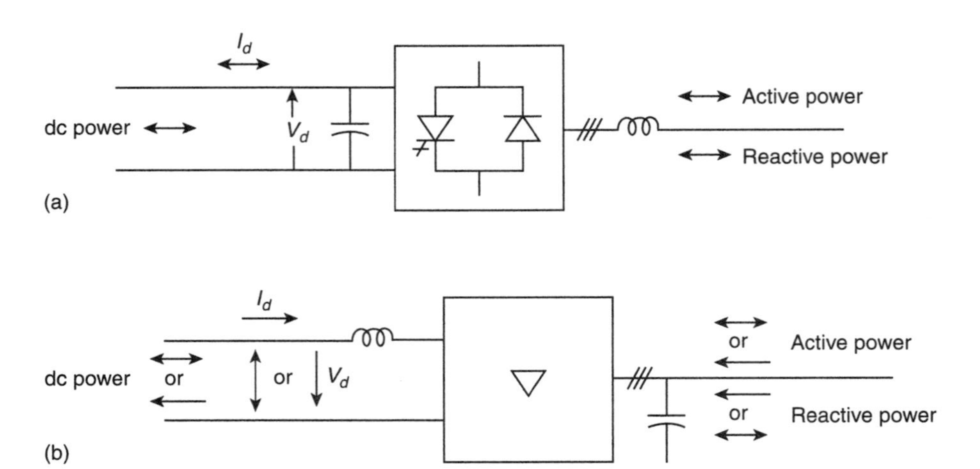

Figure 4.1 Voltage-sourced and current-sourced converter concepts: (a) voltage-sourced converter; (b) current-sourced converter.

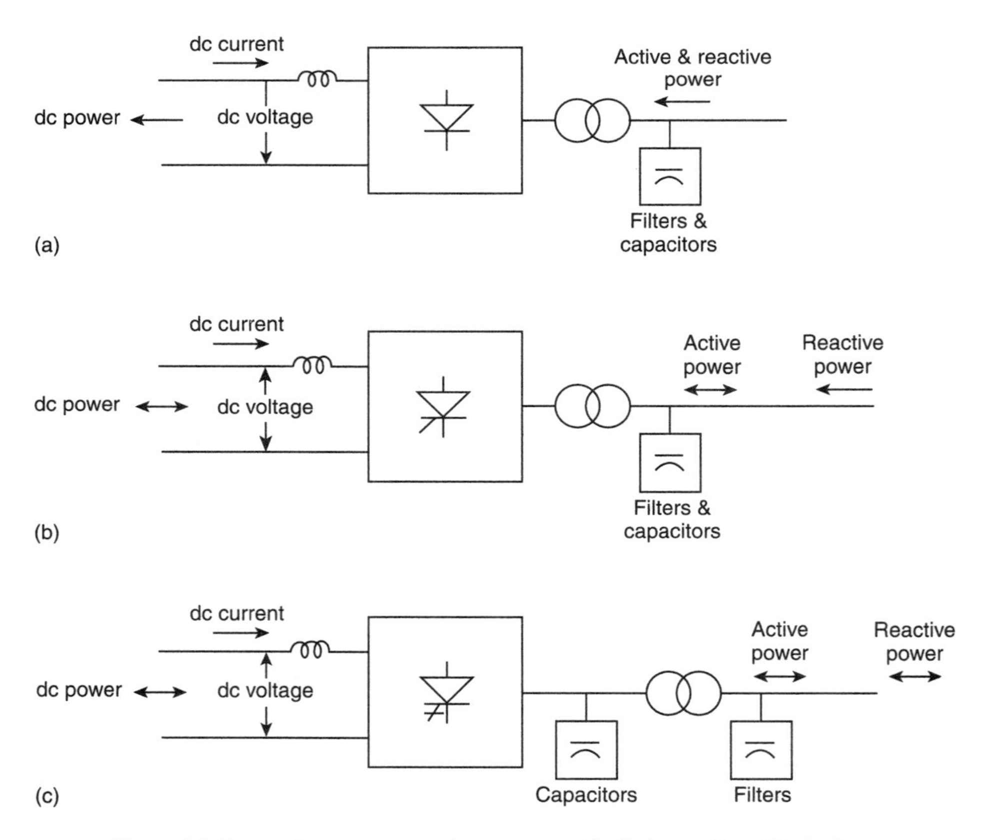

Figure 4.2 Types of current-sourced converters: (a) diode rectifier; (b) thyristor line-commutated converter; (c) self-commutated converters.

capacitor, in a self-commutated current-sourced converter, the ac capacitors provide a stiff ac bus for supplying the fast changing current pulses needed for the commutations. Apart from its capability of controlled power flow in either direction, this converter, like the voltage-sourced converter, can also supply or consume controlled reactive power. However, it is interesting to note that even though the converter can supply reactive power, sources of reactive power, i.e., capacitors and ac filters, are needed in any case. An advantage of the converters with turn-off devices (self-commutating converters) is that they offer greater flexibility including PWM mode of operation.

It must be mentioned that when the converters are based on turn-off devices, the voltage-sourced converters have been preferred over the current-sourced converters. In fact, none of the converter-based Controllers described in this book are based on current-sourced converters. However, with evolution in the device characteristics and functional details of the converters, this situation can change in the future. Therefore the current-sourced converters with turn-off devices are not discussed in much detail.

When reactive power management is not a problem, i.e., where controlled reactive power supply is not required and the reactive power consumed by the converters can be supplied from the system capacitors and/or filters, the line-commutated converters have a decisive economic advantage over self-commutated converters. For conversion of ac to dc and dc to ac, in HVDC transmission, line-commutated converters have been used almost exclusively where reactive power is managed through switched capacitors, filters and the power system. The converters for the superconducting storage can well be current-sourced converters since the superconducting reactor is itself a current source. Also the dc power supply for storage means, to drive voltage-sourced converter-based phase-angle regulators, discussed in Chapter 6, can be current-sourced converters. The economic advantage of conventional thyristor based converters arises from the fact that on a per device basis thyristors can handle two to three times the power than the next most powerful devices such as GTOs, IGCTs, MTOs, etc. In any case, an engineer should be familiar with a wide variety of converters and keep an open mind and continuously re-evaluate the converter topologies with the evolution of the power devices and other components of the system.

There are variations of the above basic types of current-sourced converters, including thyristor converters with artificial commutation, resonant converters, and hybrid converters, but they are not discussed in this book.

Since the dc voltage in a current-sourced converter can be in either direction, the converter valves must have both forward and reverse blocking capability. The conventional thyristors are usually made as symmetric devices, i.e., they have both the forward and reverse blocking capability. This is because they are easier and cheaper to make and can be made with peak blocking voltage as high as 12 kV along with a high current carrying capability. On the other hand the turn-off devices have a high on-state forward voltage drop when they are made as symmetric devices. Given the high production volumes of asymmetric turn-off devices, dictated by the industrial market, it may be advantageous to connect an asymmetric turn-off device and a diode in series to get a symmetric device combination. This again results in higher forward voltage drop and losses. Given this and other aspects, such as fast-switching characteristics of IGBTs, the industrial converter market has shifted very much towards the PWM voltage-sourced converters, discussed in Chapter 3.

4.2 THREE-PHASE FULL-WAVE DIODE RECTIFIER

Three-phase, full-wave diode rectifier is discussed in some more detail than necessary, in order to build the explanations up to the fully controlled converter. In any case the diode rectifier is very useful as a low-cost source of dc power obtained from an available ac source. Rectifiers with a rating greater than a few tens of KWs will almost always be a three-phase, full-wave circuit, shown in Figure 4.3(a), or a combination of several such circuits.

In order to first simplify the explanation and still be realistic, it is assumed that the dc side inductance is very large and therefore the dc current is constant. The circuit consists of six valves, numbered 1 to 6, the number sequence conveying the order of the current transfer and the dc output voltage. The current commutates from valve to valve to turn it into an ac current.

Figure 4.3(b) shows the three-phase ac waveforms v_a, v_b, and v_c with respect to the transformer neutral N. Assuming that the ac system impedance is zero and the transformer is ideal, the top waveforms of Figure 4.3(b) also show voltage waveforms of the two dc buses, with respect to the transformer neutral. This is followed by the waveforms for the constant dc current, ac current waveforms in relation to constant dc current, and the dc output voltage between the two dc buses.

The beginning of Figure 4.3(b) shows that from instant t_1 to t_2, valves 1 and 2 are conducting, the dc current takes the path valve 2, into phase c, out of phase a, and valve 1. The dc buses are connected to phases a and c, and the dc output voltage is that between phases a and c, as shown by the thick lines. At point t_3 the voltage of phase b becomes positive with respect to phase a, and valve 3 becomes forward biased. Being a diode, it starts conducting and takes the current over from valve 1 and the output dc voltage follows the voltage between the thick lines of phases b and c. The current waveforms show that the current is flowing through valve 2, phase c, phase b and valve 3. Next at instant t_4, phase a becomes negative with respect to phase c, and valve 4 becomes forward biased, starts conducting and takes the current over from valve 2. Then at instant t_5, valve 5 takes over from valve 3, at instant t_6 valve 6 takes over from valve 4, and at instant t_1 valve 1 takes over from valve 5 for one complete cycle. The current in the three phases is shown and is made up of 120 degree blocks of dc current, through an upper and a lower valve of each phase leg.

It should be noted that in a current-sourced converter, the commutation takes place from valve to valve among valves connected to the same dc bus, i.e., valves 1 to 3 to 5 to 1 and so on. This is different from the voltage-sourced converters in which the commutation occurs between the valves connected to the same phase leg, i.e., valve 1 to 4 to 1 and so on. As a result the waveforms of the ac voltage of a voltage-sourced converter are made up of 180 degree blocks and consequently have triplen harmonics; the waveforms of ac current of a three-phase current-sourced converter are made up of 120 degree blocks and consequently do not have triplen harmonics.

The dc output voltage as shown at the top of Figure 4.3(b), is for the two dc buses with respect to the transformer neutral. Adding the magnitude of these two waveforms gives the voltage of the upper dc bus with respect to the lower dc bus. As shown at the bottom of Figure 4.3(b), the dc output voltage has a six-pulse waveform, made up of the sum of two three-phase, half-wave circuits.

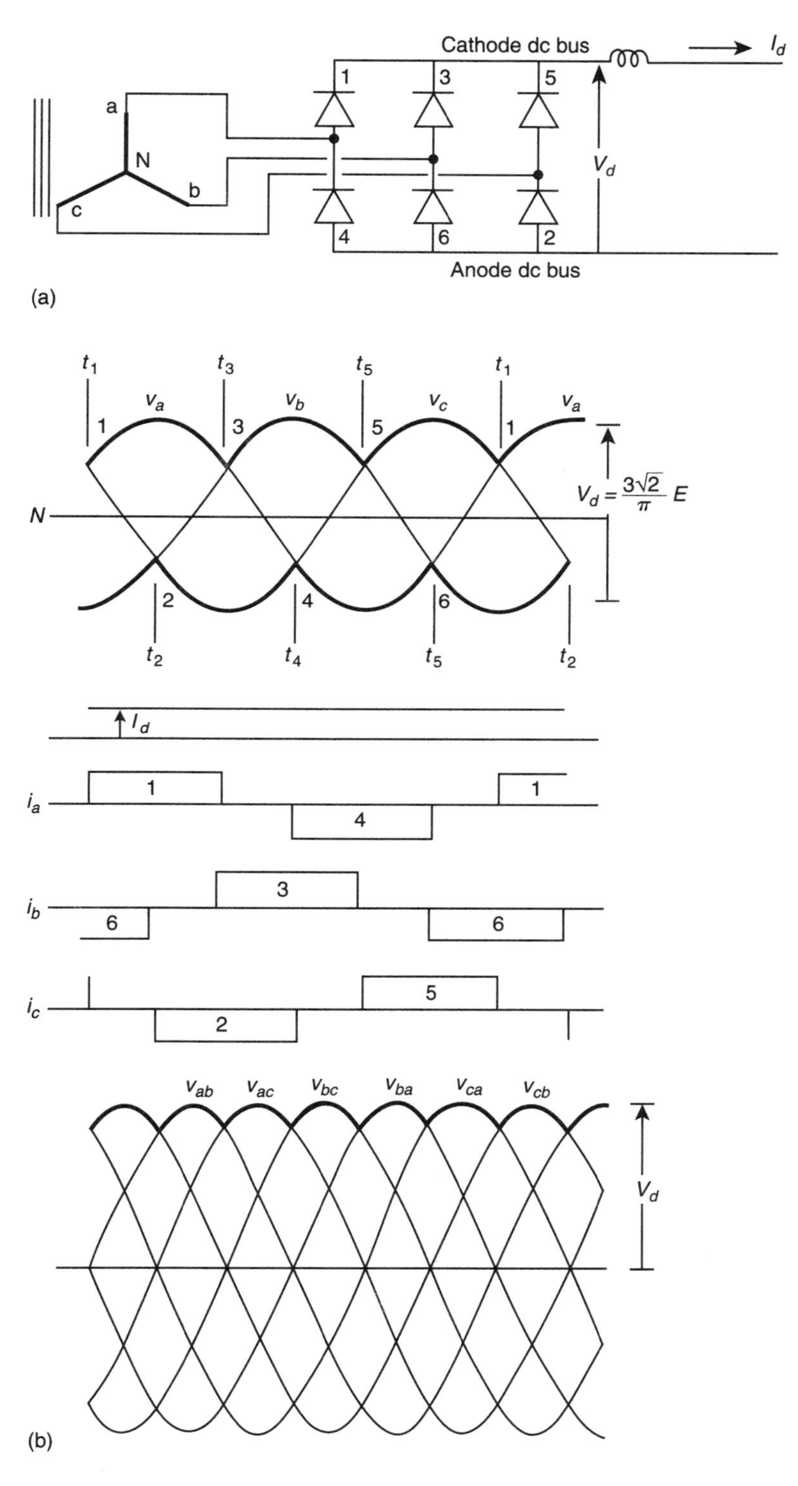

Figure 4.3 Three-phase, full-wave diode converter neglecting commutation angle: (a) three-phase full wave six-pulse diode converter; (b) current and voltage waveforms for a three-phase, full-wave diode converter.

The dc output voltage is made up of 60 degree segments and with the peak of ac voltage as a reference point, it is defined by

$$e = \sqrt{2}\, E \cos \omega t$$

where E is the phase-to-phase voltage. This ideal output voltage is given by

$$\begin{aligned} V_0 &= \frac{3}{\pi} \int_{-\pi/6}^{+\pi/6} \sqrt{2} E \cos \omega t \, d(\omega t) \\ &= \frac{3\sqrt{2}}{\pi} E [\sin \omega t]_{-\pi/6}^{+\pi/6} = \frac{3\sqrt{2}}{\pi} E = 1.35E \end{aligned} \tag{4.1}$$

The output voltage is positive with dc current flowing out of the anode bus of the converter; hence the power flow is from ac to dc (rectifier). DC output voltage contains some harmonics (discussed at the end of Section 4.3). The ac current, Figure 4.3(b), is made up of square wave blocks of 120 degree duration each half-cycle. The rms value of this phase current is given by

$$\begin{aligned} I &= \sqrt{\left(\frac{1}{\pi} \int_{-\pi/3}^{+\pi/3} I_d^2 \, d(\omega t)\right)} \\ &= \frac{I_d^2}{\pi} [\sin \omega t]_{-\pi/3}^{+\pi/3} = \frac{\sqrt{2}}{\sqrt{3}} I_d = 0.816 I_d \end{aligned} \tag{4.2}$$

AC current also has harmonics (further discussed in Section 4.3).

Equating the fundamental ac power and the dc power (neglecting losses),

$$\sqrt{3}\, EI = V_d I_d$$

and substituting V_d in terms of E from (4.1) gives the rms fundamental ac current:

$$I_1 = \frac{\sqrt{6}}{\pi} I_d = 0.78 I_d \tag{4.3}$$

The rms difference between the total rms current I (4.2), and the rms fundamental current I_1, is the total rms harmonic current:

$$I_h = \sqrt{(I^2 - I_1^2)} = 0.24 I_d \tag{4.4}$$

There was one simplifying assumption in the above discussion that the current instantaneously commutated from valve 1 to 3, 2 to 4, etc. In reality it will take a significant time. Typically it may take about 20 degrees to 30 degrees. The commutation of in-line commutated converters involves transfer of current from one phase to another through the valves in an inductive circuit of the ac system, including the transformer inductance.

Consider again the same diode circuit, Figure 4.4, which shows that at instant t_3, when valve 3 becomes forward biased starts to conduct with valve 1 carrying the full dc current. The conducting of both valves 1 and 3 represents a short circuit between phases a and b with the short-circuit current rising from phase b through valve 3 into phase a through valve 1. But once the short-circuit current equals the dc current through valve 1, its net current reaches zero, valve 1 stops conducting and the commutation is complete. The short-circuit current between the two phases for this period of commutation, angle γ_0, is defined by

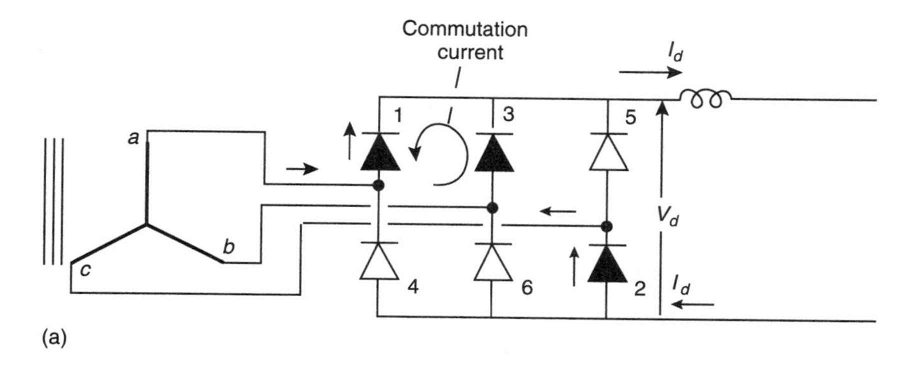

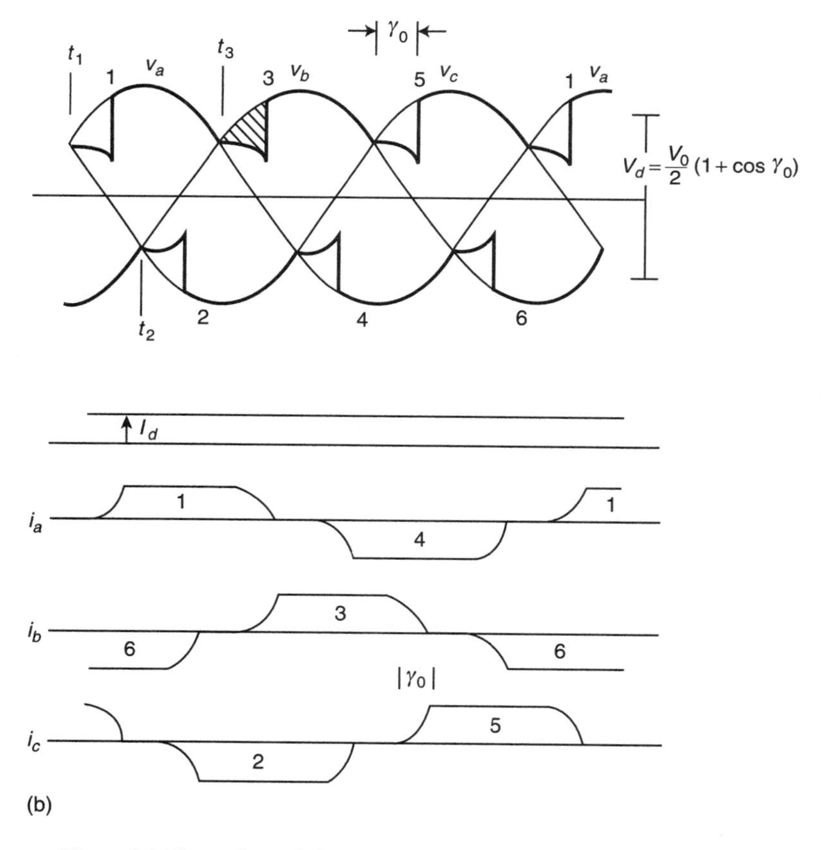

Figure 4.4 Three-phase, full-wave diode converter operation, including commutation angle: (a) six-pulse diode converter; (b) voltage and current waveforms.

$$2L\frac{di_s}{dt} = \sqrt{2}\,E\sin\omega t$$

where L is the inductance of each phase (neglecting resistance). Integration gives

$$i_s = \frac{E}{\sqrt{2}\omega L}(1 - \cos\omega t)$$

Assuming that when $I_s = I_d$, $\omega t = \gamma_0$ gives

$$I_d = \frac{E}{\sqrt{2}\omega L}(1 - \cos\gamma_0) \tag{4.5}$$

From (4.5) the commutation angle γ_0 can be calculated.

It is noted from Figure 4.4, that the output voltage is somewhat reduced compared to the output voltage corresponding to $\gamma = 0$ in Figure 4.3. During commutation the output follows the mean of the two short-circuited voltages. The lost voltage corresponds to the shaded area every 60 degrees and is given by

$$\partial V = \partial A / \frac{\pi}{3}$$

$$\partial A = \frac{1}{2}\int_0^{\gamma_0} \sqrt{2}E \sin\omega t \, d(\omega t) = \frac{1}{\sqrt{2}}E(1 - \cos\gamma_0) \tag{4.6}$$

$$\partial V = \frac{3}{\sqrt{2}\pi}E(1 - \cos\gamma_0)$$

$$\partial V = \frac{3\omega L}{\pi}I_d$$

$$V_d = V_0 - \partial V = V_0 - \frac{3\omega L}{\pi}I_d \tag{4.7}$$

Thus the dc voltage drop in the converter due to the commutation of dc current I_d is directly proportional to I_d. On the dc side the voltage drop may be simulated as a resistance, equal to $3\omega L/I_d$. This does not mean that there is a loss of power, because it is not an actual resistance. It can be visualized from the current waveform in Figure 4.4 that the current is somewhat shifted to the right by the commutation process. This means that the ac side power factor is reduced from unity to a somewhat lower value in the lagging direction, which in turn means that some reactive power is consumed. This power factor reduction corresponds to the reduction in the dc voltage.

Equating the dc and ac power

$$\sqrt{3}\,EI\cos\phi = V_d I_d = \left(V_0 - \frac{3\omega L}{\pi}I_d\right)I_d \tag{4.8}$$

which, combined with (4.1) and (4.3) gives

$$\cos\phi = 1 - \frac{1}{V_0}\frac{3\omega L}{\pi} \tag{4.9a}$$

For practical estimation, power factor angle φ may be taken as

$$\begin{aligned} \phi &= \alpha + 2\gamma/3, \quad \text{for} \quad 0 < \alpha < 30^\circ \\ \phi &= \alpha + \gamma/3, \quad \text{for} \quad 30^\circ < \alpha < 90^\circ \end{aligned} \tag{4.9b}$$

4.3 THYRISTOR-BASED CONVERTER (with gate turn-on but without gate turn-off)

4.3.1 Rectifier Operation

From the discussions on the diode rectifier, it is easy to recognize that if the devices had a turn-on control, the start of each commutation could be delayed and hence the output voltage reduced or even reversed at will.

Consider again the diode voltage waveform of Figure 4.4 when from instant t_3 valve 3 becomes forward biased and it is ready to commutate current from valve 1. This start of commutation could be delayed if the devices were thyristors. Figure 4.5 shows the output dc voltage and phase-current waveforms for the commutations delayed by an angle α. The short-circuit current of the commutation process is now defined as before by

$$2L\frac{di_s}{dt} = \sqrt{2}\,E\sin\omega t$$

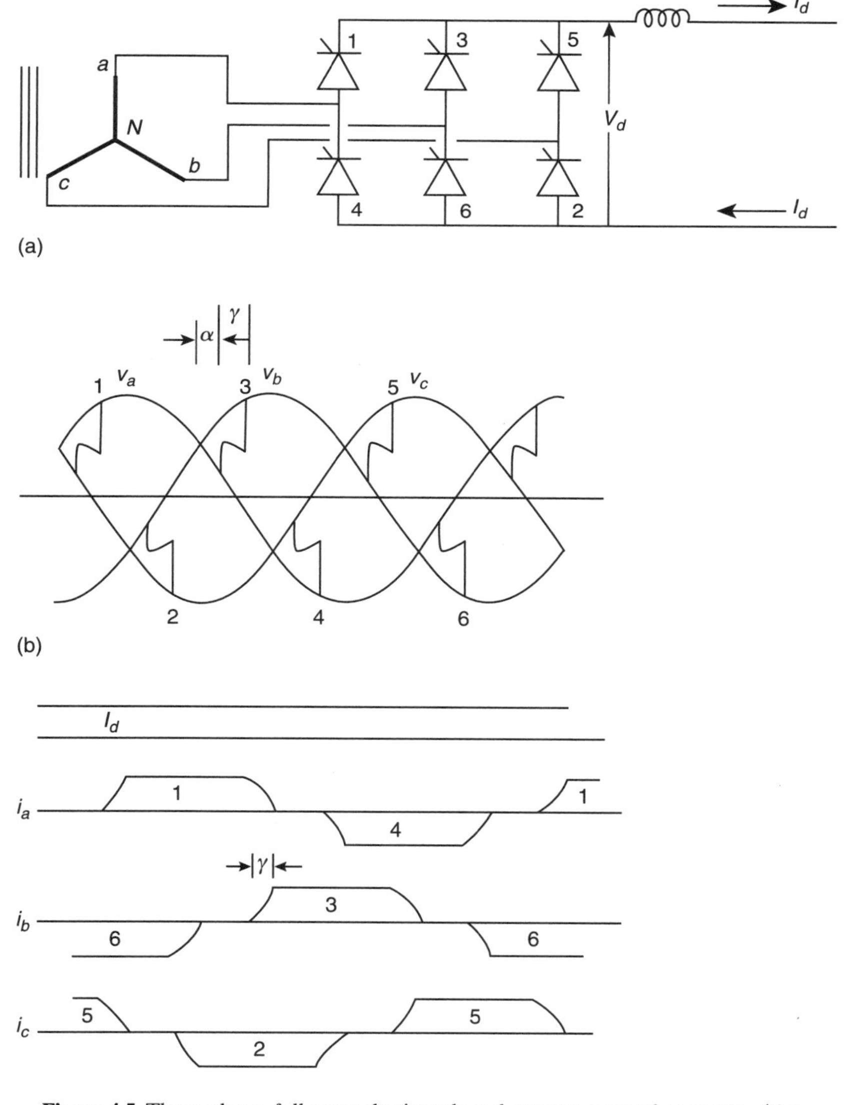

Figure 4.5 Three-phase, full-wave thyristor-based current-sourced converter (thyristors without turn-off)–rectifier operation: (a) six-pulse thyristor converter; (b) voltage and current waveforms.

Assuming now that the start of commutation is at $\omega t = \alpha$, integration for $\omega t = \alpha$ and $I_s = 0$, gives

$$i_s = \frac{E}{\sqrt{2}\omega L}(\cos\alpha - \cos\omega t)$$

and when $\omega t + \gamma$, $I_s = i_d$, gives

$$I_d = \frac{E}{\sqrt{2}\omega L}[\cos\alpha - \cos(\alpha + \gamma)] \tag{4.10}$$

From (4.10), angle γ can be calculated for a given value of angle α. It is seen that angle γ decreases with increase in angle α. This should be obvious since the rate or rise of short-circuit current is greater with increase in the short-circuited voltage.

Equation for the dc output voltage can be obtained by first calculating the output voltage with commutation angle $\gamma = 0$, and then subtracting the voltage lost due to commutations. With $\gamma = 0$, the output voltage, taking a 60 degree segment, is defined by

$$\begin{aligned} V_d &= \frac{3}{\pi}\int_{-\pi/6+\alpha}^{+\pi/6+\alpha} \sqrt{2}E\cos\omega t\, d(\omega t) \\ V_d &= \frac{3\sqrt{2}}{\pi}E\cos\alpha = V_0\cos\alpha \end{aligned} \tag{4.11}$$

where V_0 is the output voltage corresponding to $\alpha = 0$ and $\gamma_0 = 0$. Thus the output voltage with $\gamma_0 = 0$ decreases as the cosine function of α. This means that with increasing α, the output voltage decreases slowly first reaching 86.6% for $\alpha = 30$ degrees, 50% with $\alpha = 60$ degrees and 0 at $\alpha = 90$ degrees, and negative for $\alpha >$ 90 degrees.

Now taking into account the commutation angle, the area under each commutation is given by

$$\partial A = \frac{1}{2}\int_{\alpha}^{\alpha+\gamma} \sqrt{2}E\sin\omega t\, d(\omega t) = \frac{1}{\sqrt{2}}E\,(\cos\alpha - \cos\alpha + \gamma)$$

This area represents dc voltage drop every 60 degrees, therefore,

$$\partial V = \frac{3}{\sqrt{2}\pi}E[\cos\alpha - \cos(\alpha + \gamma)] \tag{4.12}$$

and

$$\begin{aligned} V_d &= V_0\cos\alpha - \partial V \\ &= \frac{V_0}{2}[\cos\alpha + \cos(\alpha + \gamma)] \end{aligned} \tag{4.13}$$

or V_d/V_0, the per unit voltage is given by

$$\frac{V_d}{V_0} = \frac{1}{2}[\cos\alpha + \cos(\alpha + \gamma)] \tag{4.14}$$

From (4.11) and (4.12),

$$\partial V = \frac{3\omega L}{\pi}I_d$$

which gives

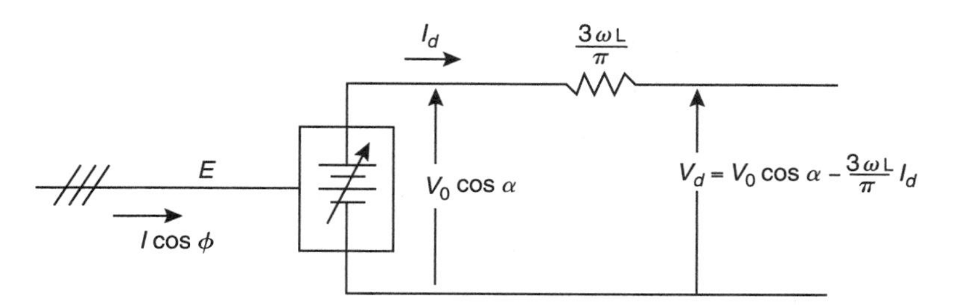

Figure 4.6 Equivalent circuit for a current-sourced converter

$$V_d = V_0 \cos \alpha - \frac{3\omega L}{\pi} I_d \tag{4.15}$$

As would be expected the voltage drop for commutation is the same $(3\omega L/\pi)I_d$ for any delay angle; i.e., the commutation voltage integral is always the same for the same current.

Thus a converter output voltage for any value of delay angle α, dc current I_d, and ac voltage E can be obtained from (4.15) and is represented by a simple equivalent circuit of a variable dc source in series with an equivalent resistor as shown in Figure 4.6. This resistor in turn is a reflection of reduced lagging reactive power on the ac side.

In per unit terms, assuming V_{on} as 1 pu V_0, inductance ωL in per unit as $X_p = (\omega L \times I)/(E/\sqrt{3})$, then combining (4.3) and (4.15) gives a per unit value of the dc output voltage:

$$V_d/V_{on} = \cos \alpha - \frac{X_p}{2} \tag{4.16}$$

Given, say, converter transformer plus the system impedance is 20% and inductive, (0.2 pu), the dc voltage will drop by about 10% from no load to full load, and the power factor on the ac side will drop correspondingly.

If it is assumed that $V_d = V_0 \cos \phi$, then the power factor $\cos \phi$ is approximately given by

$$V_d/V_0 = \cos \phi = \cos \alpha - \frac{3\omega L}{\pi} I_d = \frac{1}{2}[\cos \alpha + \cos(\alpha + \gamma)] \tag{4.17}$$

4.3.2 Inverter Operation

Consider first, converter operation assuming zero commutation angle; Figure 4.7 shows the output voltage with varying delay angle.

Figure 4.7(a) is for the normal rectifier operation with a small delay angle α and shows the positive output voltage by inclined hatched areas. Figure 4.7(b) is for the case when delay angle α is increased beyond 60 degrees and it is seen that there are some negative areas corresponding to the angle in excess of 60 degrees, shown by horizontal hatching. If the dc load on the rectifier was resistive, its operation would be discontinuous because conduction would not occur in the negative direction. But with a large inductor, the difference between the positive and negative areas will average out to a resultant output voltage on the dc side.

At 90 degrees delay, the positive and negative voltage areas become equal and the average voltage is zero as per (4.11).

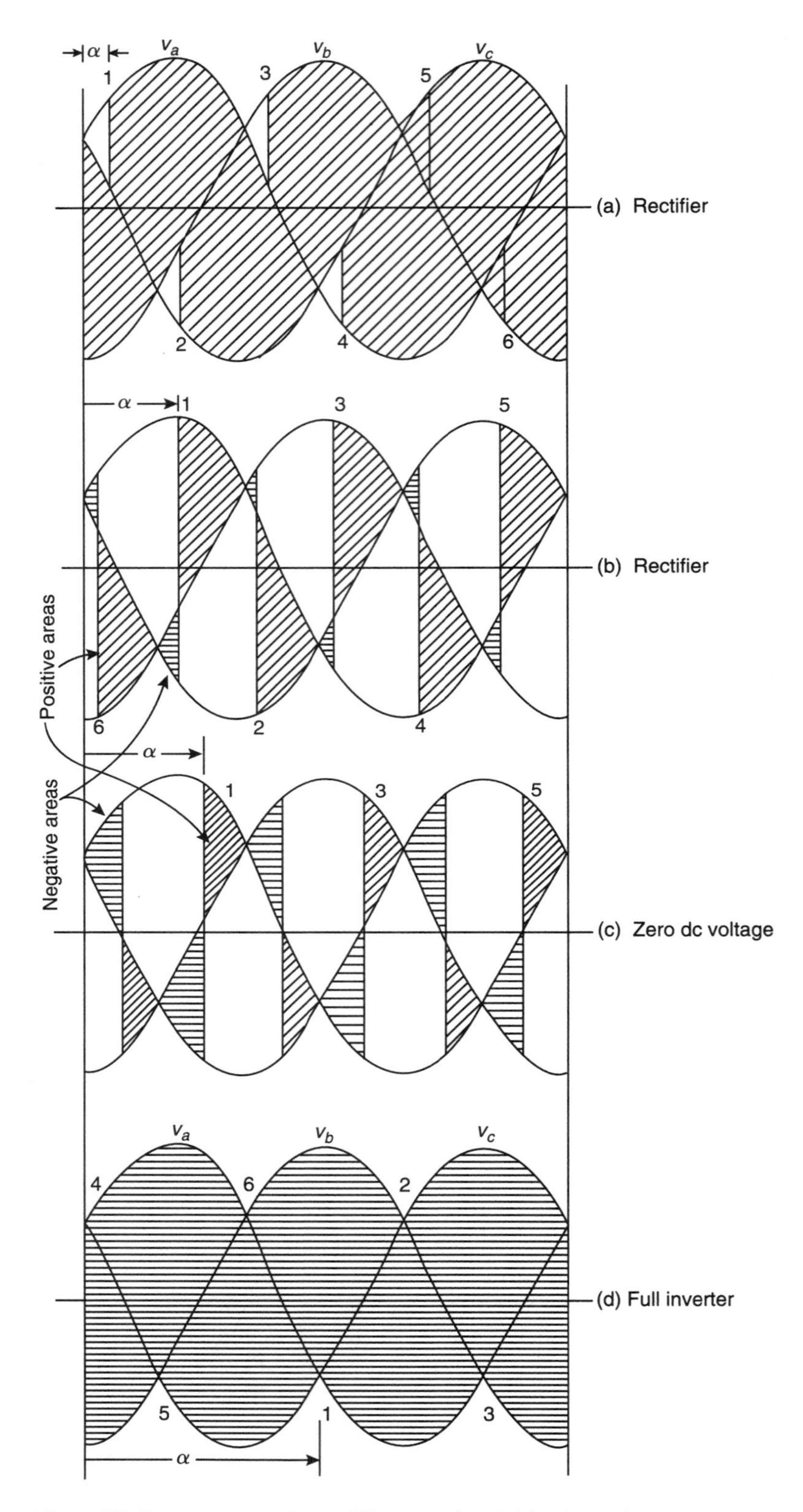

Figure 4.7 Converter operation at different angles of delay (assuming zero commutation angle).

$$V_d = V_0 \cos 90° = 0$$

As the delay angle is further increased, the average voltage becomes negative and at 180 degrees the negative voltage becomes as large as that for the rectifier with zero delay angle.

It is assumed that for an inverter to operate and feed active power into the ac system, there is another source of dc power which causes the dc current to flow in the same direction, against the converter's dc output voltage, thus feeding dc power into the inverter.

Actually, an inverter can not be operated at a delay angle of 180 degrees, because some time (angle γ) is needed for the commutation of current and further time (angle δ_0) is needed for the outgoing valve to fully recover (turn-off) before the voltage across it reverses and becomes positive. Otherwise the outgoing valve will commutate the current back and lead to a fault known as commutation failure (discussed later).

Figure 4.8 shows inverter operation with an angle α, which is somewhat less than 180 degrees. Also shown are the valve currents and various angles, dividing the range of 180 degrees angle into the delay angle α, commutation angle γ, safe turn-off angle δ, and angle of advance $\beta = \gamma + \delta$. For safe operation, angle δ must not be less than δ_0. Often for convenience, inverter equations are referred to in terms of angle of advance β, commutation angle γ, and margin angle δ. All the equations derived earlier in terms of α and γ still apply, and if preferred they can be obtained by substituting α with $\pi - \beta$ or $\pi - (\gamma + \delta)$; they are not included here.

Figure 4.9 shows the phasor diagram of ac voltage and current, indicating the range of operation of a line-commutating converter with the hatched area. It shows that the operation is limited to two quadrants; the rectifier operation limited by the commutation requirements at $\alpha = 0$, and the inverter operation limited at the other end by the commutation and valve recovery requirements. Throughout the range it consumes reactive power, which can be calculated from (4.9) and (4.17) for the power factor. For a given current and voltage, the active power decreases and reactive power increases, with increase in α from zero until it reaches 90 degrees-($\gamma/2$) when the dc output voltage reduces to zero. At this point the active power is zero and the reactive power is at its maximum. In fact, a line-commutating converter is also used as a static var compensator to consume controlled reactive power. This is done by operating the converter with a short circuit through a dc inductor on the dc side and controlling the dc current flow by controlling angle α.

It should be noted that the equations above represent the rectifier and inverter conditions for commutation angle γ less than 60 degrees; if this angle is exceeded, the operation is no longer represented by the same equations since two commutations will take place simultaneously. This situation is not likely in steady-state conditions but may occur during high dc fault current. It should also be noted that operation with delay angle α less than 30 degrees and commutation angle γ greater than 60 degrees simultaneously is not possible. If for, say, $\alpha = 0$, the current is so increased that the corresponding angle γ is greater than 60 degrees, then the firing of the valves will not take place at $\alpha = 0$ but will be automatically delayed so that $\gamma = 60°$. This situation continues until $\alpha = 30°$ and then γ can be increased. Under such conditions, the equivalent commutation resistance in (4.15) and Figure 4.6 become $9\omega L/\pi$, three times the normal value of $3\omega L/\pi$.

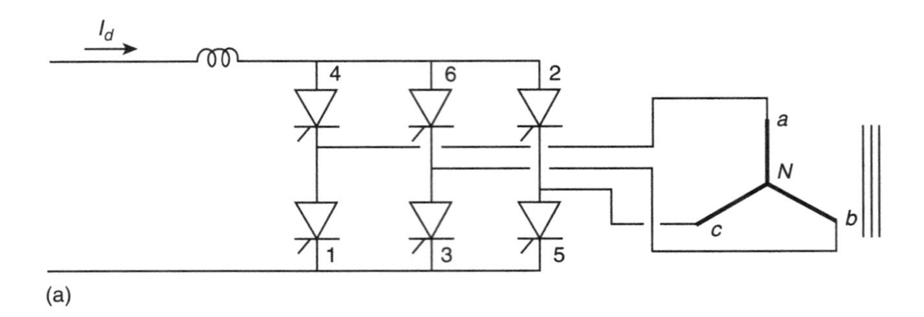

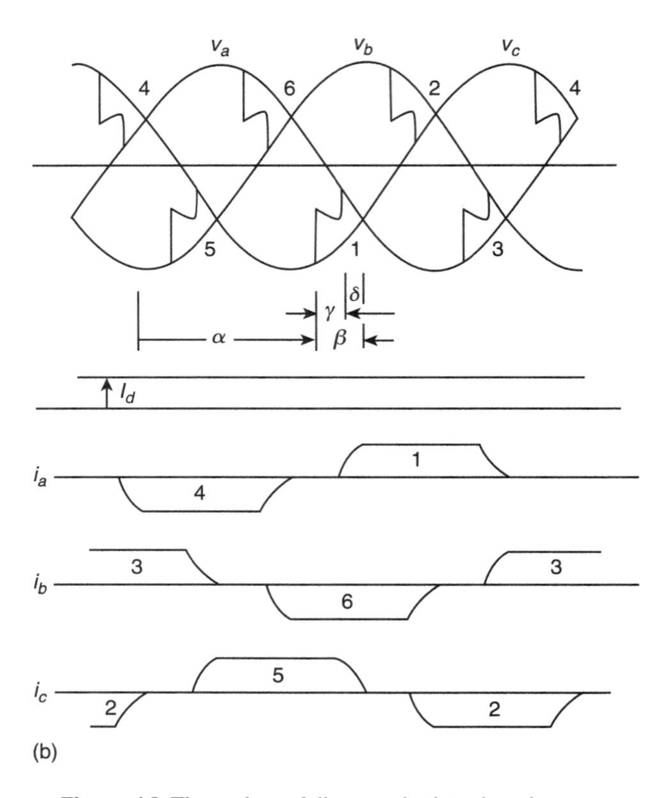

Figure 4.8 Three-phase full-wave thyristor-based current-sourced converter-inverter operation: (a) six-pulse thyristor converter; (b) voltage and current waveforms.

4.3.3 Valve Voltage

Figure 4.10 shows the valve voltage, valve 5 for example, during inverter operation. The upper waveforms show the ac voltage, valve sequence, and the voltages during commutations. As explained before, the commutations are enabled by short circuits between phases. During these short circuits the voltage of the two short-circuited ac phases follow the mean of the respective two phases. Diagram for valve 5 voltage shows that during conduction period of 120° + γ, the voltage is zero, and

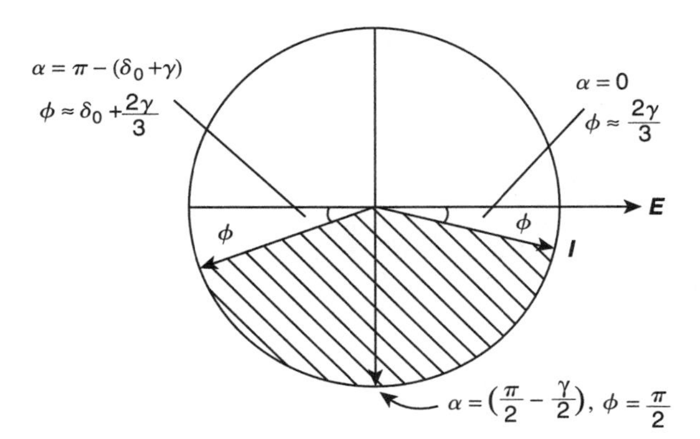

Figure 4.9 Operation zone of thyristor-based converter.

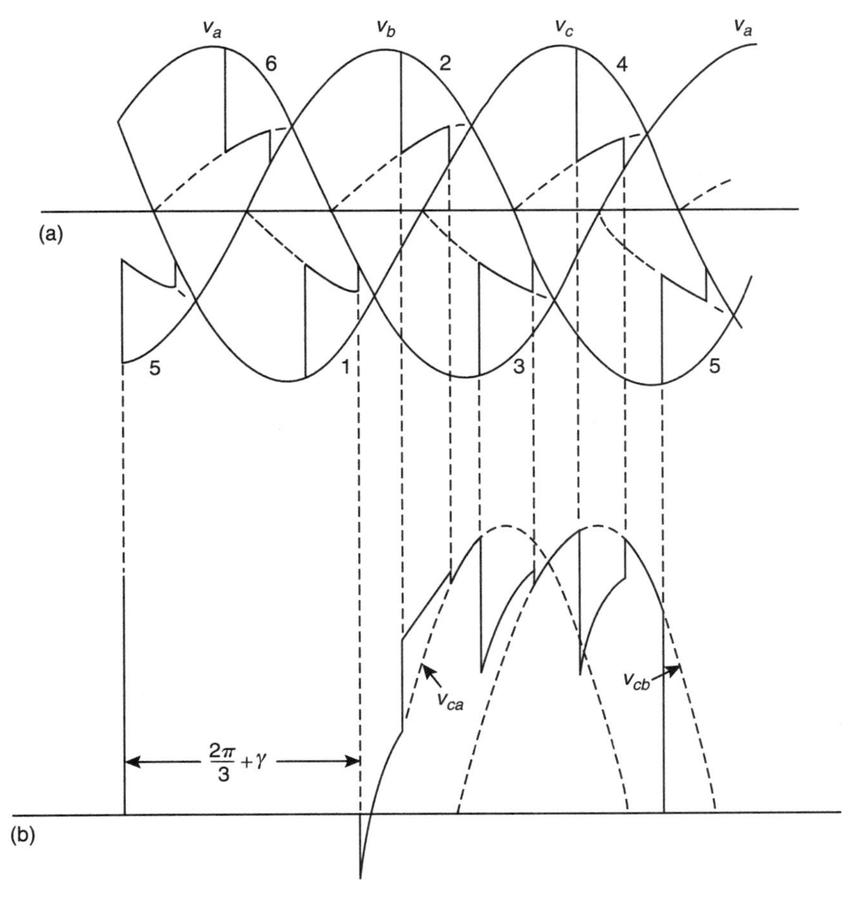

Figure 4.10 Voltage across a valve in a thyristor converter: (a) converter operation as an inverter; (b) voltage across valve 5.

at the end of commutation from valve 5 to valve 1, the voltage jumps to a negative value corresponding to the voltage v_{ca}. As mentioned before, the duration of this negative voltage needs to be sufficient for valve 5 to recover before the voltage goes positive. As the valve voltage follows voltage v_{ca}, it is distorted by change in the voltage v_{ca} during short circuit of phases b and c for commutation from valve 6 to valve 2. Following that it is again distorted by the change in voltage v_{ca} during commutation from valve 1 to valve 3; however, at the completion of this commutation, the valve voltage follows the voltage v_{cb}, which is later distorted by commutation from valve 2 to valve 4. The cycle is complete when valve 5 turns on again. Thus the voltage across valve 5, connected to phase c, follows the voltage waveforms, v_{ca} and v_{cb}. Similarly the voltage across valve 2, the other valve connected to phase c, follows the voltage waveforms v_{ac} and v_{bc}, and so on. During the inverter operation the valve voltage is positive much of the time and obviously during rectifier operation it will be negative much of the time. Given the circuit inductances and capacitances, there will be a voltage overshoot at each voltage jump, resulting in higher valve voltage when the overshoot is during the peak of the valve voltage. It is therefore necessary to provide R-C damping circuits across the valve devices.

4.3.4 Commutation Failures

It was mentioned earlier that a safe minimum angle of advance β is needed in order to allow successful commutation of current (angle γ) and for the outgoing valve to fully recover (angle $\delta > \delta_0$) before the voltage across it reverses and becomes positive. Otherwise the outgoing valve will commutate the current back and lead to a fault known as commutation failure. On one hand it is desirable to minimize the angle β when it is necessary to maximize the inverter output voltage; on the other hand, allowance has to be made for current rise and/or reduction of ac system voltage just before and during the commutation process. Some allowance has to be made for reasonable events and then it has to be accepted that some commutation failures will still occur on some statistical basis and provide for a safe ride through.

Referring to Figure 4.11, t_1 is the point where valves 1 and 2 are conducting, valve 3 turns on, and commutation from valve 1 to valve 3 is expected to take place. Suppose the commutation fails to complete by point t_2 the point of voltage crossing. The commutation will reverse, as shown, and valve 1 together with valve 2 will continue to carry the dc current. Consequently the dc voltage continues to fall as per voltage v_{ca}. Then at point t_3 valve 4 turns on to take over the current from valve 2, at which time v_{ca} is short circuited and the dc voltage falls to zero. Now assuming that the commutation from valve 2 to valve 4 is successful, the dc current is effectively bypassed through the phase-leg of valves 1 and 4. Following that, valve 5 has no chance to take back the current, because its voltage is reverse biased. The dc short circuit ends t_4 when valve 6 takes the current over from valve 4, with the dc voltage going slightly negative first and the converter continues to function correctly. This type of commutation failure is referred to as a single commutation failure.

When a commutation failure occurs, the scenario is more likely to end with a single commutation failure, given that the dc inductor is large enough to limit the rate of rise of current and the protective means advance the turn-on of the next valve with detection of a commutation failure. Again there is no guarantee that the commutation of the next valve, immediately after the commutation failure, will be successful. In the example above if the commutation from valve 2 to valve 4 also fails to complete,

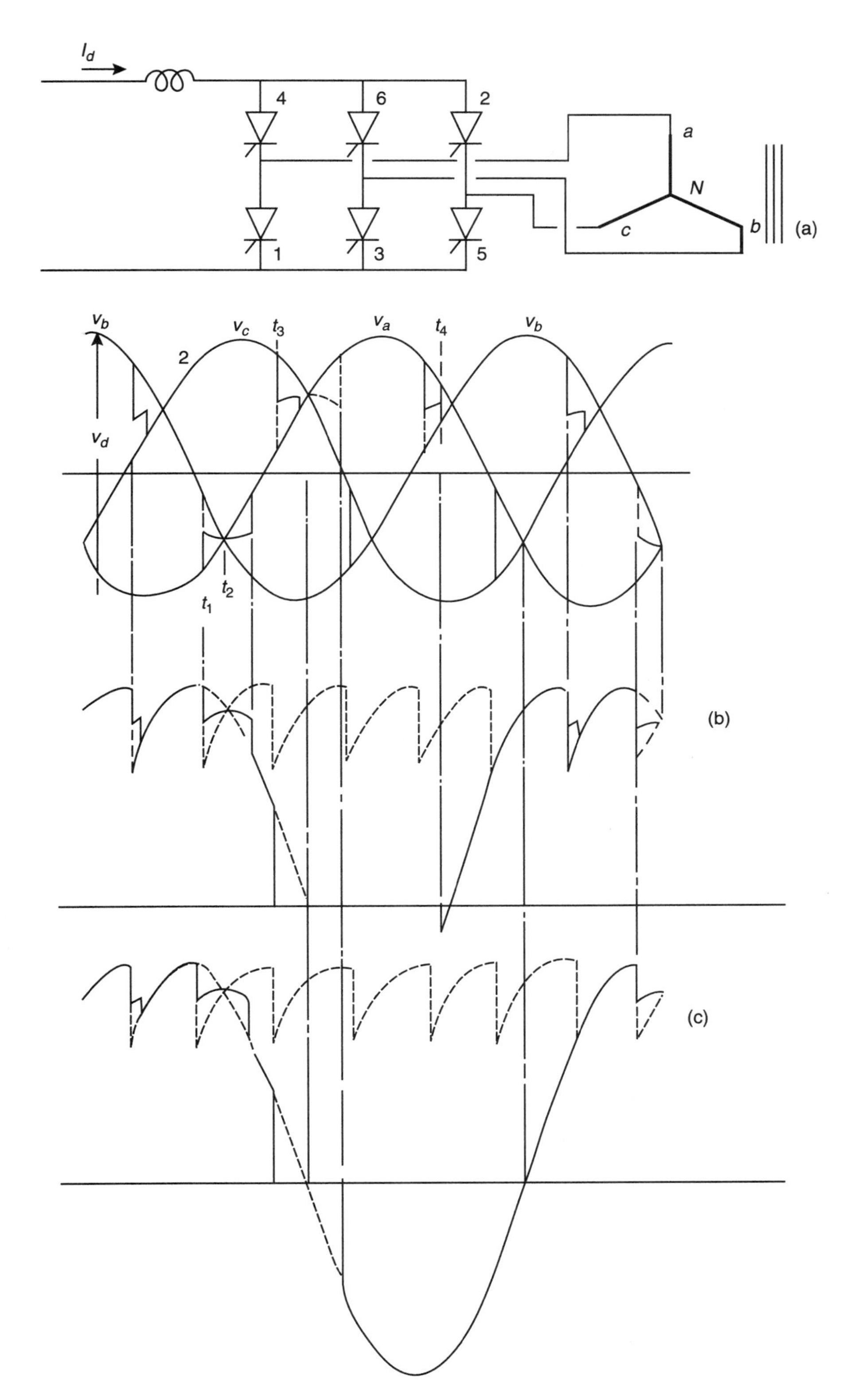

Figure 4.11 Commutation failure in a thyristor-based line-commutated converter: (a) six-pulse thyristor converter; (b) inverter voltage during a single commutation failure; (c) inverter voltage during double commutation failure.

then the dc output voltage will go through a full ac cycle voltage v_{ca}, following that if all else is well, the normal operation will continue.

In general this problem is quite manageable in power systems as is evident from a large number of HVDC projects in satisfactory operation.

4.3.5 AC Current Harmonics

Neglecting commutation angle, the current is made up of 120 degree pulses as shown in Figure 4.12(a). Analysis of ac current can be obtained by Fourier analysis and is given by

$$i_n = \frac{2\sqrt{3}}{\pi} I_d \left[\cos \omega t - \frac{1}{5}\cos 5\omega t + \frac{1}{7}\cos 7\omega t - \frac{1}{11}\cos 11\omega t + \frac{1}{13}\cos 13\omega t - \cdots \right] \quad (4.18)$$

The harmonics are of the order of $6k \pm 1$, where k is an integer. The rms value of any term in (4.18) is given by

$$I_n = I_d \frac{\sqrt{6}}{n\pi} \quad (4.19)$$

With $n = 1$, the rms value of the fundamental is given by

$$I_1 = \frac{\sqrt{6}}{\pi} I_d = 0.78 I_d \quad (4.3)$$

Assuming 1:1 ratio of a wye/wye transformer, if a delta/wye transformer with $\sqrt{3}$:1 ratio is adopted, the secondary voltages of this transformer would be displaced

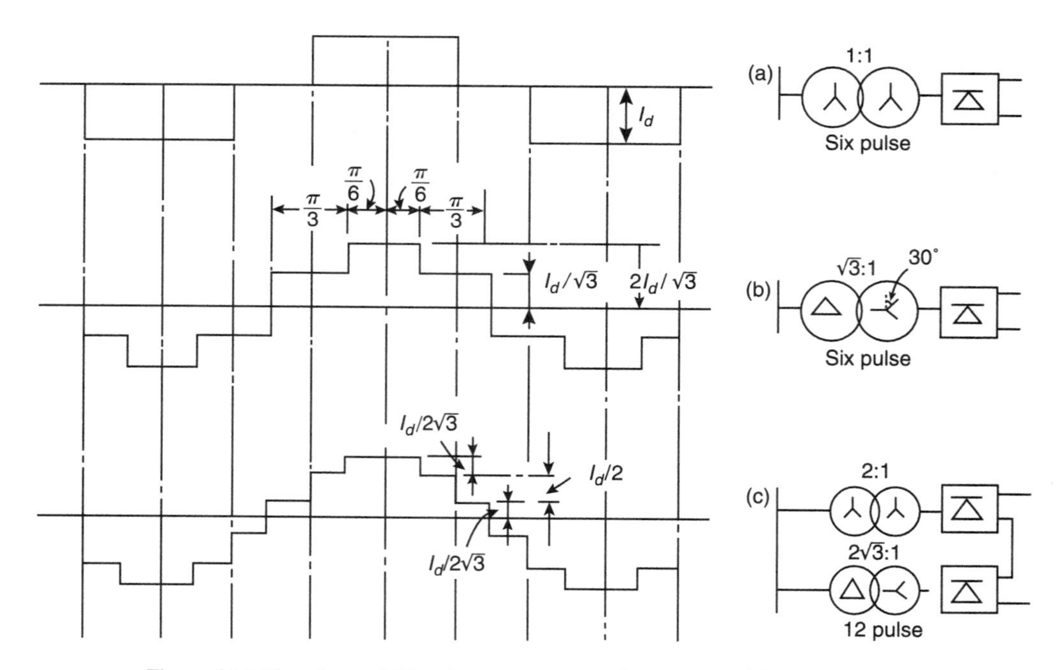

Figure 4.12 Six-pulse and 12-pulse current-sourced converters: (a) six-pulse wye/wye circuit and current waveform; (b) six-pulse wye/delta circuit and current waveform; (c) twelve-pulse circuit and current waveform.

by 30 degrees with respect to the corresponding voltages in a wye/wye transformer. The line current will now be as shown in Figure 4.12(b). The analysis will show that the harmonics are defined by the same (4.16) above, except that harmonics of the 5th, 7th, 17th, 19th, 29th, etc. orders are of opposite sign.

When two six-pulse converters of equal capacity, one with wye/wye transformer of ratio 2:1, and the other with delta/wye transformer of ratio $2\sqrt{3}:1$, are connected together in either series or parallel, the resultant current is given by

$$i_n = \frac{2\sqrt{3}}{\pi} I_d \left[\cos \omega t - \frac{1}{11} \cos 11\omega t + \frac{1}{13} \cos 13\omega t - \frac{1}{23} \cos 23\omega t + \cdots \right] \quad (4.20)$$

This equation represents the ac current of a 12-pulse converter and has a waveform shown in Figure 4.12(c). Harmonics of the orders 5th, 7th, 17th, 19th, . . . which do not enter the ac system, will circulate between the two six-pulse converters. The harmonics are of the order of $12k \pm 1$, where k is an integer. In fact a 12-pulse converter will often have one transformer with one primary and two secondaries. It follows that one can have a 24-pulse converter by combining four six-pulse converters, with 15 degrees phase-shift obtained with phase-shifting transformers.

The equations above took no account of commutation angle. However, it is important to consider the commutation angle because the harmonics decrease considerably. If this is considered, the current wave shape for analysis can be defined in three parts as in Figure 4.13.

$$i_p = I_d \frac{(\cos \alpha - \cos \omega t)}{[(\cos \alpha - \cos \alpha + \gamma)]}$$

for

$$\alpha < \omega t < (\alpha + \gamma)$$

$$I_q = I_d$$

for

$$(\alpha + \gamma) < \omega t < [(2\pi/3) + \alpha].$$

$$i_r = I_d - I_d \frac{[\cos \alpha - \cos(\omega t - 2\pi/3)]}{[(\cos \alpha - \cos \alpha + \gamma)]}$$

for

$$[(2\pi/3) + \alpha] < \omega t < [(2\pi/3) + \alpha + \gamma].$$

Based on Fourier analysis of the waveform defined above, Figures 4.14 to 4.19 show curves of the harmonic current I_n as percentage of the fundamental I (defined by equations above), against the commutation angle γ for different values of α or δ for the 5th, 7th, 11th, 13th, 17th, and 19th in order to convey the nature of these curves. Curves for specific values of the delay angle α and the margin angle δ are the same, because δ is a mirror image of α.

It is seen from these curves that, as γ increases, the magnitude of harmonics decreases, and with higher orders decreases more rapidly. Each harmonic decreases to a minimum at an angle $\gamma = 360°/n$ and then rises slightly thereafter. It is worth noting that for practical purposes γ will likely be in the range of 15 to 30 degrees at full load.

Obviously the voltage harmonics on the ac system side are a function of the ac system impedance for each harmonic current injected into the system by the converter.

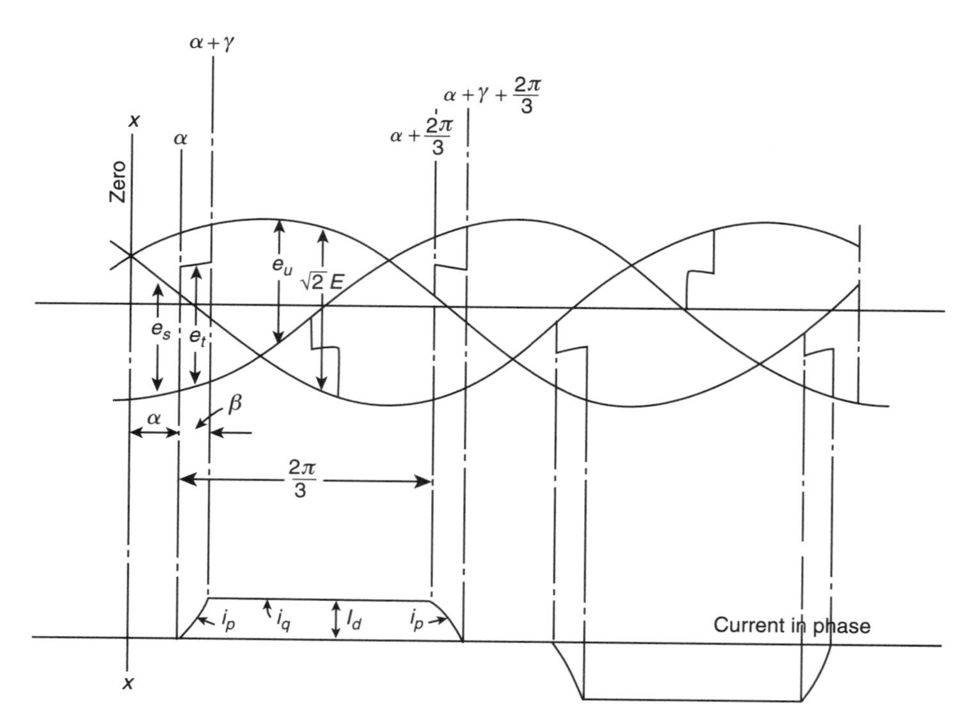

Figure 4.13 DC current and voltage waveform definition for harmonic analysis.

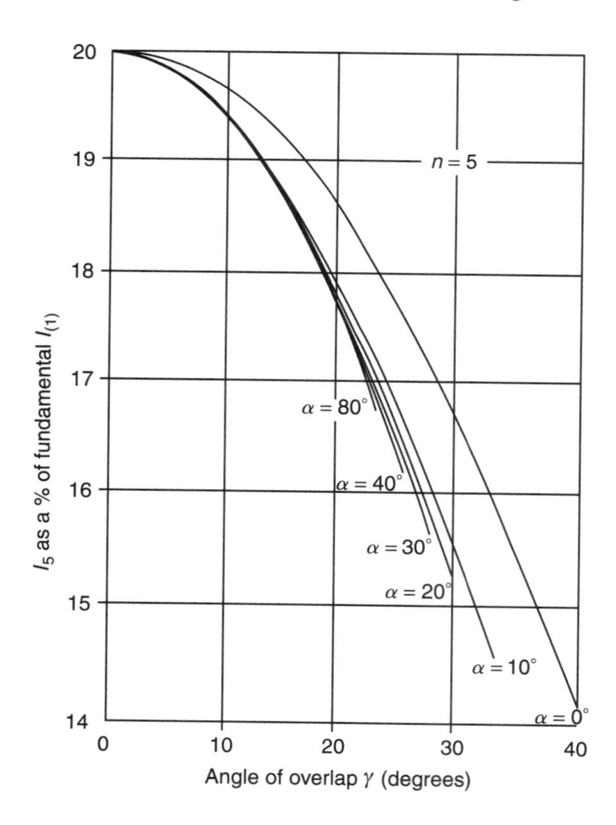

Figure 4.14 Variation of 5th harmonic current in relation to angle of delay α (or margin angle δ) and commutation angle γ.

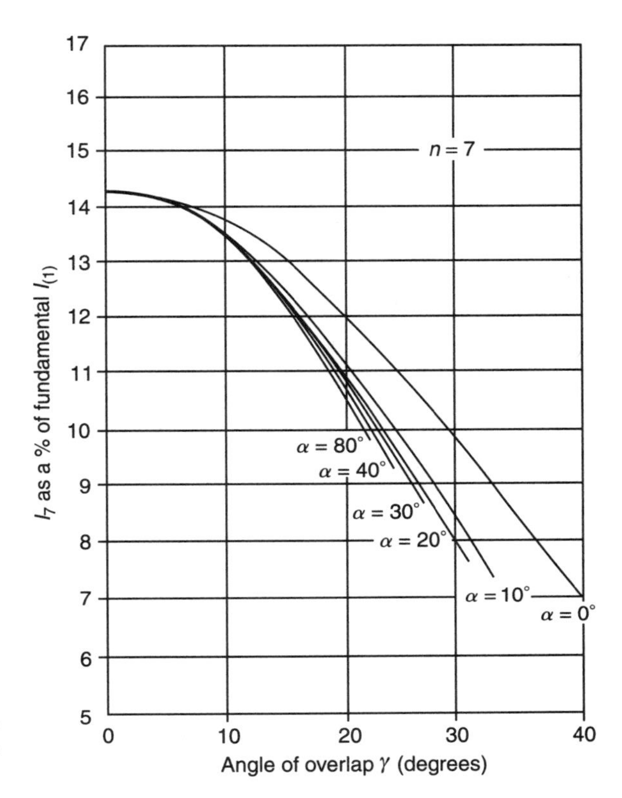

Figure 4.15 Variation of 7th harmonic current in relation to angle of delay α (or margin angle δ) and commutation angle γ.

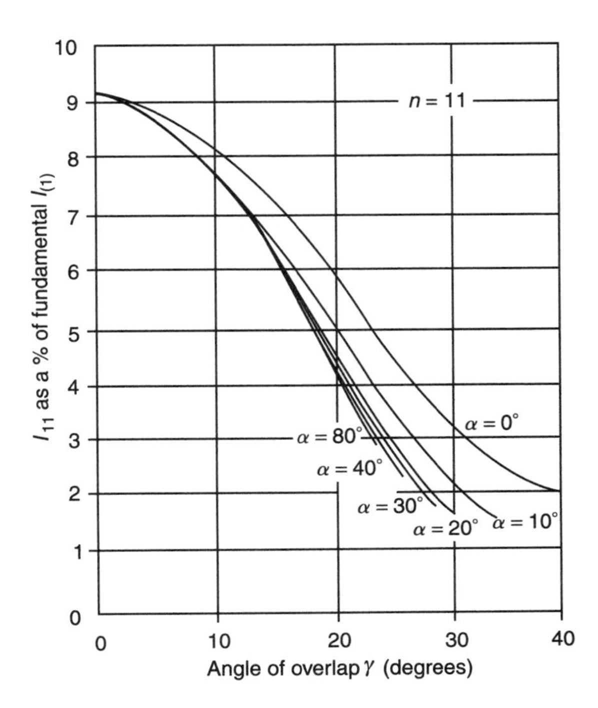

Figure 4.16 Variation of 11th harmonic current in relation to angle of delay α (or margin angle δ) and commutation angle γ.

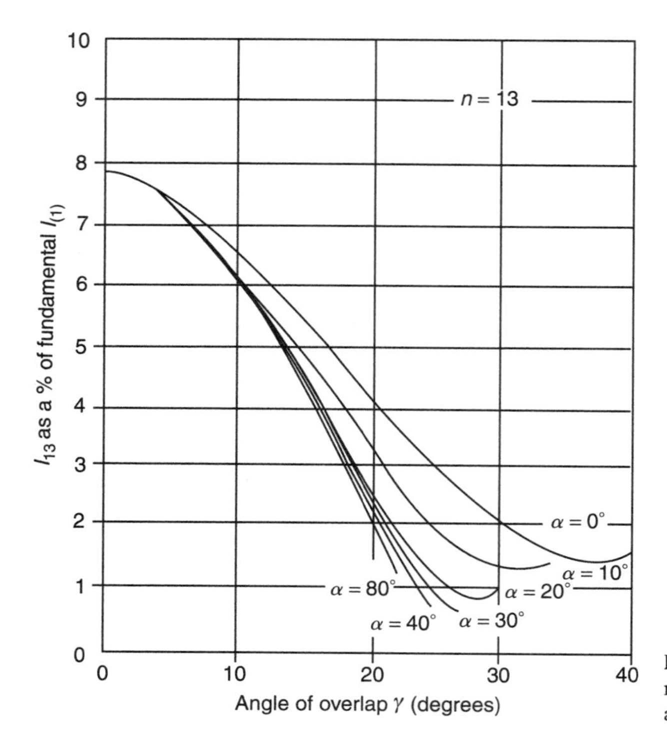

Figure 4.17 Variation of 13th harmonic current in relation to angle of delay α (or margin angle δ) and commutation angle γ.

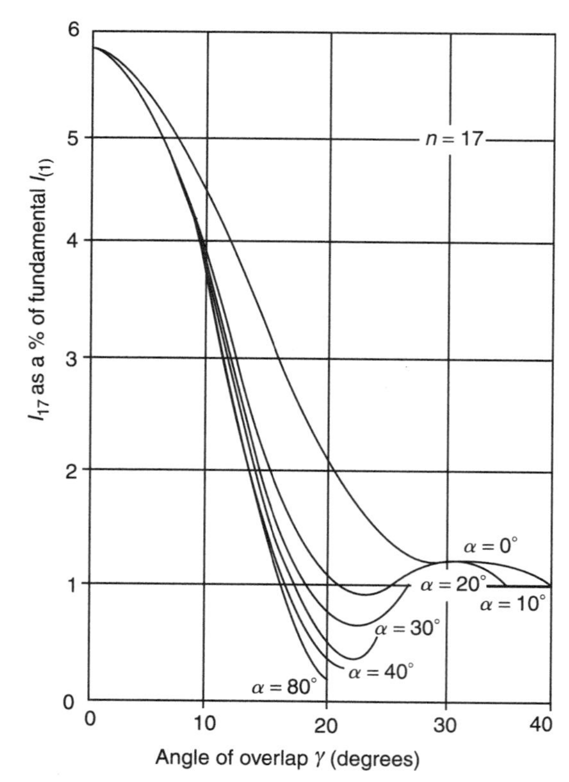

Figure 4.18 Variation of 17th harmonic current in relation to angle of delay α (or margin angle δ) and commutation angle γ.

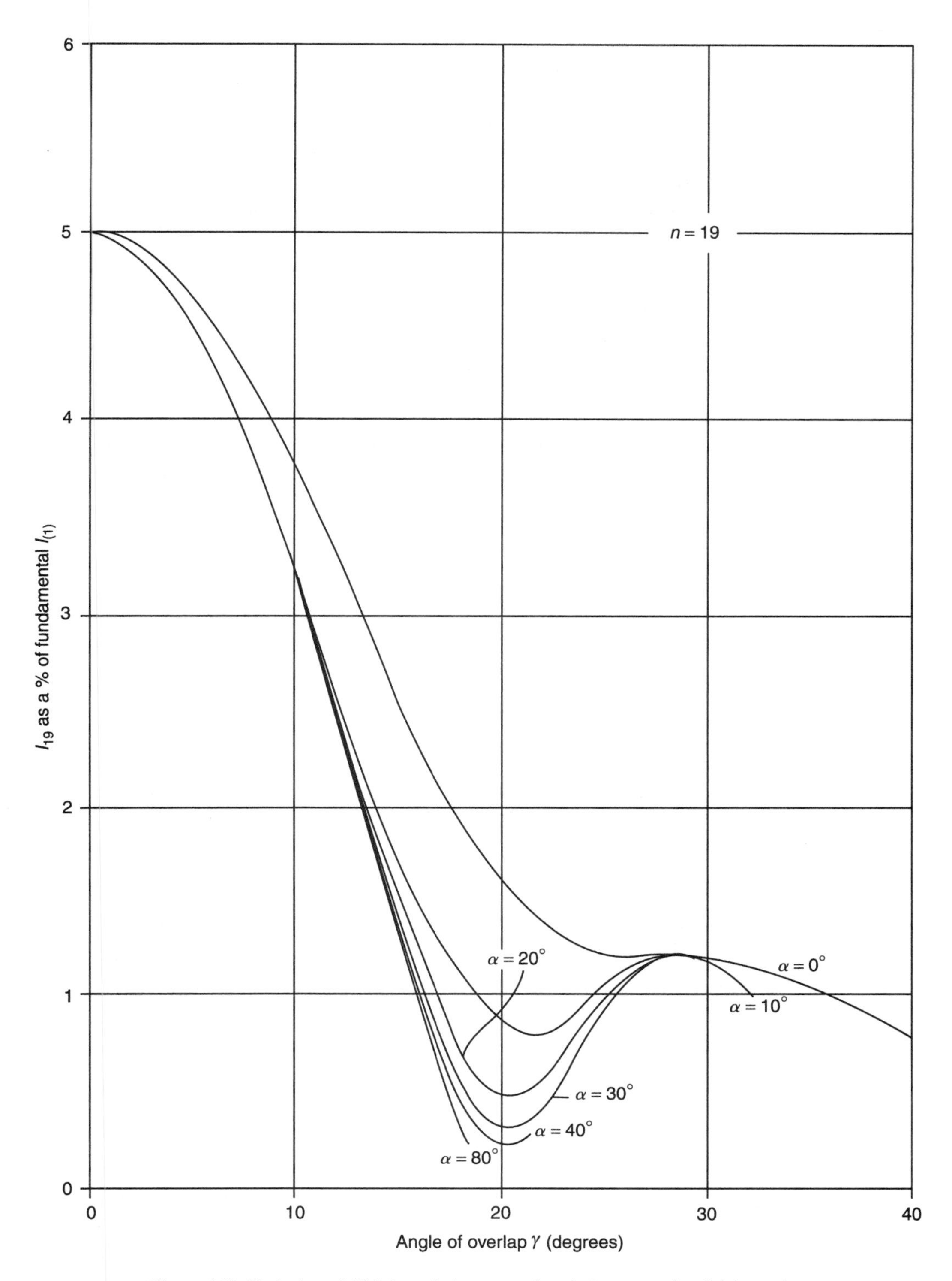

Figure 4.19 Variation of 19th harmonic current in relation to angle of delay α (or margin angle δ) and commutation angle γ.

4.3.6 DC Voltage Harmonics

On the dc side of a converter, the output voltage is of a form, which depends on the pulse order, angle of delay α, and angle of commutation γ. The output voltage consists of a dc voltage, on which are superimposed a number of harmonics; these have an order n equal to kp, where k is an integer and $p = 6$ for a six-pulse converter. With zero delay angle, and no load or neglecting commutation angle, the rms amplitude of each harmonic is given by

$$V_n = V_0 \frac{\sqrt{2}}{n^2 - 1} \tag{4.21}$$

Thus sixth harmonic has an amplitude of 4.04%, twelveth harmonic has amplitude of 0.99% of the fundamental dc voltage, etc., for zero delay and commutation angles.

With reference to Figure 4.13(a), for harmonic analysis with an angle of delay α and commutation angle γ, the 60 degree segment of the dc voltage waveform can be defined in three parts:

$$\begin{aligned} e_s &= \sqrt{2}\,E\cos\left(\omega t + \frac{\pi}{6}\right) \qquad \text{for} \quad 0 < \omega t < \alpha \\ e_t &= \frac{1}{2}\left[\sqrt{2}\,E\cos\left(\omega t - \frac{\pi}{6}\right) + \sqrt{2}\,E\cos\left(\omega t + \frac{\pi}{6}\right)\right] \quad \text{for} \quad \alpha < \omega t < (\alpha + \gamma) \\ e_u &= \sqrt{2}\,E\cos\left(\omega t - \frac{\partial}{6}\right) \qquad \text{for} \quad (\alpha + \gamma) < \omega t < (\pi/3) \end{aligned} \tag{4.22}$$

Figures 4.20, 4.21, and 4.22 show characteristics of harmonics as per unit of the ideal dc voltage V_0 against the commutation angle γ, for different values of delay angle α for 6th, 12th, and 18th harmonic, respectively. It is seen from these characteristics that

- For small values of angle γ, the harmonic magnitudes increase with the increase in angle α and the higher the harmonics, the more rapid the increase.
- For a constant angle α, the harmonics decrease (they may increase slightly at first for small angles α) and reach a first minimum at, approximately, $\gamma = \pi/n$.
- For, $\gamma = \pi/(n + 1)$ and, $\gamma = \pi/(n - 1)$, the harmonics are constant for any angle α.
- At, $\gamma = 2\pi/n$, there is a maximum and at $\gamma = 3\pi/n$ there is a further minimum.

When two six-pulse converters, one with wye/wye transformer and the other with wye/delta transformer, are connected in series to form a 12-pulse converter, the harmonics that are of the orders that are odd multiples of 6, (6th, 18th, 30th . . .), will be equal and opposite in phase and cancel out, where as harmonics of the orders that are even multiples of 6, (12th, 24th, 36th . . .) will be in phase.

Obviously the current harmonics on the dc side are a function of the dc side impedance for each harmonic voltage produced by the converter.

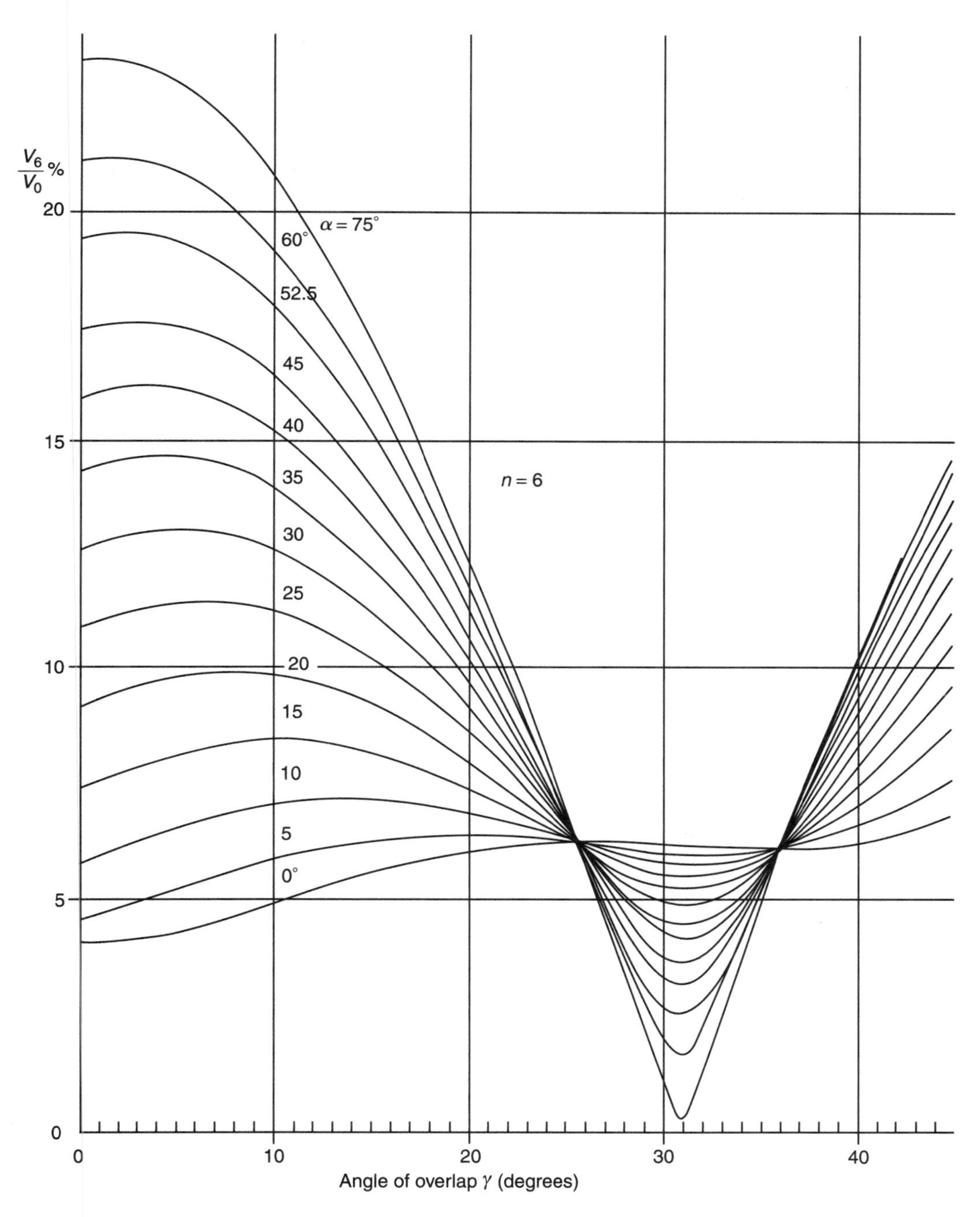

Figure 4.20 Variation of 6th harmonic voltage in relation to angle of delay α (or margin angle δ) and commutation angle γ.

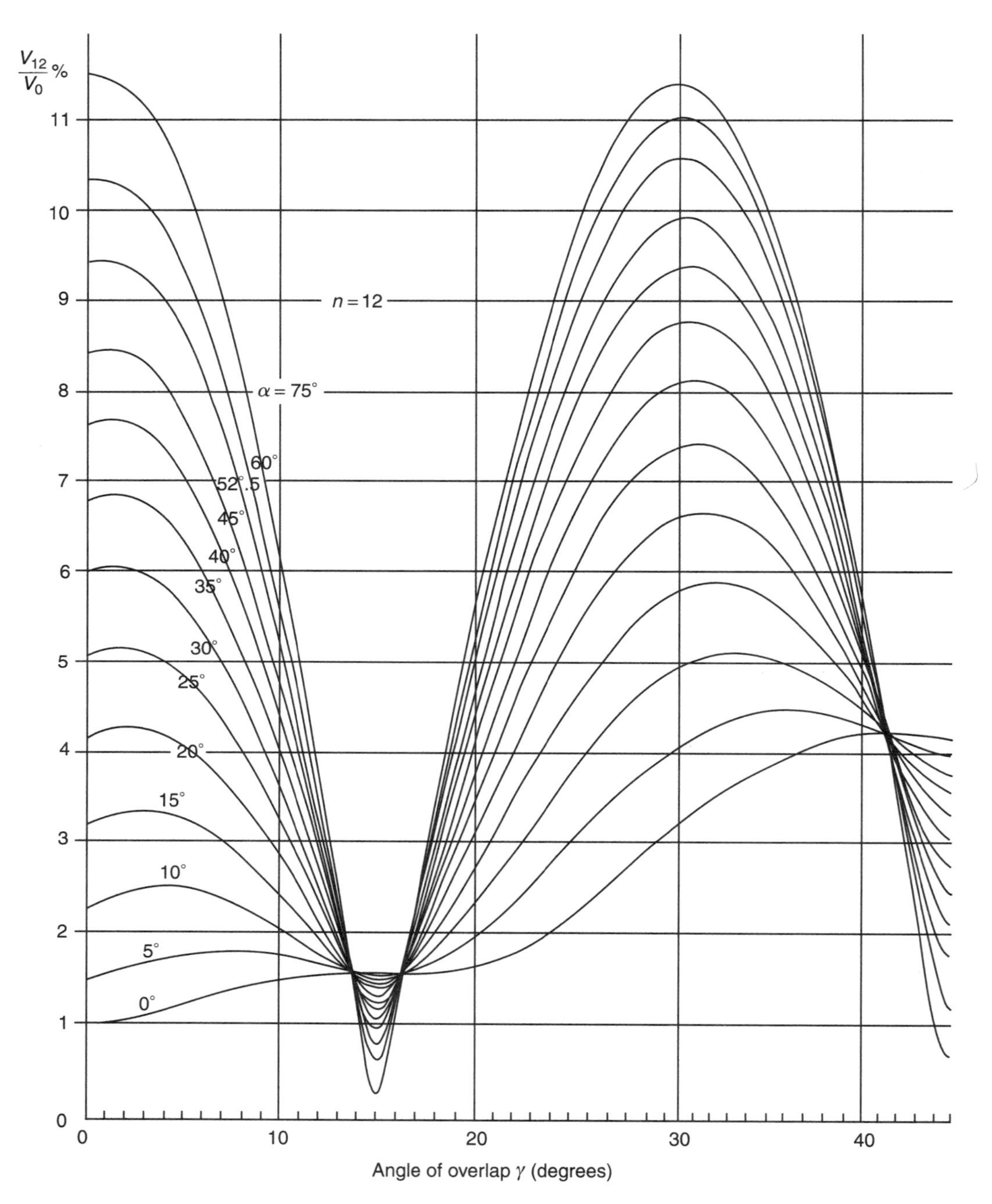

Figure 4.21 Variation of 12th harmonic voltage in relation to angle of delay α (or margin angle δ) and commutation angle γ.

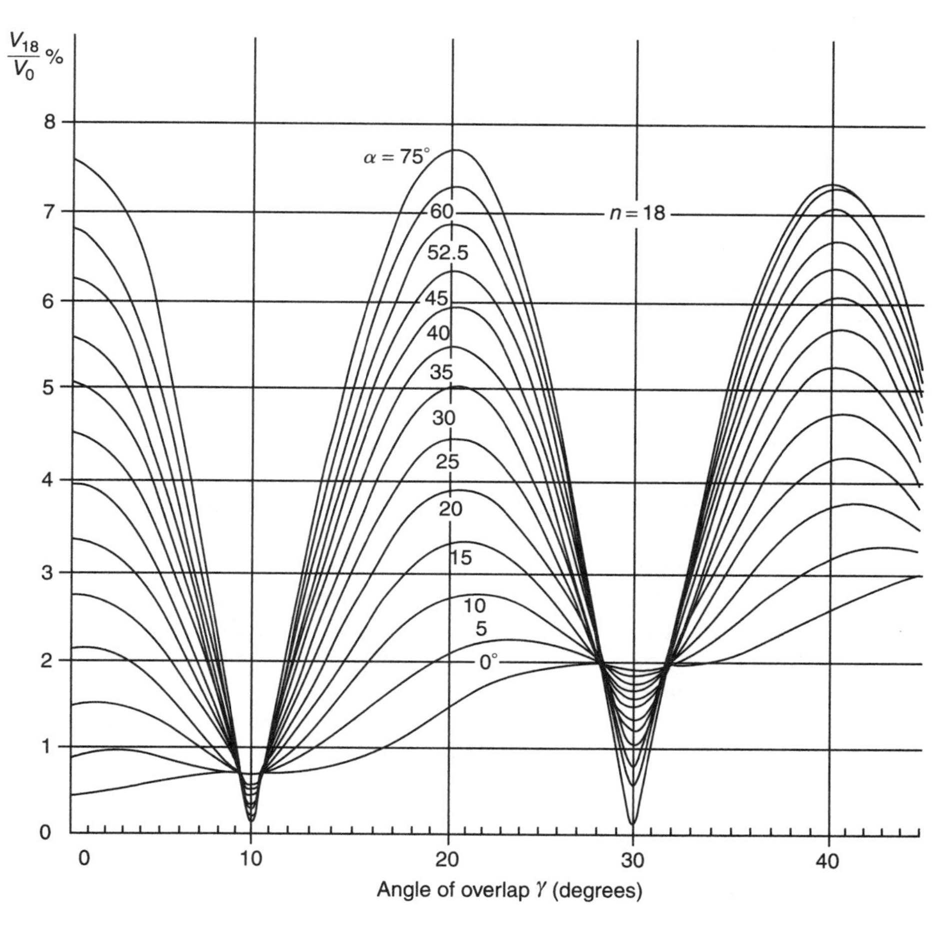

Figure 4.22 Variation of 18th harmonic voltage in relation to angle of delay α (or margin angle δ) and commutation angle γ.

4.4 CURRENT-SOURCED CONVERTER WITH TURN-OFF DEVICES (current-stiff converter)

A current-sourced converter with turn-off devices is also referred to as current-stiff converter. As mentioned previously, these turn-off devices must have reverse withstand voltage capability (symmetric devices) or have diodes in series if they are asymmetric devices.

In the current-sourced converter with conventional thyristors discussed above, operation of the converter is limited to the third and fourth quadrant (lagging power factor). This is because thyristors do not have turn-off capability and the dc current has to be commutated from one valve to another while the anode-cathode voltage of the incoming valve is still positive. Also, such a converter needs an ac voltage source for commutation.

In the voltage-sourced converters discussed in Chapter 3, there is the dc capacitor, which facilitates rapid transfer of current from an outgoing turn-off valve to the opposite valve in a phase-leg, irrespective of the direction of the ac current. The capacitor is assumed to be large enough to handle alternate charging and discharging without substantial change in dc voltage. With turn-off capability, the valves can be turned off at will. However, turn-off devices in order to turn off still require an alternate path for rapid transfer of current. Otherwise, they will have to dissipate a large amount of energy to turn off current in an inductive circuit. It can be visualized that if ac capacitors are placed between phases, on the ac side of the valves, Figure 4.23(a), they can facilitate rapid transfer of current from the outgoing turn-off valve to the incoming valve.

Commutation of current from valve 1 to valve 3 is illustrated in Figure 4.23(b). Given low inductance of the ac shunt capacitor and the bus connections, the transfer (commutation) is rapid and there is no commutation angle to speak of as far as the valves are concerned. Actually, with due respect to the turn-on di/dt limit of the devices, inductance of the capacitors and the bus connections can be duly exploited. Also it is to be noted, that when a valve turns off, valve 1 in Figure 4.23(b), its rate of rise of voltage is cushioned by the ac capacitor. Details of turn-on and turn-off in a few tens of microseconds time frame of the commutation is not discussed here. Suffice it to say that it is a complex and important matter in terms of device losses and snubber requirements. These capacitors need to handle a sustained alternating charge/discharge current of the converter valves.

Unlike the line commutated converter using conventional thyristors, this converter with turn-off valves can operate even with a leading power factor and it does not need a pre-existing ac voltage for commutation. It can in fact operate as an inverter into a passive or an active ac system.

Figure 4.23(c) shows the anode-bus current connected to the anode side of valves 1, 3, and 5 and transfer of this incoming dc current from valve 1 to 3, to 5, to 1 etc., in a closed three-valve sequence in a three-phase converter. Similarly shown are the cathode bus current, the outgoing dc current, and how it transfers from valve 2 to 4, to 6, to 2 etc., in a closed three-valve sequence. The two sequences are phase shifted by 60 degrees and they together form a three-phase, full-wave bridge converter. Consequent injected ac current in the three phases is also shown in Figure 4.23(c), which is same as for the conventional thyristor converter when neglecting commutation angle, Figure 4.3(b). The currents are injected without the support of ac system voltages, and therefore, the phase angle and the frequency of this injected ac current can be controlled.

To understand the operation of this converter, it is appropriate to visualize it as an ac current generator connected to an ac system, which is front-ended with ac capacitors [Figure 4.23(d)]. In FACTS applications, the interface beyond the ac capacitor is likely to be a transformer which may be followed by ac system and filters to limit the harmonics from entering the ac system. The ac voltage at the converter terminal is the result of the interaction of the converter-generated ac current with the ac system impedance and the system voltage. Naturally, the injected current has to be coordinated with the characteristics of the ac system in terms of its frequency and phase relationship to ensure that the consequent ac voltages are acceptable.

In the context of shunt-connected FACTS Controllers, this converter will simply inject an ac current into the system with the necessary ac voltage to force such an injection of current. This assumes of course that the dc current source is capable of driving such a current.

In the context of series-connected FACTS Controllers, the ac current in the line would flow through the valves to become a unidirectional dc current. Depending on the dc side impedance and the source, dc voltage of either polarity would appear. This dc voltage is reflected back as the converter ac voltage, which in turn will influence

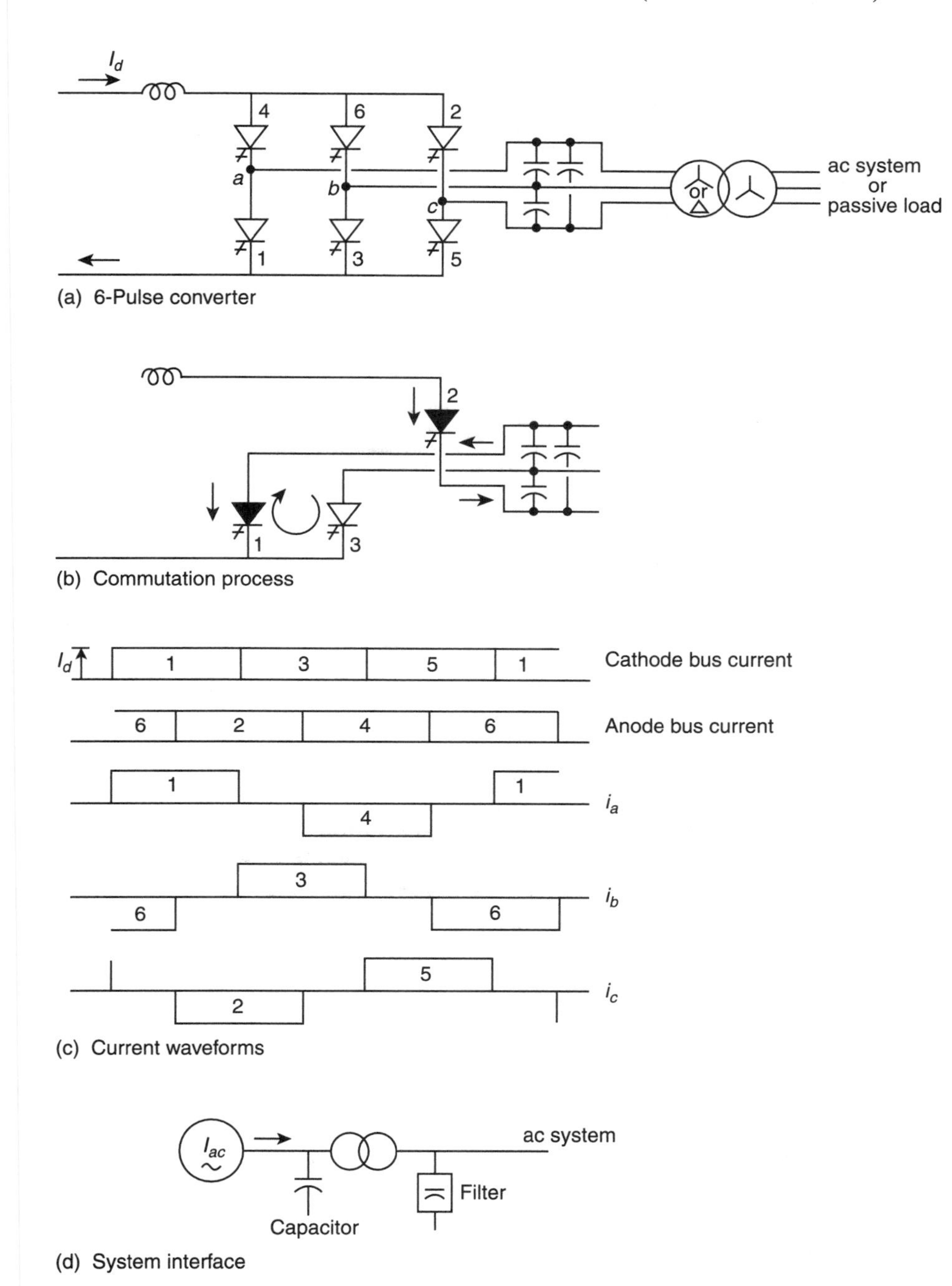

Figure 4.23 Self-commutating current-sourced converter: (a) six-pulse converter; (b) commutation process; (c) current waveforms; (d) system interface.

the line current. If for some reason the connection to the ac systems becomes open, the converter is blocked, the current bypassed by the valves and the dc side voltage can be quickly reversed to manage the bypassed current; this is a normal practice in HVDC.

Since the current waveform of this single six-pulse converter is identical to that of Figure 4.3(b), the harmonic content is given by (4.18). The harmonics are of the order of $6k \pm 1$, where k is an integer. The rms value of any term in (4.18) is given by (4.19) and the rms value of the fundamental is given by (4.3).

Two six-pulse converters of equal capacity, one with wye/wye transformer and the other with wye/delta transformer, or one transformer with two secondaries, one wye connected and the other delta connected, will result in a 12-pulse converter with harmonics defined by (4.20). The ac side fundamental and harmonic voltages are a function of the ac system and the injected current.

A dc side converter voltage will also have harmonics. Assuming sinusoidal voltage at the ac capacitor terminals, the dc voltage harmonics are defined by (4.22), but with $\gamma = 0$:

$$\begin{aligned} e_s &= \sqrt{2}\,E \cos\left(\omega t + \frac{\pi}{6}\right) \qquad \text{for} \qquad 0 < \omega t < \alpha \\ e_u &= \sqrt{2}\,E \cos\left(\omega t - \frac{\pi}{6}\right) \qquad \text{for} \qquad \alpha < \omega t < (\pi/3) \end{aligned} \tag{4.23}$$

The per-unit values of these harmonics are given by the values on the Y-axis of Figures 4.20, 4.21, and 4.22, for the 6th, 12th, and 24th harmonics.

Various PWM concepts discussed for the voltage-sourced converters in Chapter 3 are applicable to the current-sourced converters, although they are not discussed here. An advantage of the PWM operation is that the commutation capacitor size will decrease.

4.5 CURRENT-SOURCED VERSUS VOLTAGE-SOURCED CONVERTERS

There are some advantages and disadvantages of current-sourced versus voltage-sourced converters:

- Diode-based converters are the lowest cost converters, if control of active power by the converter is not required.
- If the leading reactive power is not required, then a conventional thyristor-based converter provides a low-cost converter with active power control. It can also serve as a controlled lagging reactive power load (like a thyristor-controlled reactor).
- The current-sourced converter does not have high short-circuit current, as does the voltage-sourced converter. For current-sourced converters, the rate of rise of fault current during external or internal faults is limited by the dc reactor. For the voltage-sourced converters, the capacitor discharge current would rise very rapidly and can damage the valves.
- The six-pulse, current-sourced converter does not generate third harmonic voltage, and its transformer primaries for a 12-pulse converter do not have to be connected in series for harmonic cancellation. It is also relatively simple to obtain a 24-pulse operation with phase-shifting windings.

- In a current-stiff converter, the valves are not subject to high *dv/dt*, due to the presence of the ac capacitors.
- Ac capacitors required for the current-stiff converters can be quite large and expensive, although their size can be decreased by adoption of PWM topology. In general the problem of a satisfactory interface of current-sourced converters with the ac system is more complex.
- Continuous losses in the dc reactor of a current-sourced converter are much higher than the losses in the dc capacitor. These losses can represent a significant loss penalty.
- With the presence of capacitors, which are subjected to commutation charging and discharging, this converter will produce harmonic voltages at a frequency of resonance between the capacitors and the ac system inductances. Adverse effects of this can be avoided by sizing the capacitors such that the resonance frequency does not coincide with characteristic harmonics.
- These harmonics as well as the presence of a dc reactor can result in overvoltages on the valves and transformers.
- Widespread adoption of asymmetrical devices, IGBTs and GTOs, as the devices of choice for lower on-state losses, has made voltage-sourced converters a favorable choice when turn-off capability is necessary. The device market is generally driven by high-volume industrial applications, and as a result symmetrical turn-off devices of high-voltage ratings and required operating characteristics, in particular the switching characteristics, may not be readily available until the volume of the FACTS market increases. However as the devices evolve, particularly with the evolution of advanced GTOs discussed in Chapter 2, it is important to continuously re-evaluate the converter topology of choice.

REFERENCES

Adamson, C., and Hingorani, N. G., "High Voltge Direct Current Power Transmission," *Book,* Garraway Ltd., London, U.K., 1960.

Arrillaga, J., "High Voltage Direct Current Transmission," *Book,* Published by Peter Peregrinus on behalf of IEE, London, 1983.

Cardosa, B. J., Burnet, S., and Lipo, T. "A New Control Strategy for the PWM Current Stiff Rectifier/Inverter with Resonant Snubbers," *IEEE Power Electronics Specialists Conference,* St. Louis, MO, pp. 573–579, June 1997.

Cardosa, B. J., and Lipo, T., "A Reduced Parts Count Realization of the Resonant Snubbers for High Power Current Stiff Converters," *IEEE Applied Power Electronics Conference,* Los Angeles, CA, March 1998.

Cardosa, B. J., and Lipo, T., "Current Stiff Converter Topologies with Resonant Snubbers," *IEEE Industry Applications Society Annual Meeting,* New Orleans, LA, pp. 1322–1329, October 1997.

Holt, J., Stam, M., Thur, J., and Linder, A., "High Power Pulse Width Controlled Current Source GTO Inverter for High Switching Frequency," *IEEE Industry Applications Society Annual Meeting,* New Orleans, LA, pp. 1330–1335, October 1997.

Jager, J., Herold, G., and Hosemann, G., "Current-Source Controlled Phase Shifting Transformer as a New FACTS Equipment for High Dynamical Power Flow Control," *CIGRE 1995 Symposium on Power Electronics in Electric Power Systems,* Tokyo, May 1995.

Naitoh, H., Nishikawa, H., Simamura, T., Arakawa, F., Ogimoto, K., and Saiki, K. "A Snubber

Loss Free Current Source Converter For High Power Use" *CIGRE 1995 Symposium on Power Electronics in Electric Power Systems,* Tokyo, May 1995.

Rizo, S. C., Wu, B., and Studeh, R., "Symmetric GTO and Snubber Component Characterization in PWM Current Sourced Inverters," *IEEE Power Electronics Specialists Conference,* pp. 613–619, July 1996.

Salazar, L. D., Joos, G., and Ziogas, P. D., "A Low Loss Soft-Switched PWM Current Source Inverter," *IEEE Paper No. 0-7803-0695-3/92,* 1992.

Yamaguchi, T., Matsui, K., Hayashi, K., Takase, S., and You, Y., "A Novel PWM Strategy to Minimize Surge Voltage for Current Source Converter," *IEEE Industry Applications Society Annual Meeting,* San Diego, pp. 1085–1091, October 1996.

5

Static Shunt Compensators: SVC and STATCOM

5.1 OBJECTIVES OF SHUNT COMPENSATION

It has long been recognized that the steady-state transmittable power can be increased and the voltage profile along the line controlled by appropriate reactive shunt compensation. The purpose of this reactive compensation is to chage the natural electrical characteristics of the transmission line to make it more compatible with the prevailing load demand. Thus, shunt connected, fixed or mechanically switched reactors are applied to minimize line overvoltage under light load conditions, and shunt connected, fixed or mechanically switched capacitors are applied to maintain voltage levels under heavy load conditions.

In this section, basic considerations to increase the transmittable power by ideal shunt-connected var compensation will be reviewed in order to provide a foundation for power electronics-based compensation and control techniques to meet specific compensation objectives. The ultimate objective of applying reactive shunt compensation in a transmission system is to increase the transmittable power. This may be required to improve the steady-state transmission characteristics as well as the stability of the system. Var compensation is thus used for voltage regulation at the midpoint (or some intermediate) to segment the transmission line and at the end of the (radial) line to prevent voltage instability, as well as for dynamic voltage control to increase transient stability and damp power oscillations.

5.1.1 Midpoint Voltage Regulation for Line Segmentation

Consider the simple two-machine (two-bus) transmission model in which an ideal var compensator is shunt connected at the midpoint of the transmission line, as shown in Figure 5.1(a). For simplicity, the line is represented by the series line inductance. The compensator is represented by a sinusoidal ac voltage source (of the fundamental frequency), in-phase with the midpoint voltage, V_m, and with an amplitude identical to that of the sending- and receiving-end voltages ($V_m = V_s = V_r = V$) The midpoint compensator in effect segments the transmission line into two independent parts: the

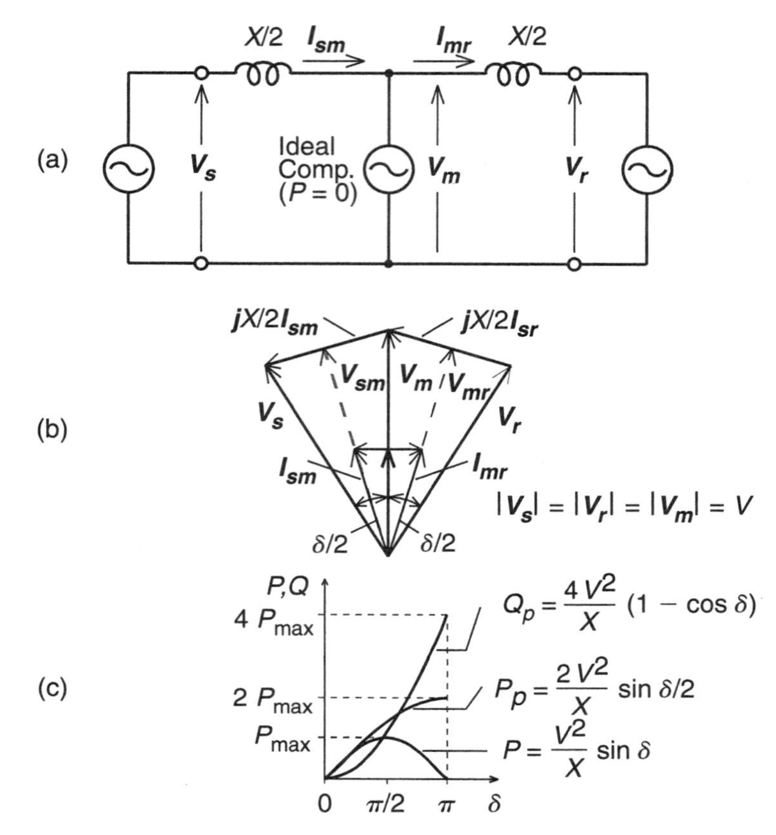

Figure 5.1 Two-machine power system with an ideal midpoint reactive compensator (a), corresponding phasor diagram (b), and power transmission vs. angle characteristic showing the variation of real power P_p and the reactive power output of the compensator Q_p with angle δ (c).

first segment, with an impedance of $X/2$, carries power from the sending end to the midpoint, and the second segment, also with an impedance of $X/2$, carries power from the midpoint to the receiving end. The relationship between voltages, $\boldsymbol{V}_s$, $\boldsymbol{V}_r$, $\boldsymbol{V}_m$, (together with $\boldsymbol{V}_{sm}$, $\boldsymbol{V}_{rm}$,), and line segment currents $\boldsymbol{I}_{sm}$ and $\boldsymbol{I}_{mr}$ is shown by the phasor diagram in Figure 5.1(b). Note that the midpoint var compensator exchanges only reactive power with the transmission line in this process.

For the lossless system assumed, the real power is the same at each terminal (sending end, midpoint, and receiving end) of the line, and it can be derived readily from the phasor diagram of Fig. 5.1(b). With

$$V_{sm} = V_{mr} = V\cos\frac{\delta}{4}; \qquad I_{sm} = I_{mr} = I = \frac{4V}{X}\sin\frac{\delta}{4} \tag{5.1}$$

the transmitted power is

$$P = V_{sm}I_{sm} = V_{mr}I_{mr} = V_mI_{sm}\cos\frac{\delta}{4} = VI\cos\frac{\delta}{4} \tag{5.2a}$$

or

$$P = 2\frac{V^2}{X}\sin\frac{\delta}{2} \tag{5.2b}$$

Similarly

$$Q = VI\sin\frac{\delta}{4} = \frac{4V^2}{X}\left(1 - \cos\frac{\delta}{2}\right) \tag{5.3}$$

The relationship between real power P, reactive power Q, and angle δ for the case of ideal shunt compensation is shown plotted in Figure 5.1(c). It can be observed that the midpoint shunt compensation can significantly increase the transmittable power (doubling its maximum value) at the expense of a rapidly increasing reactive power demand on the midpoint compensator (and also on the end-generators).

It is also evident that for the single-line system of Figure 5.1 the midpoint of the transmission line is the best location for the compensator. This is because the voltage sag along the uncompensated transmission line is the largest at the midpoint. Also, the compensation at the midpoint breaks the transmission line into two equal segments for each of which the maximum transmittable power is the same. For unequal segments, the transmittable power of the longer segment would clearly determine the overall transmission limit.

The concept of transmission line segmentation can be expanded to the use of multiple compensators, located at equal segments of the transmission line, as illustrated for four line segments in Figure 5.2. Theoretically, the transmittable power would double with each doubling of the segments for the same overall line length. Furthermore, with the increase of the number of segments, the voltage variation along the line would rapidly decrease, approaching the ideal case of constant voltage profile.

It is to be appreciated that such a distributed compensation hinges on the instantaneous response and unlimited var generation and absorption capability of the shunt compensators employed, which would have to stay in synchronism with the prevailing phase of the segment voltages and maintain the predefined amplitude of the transmis-

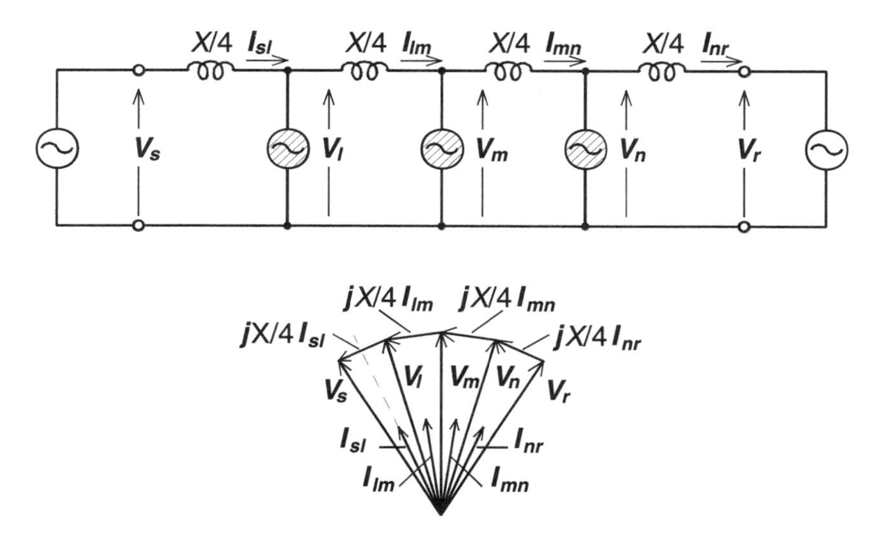

Figure 5.2 Two-machine system with ideal reactive compensators maintaining constant transmission voltage profile by line segmentation, and associated phasor diagram.

sion voltage, independently of load variation. Such a system, however, would tend to be too complex and probably too expensive, to be practical, particularly if stability and reliability requirements under appropriate contingency conditions are also considered. However, the practicability of limited line segmentation, using thyristor-controlled static var compensators, has been demonstrated by the major, 600 mile long, 735 kV transmission line of the Hydro-Quebec power system built to transmit up to 12000 MW power from the James Bay hydro-complex to the City of Montreal and to neighboring U.S. utilities. More importantly, the transmission benefits of voltage support by controlled shunt compensation at strategic locations of the transmission system have been demonstrated by numerous installations in the world.

5.1.2 End of Line Voltage Support to Prevent Voltage Instability

The midpoint voltage support of a two-machine transmission power system discussed above can easily extend to the more special case of radial transmission. Indeed, if a passive load, consuming power P at voltage V, is connected to the midpoint in place of the receiving-end part of the system (which comprises the receiving-end generator and transmission link $X/2$), the sending-end generator with the $X/2$ impedance and load would represent a simple radial system. Clearly, without compensation the voltage at the midpoint (which is now the receiving end) would vary with the load (and load power factor).

A simple radial system with feeder line reactance of X and load impedance Z, is shown in Figure 5.3(a) together with the normalized terminal voltage V_r versus power P plot at various load power factors, ranging from 0.8 lag and 0.9 lead. The "nose-point" at each plot given for a specific power factor represents the voltage instability corresponding to that system condition. It should be noted that the voltage stability limit decreases with inductive loads and increases with capacitive loads.

The inherent circuit characteristics of the simple radial structure, and the V_r versus P plots shown, clearly indicate that shunt reactive compensation can effectively increase the voltage stability limit by supplying the reactive load and regulating the terminal voltage ($V - V_r = 0$) as illustrated in Figure 5.3(b). It is evident that for a radial line, the end of the line, where the largest voltage variation is experienced, is the best location for the compensator. (Recall that, by contrast, the midpoint is the most effective location for the line interconnecting two ac system buses.)

Reactive shunt compensation is often used in practical applications to regulate the voltage at a given bus against load variations, or to provide voltage support for the load when, due to generation or line outages, the capacity of the sending-end system becomes impaired. A frequently encountered example is when a large load area is supplied from two or more generation plants with independent transmission lines. (This frequently happens when the locally generated power becomes inadequate to supply a growing load area and additional power is imported over a separate transmission link.) The loss of one of the power sources could suddenly increase the load demand on the remaining part of the system, causing severe voltage depression that could result in an ultimate voltage collapse.

5.1.3 Improvement of Transient Stability

As seen in the previous sections, reactive shunt compensation can significantly increase the maximum transmittable power. Thus, it is reasonable to expect that, with suitable and fast controls, shunt compensation will be able to change the power flow

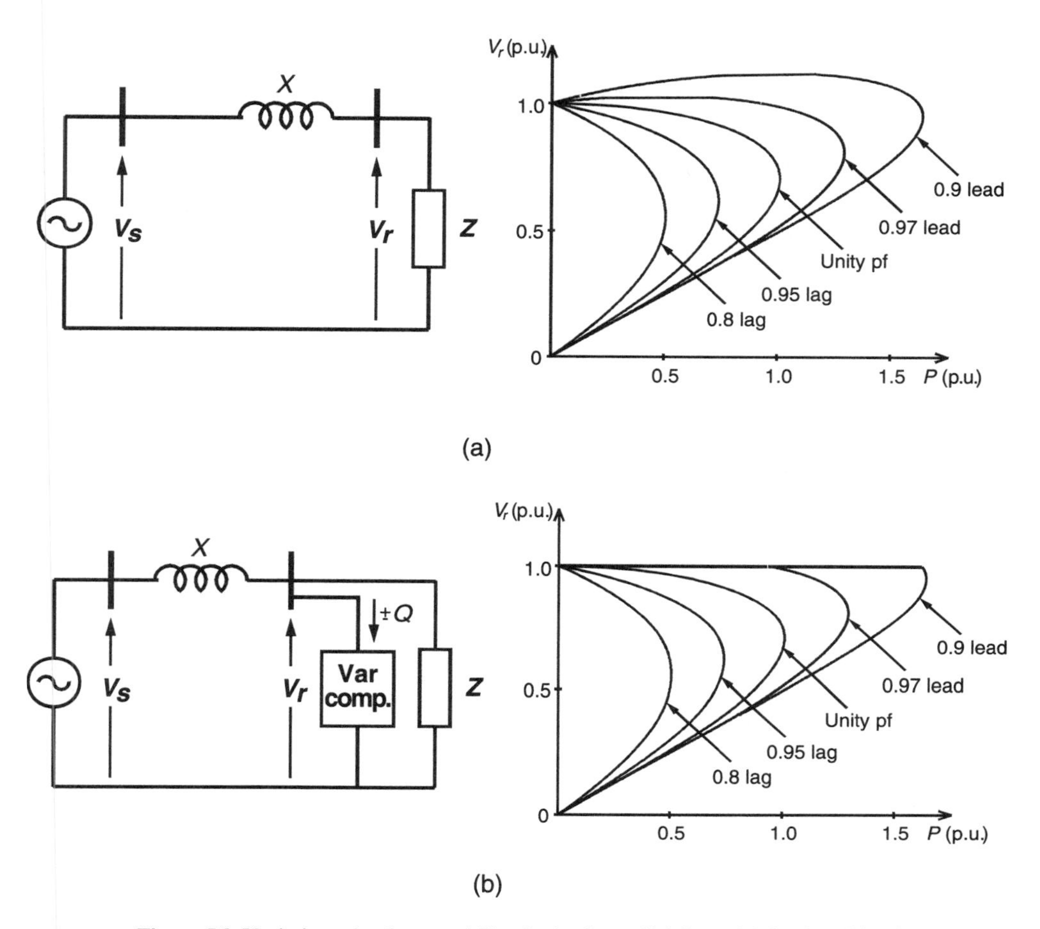

Figure 5.3 Variation of voltage stability limit of a radial line with load and load power factor (a), and extension of this limit by reactive shunt compensation (b).

in the system during and following dynamic disturbances so as to increase the transient stability limit and provide effective power oscillation damping.

The potential effectiveness of shunt (as well as other compensation and flow control techniques) on transient stability improvement can be conveniently evaluated by the *equal area criterion.* The meaning of the equal area criterion is explained with the aid of the simple two machine (the receiving end is an infinite bus), two line system shown in Figure 5.4(a) and the corresponding P versus δ curves shown in Figure 5.4(b). Assume that the complete system is characterized by the P versus δ curve "a" and is operating at angle δ_1 to transmit power P_1 when a fault occurs at line segment "1." During the fault the system is characterized by the P versus δ curve "b" and thus, over this period, the transmitted electric power decreases significantly while mechanical input power to the sending-end generator remains substantially constant corresponding to P_1. As a result, the generator accelerates and the transmission angle increases from δ_1 to δ_2 at which the protective breakers disconnect the faulted line segment "1" and the sending-end generator absorbs *accelerating* energy, represented by area "$A1$." After fault clearing, without line segment "1" the degraded system is characterized

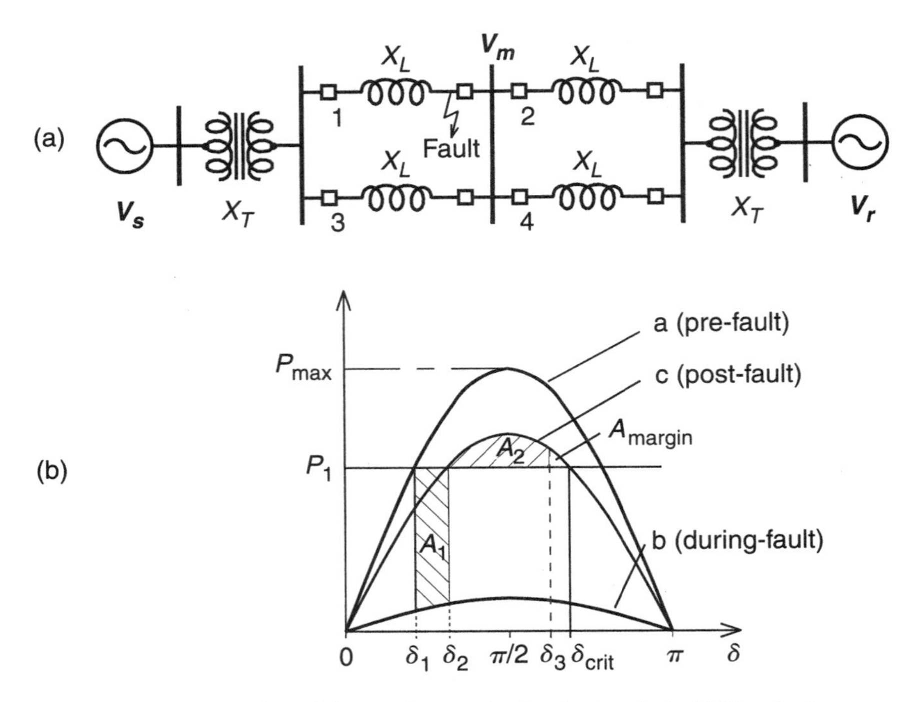

Figure 5.4 Illustration of the equal area criterion for transient stability of a two-machine, two-line power system.

by the P versus δ curve "c." At angle δ_2 on curve "c" the transmitted power exceeds the mechanical input power P_1 and the sending end generator starts to decelerate; however, angle δ further increases due to the kinetic energy stored in the machine. The maximum angle reached at δ_3, where the decelerating energy, represented by area "A_2," becomes equal to the accelerating energy represented by area "A_1". The limit of transient stability is reached at $\delta_3 = \delta_{crit}$, beyond which the decelerating energy would not balance the accelerating energy and synchronism between the sending end and receiving end could not be restored. The area "A_{margin}," between δ_3 and δ_{crit}, represent the transient stability margin of the system.

From the above general discussion it is evident that the transient stability, at a given power transmission level and fault clearing time, is determined by the P versus δ characteristic of the post-fault system. Since appropriately controlled shunt compensation can provide effective voltage support, it can increase the transmission capability of the post-fault system and thereby enhance transient stability.

For comparison, the above introduced equal-area criterion is applied here (and in subsequent chapters) in a greatly simplified manner, with the assumption that the original single line system shown in Figure 5.1(a) represents both the pre-fault and post-fault systems. (The impracticality of the single line system and the questionable validity of this assumption has no effect on this qualitative comparison, but improves the visual clarity considerably.) Suppose that this system of Figure 5.1(a), with and without the midpoint shunt compensator, transmits the same steady-state power. Assume that both the uncompensated and the compensated systems are subjected to the same fault for the same period of time. The dynamic behavior of these systems is illustrated in Figures 5.5(a) and (b). Prior to the fault both of them transmit power P_m (subscript m stands for "mechanical") at angles δ_1 and δ_{p1}, respectively. During the fault, the transmitted electric power (of the single line system considered) becomes zero while the mechanical input power to the generators remains constant (P_m). Therefore, the sending-end generator accelerates from the steady-state angles δ_1 and δ_{p1} to angles δ_2 and δ_{p2}, respectively, when the fault clears. The accelerating energies are represented by areas A_1 and A_{p1}. After fault clearing, the transmitted electric power exceeds the mechanical input power and the sending-end machine decelerates, but the accumulated kinetic energy further increases until a balance between the accelerating and decelerating energies, corresponding to areas A_1, A_{p1} and A_2, A_{p2}, respectively, is reached at δ_3 and δ_{p3}, representing the maximum angular swings for the two cases. The areas between the P versus δ curve and the constant P_m line over the intervals defined by angles δ_3 and δ_{crit}, and δ_{p3} and $\delta_{p\text{crit}}$, respectively, determine the margin of transient stability, that is, the "unused" and still available decelerating energy, represented by areas A_{margin} and A_{pmargin}.

Comparison of Figures 5.5(a) and (b) clearly shows a substantial increase in the transient stability margin the ideal midpoint compensation with unconstrained var output can provide by the effective segmentation of the transmission line. Alternatively, if the uncompensated system has a sufficient transient stability margin, shunt compensation can considerably increase the transmittable power without decreasing this margin.

In the preceding discussion, the shunt compensator is assumed to be *ideal.* The adjective "ideal" here means that the amplitude of the midpoint voltage remains constant all the time, except possibly during faults, and its phase angle follows the generator angle swings so that the compensator would not be involved in real power exchange, but it would continuously provide the necessary reactive power. As Figure 5.1(c) shows, the reactive power demand on the midpoint compensator increases rapidly with increasing power transmission, reaching a maximum value equal to four per unit at the maximum steady-state real power transmission limit of two per unit. For

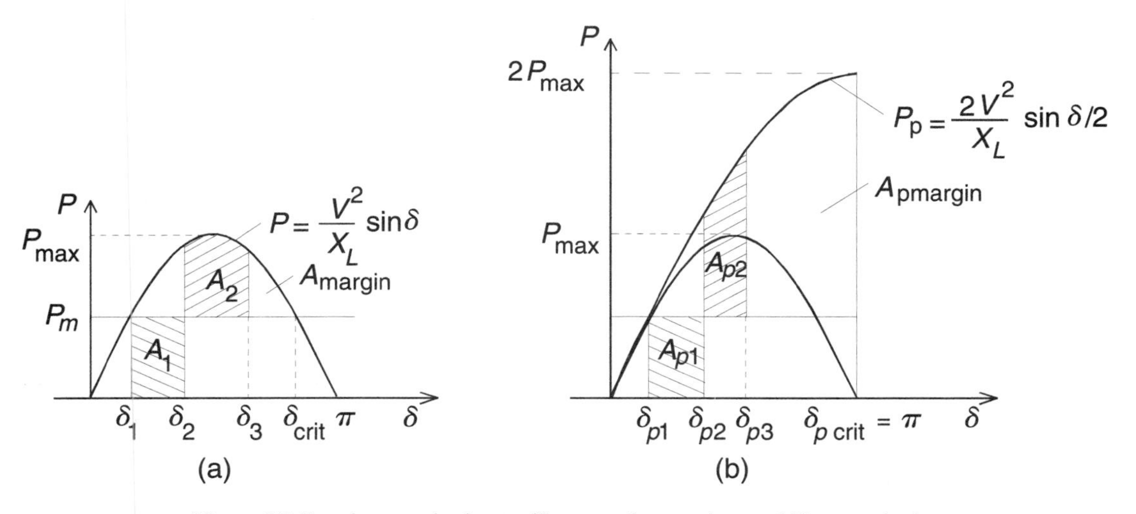

Figure 5.5 Equal area criterion to illustrate the transient stability margin for a simple two machine system without compensation (a), and with an ideal midpoint compensator (b).

obvious economic reasons, the maximum var output of a practical shunt compensator is normally considerably less than that required for full compensation. Thus, a practical compensator can perform as an ideal one only as long as the compensation var demand does not exceed its rating. Above its maximum rating the compensator either becomes a constant reactive impedance or a constant reactive current source, depending on the power circuit employed for reactive power generation. The necessary rating of the compensator is usually determined by planning studies to meet the desired power flow objectives with defined stability margins under specific system contingency conditions.

In the explanation of the equal area criterion at the beginning of this section, a clear distinction was made between the "pre-fault" and "post-fault" power system. It is important to note that from the standpoint of transient stability, and thus of overall system security, the post-fault system is the one that counts. That is, power systems are normally designed to be transiently stable, with defined pre-fault contingency scenarios and post-fault system degradation, when subjected to a major disturbance (fault). Because of this (sound) design philosophy, the actual capacity of transmission systems is considerably higher than that at which they are normally used. Thus, it may seem technically plausible (and economically savvy) to employ fast acting compensation techniques, instead of overall network compensation, specifically to handle dynamic events and increase the transmission capability of the degraded system under the contingencies encountered.

5.1.4 Power Oscillation Damping

In the case of an under-damped power system, any minor disturbance can cause the machine angle to oscillate around its steady-state value at the natural frequency of the total electromechanical system. The angle oscillation, of course, results in a corresponding power oscillation around the steady-state power transmitted. The lack of sufficient damping can be a major problem in some power systems and, in some cases, it may be the limiting factor for the transmittable power.

Since power oscillation is a sustained dynamic event, it is necessary to vary the applied shunt compensation, and thereby the (midpoint) voltage of the transmission line, to counteract the accelerating and decelerating swings of the disturbed machine(s). That is, when the rotationally oscillating generator accelerates and angle δ increases ($d\delta/dt > 0$), the electric power transmitted must be increased to compensate for the excess mechanical input power. Conversely, when the generator decelerates and angle δ decreases ($d\delta/dt < 0$), the electric power must be decreased to balance the insufficient mechanical input power. (The mechanical input power is assumed to be essentially constant in the time frame of an oscillation cycle.)

The requirements of var output control, and the process of power oscillation damping, is illustrated by the waveforms in Figure 5.6. Waveforms in Figure 5.6(a) show the undamped and damped oscillations of angle δ around the steady-state value δ_0. Waveforms in Figure 5.6(b) show the undamped and damped oscillations of the electric power P around the steady-state value P_o. (The momentary drop in power shown at the beginning of the waveform represents an assumed disturbance that initiated the oscillation.)

Waveform c shows the reactive power output Q_p of the shunt-connected var compensator. The capacitive (positive) output of the compensator increases the midpoint voltage and hence the transmitted power when $d\delta/dt > 0$, and it decreases those when $d\delta/dt < 0$.

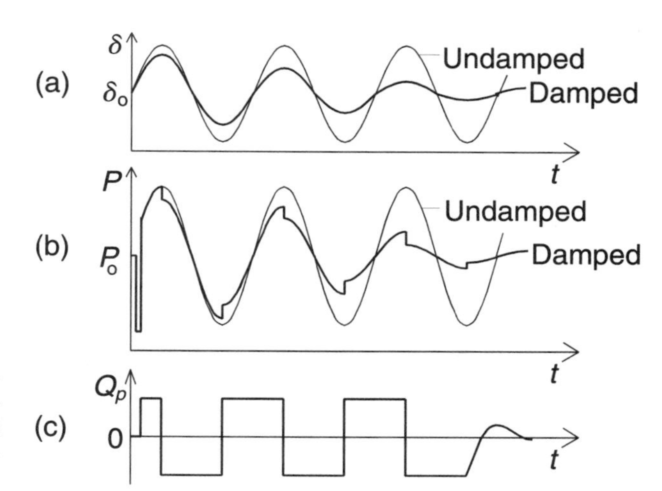

Figure 5.6 Waveforms illustrating power oscillation damping by reactive shunt compensation: (a) generator angle, (b) transmitted power, and (c) var output of the shunt compensator.

As the illustration shows, the var output is controlled in a "bang-bang" manner (output is varied between the minimum and maximum values). This type of control is generally considered the most effective, particularly if large oscillations are encountered. However, for damping relatively small power oscillations, a strategy that varies the controlled output of the compensator continuously, in sympathy with the generator angle or power, may be preferred.

5.1.5 Summary of Compensator Requirements

The functional requirements of reactive shunt compensators used for increased power transmission, improved voltage and transient stability, and power oscillation damping can be summarized as follows:

- The compensator must stay in synchronous operation with the ac system at the compensated bus under all operating conditions including major disturbances. Should the bus voltage be lost temporarily due to nearby faults, the compensator must be able to recapture synchronism immediately at fault clearing.
- The compensator must be able to regulate the bus voltage for voltage support and improved transient stability, or control it for power oscillation damping and transient stability enhancement, on a priority basis as system conditions may require.
- For a transmission line connecting two systems, the best location for var compensation is in the middle, whereas for a radial feed to a load the best location is at the load end.

As will be seen, all of the solid-state approaches for var generation and control discussed in this book can meet the steady-state and dynamic compensation requirements of power systems. However, there are considerable differences in the functional characteristics and achievable response times as well as in their evaluated capital and operating costs related to space and installation requirements, and loss evaluation criteria.

5.2 METHODS OF CONTROLLABLE VAR GENERATION

By definition, capacitors generate and reactors (inductors) absorb reactive power when connected to an ac power source. They have been used with mechanical switches for (coarsely) controlled var generation and absorption since the early days of ac power transmission. Continuously variable var generation or absorption for dynamic system compensation was originally provided by over- or under-excited rotating synchronous machines and, later, by saturating reactors in conjunction with fixed capacitors.

Since the early 1970s high power, line-commutated thyristors in conjunction with capacitors and reactors have been employed in various circuit configurations to produce variable reactive output. These in effect provide a variable shunt impedance by synchronously switching shunt capacitors and/or reactors "in" and "out" of the network. Using appropriate switch control, the var output can be controlled continuously from maximum capacitive to maximum inductive output at a given bus voltage. More recently *gate turn-off* thyristors and other power semiconductors with internal turn-off capability have been used in switching converter circuits to generate and absorb reactive power without the use of ac capacitors or reactors. These perform as ideal synchronous compensators (condensers), in which the magnitude of the internally generated ac voltage is varied to control the var output. All of the different semiconductor power circuits, with their internal control enabling them to produce var output proportional to an input reference, are collectively termed by the joint IEEE and CIGRE definition, *static var generators* (*SVG*). Thus, a *static var compensator (SVC)* is, by the IEEE CIGRE co-definition, a static var generator whose output is varied so as to maintain or control specific parameters (e.g., voltage, frequency) of the electric power system. It is important that the reader appreciate the difference between these two terms, static var generator and static var compensator, to understand the structure of this chapter. From a "black-box" viewpoint, the static var generator is a self-sufficiently functioning device that draws controllable reactive current from an alternating power source. The control input to the var generator can be an arbitrary (within the operating range) reactive current, impedance, or power reference signal that the SVG is to establish at its output. Thus, the static var generator can be viewed as a power amplifier that faithfully reproduces the reference signal at the desired power level. The functional use of the var generator is clearly defined by the reference signal provided. Consequently, according to the IEEE-CIGRE definition, a static var generator becomes a static var compensator when it is equipped with special *external* (or system) controls which derive the necessary reference for its input, from the operating requirements and prevailing variables of the power system, to execute the desired compensation of the transmission line. This means that different types of var generator can be operated with the same external control to provide substantially the same compensation functions. Evidently, the type and structure of the var generator will ultimately determine the basic operating characteristics (e.g., voltage vs. var output, response time, harmonic generation), whereas the external characteristics control the functional capabilities (e.g., voltage regulation, power factor control, power oscillation damping), of the static var compensator.

Modern static var generators are based on high-power semiconductor switching circuits. These switching circuits inherently determine some of the important operating characteristics, such as the applied voltage versus obtainable reactive output current, harmonic generation, loss versus var output, and attainable response time, setting limits for the achievable performance of the var generator and, independent of the external controls used, ultimately also that of the static var compensator. The following two main sections describe the operating principles and characteristics of the two types of static var generator presently used: those which employ thyristor-controlled reactors with fixed and/or thyristor-switched capacitors to realize a variable reactive impedance and those which employ a switching power converter to realize a controllable synchronous voltage source. Subsequent sections deal with the application requirements, structure, and operation of the external control, applicable to both types of var generator, which define the functional capabilities and operating policies of the compensator under different system conditions.

5.2.1 Variable Impedance Type Static Var Generators

The performance and operating characteristics of the impedance type var generators are determined by their major thyristor-controlled constituents: the thyristor-controlled reactor and the thyristor-switched capacitor.

5.2.1.1 The Thyristor-Controlled and Thyristor-Switched Reactor (TCR and TSR). An elementary single-phase thyristor-controlled reactor (TCR) is shown in Figure 5.7(a). It consists of a fixed (usually air-core) reactor of inductance L, and a bidirectional thyristor valve (or switch) *sw*. Currently available large thyristors can block voltage up to 4000 to 9000 volts and conduct current up to 3000 to 6000 amperes. Thus, in a practical valve many thyristors (typically 10 to 20) are connected in series to meet the required blocking voltage levels at a given power rating. A thyristor valve can be brought into conduction by simultaneous application of a gate pulse to all thyristors of the same polarity. The valve will automatically block immediately after the ac current crosses zero, unless the gate signal is reapplied.

The current in the reactor can be controlled from maximum (thyristor valve closed) to zero (thyristor valve open) by the method of firing delay angle control. That is, the closure of the thyristor valve is delayed with respect to the peak of the applied voltage in each half-cycle, and thus the duration of the current conduction intervals is controlled. This method of current control is illustrated separately for the positive and negative current half-cycles in Figure 5.7(b), where the applied voltage v and the reactor current $i_L(\alpha)$, at zero delay angle (switch fully closed) and at an arbitrary α delay angle, are shown. When $\alpha = 0$, the valve *sw* closes at the crest of the applied voltage and evidently the resulting current in the reactor will be the same as that obtained in steady state with a permanently closed switch. When the gating of the valve is delayed by an angle α ($0 \leq \alpha \leq \pi/2$) with respect to the crest of the voltage, the current in the reactor can be expressed with $v(t) = V \cos \omega t$ as follows:

$$i_L(t) = \frac{1}{L}\int_{\alpha}^{\omega t} v(t)\,dt = \frac{V}{\omega L}(\sin \omega t - \sin \alpha) \tag{5.4}$$

Since the thyristor valve, by definition, opens as the current reaches zero, (5.4) is valid for the interval $\alpha \leq \omega t \leq \pi - \alpha$. For subsequent positive half-cycle intervals the same expression obviously remains valid. For subsequent negative half-cycle intervals, the sign of the terms in (5.4) becomes opposite.

In (5.4) the term $(V/\omega L) \sin \alpha$ is simply an α dependent constant by which the sinusoidal current obtained at $\alpha = 0$ is offset, shifted down for positive, and up for negative current half-cycles, as illustrated in Figure 5.7(b). Since the valve automatically turns off at the instant of current zero crossing (which, for a lossless reactor, is symmetrical on the time axis to the instant of turn-on with respect to the peak of the current), this process actually controls the conduction interval (or angle) of the thyristor

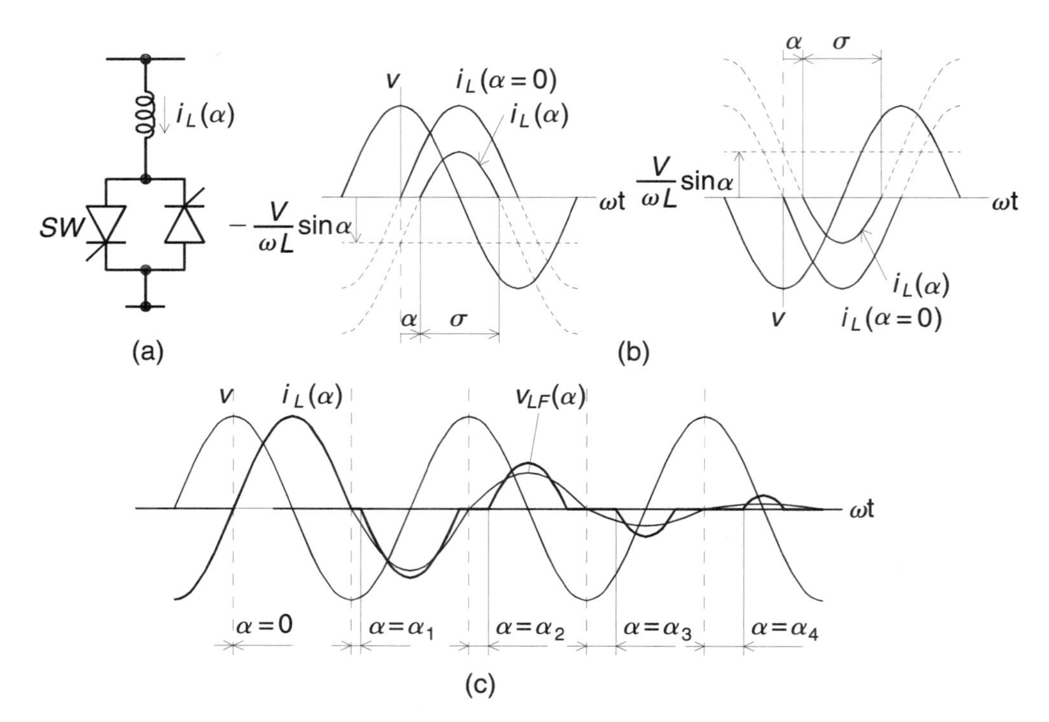

Figure 5.7 Basic thyristor-controlled reactor (a), firing delay angle control (b), and operating waveforms (c).

valve. That is, the delay angle α defines the prevailing conduction angle σ: $\sigma = \pi - 2\alpha$. Thus, as the delay angle α increases, the correspondingly increasing offset results in the reduction of the conduction angle σ of the valve, and the consequent reduction of the reactor current. At the maximum delay of $\alpha = \pi/2$, the offset also reaches its maximum of $V/\omega L$, at which both the conduction angle and the reactor current become zero. The reader should note that the two parameters, delay angle α and conduction angle σ, are equivalent and therefore TCR can be characterized by either of them; their use is simply a matter of preference. For this reason, expressions related to the TCR can be found in the literature both in terms of α and σ.

It is evident that the magnitude of the current in the reactor can be varied continuously by this method of delay angle control from maximum ($\alpha = 0$) to zero ($\alpha = \pi/2$), as illustrated in Figure 5.7(c), where the reactor current $i_L(\alpha)$, together with its fundamental component $i_{LF}(\alpha)$, are shown at various delay angles, α. Note, however, that the adjustment of current in the reactor can take place only once in each half-cycle, in the zero to $\pi/2$ interval ("gating" or "firing interval"). This restriction results in a delay of the attainable current control. The worst-case delay, when changing the current from maximum to zero (or vice versa), is a half-cycle of the applied ac voltage.

The amplitude $I_{LF}(\text{a})$ of the fundamental reactor current $i_{LF}(\alpha)$ can be expressed as a function of angle α:

$$I_{LF}(\alpha) = \frac{V}{\omega L}\left(1 - \frac{2}{\pi}\alpha - \frac{1}{\pi}\sin 2\alpha\right) \tag{5.5a}$$

where V is the amplitude of the applied ac voltage, L is the inductance of the thyristor-controlled reactor, and ω is the angular frequency of the applied voltage. The variation of the amplitude $I_{LF}(\alpha)$, normalized to the maximum current $I_{LF\max}$, ($I_{LF\max} = V/\omega L$), is shown plotted against delay angle α in Figure 5.8.

It is clear from Figure 5.8 that the TCR can control the fundamental current continuously from zero (valve open) to a maximum (valve closed) as if it was a variable reactive admittance. Thus, an effective reactive admittance, $B_L(\alpha)$, for the TCR can be defined. This admittance, as a function of angle α, can be written directly from (5.5a), i.e.,

$$B_L(\alpha) = \frac{1}{\omega L}\left(1 - \frac{2}{\pi}\alpha - \frac{1}{\pi}\sin 2\alpha\right) \tag{5.5b}$$

Evidently, the admittance $B_L(\alpha)$ varies with α in the same manner as the fundamental current $I_{LF}(\alpha)$.

The meaning of (5.5b) is that at each delay angle α an effective admittance $B_L(\alpha)$ can be defined which determines the magnitude of the fundamental current, $I_{LF}(\alpha)$, in the TCR at a given applied voltage V. In practice, the maximal magnitude of the applied voltage and that of the corresponding current will be limited by the ratings of the power components (reactor and thyristor valve) used. Thus, a practical TCR can be operated anywhere in a defined V-I area, the boundaries of which are determined by its maximum attainable admittance, voltage, and current ratings, as illustrated in Figure 5.9(a). The TCR limits are established by design from actual operating requirements.

If the TCR switching is restricted to a fixed delay angle, usually $\alpha = 0$, then it becomes a thyristor-switched reactor (TSR). The TSR provides a fixed inductive admittance and thus, when connected to the ac system, the reactive current in it will be proportional to the applied voltage as the V-I plot in Figure 5.9(b) indicates. Several

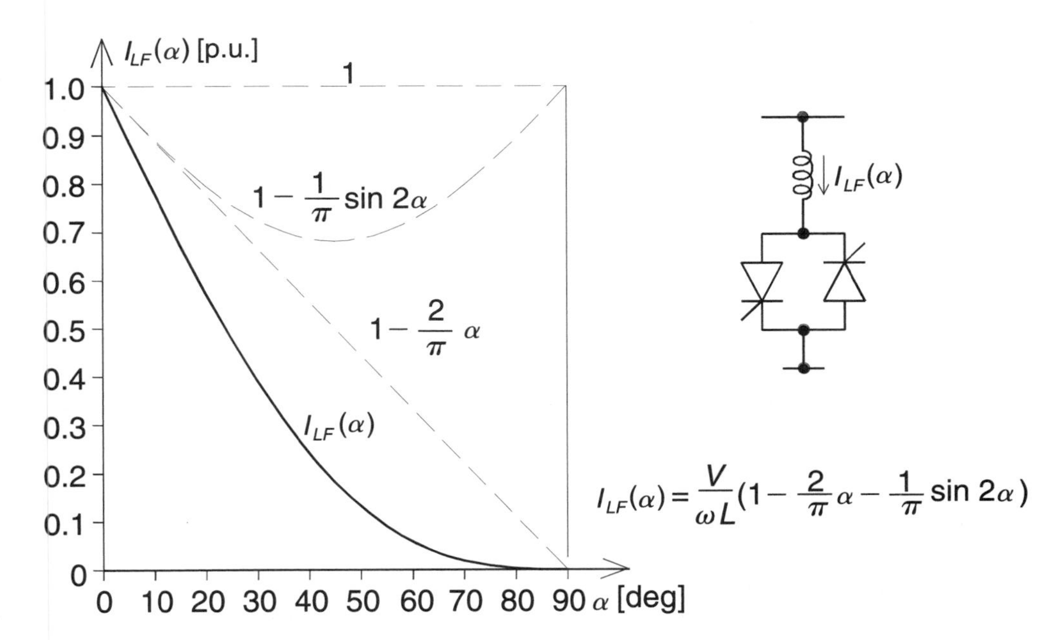

Figure 5.8 Amplitude variation of the fundamental TCR current with the delay angle α.

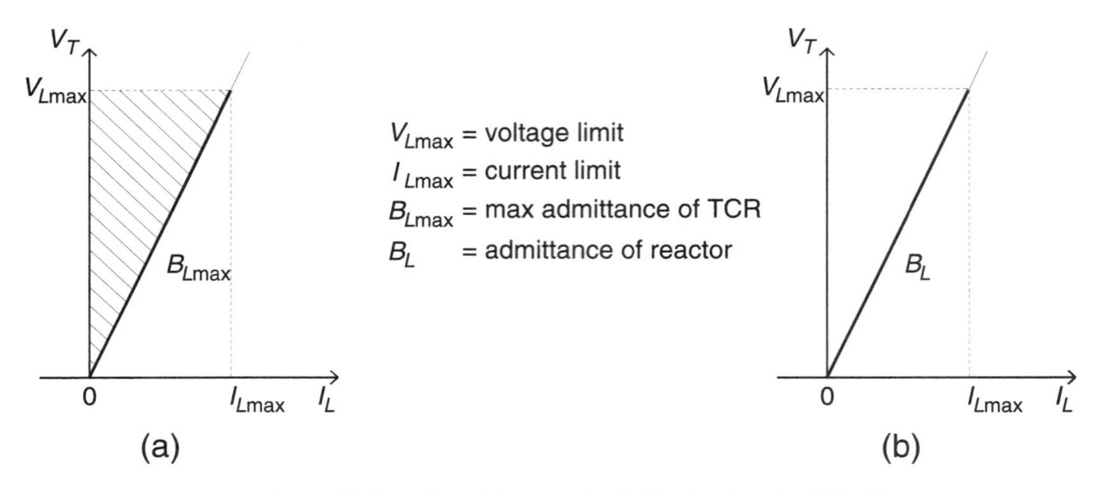

Figure 5.9 Operating *V-I* area of the TCR (a) and of the TSR (b).

TSRs can provide a reactive admittance controllable in a step-like manner. If the TSRs are operated at $\alpha = 0$, the resultant steady-state current will be sinusoidal.

Inspection of Figure 5.7(b) shows that the conduction angle control, characterizing the operation of the TCR, results in a nonsinusoidal current waveform in the reactor. In other words, the thyristor-controlled reactor, in addition to the wanted fundamental current, also generates harmonics. For identical positive and negative current half-cycles, only odd harmonics are generated. The amplitudes of these are a function of angle α, as expressed by the following equation:

$$I_{Ln}(\alpha) = \frac{V}{\omega L}\frac{4}{\pi}\left\{\frac{\sin\alpha\cos(n\alpha) - n\cos\alpha\sin(n\alpha)}{n(n^2-1)}\right\} \tag{5.6}$$

where $n = 2k + 1$, $k = 1, 2, 3, \ldots$

The amplitude variation of the harmonics, expressed as percent of the maximum fundamental current, is shown plotted against α in Figure 5.10.

In a three-phase system, three single-phase thyristor-controlled reactors are used, usually in delta connection. Under balanced conditions, the triple-n harmonic currents (3rd, 9th, 15th, etc.) circulate in the delta connected TCRs and do not enter the power system. The magnitudes of the other harmonics generated by the thyristor-controlled reactors can be reduced by various methods.

One method, particularly advantageous for high power applications, employs m ($m \geq 2$) parallel-connected TCRs, each with $1/m$ of the total rating required. The reactors are "sequentially" controlled, that is, only one of the m reactors is delay angle controlled, and each of the remaining $m - 1$ reactors is either fully "on" or fully "off," depending on the total reactive power required, as illustrated for four reactors in Figure 5.11. In this way, the amplitude of every harmonic is reduced by the factor m with respect to the maximum rated fundamental current. Furthermore, losses associated with this scheme are generally lower than those characterizing a TCR with equivalent rating due to the reduction in switching losses.

Another method employs a 12-pulse TCR arrangement. In this, two identical three-phase delta connected thyristor-controlled reactors are used, one operated from wye-connected windings, the other from delta-connected windings of the secondary

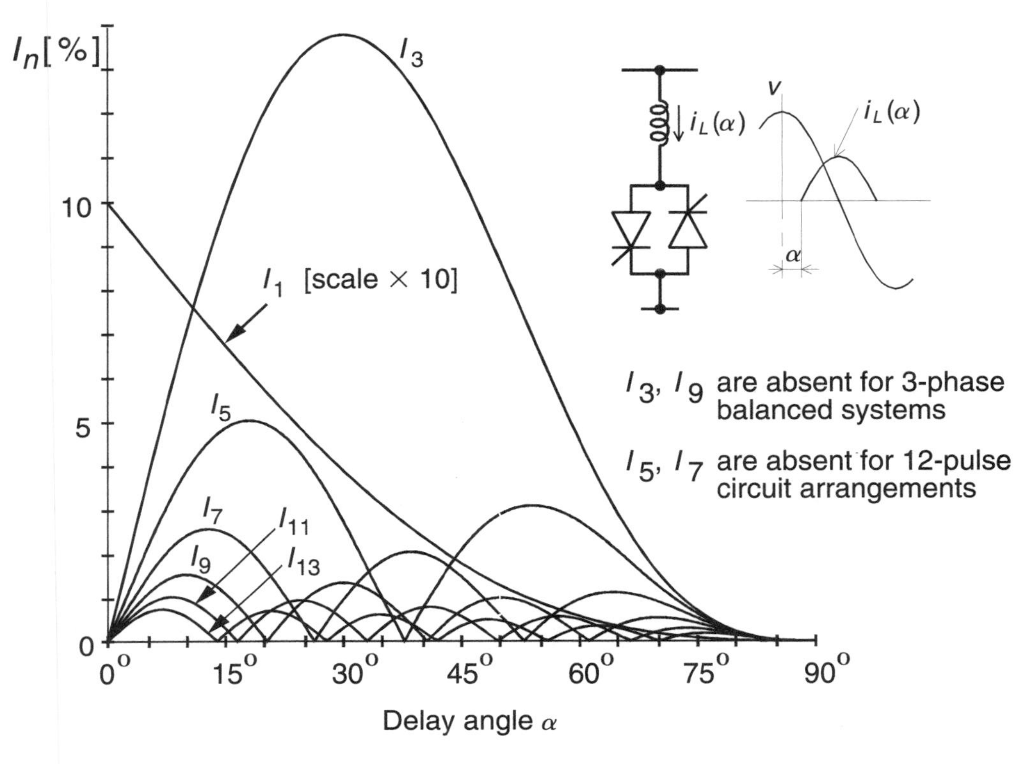

Figure 5.10 Amplitudes of the harmonic components in the current of the TCR versus the delay angle α.

of a coupling transformer. (Other types of transformer arrangements providing two sets of three-phase voltages with 30-degree phase shift can, of course, also be used.) Because of the 30-degree phase shift between the related voltages of the two transformer windings, the 5th, 7th, 17th, 19th, generally the harmonic currents of order $6(2k-1)-1$ and $6(2k-1)+1$, $k=1,2,3,\ldots$ cancel, resulting in a nearly sinusoidal output current, at all delay angles, as illustrated by the current waveforms in Figure 5.12.

Further harmonic cancellation is possible by operating three or more delta connected TCRs from appropriately phase shifted voltage sets. In practice, however, these 18 and higher pulse circuit arrangements tend to be too complex and expensive. Also, it becomes increasingly difficult to meet the requirements for symmetry, due to possible unbalance in the ac system voltages, to achieve significant reduction in the amplitudes of higher order harmonics. For these reasons, higher than 12-pulse circuit configurations are seldom used. (The reader should note that this statement generally applies only to line commutated circuits employing conventional thyristors. The output voltage waveform construction of the self-commutated power circuits is largely independent of the ac system voltages and high, typically 48- or 24-pulse voltage-sourced converter structures have been used in all existing high-power compensator installations.)

If the TCR generated harmonics cannot be reduced sufficiently by circuit arrangements, such as the four-reactor system of Figure 5.11 or the 12-pulse structure of Figure 5.12, to meet specification requirements for economic or other practical reasons (which is often the case), harmonic filters are employed. Normally, these filters are series LC and LCR branches in parallel with the TCR and are tuned to the dominant

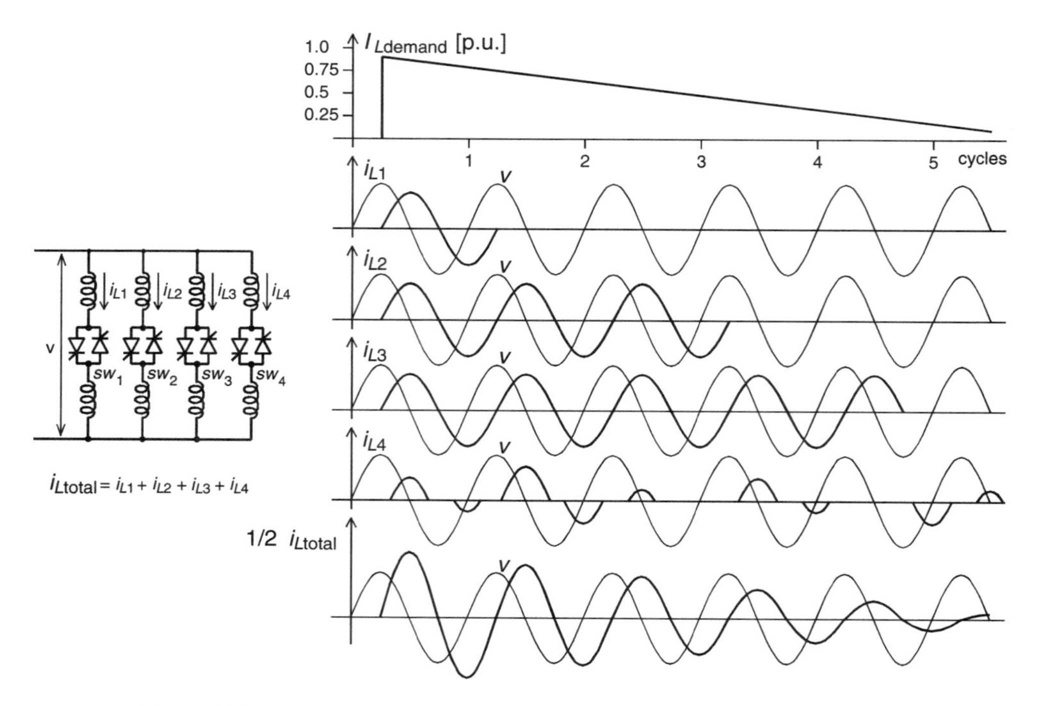

Figure 5.11 Waveforms illustrating the method of controlling four TCR banks "sequentially" to achieve harmonic reduction.

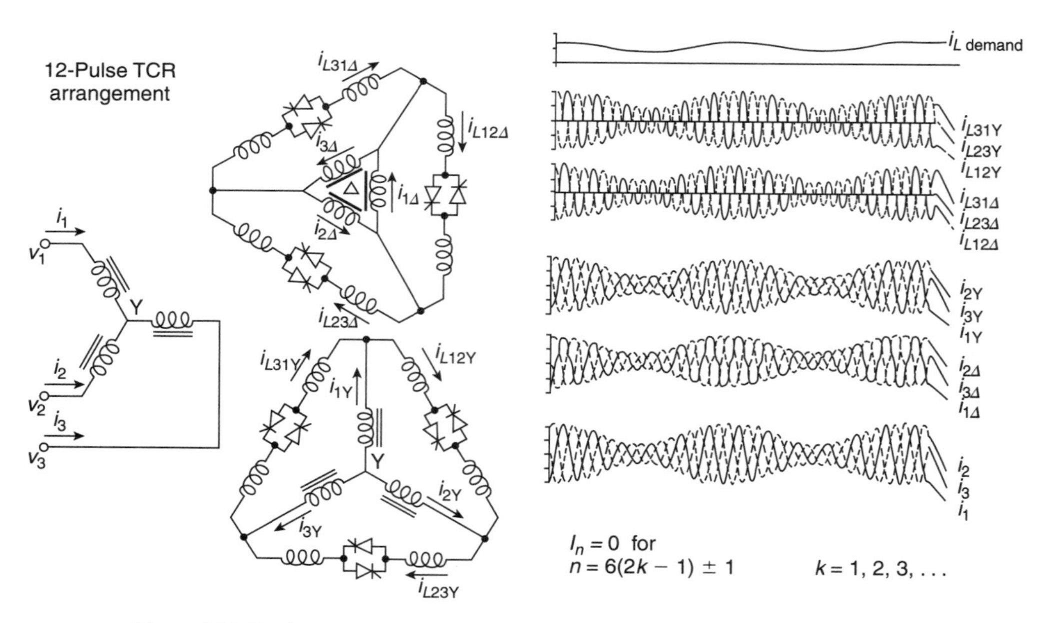

Figure 5.12 Twelve-pulse arrangement of two sets of thyristor-controlled reactors and associated current waveforms.

harmonics, such as the 5th, 7th, and occasionally, the 11th and 13th, usually with an additional high-pass branch. The high-pass filter is sometimes implemented by shunting the reactor of one of the LC filter branches with a resistor in order to maintain reasonable attenuation at higher frequencies where tuned filters are not effective. In many practical applications, due to unbalances, resonant conditions in the ac system network, or independent (single-phase) control of the three TCRs, a tuned filter branch at the third harmonic frequency may also be required.

5.2.1.2 The Thyristor-Switched Capacitor (TSC). A single-phase thyristor-switched capacitor (TSC) is shown in Figure 5.13(a). It consists of a capacitor, a bidirectional thyristor valve, and a relatively small surge current limiting reactor. This reactor is needed primarily to limit the surge current in the thyristor valve under abnormal operating conditions (e.g., control malfunction causing capacitor switching at a "wrong time," when transient free switching conditions are not satisfied); it may also be used to avoid resonances with the ac system impedance at particular frequencies.

Under steady-state conditions, when the thyristor valve is closed and the TSC branch is connected to a sinusoidal ac voltage source, $v = V \sin \omega t$, the current in the branch is given by

$$i(\omega t) = V \frac{n^2}{n^2 - 1} \omega C \cos \omega t \tag{5.7}$$

where

$$n = \frac{1}{\sqrt{\omega^2 LC}} = \sqrt{\frac{X_C}{X_L}} \tag{5.8}$$

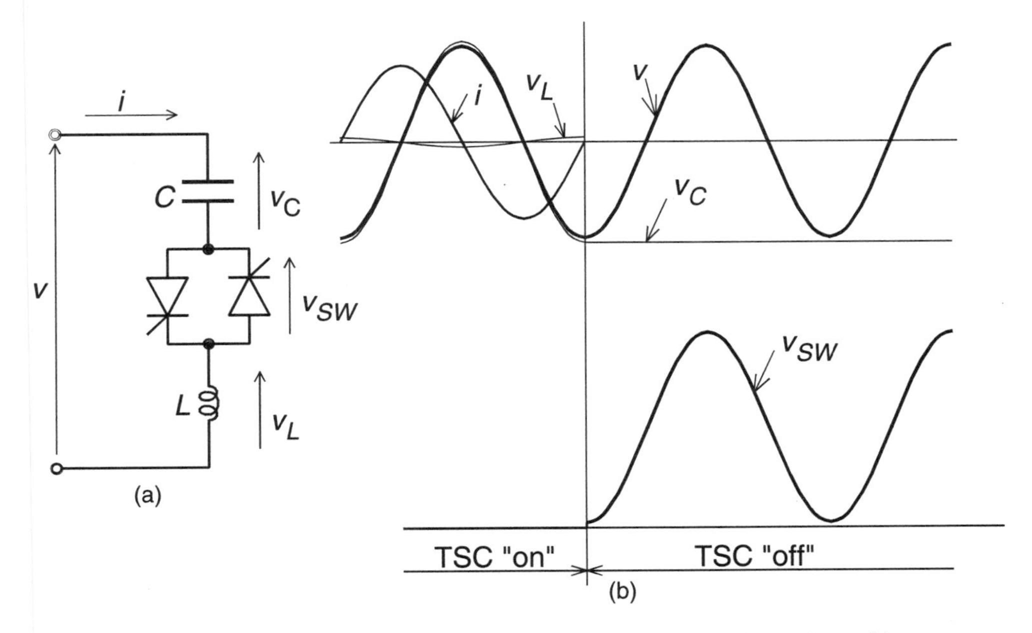

Figure 5.13 Basic thyristor-switched capacitor (a) and associated waveforms (b).

The amplitude of the voltage across the capacitor is

$$V_C = \frac{n^2}{n^2 - 1} V \tag{5.9}$$

The TSC branch can be disconnected ("switched out") at any current zero by prior removal of the gate drive to the thyristor valve. At the current zero crossing, the capacitor voltage is at its peak value, $v_{C,i=0} = Vn^2/(n^2 - 1)$. The disconnected capacitor stays charged to this voltage and, consequently, the voltage across the non-conducting thyristor valve varies between zero and the peak-to-peak value of the applied ac voltage, as illustrated in Figure 5.13(b).

If the voltage across the disconnected capacitor remained unchanged, the TSC bank could be switched in again, without any transient, at the appropriate peak of the applied ac voltage, as illustrated for a positively and negatively charged capacitor in Figure 5.14(a) and (b), respectively. Normally, the capacitor bank is discharged after disconnection. Thus, the reconnection of the capacitor may have to be executed at some residual capacitor voltage between zero and $Vn^2/(n^2 - 1)$. This can be accomplished with the minimum possible transient disturbance if the thyristor valve is turned on at those instants at which the capacitor residual voltage and the applied ac voltage are equal, that is, when the voltage across the thyristor valve is zero. Figure 5.15(a) and (b) illustrate the switching transients obtained with a fully and a partially discharged capacitor. These transients are caused by the nonzero dv/dt at the instant of switching, which, without the series reactor, would result in an instantaneous current of $i_C = Cdv/dt$ in the capacitor. (This current represents the instantaneous value of the steady-

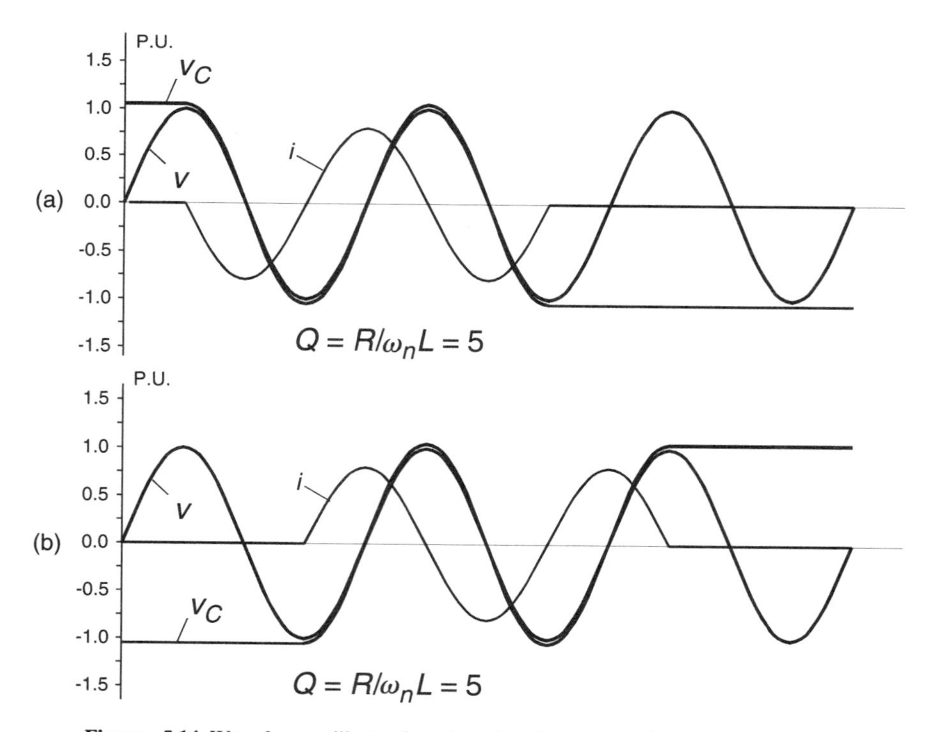

Figure 5.14 Waveforms illustrating transient-free switching by a thyristor-switched capacitor.

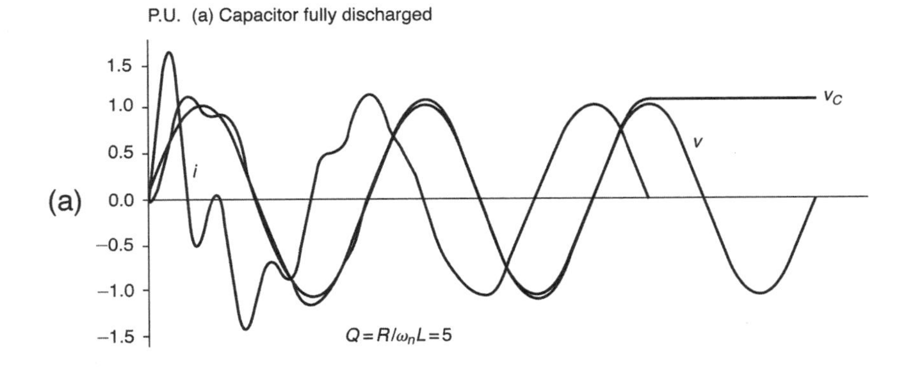

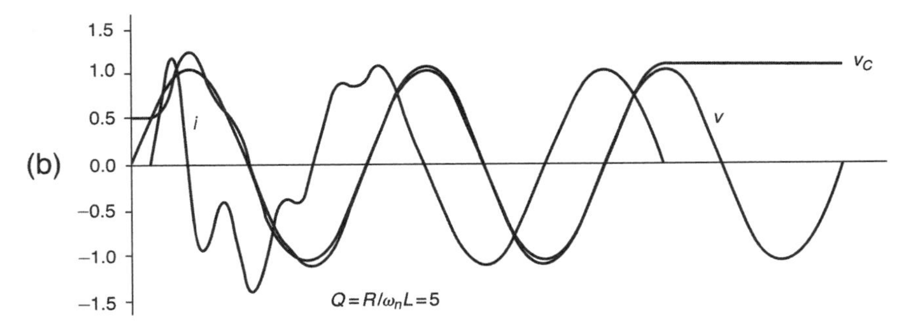

Figure 5.15 Waveforms illustrating the switching transients with the thyristor-switched capacitor fully (a) or partially discharged (b).

state capacitor current at the time of the switching.) The interaction between the capacitor and the current (and *di*/*dt*) limiting reactor, with the damping resistor, produces the oscillatory transients visible on the current and voltage waveforms. (Note that the switching transient is greater for the fully discharged than for the partially discharged capacitor because the *dv*/*dt* of the applied (sinusoidal) voltage has its maximum at the zero crossing point.)

The conditions for "transient-free" switching of a capacitor are summarized in Figure 5.16. As seen, two simple rules cover all possible cases: (1) if the residual capacitor voltage is lower than the peak ac voltage ($V_C < V$), then the correct instant of switching is when the instantaneous ac voltage becomes equal to the capacitor voltage; and (2) if the residual capacitor voltage is equal to or higher than the peak ac voltage ($V_C \geq V$), then the correct switching is at the peak of the ac voltage at which the thyristor valve voltage is minimum.

From the above, it follows that the maximum possible delay in switching in a capacitor bank is one full cycle of the applied ac voltage, that is, the interval from one positive (negative) peak to the next positive (negative) peak. It also follows that firing delay angle control is not applicable to capacitors; the capacitor switching must take place at that specific instant in each cycle at which the conditions for minimum transients are satisfied, that is, when the voltage across the thyristor valve is zero or minimum. For this reason, a TSC branch can provide only a step-like change in the reactive current it draws (maximum or zero). In other words, the TSC branch represents

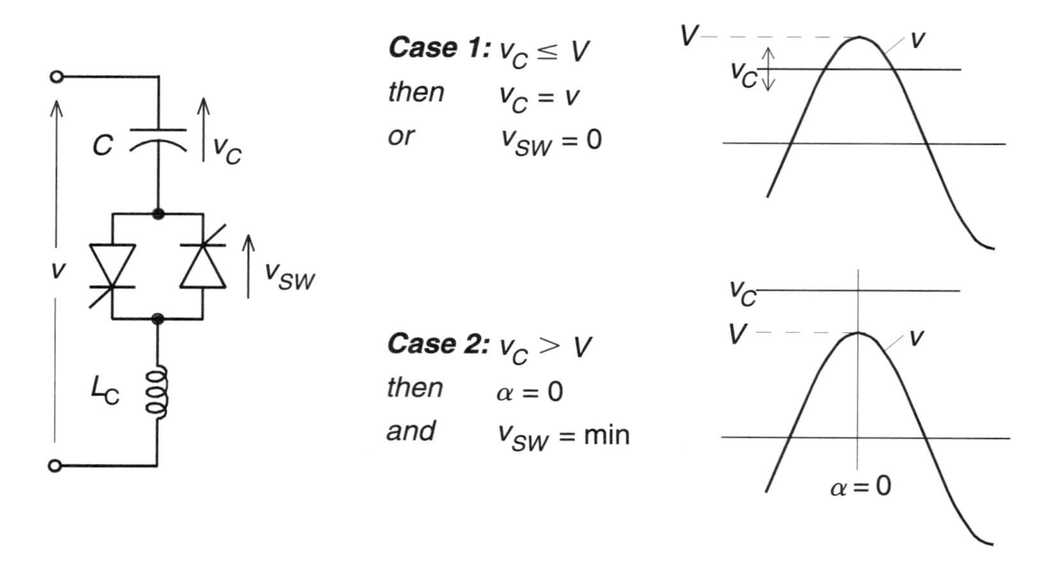

Figure 5.16 Conditions for "transient-free" switching for the thyristor-switched capacitor with different residual voltages.

a single capacitive admittance which is either connected to, or disconnected from the ac system. The current in the TSC branch varies linearly with the applied voltage according to the admittance of the capacitor as illustrated by the *V-I* plot in Figure 5.17. The maximum applicable voltage and the corresponding current are limited by the ratings of the TSC components (capacitor and thyristor valve).

To approximate continuous current variation, several TSC branches in parallel (which would increase in a step-like manner the capacitive admittance) may be employed, or, as is explained later, the TSC branches have to be complemented with a TCR.

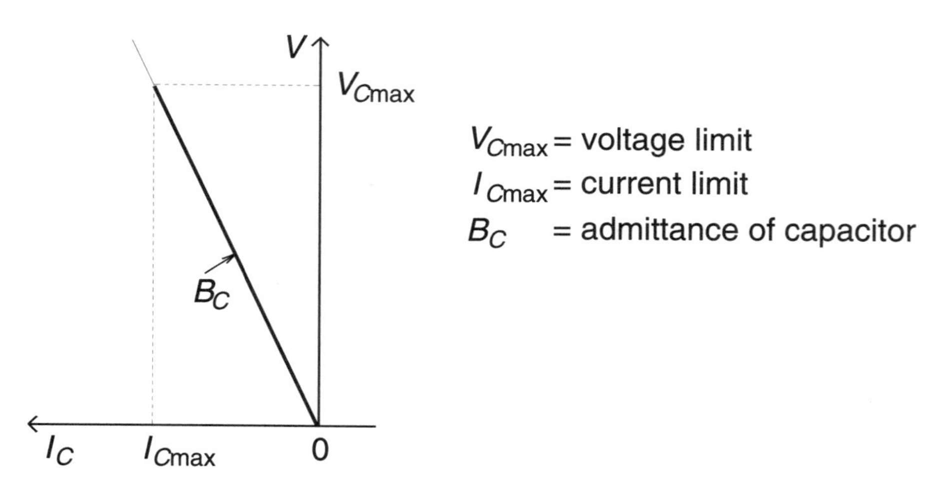

Figure 5.17 Operating *V-I* area of a single TSC.

5.2.1.3 Fixed Capacitor, Thyristor-Controlled Reactor Type Var Generator. A basic var generator arrangement using a fixed (permanently connected) capacitor with a thyristor-controlled reactor (FC-TCR) is shown functionally in Figure 5.18(a). The current in the reactor is varied by the previously discussed method of firing delay angle control. The fixed capacitor in practice is usually substituted, fully or partially, by a filter network that has the necessary capacitive impedance at the fundamental frequency to generate the reactive power required, but it provides a low impedance at selected frequencies to shunt the dominant harmonics produced by the TCR.

The fixed capacitor, thyristor-controlled reactor type var generator may be considered essentially to consist of a variable reactor (controlled by delay angle α) and a fixed capacitor, with an overall var demand versus var output characteristic as shown in Figure 5.18(b). As seen , the constant capacitive var generation (Q_C) of the fixed capacitor is opposed by the variable var absorption (Q_L) of the thyristor-controlled reactor, to yield the total var output (Q) required. At the maximum capacitive var output, the thyristor-controlled reactor is off ($\alpha = 90°$). To decrease the capacitive output, the current in the reactor is increased by decreasing delay angle α. At zero var output, the capacitive and inductive currents become equal and thus the capacitive and inductive vars cancel out. With a further decrease of angle α (assuming that the rating of the reactor is greater than that of the capacitor), the inductive current becomes larger than the capacitive current, resulting in a net inductive var output. At zero delay angle, the thyristor-controlled reactor conducts current over the full 180 degree

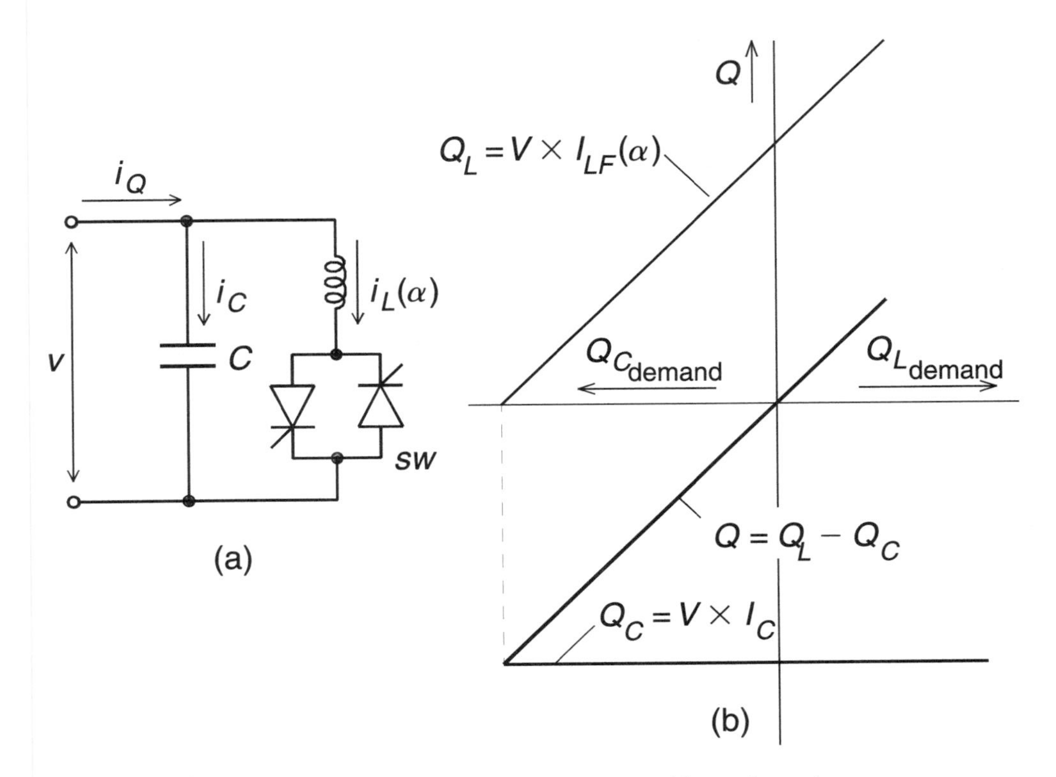

Figure 5.18 Basic FC-TCR type static var generator and its var demand versus var output characteristic.

interval, resulting in maximum inductive var output that is equal to the difference between the vars generated by the capacitor and those absorbed by the fully conducting reactor.

The control of the thyristor-controlled reactor in the FC-TCR type var generator needs to provide four basic functions, as shown in Figure 5.19(a).

One function is *synchronous timing.* This function is usually provided by a phase-locked loop circuit that runs in synchronism with the ac system voltage and generates appropriate timing pulses with respect to the peak of that voltage. (In a different approach, the ac voltage itself may be used for timing. However, this seemingly simple approach presents difficult problems during system faults and major disturbances when the voltage exhibits wild fluctuations and large distortion.)

The second function is the *reactive current* (or *admittance*) *to firing angle conversion.* This can be provided by a real time circuit implementation of the mathematical relationship between the amplitude of the fundamental TCR current $I_{LF}(\alpha)$ and the delay angle α given by (5.5). Several circuit approaches are possible. One is an analog function generator producing in each half-cycle a scaled electrical signal that represents the $I_{LF}(\alpha)$ versus α relationship. [This approach is illustrated in Figure 5.19(b).] Another is a digital "look-up table" for the normalized $I_{LF}(\alpha)$ versus α function which is read at regular intervals (e.g., at each degree) starting from $\alpha = 0$ (peak of the voltage) until the requested value is found, at which instant a firing pulse is initiated. A third approach is to use a microprocessor and compute, prior to the earliest firing angle ($\alpha = 0$), the delay angle corresponding to the required $I_{LF}(\alpha)$. The actual firing instant is then determined simply by a timing circuit (e.g., a counter) "measuring" α from the peak of the voltage.

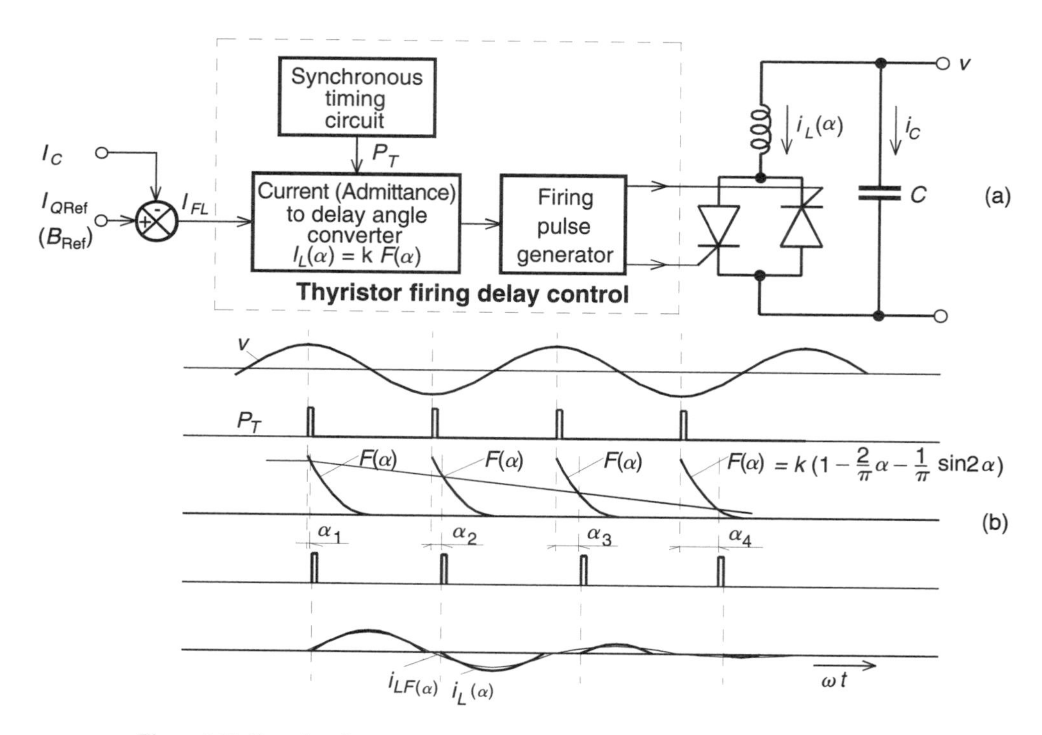

Figure 5.19 Functional control scheme for the FC-TCR type static var generator (a), and associated waveforms illustrating the basic operating principles (b).

The third function is the *computation of the required fundamental reactor current* I_{FL}, from the requested total output current I_Q (sum of the fixed capacitor and the TCR currents) defined by the amplitude reference input I_{QRef} to the var generator control. This is simply done by subtracting the (scaled) amplitude of the capacitor current, I_C from I_{QRef}. (Positive polarity for I_{QRef} means inductive output current, and negative polarity means capacitive output current.)

The fourth function is the *thyristor firing pulse generation.* This is accomplished by the firing pulse generator (or gate drive) circuit which produces the necessary gate current pulse for the thyristors to turn on in response to the output signal provided by the reactive current to firing angle converter. The gate drive circuits are sometimes at ground potential with magnetic coupling to the thyristor gates; more often, however, they are at the (high) potential level of the thyristors. In the latter case, in order to provide sufficient insulation between the ground level control and the gate drive circuits, the gating information is usually transmitted via optical fibers ("light pipes"). The operation of the FC-TCR type var generator is illustrated by the waveforms in Figure 5.19(b).

Taking a "black box" viewpoint, the FC-TCR type var generator can be considered as a controllable reactive admittance which, when connected to the ac system, faithfully follows (within a given frequency band and within the specified capacitive and inductive ratings) an arbitrary input (reactive admittance or current) reference signal. The *V-I* operating area of the FC-TCR var generator is defined by the maximum attainable capacitive and inductive admittances and by the voltage and current ratings of the major power components (capacitor, reactor, and thyristor valve), as illustrated in Figure 5.20. The ratings of the power components are derived from application requirements.

The dynamic performance (e.g., the frequency band) of the var generator is limited by the firing angle delay control, which results in a time lag or *transport lag* with respect to the input reference signal. The actual transfer function of the FC-TCR type var generator can be expressed with the transport lag in the following form:

$$G(s) = ke^{-T_d s} \tag{5.10}$$

where s is the Laplace transform operator, k is a gain constant, and T_d is the transport lag corresponding to firing delay angle α.

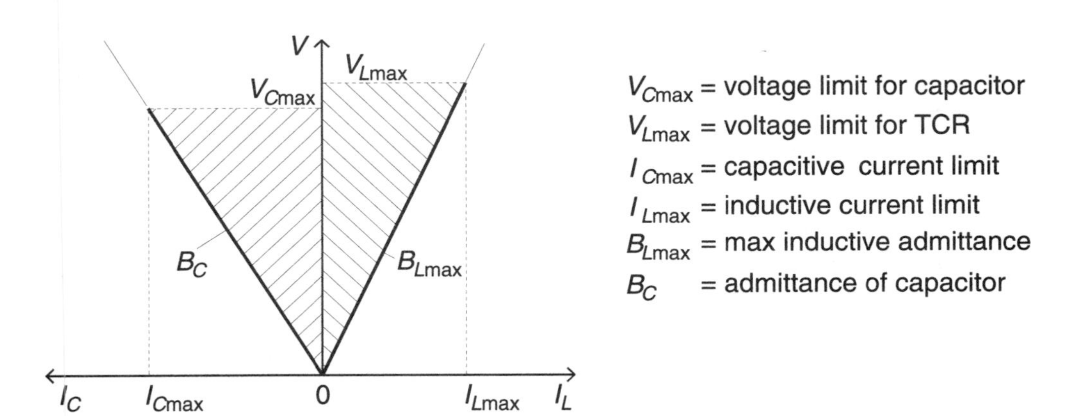

Figure 5.20 Operating *V-I* area of the FC-TCR type var generator.

The transfer function $G(s)$ can be written in a simpler from by using a first order approximation for the exponential term, that is,

$$G(s) \approx k \frac{1}{1 + T_d s} \tag{5.11}$$

Note that this approximation is optimistic and could be misleading by yielding a considerably wider bandwidth than that obtainable with the exact term (and with the actual var generator).

For a single-phase (or individually controlled three-phase) TCR, the maximum transport lag, T_d, is $1/2f = T/2$ (where f is the frequency and T is the period time of the applied ac voltage). For a three-phase, six-pulse TCR arrangement (three identical TCRs in delta) under balanced operating conditions the maximum average transport lag is $T/6$ for the increase, and it is $T/3$ for the decrease, of the reactive current. The difference in transport lag characterizing the turn-on and the turn-off delays of the TCR is due to the fact that only the start of the current conduction for the TCR is controllable. Once the conduction is initiated, the magnitude of the current cannot be changed, indeed the flow of current cannot be stopped, before the current naturally decays to zero. Thus, assuming full current in the three-phase TCR and stipulating a turn-off command immediately after the zero crossing of one of the TCR currents, it can readily be established that the three individual TCR branches would turn off in sequence after $T/6$, $T/3$, and $T/2$ successive time delays, yielding an average transport lag delay of $T/3$. With similar reasoning it can be shown easily that for a 12- (or higher) pulse TCR structure, the overall worst case transport lags for current increase and decrease would not significantly change because the turn-on and turn-off delays of the individual TCR branches would stretch over the same maximum time intervals established for the basic six-pulse TCR. However, the smoothness and continuity of transiting from one current level to another would progressively improve with the pulse number. In system studies, for simplicity, single transport lag of $T/6$ for the TCR is usually assumed which is sufficiently representative for planning and general performance evaluation.

In addition to dynamic performance, the loss versus var output characteristic of a var generator in practical applications is of major importance. In the FC-TCR type var generator, there are three major constituents of the losses encountered: (1) the capacitor (or capacitive filter) losses (these are relatively small but constant), (2) the reactor losses (these increase with the square of the current), and (3) thyristor losses (these increase almost linearly with the current). Thus, the total losses increase with increasing TCR current and, consequently, decrease with increasing capacitive var output. With reference to Figure 5.18, in the FC-TCR arrangement zero output is obtained by canceling the fixed capacitive vars with inductive vars. This, of course, means that the current in the capacitor is circulated through the reactor via the thyristor valve, resulting in appreciable no load or standby losses (about one percent of rated capacitive output). These losses decrease with increasing capacitive var output (reduced current) in the TCR and, conversely, further increase with increasing inductive var output (the TCR current is made larger than the capacitive current to produce a net inductive var output), as shown in Figure 5.21. This type of loss characteristic is advantageous when the average capacitive var output is relatively high as, for example, in industrial applications requiring power factor correction, and it is disadvantageous when the average var output is low, as for example, in the case of dynamic compensation of power transmission systems.

5.2.1.4 Thyristor-Switched Capacitor, Thyristor-Controlled Reactor Type Var Generator. The thyristor-switched capacitor, thyristor-controlled reactor (TSC-

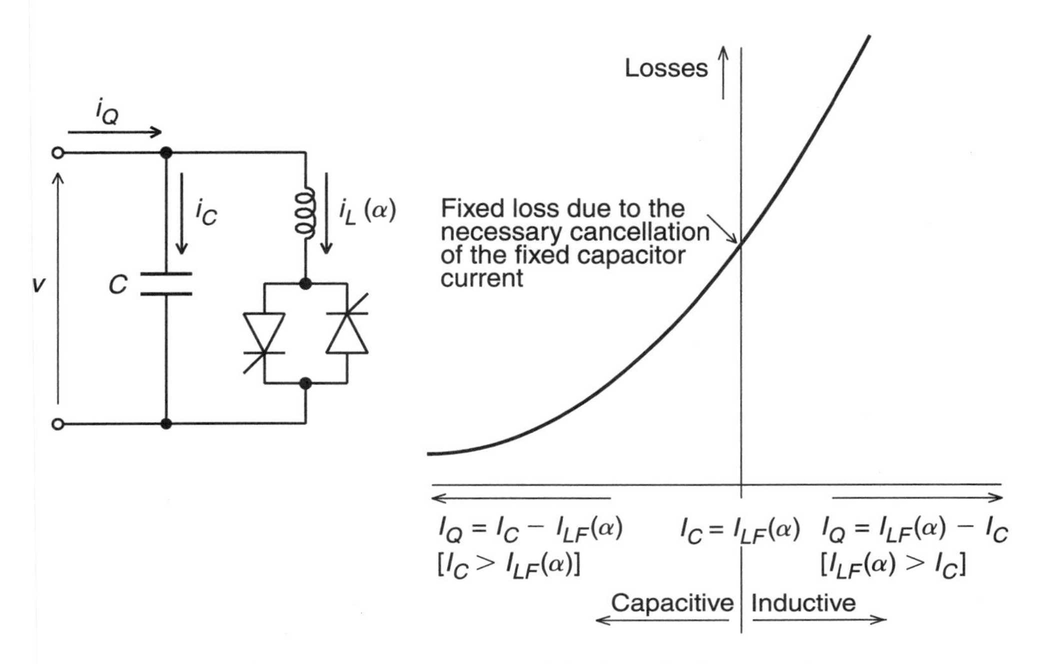

Figure 5.21 Loss versus var output characteristic of the FC-TCR type static var generator.

TCR) type compensator was developed primarily for dynamic compensation of power transmission systems with the intention of minimizing standby losses and providing increased operating flexibility.

A basic single-phase TSC-TCR arrangement is shown in Figure 5.22(a). For a given capacitive output range, it typically consists of n TSC branches and one TCR. The number of branches, n, is determined by practical considerations that include the operating voltage level, maximum var output, current rating of the thyristor valves, bus work and installation cost, etc. Of course, the inductive range also can be expanded to any maximum rating by employing additional TCR branches.

The operation of the basic TSC-TCR var generator shown in Figure 5.22(a) can be described as follows:

The total capacitive output range is divided into n intervals. In the first interval, the output of the var generator is controllable in the zero to Q_{Cmax}/n range, where Q_{Cmax} is the total rating provided by all TSC branches. In this interval, one capacitor bank is switched in (by firing, for example, thyristor valve SW_1,) and, simultaneously, the current in the TCR is set by the appropriate firing delay angle so that the sum of the var output of the TSC (negative) and that of the TCR (positive) equals the capacitive output required.

In the second, third, . . ., and nth intervals, the output is controllable in the Q_{Cmax}/n to $2Q_{Cmax}/n$, $2Q_{Cmax}/n$ to $3Q_{Cmax}/n$, . . ., and $(n-1)Q_{Cmax}/n$ to Q_{Cmax} range by switching in the second, third, . . ., and nth capacitor bank and using the TCR to absorb the surplus capacitive vars.

By being able to switch the capacitor banks in and out within one cycle of the applied ac voltage, the maximum surplus capacitive var in the total output range can be restricted to that produced by one capacitor bank, and thus, theoretically, the TCR

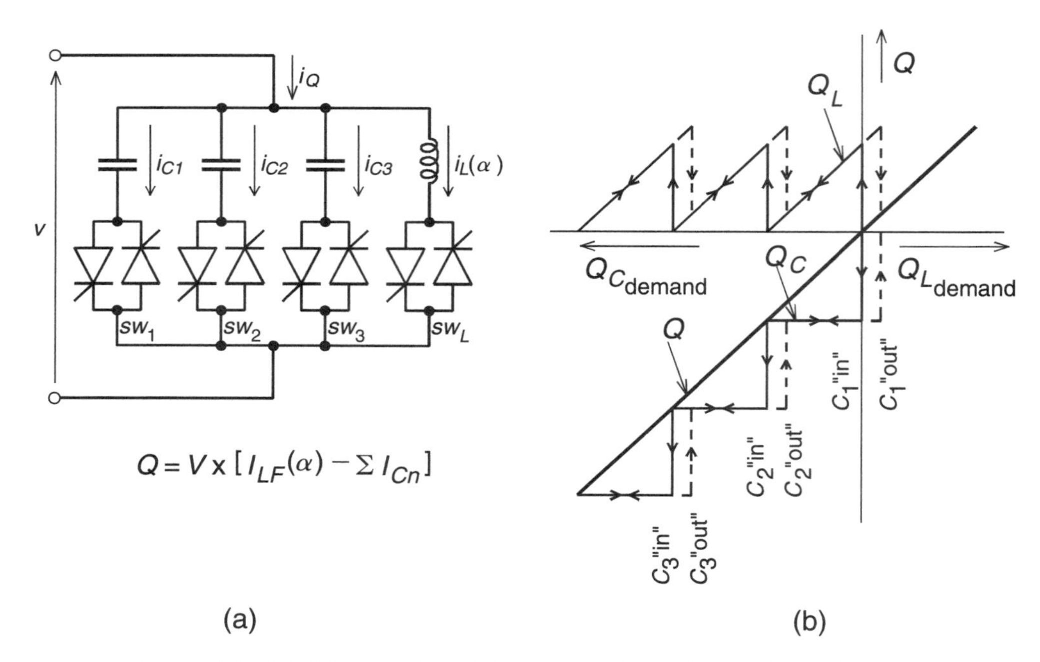

Figure 5.22 Basic TSC-TCR type static var generator and its var demand versus var output characteristic.

should have the same var rating as the TSC. However, to ensure that the switching conditions at the endpoints of the intervals are not indeterminate, the var rating of the TCR has to be somewhat larger in practice than that of one TSC in order to provide enough overlap (hysteresis) between the "switching in" and "switching out" var levels.

The var demand versus var output characteristic of the TSC-TCR type var generator is shown in Figure 5.22(b). As seen, the capacitive var output, Q_C, is changed in a step-like manner by the TSCs to approximate the var demand with a net capacitive var surplus, and the relatively small inductive var output of the TCR, Q_L, is used to cancel the surplus capacitive vars.

In a way, this scheme could be considered as a special fixed capacitor, thyristor-controlled reactor arrangement, in which the rating of the reactor is kept relatively small ($1/n$ times the maximum capacitive output), and the rating of the capacitor is changed in discrete steps so as to keep the operation of the TCR within its normal control range.

A functional control scheme for the TSC-TCR type var generator is shown in Figure 5.23. It provides three major functions:

1. Determines the number of TSC branches needed to be switched in to approximate the required capacitive output current (with a positive surplus), and computes the amplitude of the inductive current needed to cancel the surplus capacitive current.
2. Controls the switching of the TSC branches in a "transient-free" manner.
3. Varies the current in the TCR by firing delay angle control.

The first function is relatively simple. The input current reference I_{QRef} representing the magnitude of the requested output current is divided by the (scaled) amplitude

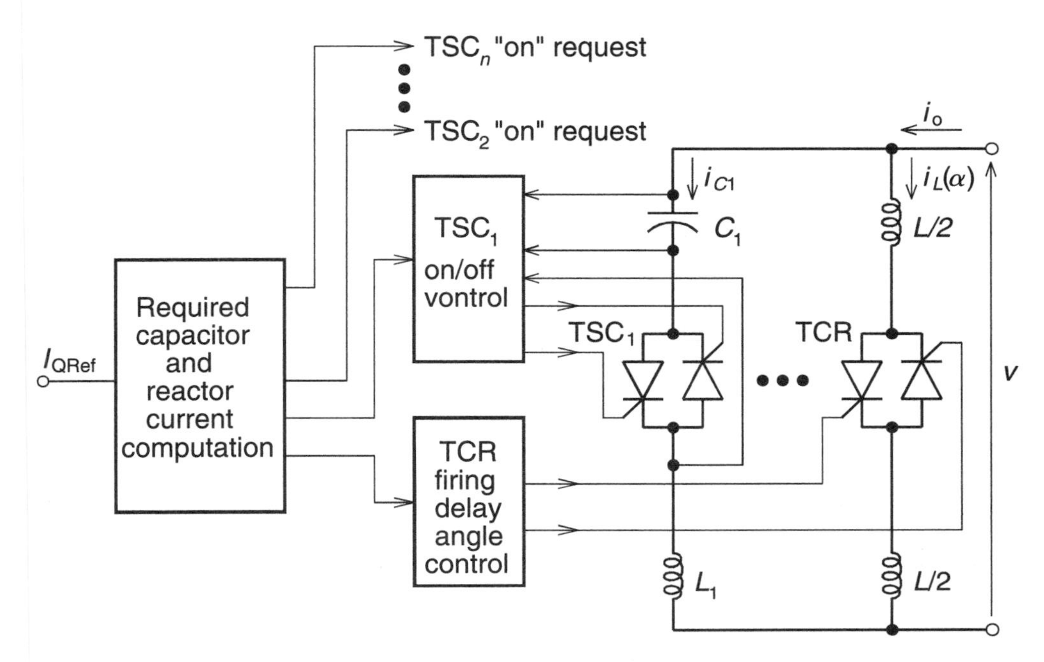

Figure 5.23 Functional control scheme for the TSC-TCR type static var generator.

I_C of the current that a TSC branch would draw at the given amplitude V of the ac voltage. The result, rounded to the next higher integer, gives the number of capacitor banks needed. The difference in magnitude between the sum of the activated capacitor currents, ΣI_{Cn}, and the reference current, I_{QRef}, gives the amplitude, I_{LF}, of the fundamental reactor current required.

The basic logic for the second function (switching of the TSC branches) is detailed in Figure 5.24. This follows the two simple rules for "transient-free" switching summarized in Figure 5.16. That is, either switch the capacitor bank when the voltage across the thyristor valve becomes zero or when the thyristor valve voltage is at a minimum. (The first condition can be met if the capacitor residual voltage is less than the peak ac voltage and the latter condition is satisfied at those peaks of the ac voltage which has the same polarity as the residual voltage of the capacitor.) The actual firing pulse generation for the thyristors in the TSC valve is similar to that used for the TCR with the exception that a continuous gate drive is usually provided to maintain continuity in conduction when the current is transferred from one thyristor string carrying current of one polarity (e.g., positive) to the other string carrying current of opposite polarity (e.g., negative).

The third function (TCR firing delay angle control) is identical to that used in the fixed-capacitor, thyristor-controlled reactor scheme (refer to Figure 5.19). The operation of the TSC-TCR type var generator with three capacitor banks is illustrated by the oscillograms in Figure 5.25. The oscillograms show the reactive current reference signal I_{QRef}, the total output current $i_Q(= i_C + i_L)$, the current i_C drawn by the thyristor-switched capacitor banks, and the current i_L drawn by the thyristor-controlled reactor.

From the "black box" viewpoint, the TSC-TCR type var generator, similarly to its FC-TCR counterpart, can be considered as a controllable reactive admittance, which, when connected to the ac system, faithfully follows an arbitrary input reference (reactive admittance or current) signal. An external observer monitoring the output

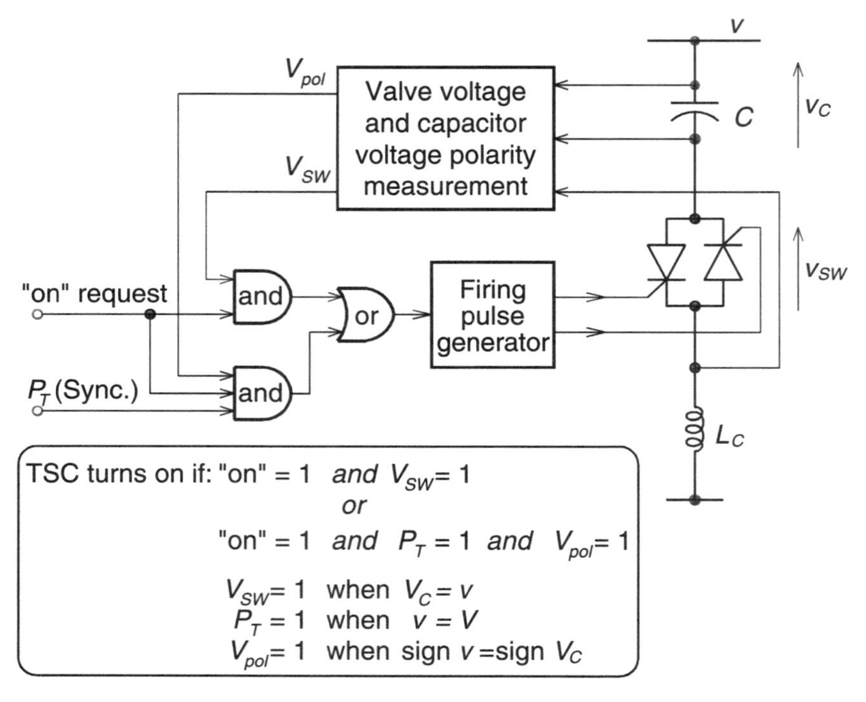

Figure 5.24 Functional logic for the implementation of "transient-free" switching strategy for the TSC.

current generally would not be able to detect (when the conditions for transient-free switching are satisfied) the internal capacitor switching; indeed, would not be able to tell whether the var generator employs fixed or thyristor-switched capacitors. The *V-I* characteristic of the TSC-TCR type generator, shown for two TSCs in Figure 5.26, is also identical to that of its FC-TCR counterpart.

The response of the TSC-TCR type var generator, depending on the number of TSC branches used, may be somewhat slower than that of its FC-TCR counterpart. This is because the maximum delay of switching in a single TSC, with a charged capacitor, is one full cycle, whereas the maximum delay of the TCR is only half of a cycle. (Note that the maximum switching out delay for both the TSC and TCR is a half-cycle.) However, if two or more TSC branches are employed, then there is a reasonable chance that, on the average, one or more capacitor banks will be available

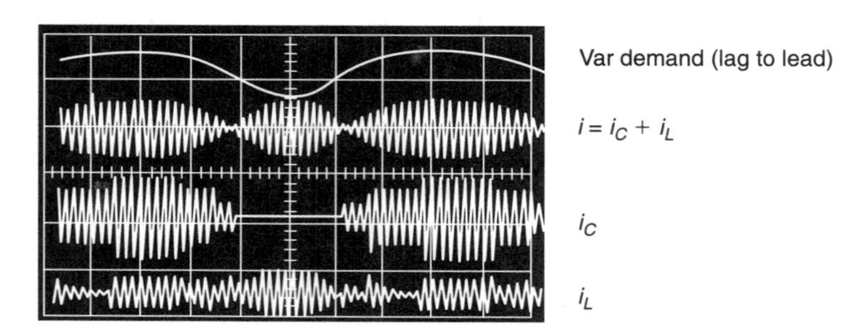

Figure 5.25 Oscillographic waveforms illustrating the operation of the TSC-TCR type static var generator.

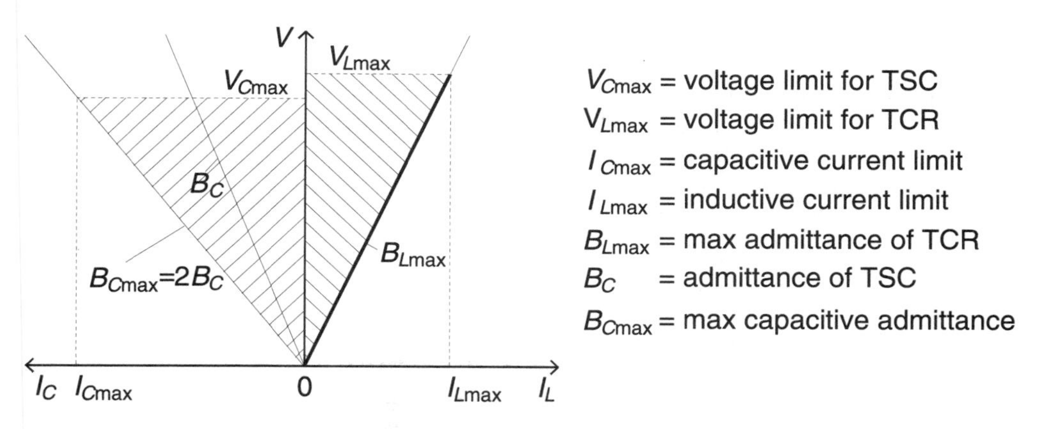

Figure 5.26 Operating *V-I* area of the TSC-TCR type var generator with two thyristor-switched capacitor banks.

with the charge of the desired polarity (or without any charge) at the instant at which an increase in capacitive output is required.

The transfer function of the TSC-TCR type var generator is the same as that of its FC-TCR counterpart [see (5.10)] except that the maximum transport lag T_d, encountered when the capacitive output is to be increased, is theoretically twice as large: it is $1/f = T$ for single phase, and $1/3f = T/3$ for balanced three-phase operation.

From the practical viewpoint, in the linear operating range the dynamic performance of the TSC-TCR type var generator in power transmission system applications is generally indistinguishable from that of its FC-TCR counterpart.

The loss versus var output characteristic of the TSC-TCR type var generator follows from its basic operating principle (refer to Figure 5.22). At or slightly below zero var output, all capacitor banks are switched out, the TCR current is zero or negligibly small, and consequently, the losses are zero or almost zero. As the capacitive output is increased, an increasing number of TSC banks are switched in with the TCR absorbing the surplus capacitive vars. Thus, with each switched-in TSC bank, the losses

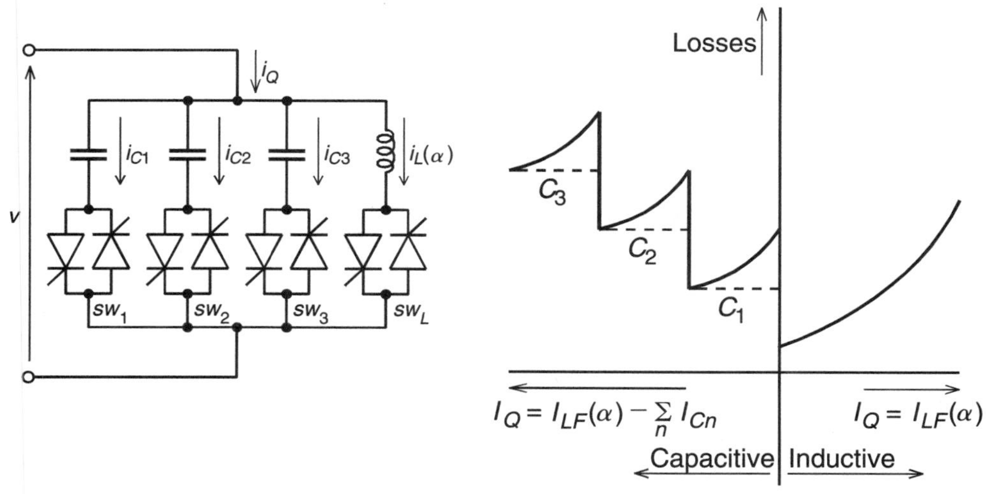

Figure 5.27 Loss versus var output characteristic of the TSC-TCR type static var generator.

increase by a fixed amount. To this fixed loss, there are the added losses of the TCR, which vary from maximum to zero between successive switchings of the TSC banks, as illustrated in Figure 5.27. Overall, the losses of the TSC-TCR type var generator vary, on the average, in proportion with the var output. This type of loss characteristic is clearly advantageous in those applications in which the var generator is used for dynamic compensation and is not required to provide high average var output for the normally functioning power system.

In order to reduce the losses of the TSC-TCR type var generator at high capacitive output, the replacement of the thyristor valves with mechanical breakers seems, at first sight, plausible. Some of the technical literature actually uses the term "mechanically-switched capacitor, thyristor-controlled reactor (MSC-TCR)." Unfortunately, although mechanically-switched capacitors can play a significant part in an overall var compensation system, the MSC-TCR arrangement does not have the response, nor the repeatability of operation that are generally needed for the dynamic compensation of power systems. In the final analysis, the response of the mechanical breakers employed will determine mostly the elapsed time between the capacitive var demand and the actual capacitive var output. Since precise and consistent control of the mechanical switch closure is not possible, the capacitor bank must be switched without any appreciable residual charge to avoid high and, possibly, damaging transients. For this reason, whenever the capacitor is switched out, it is discharged (usually via a saturating reactor) before the next switching is allowed to take place. Considering a practical discharge time of about 3–4 cycles, a typical breaker closing time of about 3–7 cycles, the MSC delay time may be 6–11 cycles, which is 6 to 11 times higher than that achievable with a TSC under worst case conditions.

The switched capacitors in a compensator are occasionally subjected to repeated switching operations. This would be the case, for example, if the var demand would repeatedly change above and below those levels at which a capacitor bank is switched in and out. Considering a typical life of 2000 to 5000 operations for mechanical breakers or switches, repeated switching of the capacitor banks in practice could not be permitted and therefore the actual var output would have to be allowed to settle (by adjusting the reactive current reference) above or below the value needed for proper compensation.

The above considerations lead to the overall conclusion that, because of the much slower response and limitations in the execution of capacitor switchings, the output of the MSC-TCR combination would not be able to follow a changing reactive current reference unless the rate of change is very low, or unless it is a single step-like change at a time when the capacitor is already discharged and ready for switching. For these reasons, a mechanically-switched arrangement—in agreement with the appropriate IEEE and CIGRE definitions—is not considered as a static var generator meeting the general requirements of dynamic compensation, and therefore it is not considered any further here. However, as already indicated, mechanically-switched reactive branches play a significant part in overall *static var systems* and will be further treated in Section 5.5 (page 205).

5.2.2 Switching Converter Type Var Generators

Static var generators discussed in the previous section generate or absorb controllable reactive power (var) by synchronously switching capacitor and reactor banks "in" and "out" of the network. The aim of this approach is to produce a variable reactive shunt impedance that can be adjusted (continuously or in a step-like manner) to meet the compensation requirements of the transmission network. The possibility of generating controllable reactive power directly, without the use of ac capacitors or reactors, by various switching power converters was disclosed by Gyugyi in 1976. These (dc to ac or ac to ac) converters are operated as voltage and current sources and they produce reactive power essentially without reactive energy storage components by circulating alternating current among the phases of the ac system. Functionally, from the standpoint of reactive power generation, their operation is similar to that of an ideal synchronous machine whose reactive power output is varied by excitation control. Like the mechanically powered machine, they can also exchange real power with the ac system if supplied from an appropriate, usually dc energy source. Because of these similarities with a rotating synchronous generator, they are termed *Static Synchronous Generators* (*SSGs*). When an SSG is operated without an energy source, and with appropriate controls to function as a shunt-connected reactive compensator, it is termed, analogously to the rotating synchronous compensator (condenser), a *Static Synchronous Compensator* (*Condenser*) or *STATCOM* (*STATCON*).

Controllable reactive power can be generated by all types of dc to ac and ac to ac switching converters. The former group is generally called *dc to ac converters* or just *converters*, whereas the latter one is referred to as *frequency changers* or *frequency converters* or *cycloconverters*. The normal function of converters is to change dc power to ac and that of frequency changers to change ac power of one frequency to ac power of another frequency. A power converter of either type consists of an array of solid-state switches which connect the input terminals to the output terminals. Consequently, a switching power converter has no internal energy storage and therefore the instantaneous input power must be equal to the instantaneous output power. Also, the termination of the input and output must be complementary, that is, if the input is terminated by a voltage source (which can be an active voltage source like a battery or a passive one like a capacitor) then the output must be terminated by a current source (which in practice would always mean a voltage source with an inductive source impedance or a passive inductive impedance) and vice versa. In the case of dc to ac converters the dc terminals are usually considered as "input" and therefore *voltage-sourced* and *current-sourced* converters are distinguished according to whether these are shunted by a voltage source (capacitor) or by a current source (inductor). Due to their practical importance in transmission applications, voltage-sourced and current-sourced converters are treated in detail in Chapters 3 and 4 to provide the necessary technical foundation for their use in FACTS Controllers. Although ac to ac power converters have application potential in FACTS Controllers, they are not considered in this book because, without unforeseen major technological advances, they are not economically viable for high power applications.

Converters presently employed in FACTS Controllers are the voltage-sourced type, but current-sourced type converters may also be used in the future. As explained in Chapters 3 and 4, and summarized here for the convenience of the reader, the major reasons for the preference of the voltage-sourced converter are: (1) Current-sourced converters require power semiconductors with bi-directional voltage blocking capability. The available high power semiconductors with gate turn-off capability (GTOs, IGBTs) either cannot block reverse voltage at all or can only do it with detrimental effect on other important parameters (e.g., increased conduction losses). (2) Practical current source termination of the converter dc terminals by a current-charged reactor is much lossier than complementary voltage source termination by a voltage-charged capacitor. (3) The current-sourced converter requires a voltage source termination at ac terminals, usually in the form of a capacitive filter. The voltage-sourced converter requires a current source termination at the ac terminals that is naturally provided by the leakage inductance of the coupling transformer. (4) The voltage source termination (i.e., a large dc capacitor) tends to provide an automatic protection of the power semiconductors against transmission line voltage transients. Current-sourced converters may require additional overvoltage protection or higher voltage rating for the semiconductors. However, the current-sourced converters have one major advantage over their voltage-sourced counterpart in that they are almost totally immune to terminal shorts due to their inherent output current limitation provided by the dc current source.

Because of the practical advantages voltage-sourced converters offer, in this and all other chapters dealing with FACTS Controllers, only this type of converter is considered, and treated only to the extent necessary to convey the operating principles and characteristics of the particular Controllers discussed. (For operating and circuit details the reader is referred to Chapter 3.) However, the operating principles and characteristics of the Controllers discussed would generally remain valid if their future implementation would be based upon current-sourced converters discussed in Chapter 4.

5.2.2.1 Basic Operating Principles. The basic principle of reactive power generation by a voltage-sourced converter is akin to that of the conventional rotating synchronous machine shown schematically in Figure 5.28. For purely reactive power flow, the three-phase induced electromotive forces (EMFs), e_a, e_b, and e_c, of the synchronous rotating machine are in phase with the system voltages, v_a, v_b, and v_c. The reactive current I drawn by the synchronous compensator is determined by the magnitude of

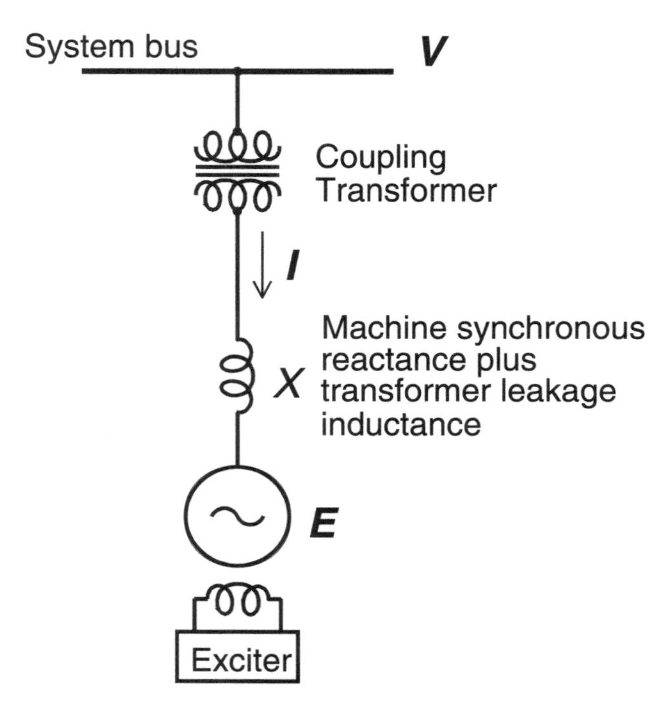

Figure 5.28 Reactive power generation by a rotating synchronous compensator (condenser).

the system voltage V, that of the internal voltage E, and the total circuit reactance (synchronous machine reactance plus transformer leakage reactance plus system short-circuit reactance) X:

$$I = \frac{V - E}{X} \tag{5.12}$$

The corresponding reactive power Q exchanged can be expressed as follows:

$$Q = \frac{1 - \frac{E}{V}}{X} V^2 \tag{5.13}$$

By controlling the excitation of the machine, and hence the amplitude E of its internal voltage relative to the amplitude V of the system voltage, the reactive power flow can be controlled. Increasing E above V (i.e., operating over-excited) results in a leading current, that is, the machine is "seen" as a capacitor by the ac system. Decreasing E below V (i.e., operating under-excited) produces a lagging current, that is, the machine is "seen" as a reactor (inductor) by the ac system. Under either operating condition a small amount of real power of course flows from the ac system to the machine to supply its mechanical and electrical losses. Note that if the excitation of the machine is controlled so that the corresponding reactive output maintains or varies a specific parameter of the ac system (e.g., bus voltage), then the machine (rotating var generator) functions as a rotating synchronous compensator (condenser).

The basic voltage-sourced converter scheme for reactive power generation is shown schematically, in the form of a single-line diagram, in Figure 5.29. From a dc

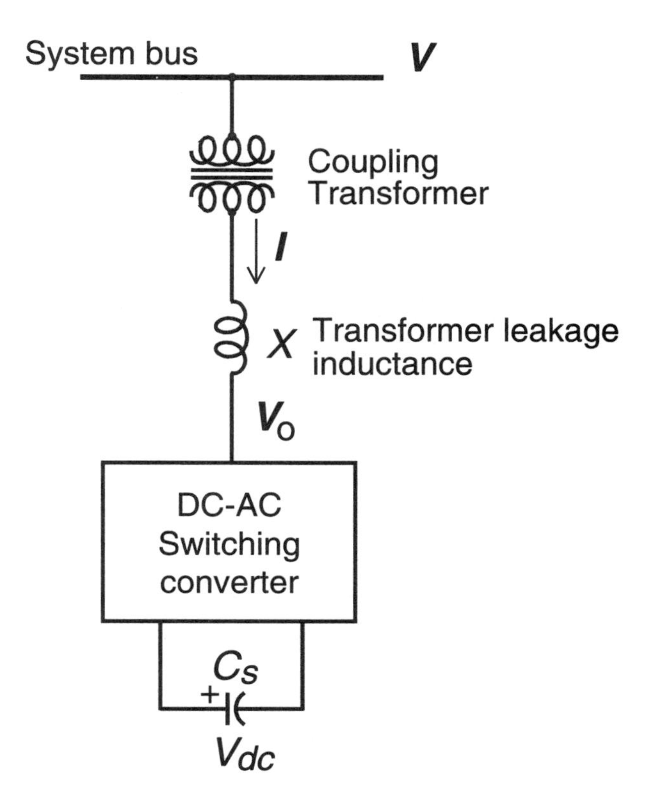

Figure 5.29 Reactive power generation by rotating a voltage-sourced switching converter.

input voltage source, provided by the charged capacitor C_S, the converter produces a set of controllable three-phase output voltages with the frequency of the ac power system. Each output voltage is in phase with, and coupled to the corresponding ac system voltage via a relatively small (0.1–0.15 p.u.) tie reactance (which in practice is provided by the per phase leakage inductance of the coupling transformer). By varying the amplitude of the output voltages produced, the reactive power exchange between the converter and the ac system can be controlled in a manner similar to that of the rotating synchronous machine. That is, if the amplitude of the output voltage is increased above that of the ac system voltage, then the current flows through the tie reactance from the converter to the ac system, and the converter generates reactive (capacitive) power for the ac system. If the amplitude of the output voltage is decreased below that of the ac system, then the reactive current flows from the ac system to the converter, and the converter absorbs reactive (inductive) power. If the amplitude of the output voltage is equal to that of the ac system voltage, the reactive power exchange is zero.

The three-phase output voltage is generated by a voltage-sourced dc to ac converter operated from an energy storage capacitor. All of the practical converters so far employed in actual transmission applications are composed of a number of elementary converters, that is, of single-phase H-bridges, or three-phase, two-level, six-pulse bridges, or three-phase, three-level, 12-pulse bridges, shown in Figure 5.30. The valves used in the elementary converters usually comprise a number (3 to 10) of series connected power semiconductors, e.g., GTO thyristors with reverse-parallel diodes. (In the case of the single-phase bridge, complete elementary converters rather than individual switching devices may be connected in series in a so-called "chain-link" circuit.) Each elementary converter, as discussed in Chapter 3, produces a square or a quasi-square or a pulse-width modulated output voltage waveform. These component voltage waveforms are phase-shifted from each other (or otherwise made complementary to each other) and then combined, usually with the use of appropriate magnetic components, to produce the final output voltage of the total converter. With sufficient

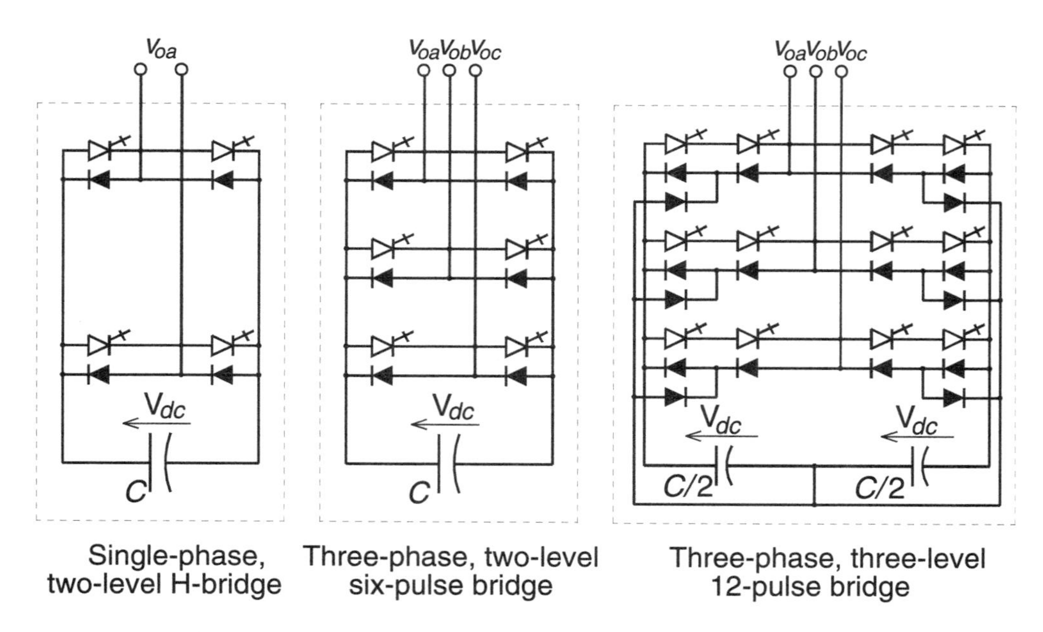

Figure 5.30 Basic converter schemes used for reactive power generation.

design, this final output voltage can be made to approximate a sine wave closely enough so that no (or a very small amount) filtering is required. For example, Figure 5.31 shows a typical 48-pulse output voltage waveform generated by the combined outputs of either eight two-level, six-pulse, or four three-level, 12-pulse converters. In this and subsequent FACTS Controller chapters, the waveform construction aspects of the converter will not be any further discussed (the interested reader should refer back to Chapter 3) and the term "converter" here on will imply a complete converter structure capable of producing a substantially harmonic free output voltage.

The operation of the voltage-sourced converter, used as a controllable static var generator, can be explained without considering the detailed operation of the converter valves by basic physical laws governing the relationship between the output and input powers. The key to this explanation resides in the physical fact that, like in all switching power converters, the net instantaneous power at the ac output terminals must always be equal to the net instantaneous power at the dc input terminals (neglecting the losses in the semiconductor switches).

Since the converter supplies only reactive output power (its output voltages are controlled to be in phase with the ac system voltages), the real input power provided by the dc source (charged capacitor) must be zero (as the total instantaneous power on the ac side is also zero). Furthermore, since reactive power at zero frequency (at the dc capacitor) by definition is zero, the dc capacitor plays no part in the reactive power generation. In other words, the converter simply interconnects the three ac terminals in such a way that the reactive output currents can flow freely between them. Viewing this from the terminals of the ac system, one could say that the converter establishes a circulating current flow among the phases with zero net instantaneous power exchange.

The need for the dc storage capacitor is theoretically due to the above stipulated equality of the instantaneous output and input powers. The output voltage waveform of the dc to ac converter is not a perfect sine wave. For this reason, the net instantaneous output power (VA) has a fluctuating component even if the converter output currents

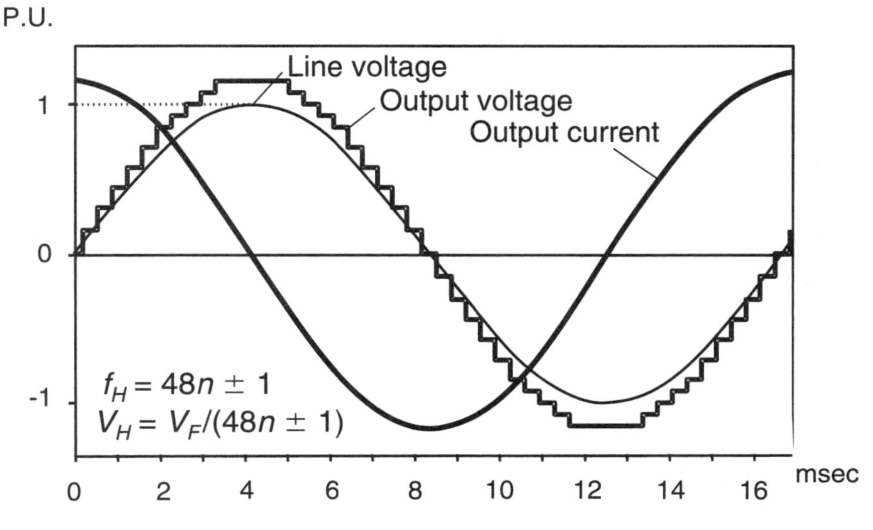

Figure 5.31 Typical output voltage and current waveforms of a 48-pulse converter generating reactive power.

were pure sine waves (which condition is approximated quite closely in practical systems). Thus, in order not to violate the equality of instantaneous output and input powers, the converter must draw a fluctuating ("ripple") current from the dc storage capacitor that provides a constant voltage termination at the input.

The presence of the input ripple current components is thus entirely due to the ripple components of the output voltage, which are a function of the output waveform fabrication method used. In a practical var generator, as explained above, the elementary two- or three-level converters would not meet practical harmonic requirements either for the output voltage or for the input (dc capacitor) current. However, by combining a number of these basic converters into a multi-pulse structure (and/or using appropriate pulse-width modulation—PWM—or other wave shaping techniques), the output voltage distortion and capacitor ripple current can be theoretically reduced to any desired degree. Thus, a static (var) generator, employing a perfect voltage-sourced converter, would produce sinusoidal output voltages, would draw sinusoidal reactive currents from the ac system and zero input current from the dc capacitor. In practice, due to system unbalance and other imperfections (which could increase the ac power fluctuation considerably), as well as economic restrictions, these ideal conditions are not achieved, but can be approximated quite satisfactorily by appropriate converter structures and wave shaping techniques so that the size of the dc capacitor in normal transmission applications remains relatively small.

In a practical converter, the semiconductor switches are not lossless, and therefore the energy stored in the dc capacitor would be used up by the internal losses. However, these losses can be supplied from the ac system by making the output voltages of the converter lag the ac system voltages by a small angle. In this way the converter absorbs a small amount of real power from the ac system to replenish its internal losses and keep the capacitor voltage at the desired level. The mechanism of phase angle adjustment can also be used to control the var generation or absorption by increasing or decreasing the capacitor voltage, and thereby the amplitude of the output voltage produced by the converter. (Recall that the amplitude difference between the converter output voltage and ac system voltage determines completely the magnitude and direction of the reactive current flow, and thus the var generation or absorption produced.) The dc capacitor also has a vital function, even in the case of a perfect converter, in establishing the necessary energy balance between the input and output during the dynamic changes of the var output.

It is, of course, also possible to equip the converter with a dc source (e.g., a battery) or with an energy storage device of significant capacity (e.g., a large dc capacitor, or a superconducting magnet). In this case the converter can control both reactive and real power exchange with the ac system, and thus it can function as a static synchronous generator. The capability of controlling real as well as reactive power exchange is a significant feature which can be used effectively in applications requiring power oscillation damping, leveling peak power demand, and providing uninterrupted power for critical loads. This capability is unique to the switching converter type var generator and it fundamentally distinguishes it from its conventional thyristor-controlled counterpart.

5.2.2.2 Basic Control Approaches. A static (var) generator converter comprises a large number of gate-controlled semiconductor power switches (GTO thyristors). The gating commands for these devices are generated by the *internal* converter control (which is part of the var generator proper) in response to the demand for reactive and/or real power reference signal(s). The reference signals are provided by the *external* or *system control*, from operator instructions and system variables, which determine the functional operation of the STATCOM.

The internal control is an integral part of the converter. Its main function is to operate the converter power switches so as to generate a fundamental output voltage waveform with the demanded magnitude and phase angle in synchronism with the ac system. In this way the power converter with the internal control can be viewed as a sinusoidal, synchronous voltage source behind a tie reactor (provided by the leakage inductance of the coupling transformer), the amplitude and phase angle of which are controlled by the external (STATCOM system) control via appropriate reference signal(s).

There are many tasks the internal control executes in operating the power switches of the converter which are not important from the application viewpoint. These are mostly related to keeping the individual power semiconductors within their maximum voltage and current limits, and otherwise to maintaining the safe operation of the converter under all system conditions

The main function of the internal control, as stated above, is to operate the converter power switches so as to produce a synchronous output voltage waveform that forces the reactive (and real) power exchange required for compensation. As illustrated schematically in Figure 5.32, the internal control achieves this by computing the magnitude and phase angle of the required output voltage from I_{QRef} (and I_{PRef}) provided by the external control and generating a set of coordinated timing waveforms ("gating pattern"), which determines the on and off periods of each switch in the converter corresponding to the wanted output voltage. These timing waveforms have a defined phase relationship between them, determined by the converter pulse number, the method used for constructing the output voltage waveform, and the required angular phase relationship between the three outputs (normally 120 degrees).

The magnitude and angle of the output voltage are those internal parameters which determine the real and reactive current the converter draws from, and thereby the real and reactive power it exchanges with the ac system. If the converter is restricted for reactive power exchange, i.e., it is strictly operated as a static var generator, then the reference input to the internal control is the required reactive current. From this the internal control derives the necessary magnitude and angle for the converter output voltage to establish the required dc voltage on the dc capacitor since the magnitude of the ac output voltage is directly proportional to the dc capacitor voltage. Because of this proportionality, the reactive output current, as one approach, can be controlled

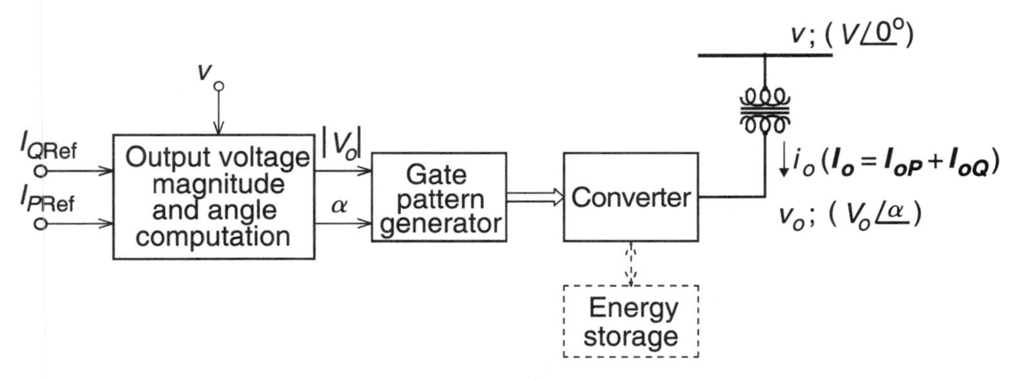

Figure 5.32 Main functions of the internal converter control.

indirectly via controlling the dc capacitor voltage (which in turn is controlled by the angle of the output voltage) or, as another approach, directly by the internal voltage control mechanism (e.g., PWM) of the converter in which case the dc voltage is kept constant (by the control of the angle). These two basic approaches to output voltage, and thus to var control are illustrated in Figures 5.33 and 5.34 for an assumed two-level converter ("indirect" control) and for a three-level converter ("direct" control), respectively. Note that if the converter is equipped with an energy storage device, then the internal control can accept an additional real current reference, which would control the angle of the output voltage so as to establish a real component of current in the output as demanded by this reference.

From the "black box" viewpoint the converter-based static var generator can be viewed as a synchronous voltage source that can be controlled to draw either capacitive or inductive current up to a maximum value determined by its MVA rating. It is important to note that the maximum reactive current can be maintained even if the system voltage is significantly depressed from its nominal value.

A simplified block diagram of the internal control for purely reactive compensation, based on the indirect approach of dc capacitor voltage control as illustrated in Figure 5.33, is shown in Figure 5.35. The inputs to the internal control are: the ac system bus voltage, v, the output current of the converter, i_o, and the reactive current reference, I_{QRef}. Voltage v operates a *phase-locked loop* that provides the basic synchronizing signal, angle θ. The output current, i_o, is decomposed into its reactive and real components, and the magnitude of the reactive current component, I_{oQ}, is compared to the reactive current reference, I_{QRef}. The error thus obtained provides, after suitable amplification, angle α, which defines the necessary phase shift between the output

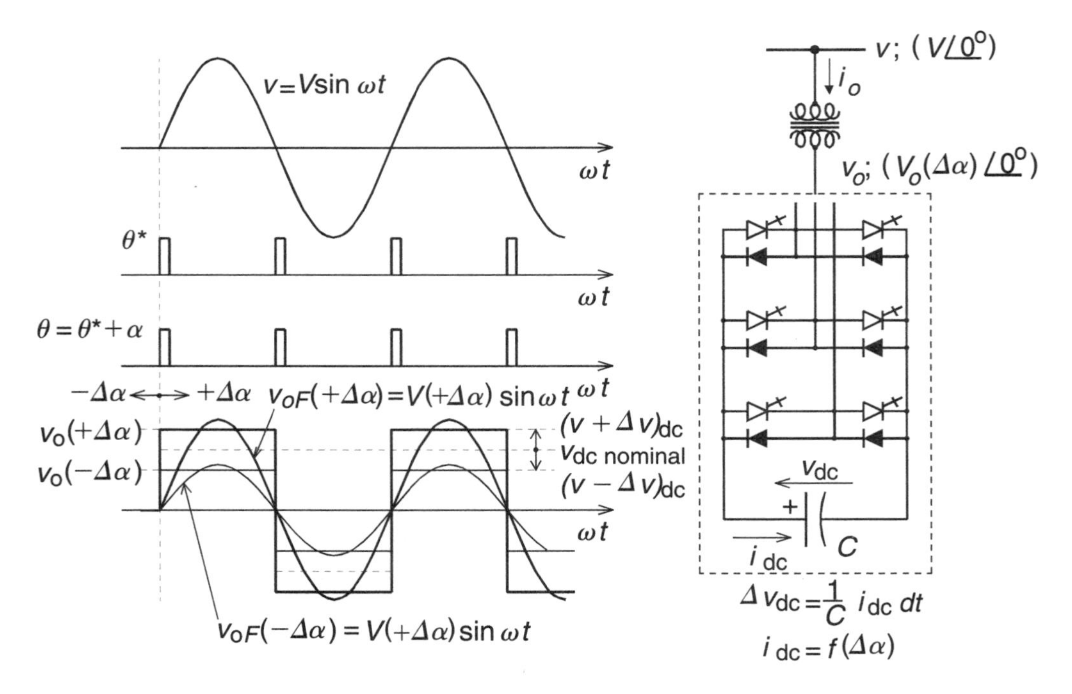

Figure 5.33 "Indirect" output voltage control by varying the dc capacitor voltage through the temporary phase shift of the output voltage.

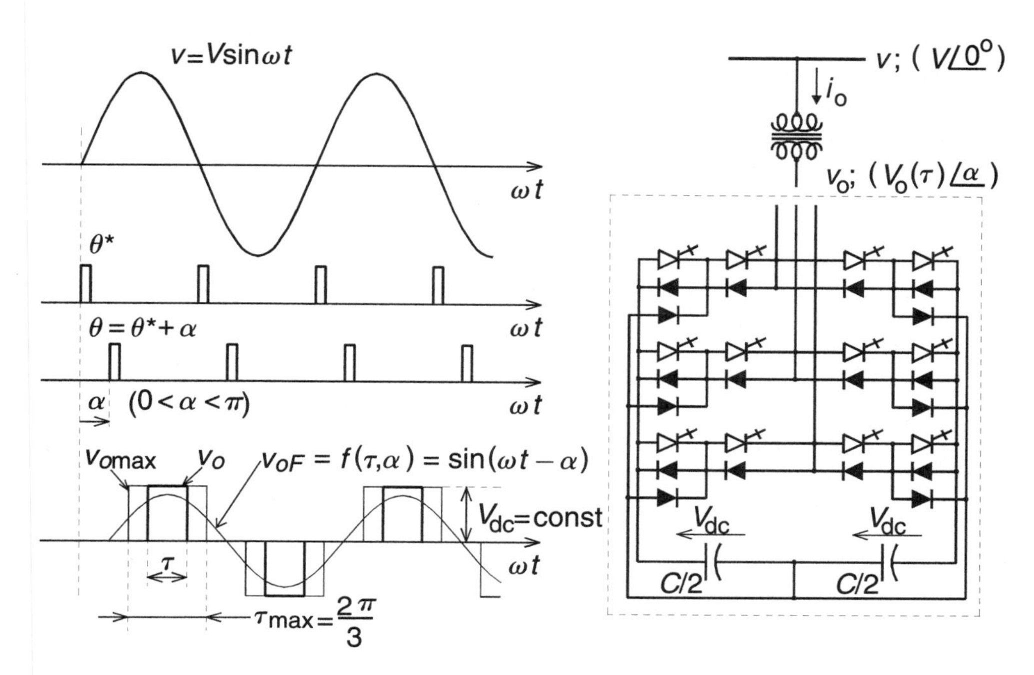

Figure 5.34 "Direct" output voltage control by the variation of the midpoint (zero-level) intervals of a three-level converter.

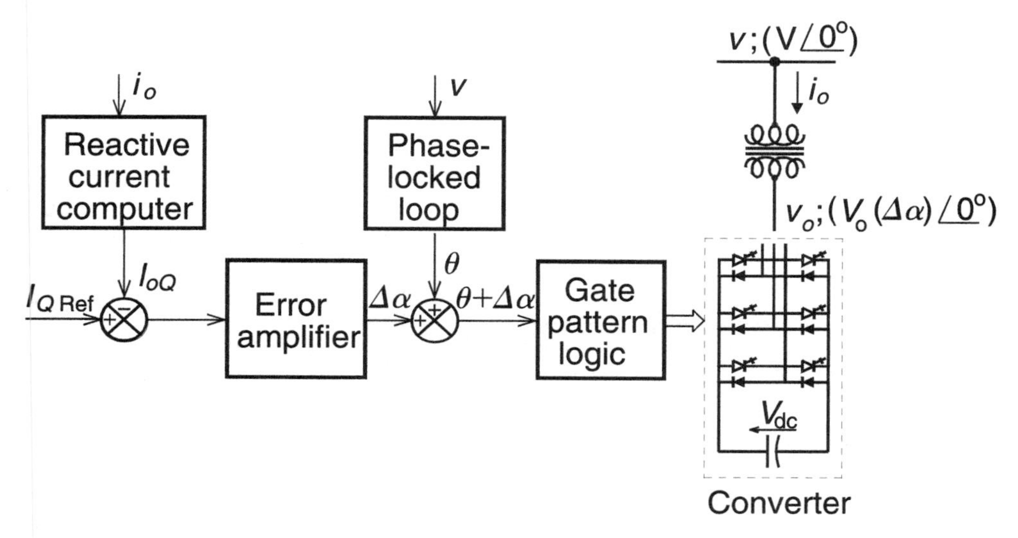

Figure 5.35 Basic control scheme for the voltage-sourced converter type var generator controlling the reactive output by the variation of the dc capacitor voltage ("indirect" output voltage control).

voltage of the converter and the ac system voltage needed for charging (or discharging) the storage capacitor to the dc voltage level required. Accordingly, angle α is summed to θ to provide angle $\theta + \alpha$, which represents the desired synchronizing signal for the converter to satisfy the reactive current reference. Angle $\theta + \alpha$ operates the gate pattern logic (which may be a digital look-up table) that provides the individual gate drive logic signals to operate the converter power switches.

A simplified block diagram of the internal control for a converter with internal voltage control capability, such as the three-level converter illustrated in Figure 5.34, is shown in Figure 5.36. The input signals are again the bus voltage, v, the converter output current, i_o, and the reactive current reference, I_{QRef}, plus the dc voltage reference V_{dc}. This dc voltage reference determines the real power the converter must absorb from the ac system in order to supply its internal losses. As the block diagram illustrates, the converter output current is decomposed into reactive and real current components. These components are compared to the external reactive current reference (determined from compensation requirements) and the internal real current reference derived from the dc voltage regulation loop. After suitable amplification, the real and reactive current error signals are converted into the magnitude and angle of the wanted converter output voltage, from which the appropriate gate drive signals, in proper relationship with the phase-locked loop provided phase reference, are derived. Note that this internal control scheme could operate the converter with a dc power supply or energy storage as a *static synchronous generator*. In this case the internal real current reference would be summed to an externally provided real current reference that would indicate the desired real power exchange (either positive or negative) with the ac system. The combined internal and external real current references (for converter losses and active power compensation), together with the prevailing reactive current demand, would determine the magnitude and angle of the output voltage generated, and thus the real and reactive power exchanged with the ac system.

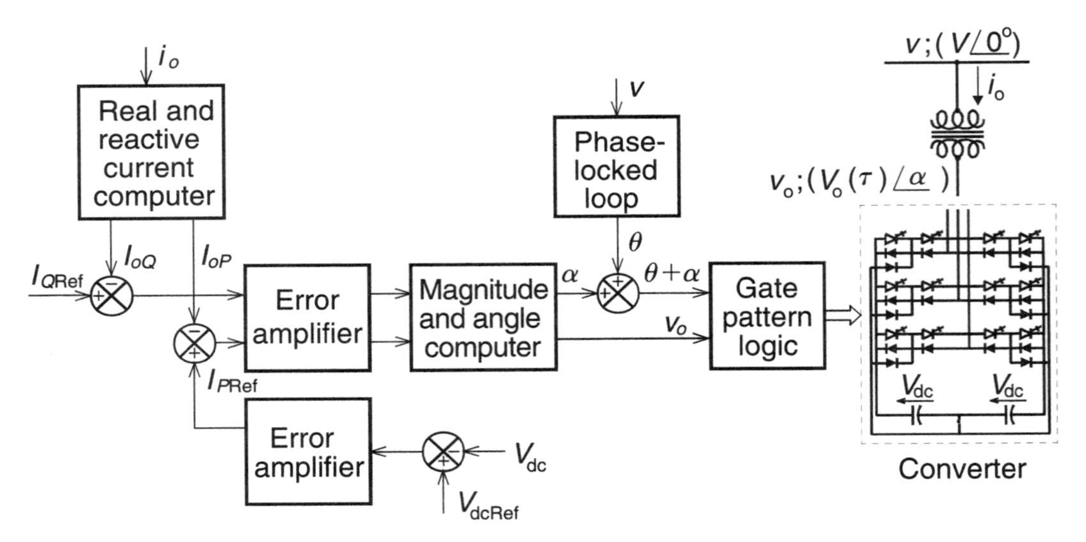

Figure 5.36 Basic control scheme for the voltage-sourced converter type var generator controlling the reactive output by internal voltage (magnitude and angle) control at a sustained dc capacitor voltage ("direct" output voltage control).

From the "black box" viewpoint, the voltage-sourced converter type var generator can be considered a synchronous voltage source which will draw reactive current from the ac system according to an external reference which may vary in a defined range between the same capacitive and inductive maxima, independent of the ac system voltage. The *V-I* operating area of this var generator is limited only by the maximum voltage and current ratings of the converter, as illustrated in Figure 5.37. (Note that there would also be a low voltage—about 0.2 p.u.—limit at which the converter still would be able to absorb the necessary real power from the ac system to supply its operating losses.)

The dynamic response of this type of var generator, due to its almost negligible transport lag, is generally much faster (about an order of magnitude) than that attainable with its variable impedance (FC-TCR and TSC-TCR) counterparts. There are several reasons for this. The main reason is that both the turn-on and turn-off instants of the semiconductors (GTOs) used in the converter, and thus both the starting and ending edges of the converter produced waveform, can be controlled, whereas only the turn-on instants of the thyristors in the TCR and TSC banks, and thus only the starting edge of the current waveform they draw, are controllable. (In the case of the TSC, even the turn-on is restricted to a single instant in each cycle.) The turn-on and turn-off capability of the semiconductors also facilitates optional operating modes for the converter, e.g., pulse-width modulation, waveform "notching," multilevel waveform construction, which are not possible with the naturally commutated TCRs and TSCs. Another significant reason is that only about 15% change in the magnitude of the converter output voltage changes the reactive output current 100%. (The TCR and TSC must transit from full on to full off, or vice versa, to achieve this.) In addition, the converter-based var generator also has the unique capability to increase or decrease the voltage of the dc capacitor (and thereby the magnitude of all three phases of the output voltage) by a single "shot" of short duration real power exchange with the ac system, in addition to using its internal voltage control. It is also helpful that for high power transmission applications the voltage-sourced converters for a variety of technical and economic reasons are of multipulse (24 or higher) construction, which, because of the above reasons (the large effect of small voltage change on reactive output current), can also significantly contribute to the improvement of the attainable

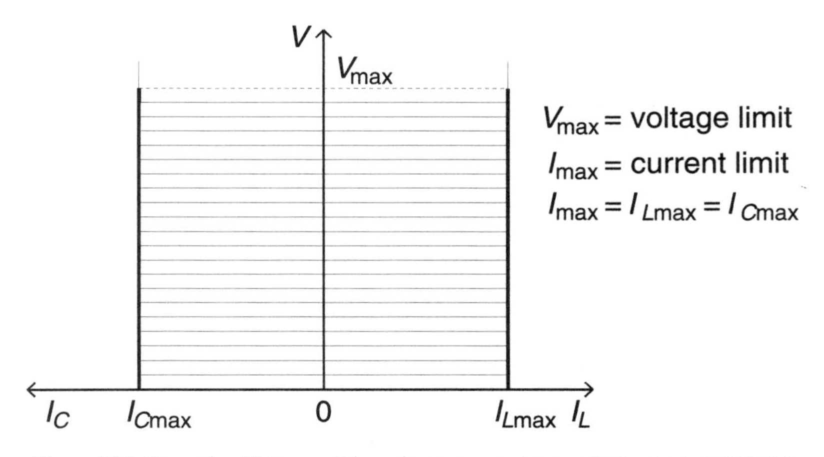

Figure 5.37 Operating *V-I* area of the voltage-sourced converter type var generator.

response time. (The TCR/TSC-based var generators, also for economic and technical reasons, are usually of six-pulse, or occasionally of 12-pulse structure. But even if the pulse number of these would be increased, the improvement in response time, as discussed previously in Section 5.2.1.3, would be marginal because the maximum overall transport lag of the TCR/TSC var generator does not significantly change with pulse number.)

The very fast response time, per se, attainable by the converter-based var generator, is not generally required for transmission line compensation. However, as will be seen, the small transport lag allows a wide frequency closed-loop bandwidth, which provides stable operation for the STATCOM over a much wider variation of the ac system impedance than is possible with the SVC.

The loss versus reactive output current characteristic of an actual 100 MVA, 48-pulse voltage-sourced converter, operated as a STATCOM, is shown in Figure 5.38. The total loss is the sum of the losses consumed by the eight constituent six-pulse, two-level converters of the multipulse arrangement, by the main coupling transformer, and by the interface magnetics used to combine the output voltages of the constituent converters.

The converter losses are due to semiconductor conduction and switching losses, as well as to "snubber" losses (consumed by *dv/dt* and *di/dt* limiting circuits). These losses are greatly dependent on the characteristics of the power semiconductors employed in the converter and the number of switchings they have to execute during each fundamental cycle. The loss characteristic shown is valid for a GTO-based converter, employing devices with a voltage rating of 4.5 kV and a peak turn-off current capability of 4 kA, operated at 60 Hz switching frequency with 6 μF snubber capacitor. Due to the low switching frequency, only about one-third of the converter losses is due to

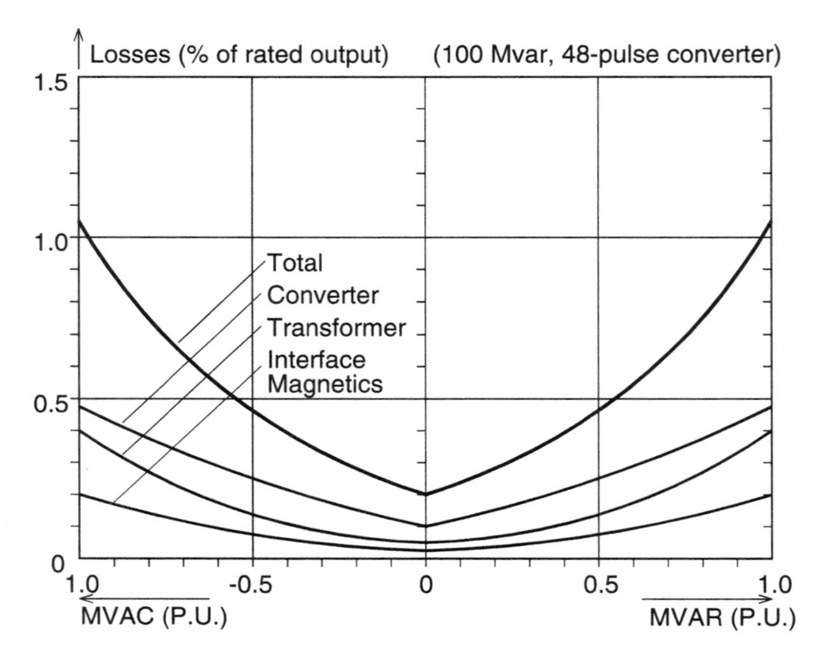

Figure 5.38 Loss versus var output characteristic of a 48-pulse, ±100 Mvar voltage-sourced converter type static var generator.

(semiconductor + snubber) switching losses, the other two-thirds is due to conduction losses.

The total magnetic losses are due to the main coupling transformer and the interface magnetics. The main transformer losses are, of course, unavoidable in high-voltage applications. The losses of the interface magnetics are dependent on the overall converter structure and operating mode employed.

The above loss characteristic gives an indication for the restrictions imposed on the design of a high-power converter. The application of PWM would allow some simplification of the interface magnetics. However, the impracticality of using pulse-width modulation to improve the frequency spectrum of the basic converter with presently available semiconductors is evident from the above data. As explained in Chapter 3, the elimination of each harmonic from the output waveform would require one "notch" in the waveform. Thus, for example, the elimination of the fifth and seventh harmonics would increase the switching rate of the power semiconductors by a factor of three. This would mean that, in the above 100 MVA converter considered, the losses (not including magnetics) would go up by about 65% (with a corresponding increase in cooling requirements) and the total installation losses would increase by about 30%. There are on-going development efforts to improve the switching characteristics of GTO thyristors and to devise a simple and reliable method to recover (into the dc storage capacitor) the energy stored in the snubber capacitor at each turn-off. The results of these efforts would allow the application of limited PWM techniques, requiring only modest switching frequencies and aiming at the elimination of specific harmonics (such as the 5th and 7th) from the output voltage waveform. As indicated in Chapter 2, major parallel efforts are also being expended in the development of a new type of semiconductor switching devices that would be free of the switching limitation of the GTO thyristor. Such devices are a new generation of high power MOS Turn-Off Thyristors (MTOs), Emitter Turn-Off Thyristors (ETOs), Integrated Gate Commutated Thyristors (IGCTs), Insulated Gate Bipolar Transistors (IGBTs), and MOS-Controlled Thyristors (MCTs). These devices would, in principle, allow an increase in the switching frequency, and possibly the elimination of harmonics from the output waveform at least up to the 13th, without any energy recovery circuit. However, PWM with high switching frequency for high-power converters may remain a challenge for some years to come because of the problems associated with the trapped energy stored in the leakage inductance of the converter structure, associated transient voltage appearing across opening semiconductor switches, the filtering of the high-frequency and high-energy carrier component, and, in general, the containment of high-level, high-frequency electrical noise.

Overall, even at the present state of the art, the loss versus reactive output characteristic, as well as the actual operating losses, of a converter-based var generator are comparable to those attainable with its conventional counterpart employing both thyristor-controlled reactors and thyristor-switched capacitors. With more advanced GTOs and other power semiconductors presently under development, and with the judicious combination of multipulse, PWM and other wave shaping (e.g., multilevel) techniques, both the complexity of the magnetic structure and the overall converter losses are expected to be further reduced in the future.

5.2.3 Hybrid Var Generators: Switching Converter with TSC and TCR

The converter-based var generator can generate or absorb the same amount of maximum reactive power; in other words, it has the same control range for capacitive and inductive var output. However, many applications may call for a different var generation and absorption range. This can simply be achieved by combining the converter with either fixed and/or thyristor-switched capacitors and/or reactors.

The combination of a converter-based var generator with a fixed capacitor is shown in Figure 5.39(a). This arrangement can generate vars in excess of the rating of the converter, shifting the operating range into the capacitive region, as illustrated by the associated *V-I* characteristic shown in Figure 5.39(b).

It is, of course, also possible to move the operating characteristic of a static var generator further into the absorption region by combining the converter with a shunt reactor as shown in Figure 5.40(a). The *V-I* characteristic of this arrangement is shown in Figure 5.40(b).

Whereas fixed capacitors or reactors shift the operating range of the converter-based var generator more into the capacitive or into the inductive region without changing the amount of controllable Mvars, thyristor-switched capacitors and reactors actually increase the total control range of var output. A converter-based var generator combined with a TSC and a TCR, together with the associated *V-I* characteristic are shown in Figures 5.41(a) and 1(b).

Note that the addition of fixed or switched reactive admittances to the converter-based var generator undesirably changes the *V-I* characteristic in that the output current becomes a function of the applied voltage. The change in the *V-I* characteristic is, clearly, dependent on the MVA rating of the converter relative to the total controlled Mvar range.

Apart from the shift or extension of the controlled var range, hybrid var generator arrangements, employing a converter with fixed and/or thyristor-controlled capacitor and reactor banks, may be used for the purpose of providing an optimal loss versus var output characteristic to a given application.

The generalized hybrid var generator scheme, employing a switching converter with TSCs, TCRs, and possibly fixed or mechanically-switched capacitors, provides a useful possibility for the designer to optimize the compensator for defined var range, loss versus var output characteristics, performance, and cost.

5.2.4 Summary of Static Var Generators

The basic characteristics of the main static var generator schemes are summarized in Table 5.1 and representative loss versus var output characteristics are shown in Figure 5.42.

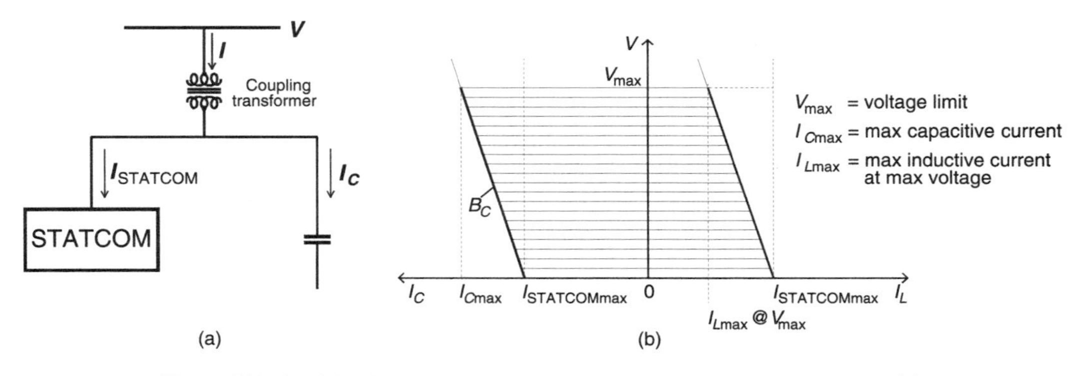

Figure 5.39 Combined converter-based and fixed capacitor type var generator (a), and associated operating *V-I* area (b).

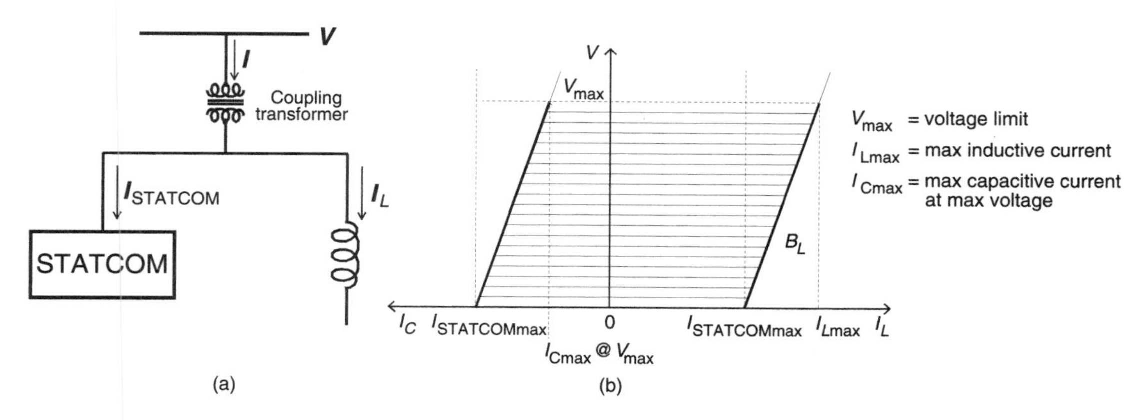

Figure 5.40 Combined converter-based and fixed reactor type var generator (a), and associated operating V-I area (b).

5.3 STATIC VAR COMPENSATORS: SVC AND STATCOM

Static Var Compensator (SVC) and Static Synchronous Compensator (*STATCOM*) are static var generators, whose output is varied so as to maintain or control specific parameters of the electric power system. It has been discussed in the previous sections that a static (var) generator may be of a controlled reactive impedance type, employing thyristor-controlled and switched reactors and capacitors, or synchronous voltage source type, employing a switching power converter, or a hybrid type, which employs a combination of these elements. Although the operating principles of these var generators are disparate and their V-I and loss versus var output characteristics, as well as their speed of response and attainable frequency bandwidth, are quite different, they all can provide the controllable reactive shunt compensation, exhibiting similar overall functional capabilities within their linear operating range. This means that the basic external control structure that defines the functional operation of the compensator, and to this end derives the necessary reference inputs for the var generator, is substantially the same independent of the type of var generator used. (Note,

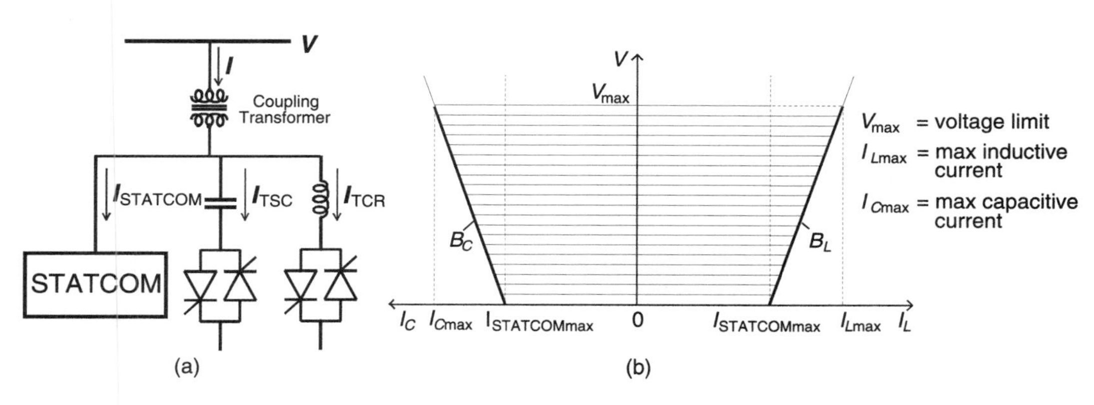

Figure 5.41 Combined converter-based and TSC-TCR type var generator (a), and associated operating V-I area (b).

TABLE 5.1 Comparison of Static Var Generators

VAR generator	TCR-FC	TSC-(TSR)	TCR-TSC	Converter-based
Type	Controlled Impedance	Controlled Impedance	Controlled Impedance	Synch. Voltage Source
V-I and V-Q characteristics	Max comp. current is proportional to system voltage. Max. cap. var output decreases with the square of voltage decrease.	Max comp. current is proportional to system voltage. Max. cap. var output decreases with the square of voltage decrease.	Max comp. current is proportional to system voltage. Max. cap. var output decreases with the square of voltage decrease.	Max comp. current is maintained independent of system voltage. Max. cap. var output decreases linearly with voltage decrease.
Loss vs. var output	High losses at zero output. Losses decrease smoothly with cap. output, increase with inductive output.	Low losses at zero output. Losses increase step-like with cap. output.	Low losses at zero output. Losses increase step-like with cap. output, smoothly with ind. output.	Low losses at zero output. Losses increase smoothly with both cap. and inductive output.
Harmonic generation	Internally high (large p.u. TCR). Requires significant filtering.	Internally very low. Resonances (and current limitation) may necessitate tuning reactors.	Internally low (small p.u. TCR). Requires filtering.	Can be internally very low (multipulse, multilevel converters). May require no filtering.
Max. theoret. delay	Half-cycle	One cycle	One cycle	Negligible
Transient behavior under system voltage disturbances	Poor. (FC causes transient over-voltages in response to step disturbances.)	Can be neutral. (Capacitors can be switched out to minimize transient over-voltages.)	Can be neutral. (Capacitors can be switched out to minimize transient over-voltages.)	Trends to damp transients. (Low impedance voltage source.)

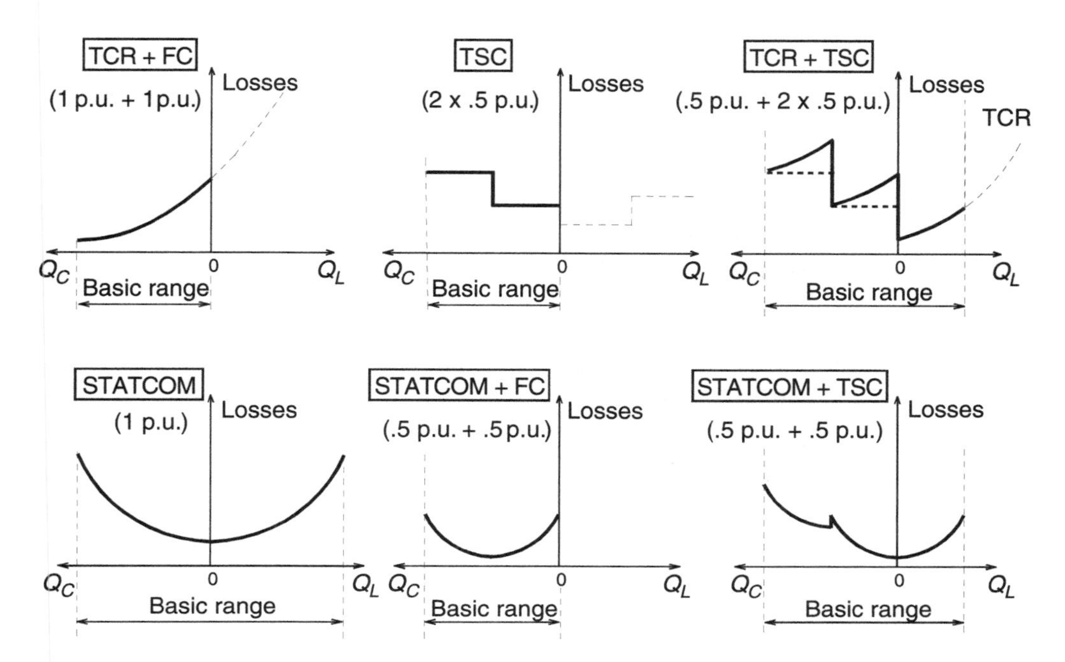

Figure 5.42 Summary of loss versus var output characteristics of different static var generator schemes.

however, that the converter-based var generator may be equipped with a suitable energy storage in order to provide active as well as reactive compensation in which case the compensator control would have to be complemented with additional control loops for managing the real power exchange between the ac system and the converter.)

The primary objective of applying a static compensator (this term or the shorter term compensator will be used in a general sense to refer to an SVC as well as to a STATCOM) in a power system is to increase the power transmission capability, with a given transmission network, from the generators to the loads. Since static compensators cannot generate or absorb real power (neglecting the relatively low internal losses of the SVC and assuming no energy storage for the STATCOM), the power transmission of the system is affected indirectly by voltage control. That is, the reactive output power (capacitive or inductive) of the compensator is varied to control the voltage at given terminals of the transmission network so as to maintain the desired power flow under possible system disturbances and contingencies.

The control requirements for the compensator (which define the way the output of the var generator has to be varied to increase power flow and to stabilize specific parameters of the power system in face of network contingencies and dynamic disturbances) has been derived from the functional compensation considerations in Section 5.1 of this chapter. As shown there, the basic compensation needs usually fall into one of the following two main categories: (1) direct voltage support (to maintain sufficient line voltage for facilitating increased power flow under heavy loads and for preventing voltage instability) and (2) transient and dynamic-stability improvements (to increase the first swing stability margin and provide power oscillation damping). In that section it was shown that terminal voltage control can enhance significantly the power transmission capability of the power system. Specifically, the regulation of the voltage at particular intermediate points and selected load terminals of the transmission system, limits voltage variation, prevents voltage instability (voltage collapse), and increases transient (first swing) stability limits, whereas appropriate variation of the terminal voltage can further enhance transient stability and provide effective power oscillation damping (dynamic stability).

In order to meet the general compensation requirements of the power system, the output of the static var generator is to be controlled to maintain or vary the voltage at the point of connection to the transmission system. A general control scheme, converting a static var generator (either a controlled impedance type or a converter-based type) into a transmission line compensator, is shown in Figure 5.43.

The power system, at the terminal of the compensator, is represented by a generator with a generally varying rotor angle δ, internal voltage v, and source impedance Z (including the generator and transmission line impedance) that is a function of the angular frequency ω and time t. (The impedance variation in time is due to faults, line switching, etc.) The terminal voltage v_T of the power system can be characterized by a generally varying amplitude V_T and angular frequency ω.

The output of the static var generator is controlled so that the amplitude I_o of the reactive current i_o drawn from the power system follows the current reference I_{QRef}. With the basic static compensator control, the var generator is operated as a *perfect* terminal voltage regulator: the amplitude V_T of the terminal voltage v_T is measured and compared with the voltage reference V_{Ref}; the error ΔV_T is processed and amplified by a PI (proportional integral) controller to provide the current reference I_{QRef} for the var generator. In other words, I_o is closed-loop controlled via I_{QRef} so that V_T is maintained precisely at the level of the reference voltage V_{Ref} in face of power system and load changes.

If the proper compensation of the ac power system requires some specific variation in the amplitude of the terminal voltage with time or some other variable, then an appropriate correcting signal V_{Rc}, derived from the auxiliary inputs, is summed to

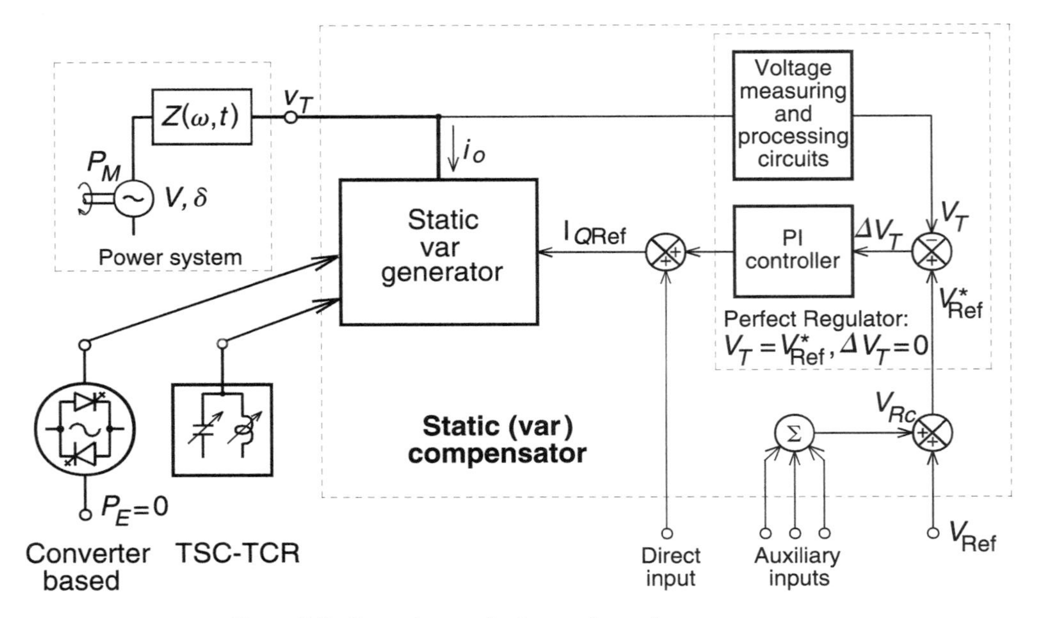

Figure 5.43 General control scheme of a static var generator.

the fixed reference V_{Ref} in order to obtain the desired effective (variable) reference signal V^*_{Ref} that closed-loop controls the terminal voltage V_T. In the following sections, some practical auxiliary control loops and the corresponding characteristics of the static compensator are described.

5.3.1 The Regulation Slope

In many applications, the static compensator is not used as a perfect terminal voltage regulator, but rather the terminal voltage is allowed to vary in proportion with the compensating current. There are several reasons for this:

1. The linear operating range of a compensator with given maximum capacitive and inductive ratings can be extended if a regulation "droop" is allowed. Regulation "droop" means that the terminal voltage is allowed to be smaller than the nominal no load value at full capacitive compensation and, conversely, it is allowed to be higher than the nominal value at full inductive compensation.
2. Perfect regulation (zero droop or slope) could result in poorly defined operating point, and a tendency of oscillation, if the system impedance exhibited a "flat" region (low impedance) in the operating frequency range of interest.
3. A regulation "droop" or slope tends to enforce automatic load sharing between static compensators as well as other voltage regulating devices normally employed to control transmission voltage.

The desired terminal voltage versus output current characteristic of the compensator can be established by a minor control loop using the previously defined auxiliary input as shown schematically in Figure 5.44. A signal proportional to the amplitude of the compensating current κI_Q with an ordered polarity (capacitive current is negative and

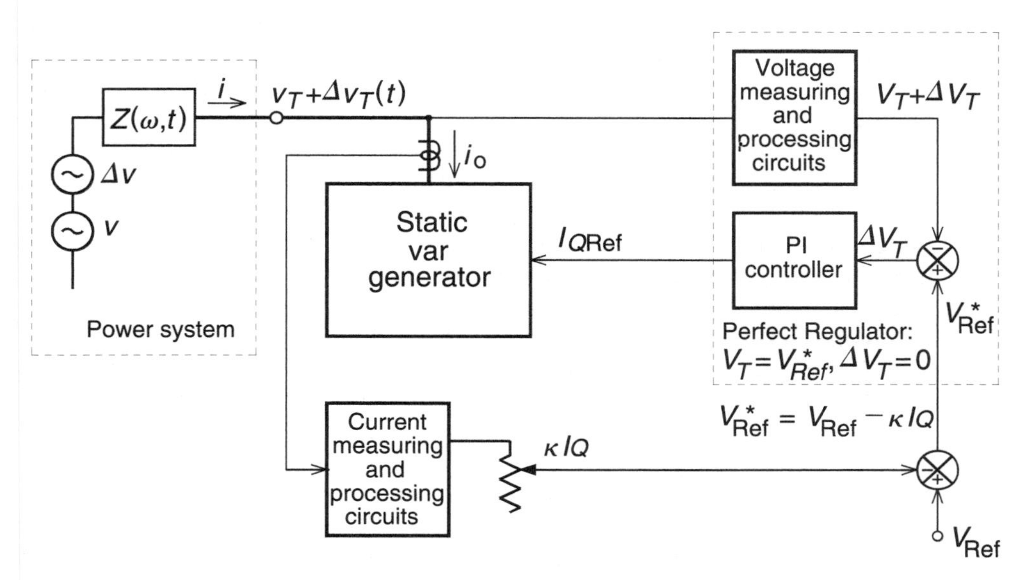

Figure 5.44 Implementation of the *V-I* slope by a minor control loop changing the reference voltage in proportion to the line current.

inductive current is positive) is derived and summed to the reference V_{Ref}. The effective reference V^*_{Ref} controlling the terminal voltage thus becomes

$$V^*_{Ref} = V_{Ref} + \kappa I_Q \tag{5.14}$$

In (5.14), κ, the regulation slope is defined by

$$\kappa = \frac{\Delta V_{Cmax}}{I_{Cmax}} = \frac{\Delta V_{Lmax}}{I_{Lmax}} \tag{5.15}$$

where ΔV_{Cmax} is the deviation (decrease) of the terminal voltage from its nominal value at maximum capacitive output current ($I_{Qmax} = I_{Cmax}$), ΔV_{Lmax} is the deviation (increase) of the terminal voltage from its nominal value at maximum inductive output current ($I_{Qmax} = I_{Lmax}$), I_{Cmax} is the maximum capacitive compensating current, and I_{Lmax} is the maximum inductive compensating current.

Equation (5.14) indicates that V^*_{Ref} is controlled to decrease from the nominal (set) value (no compensation) with increasing capacitive compensating current (as determined by the selected slope κ), and, conversely, it is controlled to increase with increasing inductive compensating current until the maximum capacitive or, respectively, inductive compensating current is reached. Consequently, the amplitude of the terminal voltage, V_T, is regulated along a set linear slope over the control range of the compensator. For terminal voltage changes outside of the linear control range, the output current of the compensator is determined by the basic *V-I* characteristic of the var generator used. That is, the compensating current will stay at the maximum capacitive or inductive value in the case of a converter-based var generator (i.e., if the compensator considered is STATCOM) and, in contrast, it will change in the manner of a fixed capacitor or inductor in the case of a variable impedance (TCR/TSC) type var generator (i.e., if the compensator is an SVC).

A typical terminal voltage versus output current characteristic of a static compensator with a specific slope is shown in Figure 5.45, together with particular "load lines" (voltage versus reactive current characteristic) of the ac system. Load line 1 intersects the compensator *V-I* characteristic at the nominal (reference) voltage, thus the output current of the compensator is zero. Load line 2 is below load line 1 due to a decrease in the power system voltage (e.g., generator outage). Its intersection with the compensator *V-I* characteristic calls for the capacitive compensating current I_{C2}. Load line 3 is above load line 1 due to an increase in the power system voltage (e.g., load rejection). Its intersection with the compensator *V-I* characteristic defines the inductive compensating current I_{L3}. The intersection points of the load lines 2 and 3 with the vertical (voltage) axis define the terminal voltage variation without any compensation. The terminal voltage variation with compensation, in the linear operating range under steady-state conditions and slow system changes is entirely determined by the regulation slope, independent of the type of var generator used, as observable in Figure 5.45. Outside of the linear operating range the STATCOM and SVC act differently, as indicated in the figure. The dynamic performance of the two types of compensator is also different.

5.3.2 Transfer Function and Dynamic Performance

The *V-I* characteristic of the static compensator shown in Figure 5.45 represents a steady-state relationship. The dynamic behavior of the compensator in the normal compensating range can be characterized by the basic transfer function block diagram

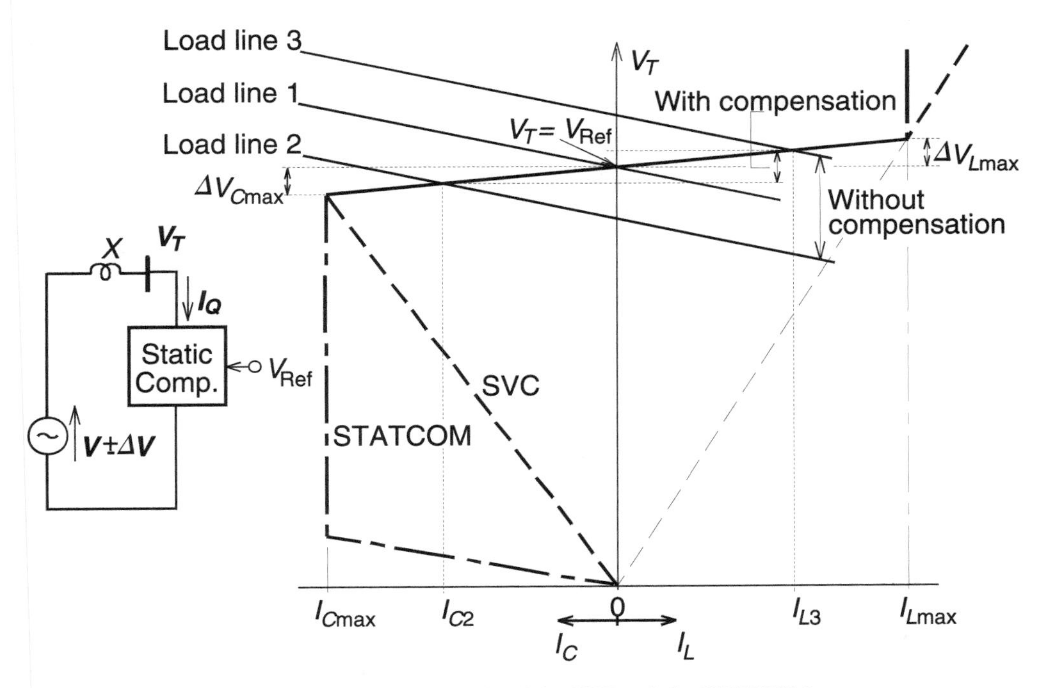

Figure 5.45 *V-I* characteristic of the SVC and the STATCOM.

shown in Figure 5.46. This block diagram is derived directly from the basic control scheme shown in Figure 5.44 and the transfer functions established for the variable impedance (FC-TCR and TSC-TCR) type and converter-based var generators in Sections 5.2.1 and 5.2.2, respectively.

In the linear operating range of the compensator, the terminal voltage V_T can be expressed from Figure 5.46 in terms of the internal voltage V and the reference voltage V_{Ref} as follows:

$$V_T = V\frac{1}{1 + G_1G_2HX} + V_{\text{Ref}}\frac{G_1G_2X}{1 + G_1G_2HX} \tag{5.16}$$

Since the objective is to establish how well the terminal voltage is regulated

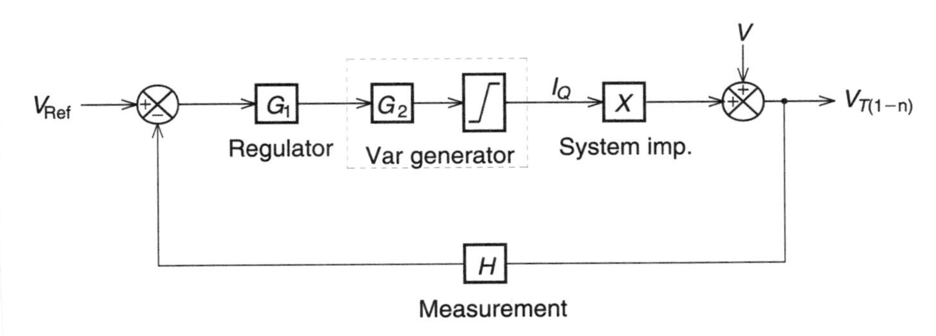

Figure 5.46 Basic transfer function block diagram of the static (var) compensator.

against the (varying) system voltage, let $V_{Ref} = 0$ and consider small variation only. Then the amplitude variation of the terminal voltage ΔV_T against the amplitude variation of the power system voltage ΔV can be expressed in the following form:

$$\frac{\Delta V_T}{\Delta V} = \frac{1}{1 + G_1 G_2 H X} = \frac{1}{1 + GHX} \tag{5.17}$$

where

$$G_1 = \frac{1/\kappa}{1 + T_1 s} \tag{5.18}$$

$$G_2 = e^{-T_d s} \tag{5.19}$$

$$G = G_1 G_2 = \frac{1/\kappa}{1 + T_1 s} e^{-T_d s} \tag{5.20}$$

$$H = \frac{1}{1 + T_2 s} \tag{5.21}$$

and T_1 = main time constant of the PI controller (typically about 10–50 ms depending on the var generator transport lag), T_2 = amplitude measuring circuit time constant (typically about 8–16 ms), T_d = transport lag of the var generator (typically 2.5 ms for TCR, 5.0 ms for TSC and 0.2–0.3 ms for converter), X = Im Z (reactive part of the system impedance), κ = regulation slope (typically 1–5%) given by (5.15), and s = Laplace operator.

It should be pointed out that practical compensator controls often employ filters in the signal processing circuits which may introduce additional time constants in the transfer functions affected. Sometimes phase correcting (lead/lag) circuits are also employed.

Under steady-state conditions ($s \to 0$) (5.17) becomes

$$\frac{\Delta V_T}{\Delta V} = \frac{1}{1 + \dfrac{X}{\kappa}} \tag{5.22}$$

Equation (5.22) confirms that as the slope becomes smaller ($\kappa \to 0$), the terminal voltage remains constant, independent of the system voltage variation ($\Delta V_T/\Delta V \to 0$). Similarly, with increasing slope ($\kappa \gg X$) the terminal voltage becomes unregulated ($\Delta V_T/\Delta V \to 1$).

It is to be noted too that (5.17) defining the dynamic behavior of the compensator is a *function of the power system impedance*, that is, the system impedance is an integral part of the feedback loop. This means that the time response, and thus the stability of the control, is dependent on the system impedance. For this reason, the control is normally optimized for the maximum system impedance (minimum short-circuit capacity) expected. This, of course, means that the response time of the system will be somewhat longer if the system impedance is decreased (short-circuit capacity is increased). With practical static compensators, the worst case response time is typically in the range of 30 to 70 ms (2 to 4 cycles at the power system frequency). The tolerance to system impedance variation and the attainable worst case response time are considerably dependent on the achievable frequency bandwidth of the compensator, which is ultimately limited by the transport lag of the var generator employed.

In order to estimate the attainable bandwidth and the stability of the closed voltage regulation loop of the two types of static compensator (SVC and STATCOM), consider the general expression given by (5.17). This equation equally characterizes both types, except for the "transport lag" time constant T_d in transfer function G_2, which is about an order of magnitude smaller for the STATCOM than it is for the SVC. Recognizing the very undesirable phase characteristic of the $e^{-T_d s}$ term, the improvement the STATCOM offers is very significant from the standpoint of attainable frequency bandwidth. Refer to Figure 5.47 where the phase angle characteristic of the $e^{-T_d s}$ term (with $s = j\omega$) versus frequency for the previously established transport lag delay times, $T_d = 5.55$ ms (TSC), $T_d = 2.77$ ms (TCR), and $T_d = 0.5$ ms (converter), are plotted. As can be seen, the frequency at which the phase shift reaches 180 degrees (where the closed-loop gain of the voltage regulator for stability must be less than unity) is typically more than ten times greater for the converter-based var generator of the STATCOM than it is for the TSC and TCR based var generator of the SVC.

It follows from 5.17 that, if the system impedance varies (as it always does in practice due to line switchings, generator outages, etc.), the compensator will remain stable only if the overall loop gain versus frequency, determined by the product G_1G_2HX, is less than unity with the maximum system impedance (weakest system) before the angle of G_1G_2HX versus frequency reaches 180 degrees. This is usually achieved by setting the time constant of the main error amplifier appropriately large. To illustrate this, consider the basic compensator-based voltage regulator shown in Figure 5.43, with the transfer function representation and corresponding data defined in Figure 5.48. If the compensator is a STATCOM with $T_d = 0.5$ ms, and thus with $G_2 = e^{-0.0005s}$, then stable operation with a regulation bandwidth of 65 Hz ($T_1 = 14$ ms) is maintained over the total two to one system impedance range. As the Bode plots shown in Figures 5.49(a) and (b) illustrate, with the strong system ($X = 4.761$) the gain margin (i.e., the gain at 180° phase angle) is almost −20 dB and even with the weak system ($X = 9.522$) a comfortable gain margin of about −10 dB is maintained. By contrast, as the Bode plots in Figures 5.50(a) and (b) illustrate, the TSC/TCR-based SVC ($T_2 = 5.55$ ms), even if the bandwidth is limited to 35 Hz ($T_1 = 35$ ms) would become unstable (the gain becomes positive at 180 degree phase angle) as

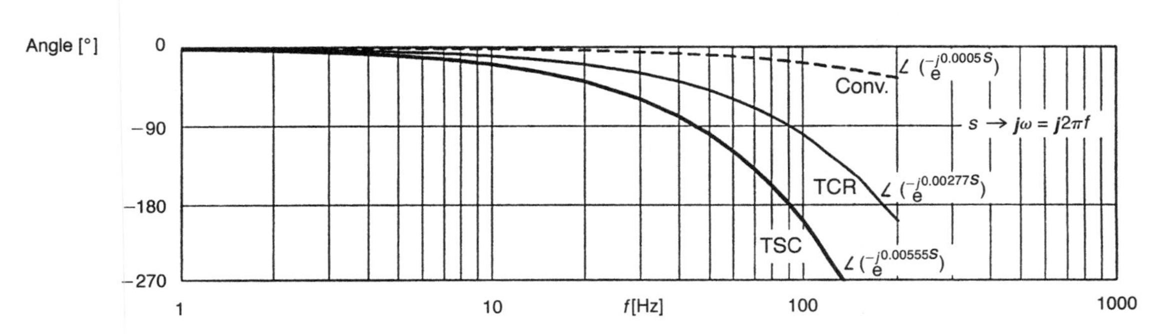

TSC: thyristor-switched capacitor
TCR: thyristor-controlled reactor
Conv.: multipulse, voltage-sourced converter

Figure 5.47 Bode-plot representation of the transport lag characterizing the controlled elements of a static var generator: the TSC, the TCR, and the converter.

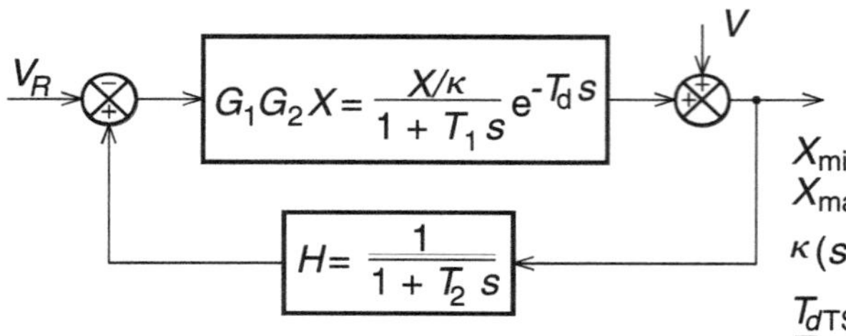

Data:

X_{min} = 4.761 (strong system)
X_{max} = 9.522 (weak system)
$\kappa(slp.)$ = 0.846 (with strong system)
T_{dTSC} = 5.55 ms (TSC/TCR SVC)
T_{dConv} = 0.5 ms (STATCOM)
T_2 = 4 ms

Steady-state performance:

$$\left.\frac{\Delta V_T}{\Delta V}\right|_{s=0} = \frac{1}{1 + G_1 G_2 HX} = \frac{1}{1 + \frac{4.761}{0.846}} = 0.151$$

Determine:

Maximum bandwidth for stable operation with both strong and weak systems, i.e., maximum T_{1Conv} for STATCOM and T_{1TSC} for TSC/TCR type SVC

Figure 5.48 Compensation example to demonstrate the effect of transport lag on the stability of voltage regulation in face of varying system impedance.

the system impedance approaches its maximum value. For acceptable stability, the bandwidth of the SVC would have to be reduced further, to about 20 Hz. The reduced bandwidth would, of course, result in a reduced response to system disturbances. If the wide impedance variation in the application is encountered, and fast response is required, then the TSC/TCR-based SVC would have to be equipped with an automatic gain adjustment that would change the regulation gain either according to the prevailing system impedance or with the sensing of an oscillatory trend in the control.

5.3.3 Transient Stability Enhancement and Power Oscillation Damping

As established in Section 5.1.5, transient stability enhancement and power oscillation damping require the appropriate variation of the transmission line (terminal) voltage in order to control the transmitted power so as to counteract the prevailing acceleration or deceleration of the disturbed machine(s).

5.3.3.1 Transient Stability Enhancement. The transient stability indicates the capability of the power system to recover following a major disturbance. A major disturbance, for example a severe fault on a heavily loaded line, can result in a large step-like decrease in the transmitted electric power while the generators feeding the line receive constant mechanical input power. The difference between mechanical input and electrical output power causes the machines to accelerate. As explained in Section 5.1.3, transient stability at a given power level and fault clearing time is primarily determined by the P versus δ characteristic of the post-fault system that controls the electric power transmission and thereby the rate of energy absorption from the machine.

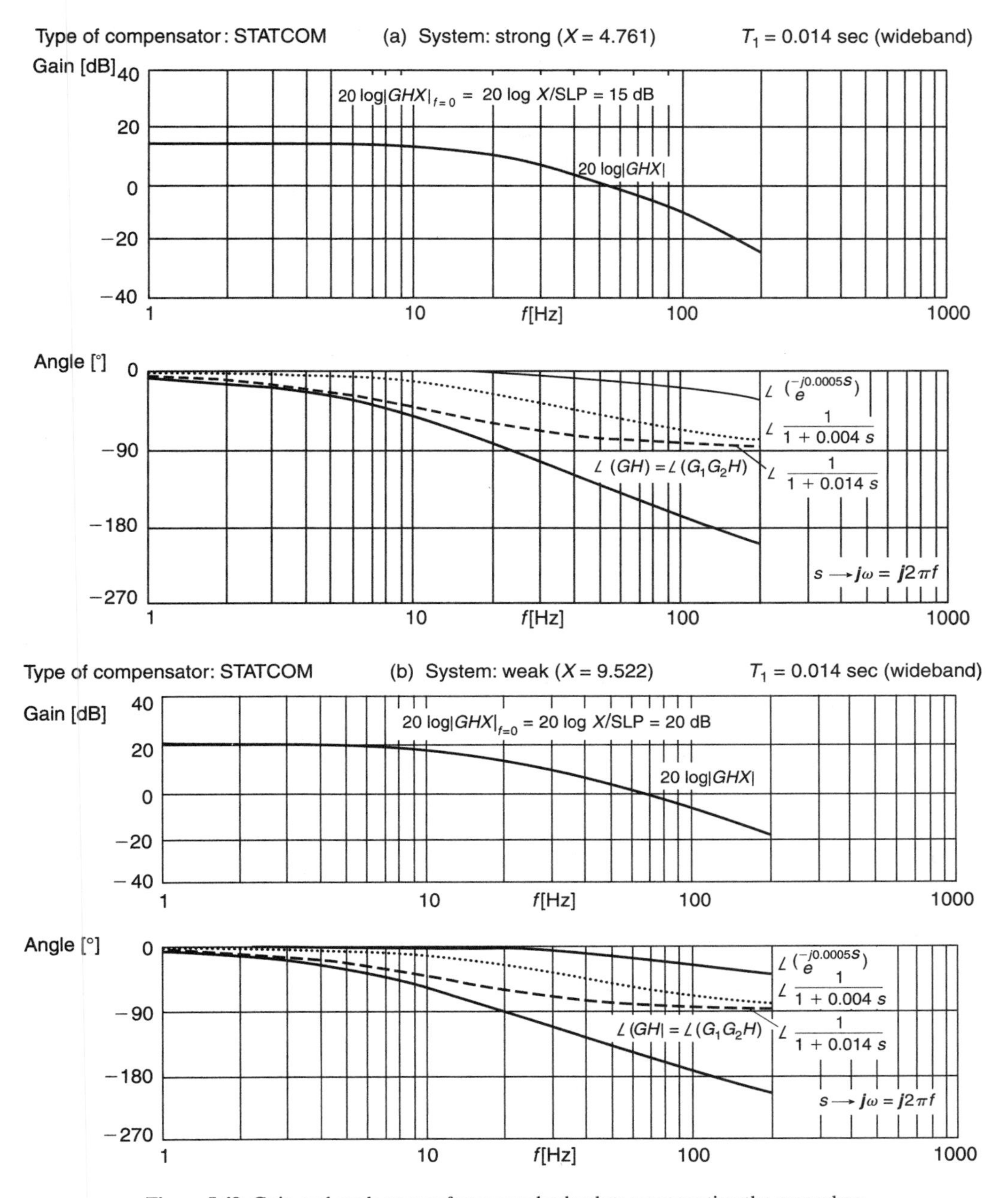

Figure 5.49 Gain and angle versus frequency bode plots representing the case when the STATCOM regulates the voltage of a strong (low impedance) system (a) and a weak (high impedance) system (b).

As was shown in that section, a static compensator, controlled to regulate the terminal voltage, can increase the transient stability by maintaining the transmission voltage (at the midpoint or some appropriate intermediate point) in face of the increased power flow encountered immediately after fault clearing. However, the transient stability can be increased further by temporarily increasing the voltage above the regulation reference for the duration of the first acceleration period of the machine.

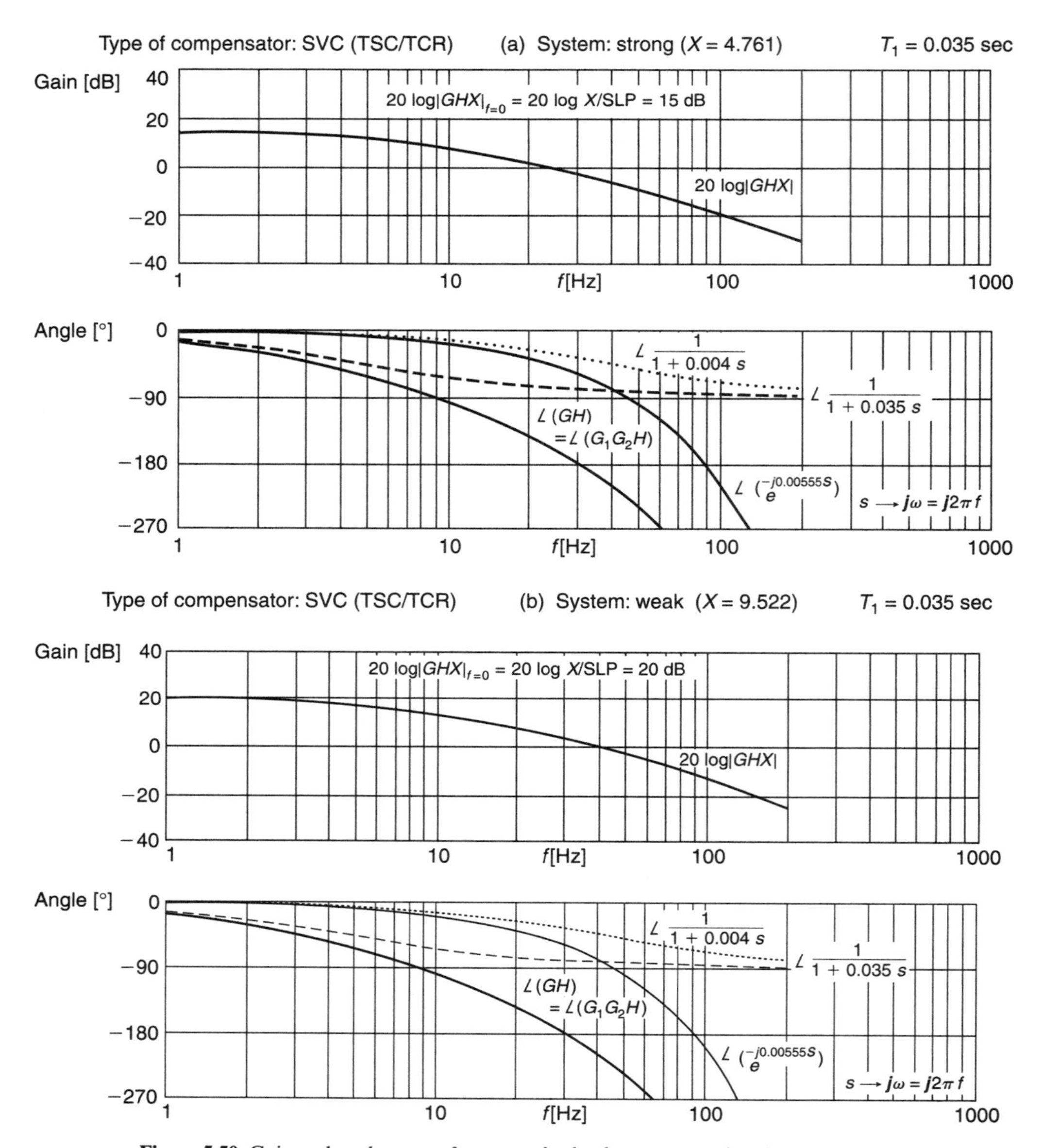

Figure 5.50 Gain and angle versus frequency bode plots representing the case when the SVC regulates the voltage of a strong (low impedance) system (a) and a weak (high impedance) system (b).

The voltage increased above its nominal value will increase the electric power transmitted and thus will increase also the deceleration of the machine. This is illustrated in Figure 5.51, where the P versus δ plots of a simple two-machine system with different midpoint compensations are shown. The plot marked $V_m = V$ represents the P versus δ plot obtained with an ideal compensator holding the midpoint voltage constant (refer back to Figure 5.1). The plots marked STATCOM and SVC represent these

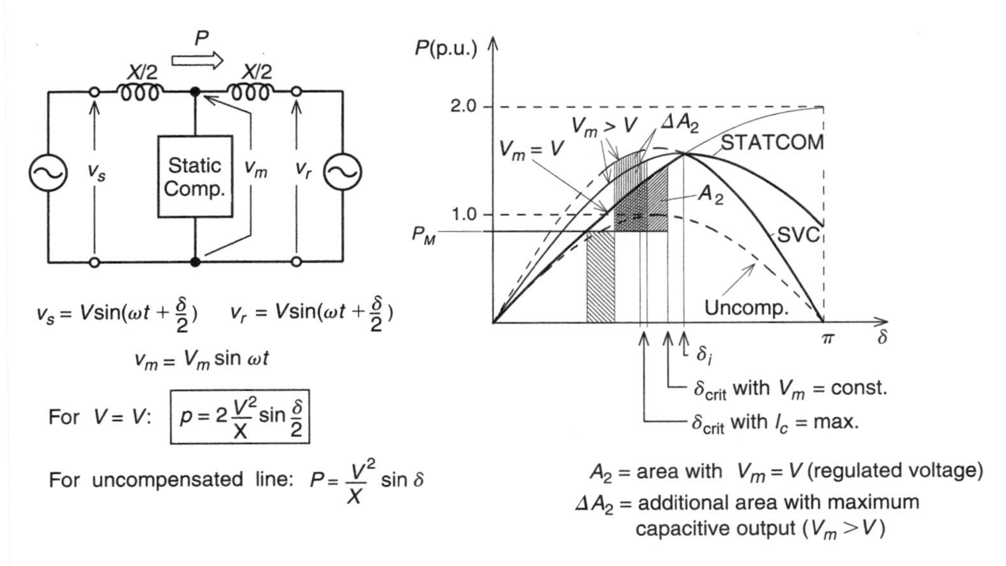

Figure 5.51 Attainable enhancement of transient stability by the SVC and STATCOM increasing temporarily the (midpoint) transmission voltage.

compensators with a given rating insufficient to maintain constant midpoint voltage over the total range of δ. Thus, these P versus δ plots are identical to that of the ideal compensator up to a specific δ ($\delta = \delta_i$) at which the SVC becomes a fixed capacitor and the STATCOM a constant current source. In the interval between δ_i and π, the P versus δ plots are those which correspond to a fixed midpoint capacitor and a constant reactive current source. The continuations of these plots in the δ_i to zero interval show the P versus δ characteristic of the two-machine system with the maximum capacitive admittance of the SVC and with the maximum capacitive output current of the STATCOM. That is, at angles smaller than δ_i the transmission line is overcompensated (and, of course, for angles greater, it is undercompensated). This overcompensation capability of the compensator can be exploited to enhance the transient stability by increasing the var output to the maximum value after fault clearing and thereby match the area of accelerating energy (A_1) by that of the decelerating energy (A_2) with a smaller δ_{crit}, as indicated in the figure. Depending on the rating of the compensator, and the allowed voltage increase, the attainable increase in transient stability margin (represented by ΔA_2) can be significant.

The implementation of the transient stability enhancement in the basic control scheme shown in Figure 5.43 can be accomplished simply by summing a signal δV to the fixed voltage reference signal V_{Ref}, as illustrated in Figure 5.52. The signal δV can be derived from the rate of change of transmitted power, line current or system frequency, indicating the angular change of the disturbed machines.

5.3.3.2 Power Oscillation Damping. As discussed in Section 5.1.4, power oscillation damping generally requires the variation of the voltage at the terminal of the compensator in proportion to the rate of change of the effective rotor (or power transmission) angle. Rotor angle changes, of course, result in frequency and real power variations. In practice, usually the variation of the transmitted real power or the system

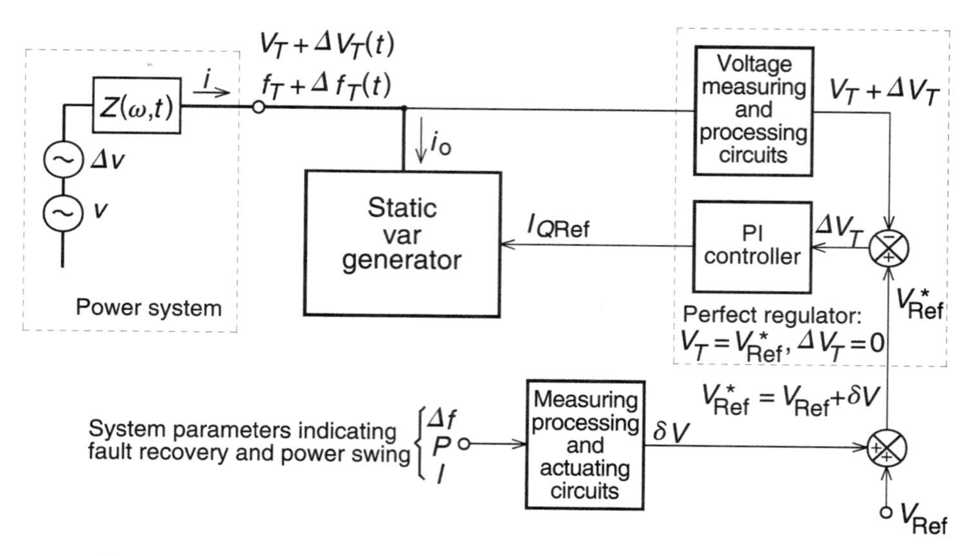

Figure 5.52 Implementation of the transient stability enhancement concept by increasing the reference voltage during the first swing of a major disturbance.

frequency is measured and used for controlling the var output to produce the terminal voltage variation desired.

The functional control scheme for damping power oscillations (and for providing terminal voltage regulation when power oscillation is absent) is shown in Figure 5.53. In this scheme, the same general idea of modifying the fixed voltage reference by an auxiliary control signal to derive the effective voltage reference that controls the terminal voltage is followed again. Accordingly, a signal corresponding to the variation

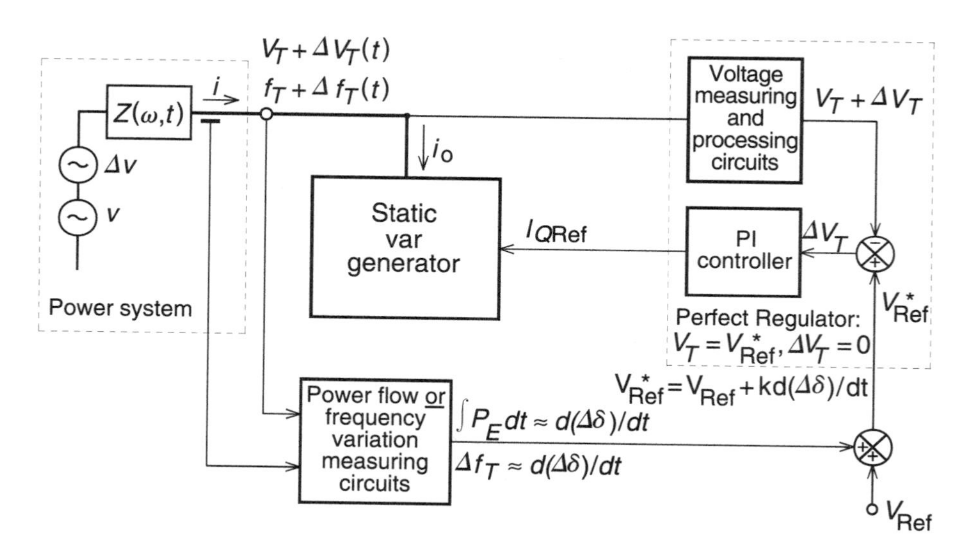

Figure 5.53 Implementation of power oscillation damping by modulating the reference voltage according to frequency or power flow variations.

of the real power or that of the system frequency is summed to the fixed reference voltage signal V_{Ref}. The added signal causes the output current of the compensator to vary (oscillate) around the fixed operating point to control the terminal voltage so as to aid system damping. That is, the terminal voltage is increased when, for example, the frequency deviation $\Delta f \approx d(\Delta\delta)/dt$ is positive (in order to increase the transmitted electrical power and thereby to oppose the acceleration of the generators) and it is decreased when Δf is negative (to reduce the transmitted electrical power and thereby oppose the deceleration of the generators).

Effective power oscillation damping can be achieved also by the alternative control scheme shown in Figure 5.54. In this scheme, the signal representing power oscillation is fed directly to the input of the static var generator. This may be done to avoid the time constants associated with the terminal voltage regulator loop and to improve the overall response time, or to simplify the control system if the sole purpose of the compensator is power oscillation damping. This arrangement is particularly suitable to the implementation of the "bang-bang" type control strategy that alternates the var output between the attainable maximum positive and negative values.

The control scheme shown in Figure 5.54 can also be used successfully to dampen subsynchronous oscillations encountered with transmission lines employing series capacitive compensation. The frequency of subsynchronous oscillation is normally significantly higher than that of normal power oscillation. For subsynchronous oscillation damping, the angular velocity of the generator is usually measured directly by using, for example, a toothed-wheel magnetic transducer located at the shaft of the rotor.

5.3.4 Var Reserve (operating point) Control

As evident from the previous sections, a static compensator has the functional capability to handle dynamic system conditions, such as transient stability and power oscillation damping in addition to providing voltage regulation. Even in the area of

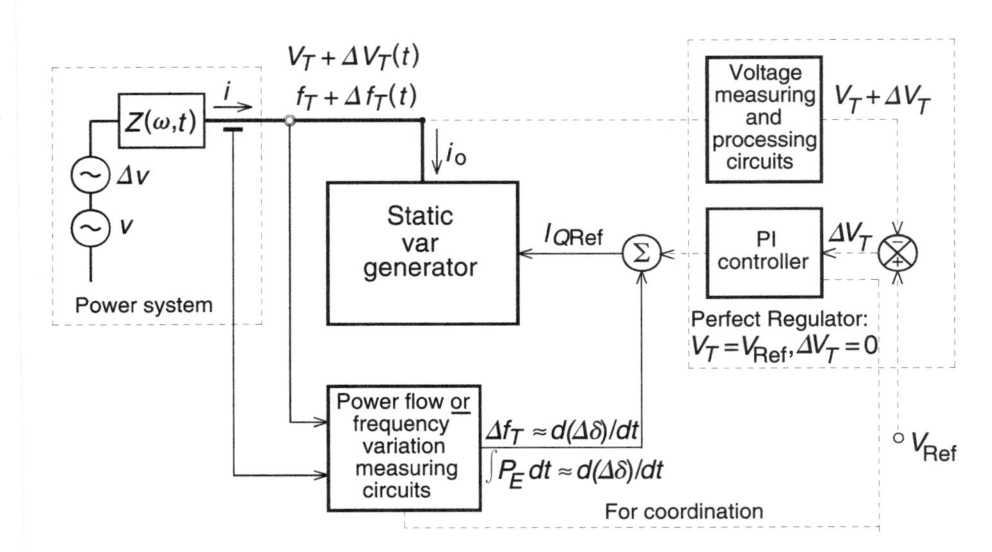

Figure 5.54 Alternative method of power oscillation damping by modulating the reference current to the static var generator according to frequency or power flow variations.

voltage regulation, the static compensator is viewed primarily as a fast var source to counteract rapid and unexpected voltage disturbances due to faults, line and load switching, generator outages, etc. In order to fulfill these applications requirements it is necessary to ensure that the compensator will have sufficient var capacity to handle unpredictable dynamic disturbances. This is often accomplished with an automatic control that maintains predetermined var reserve by adjusting the operating point of the compensator.

The objective of this control is to limit the steady-state reactive power output of the compensator to a given reference value. The basic concept is to allow the compensator to change its output rapidly to counteract transient disturbances. However, when a disturbance results in a new operating point, with a steady var output, the var reserve control effectively changes the voltage reference in order to bring back the var output slowly to the set reference value, and thereby activating "slow" var sources (e.g., mechanically-switched capacitors) and other compensating means (e.g., generator excitation) to pick up the steady-state var load. The response time of this control loop is slow in order not to interfere with the rapid voltage regulation or any fast stabilizing or auxiliary functions that might be included in the overall var output control.

A possible scheme to implement a basic var reserve control is shown in Figure 5.55. The magnitude of the output current of the compensator is measured and compared against the reference I_Q^*. The error signal ΔI_Q is processed by an integrator of large time constant and added to the fixed voltage reference V_{Ref}. This forces the input signal to the voltage regulator to change until the difference between the actual output current of the compensator and the steady-state output current reference I_Q^* become equal.

The operation of the simple var reserve control described is illustrated in Figure 5.56. Assume that the compensator is operating at point 1 ($I_Q = I_{C1}$) on the V-I curve when a disturbance in the form of a sudden ΔV_T drop in the amplitude of the terminal voltage occurs. The voltage change ΔV_T forces, via the fast voltage regulating loop,

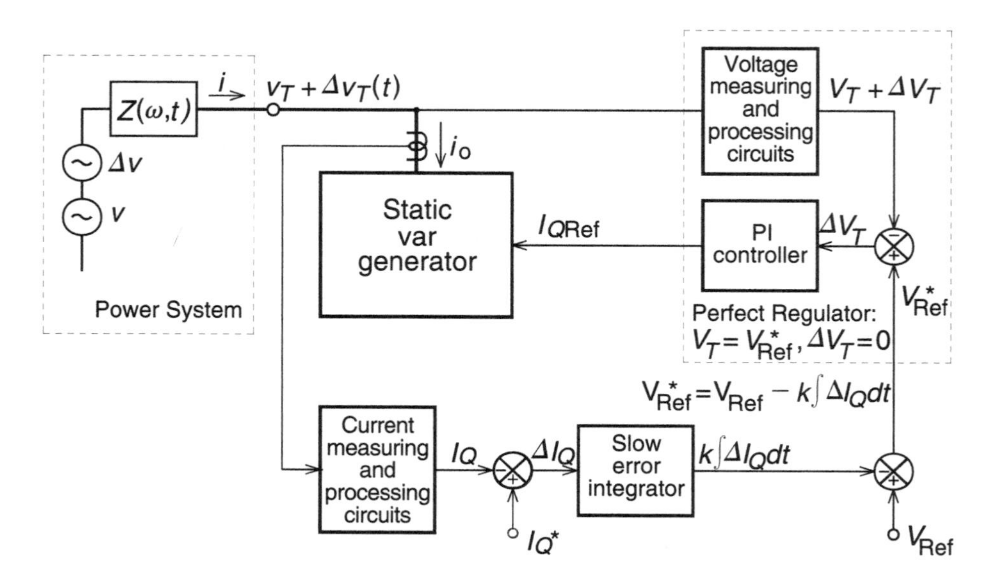

Figure 5.55 Implementation of the var reserve control.

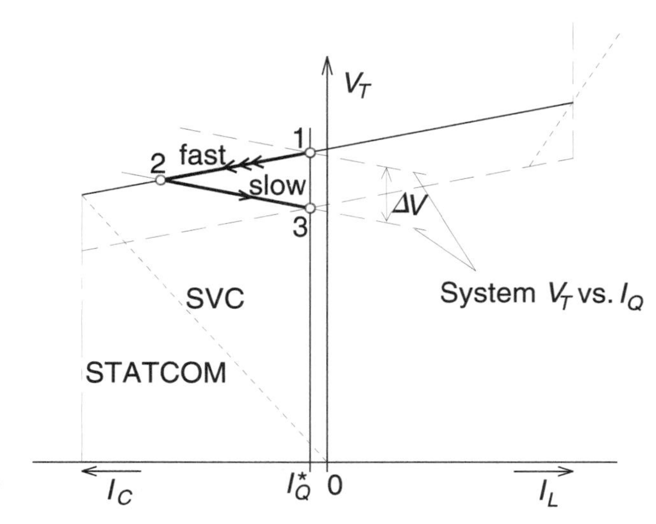

Figure 5.56 Diagrammatic illustration of the concept of var reserve control.

the output current to increase from the steady-state value I_Q^* to I_{C2} and the compensator assumes working point 2 on the *V-I* curve. However, since $I_{C2} > I_Q^*$, an error signal ΔI_Q is generated within the var reserve control loop, which, via the slow integrator, changes the reference signal to the voltage regulator, forcing the compensator to reduce slowly its output current. The compensator finally assumes a new steady state in working point 3 on the *V-I* curve.

5.3.5 Summary of Compensator Control

The structure of static compensator control illustrating the underlying principle of superimposing auxiliary input signals on the basic voltage reference, to carry out specific compensation functions automatically as required by system conditions, is shown in Figure 5.57. With this principle, the compensator, within its MVA rating and operating frequency band, acts as a perfect amplifier forcing the magnitude of the regulated terminal voltage to follow the effective voltage reference, which is the sum of the fixed voltage magnitude reference and auxiliary signals. The effective reference thus defines the operating modes and characteristics (e.g., voltage regulation with the corresponding steady-state operating point and regulation slope) as well as the desired actions in response to dynamic changes of selected system variables (e.g., transient stability enhancement and oscillation damping).

Apart from the real-time control functions illustrated in Figure 5.57, the total control system of a modern static compensator has many other elements to manage the proper and safe operation of the equipment with high reliability and availability, as well as to accommodate proper interface with local and remote operators, as shown in a glossary manner in Figure 5.58. The main elements of this overall compensator control system include:

1. Interface between high-power, high-voltage semiconductor valves of the overall switching converter and/or TSC and TCR structure and a highly-sophisticated real-time control required for the internal operation of the var generator and for the desired system compensation functions. This interface,

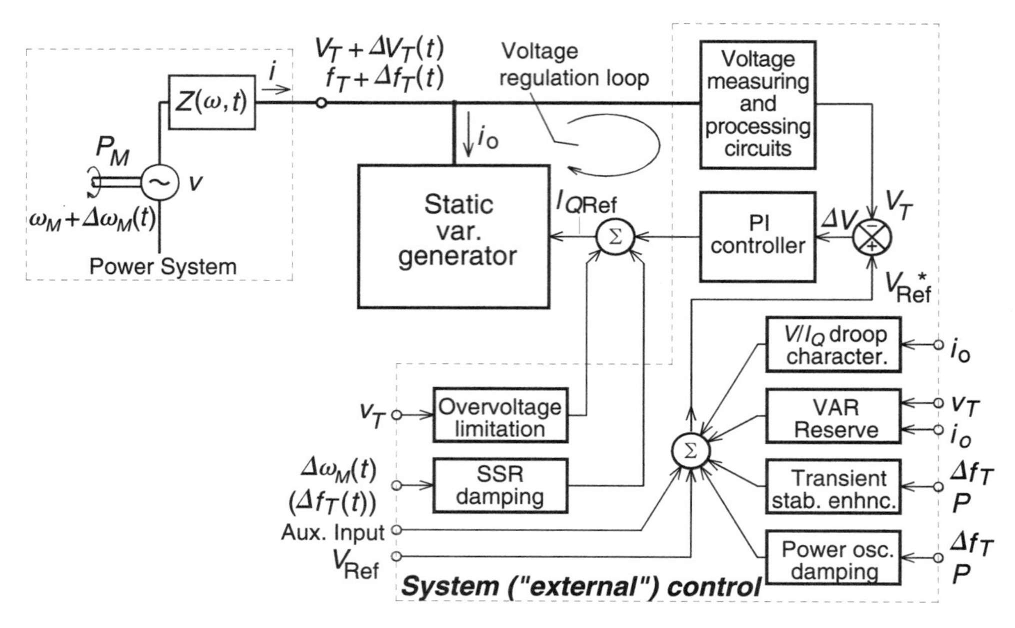

Figure 5.57 Structure of the basic compensator control for multifunctional power system compensation.

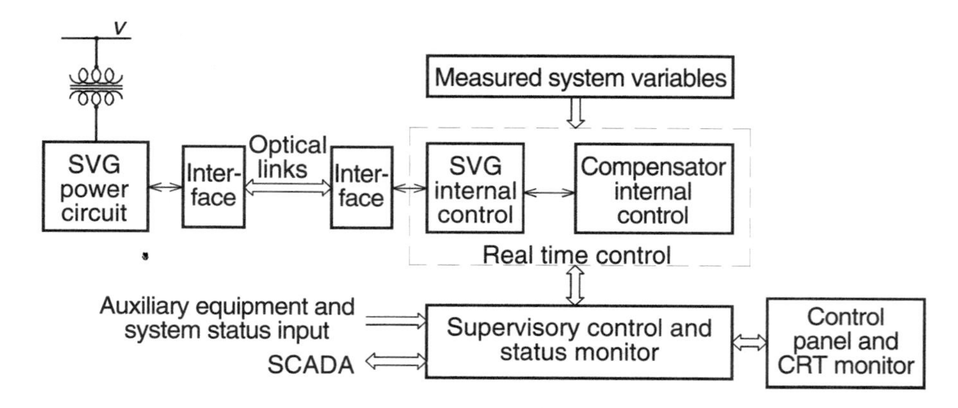

Figure 5.58 Main elements of the complete control operating a static compensator.

transmitting gating commands from the control to the valves and status information from the valves to the control is usually implemented by optical links.

2. Signal measuring and processing circuits for system and equipment variables. The real-time control and protection relays (and operator displays) need as inputs certain system variables, such as terminal voltage and compensator output current, as well as appropriate internal voltages and currents of the equipment, from which magnitude, phase, frequency, and other relevant information can be derived to follow in real-time system conditions and monitor equipment operation.
3. Supervisory control and status monitor which interfaces with the all parts of

the compensator, including all essential components of the compensator proper and its support equipment (e.g., cooling system, power supplies, breakers, switches, interlocks, etc.). It collects status information from every part of the system, usually via serial communication links, organizes and interprets the status data to determine the operational integrity of the compensator and to provide diagnostics for possible malfunctions and failures. It also carries out the start-up and shutdown sequencing and other operating routines of the compensator and provides appropriate communication links for the local and remote operators.

4. User interface with CRT graphical displays is usually provided by a standalone computer with an appropriate CRT monitor, keyboard, and pointing device for data entry. This computer usually has a serial link to the status processor and runs a graphical display and control software. Through the interface a large amount of information is available for the operation, diagnostic, and maintenance purposes in graphical and numerical form. The information includes: status information from the valves, identifying failed power semiconductors and other components and associated circuits; selected operating modes of the compensator and associated control and operational parameter settings; control operating and redundancy; and status of support equipment such as cooling system, auxiliary power supplies, breakers, switches, etc., and building climate status (temperature, humidity, etc.).

5.4 COMPARISON BETWEEN STATCOM AND SVC

On the basis of explanations provided in the previous sections it should be clear to the reader that, on the one hand, in the linear operating range the *V-I* characteristic and functional compensation capability of the STATCOM and the SVC are similar. However, the basic operating principles of the STATCOM, which, with a converter-based var generator, functions as a shunt-connected synchronous voltage source, are fundamentally different from those of the SVC, which, with thyristor-controlled reactors and thyristor-switched capacitors, functions as a shunt-connected, controlled reactive admittance. This basic operational difference (voltage source versus reactive admittance) accounts for the STATCOM's overall superior functional characteristics, better performance, and greater application flexibility than those attainable with the SVC. These operational and performance characteristics are summarized here, with the underlying physical reasons behind them, and with the corresponding application benefits.

5.4.1 *V-I* and *V-Q* Characteristics

The STATCOM is essentially an alternating voltage source behind a coupling reactance with the corresponding *V-I* and *V-Q* characteristics shown in Figures 5.59(a) and 5.60(a), respectively. These show that the STATCOM can be operated over its full output current range even at very low (theoretically zero), typically about 0.2 p.u. system voltage levels. In other words, the maximum capacitive or inductive output current of the STATCOM can be maintained independently of the ac system voltage, and the maximum var generation or absorption changes linearly with the ac system voltage.

In contrast to the STATCOM, the SVC, being composed of (thyristor-switched capacitors and reactors, becomes a fixed capacitive admittance at full output. Thus,

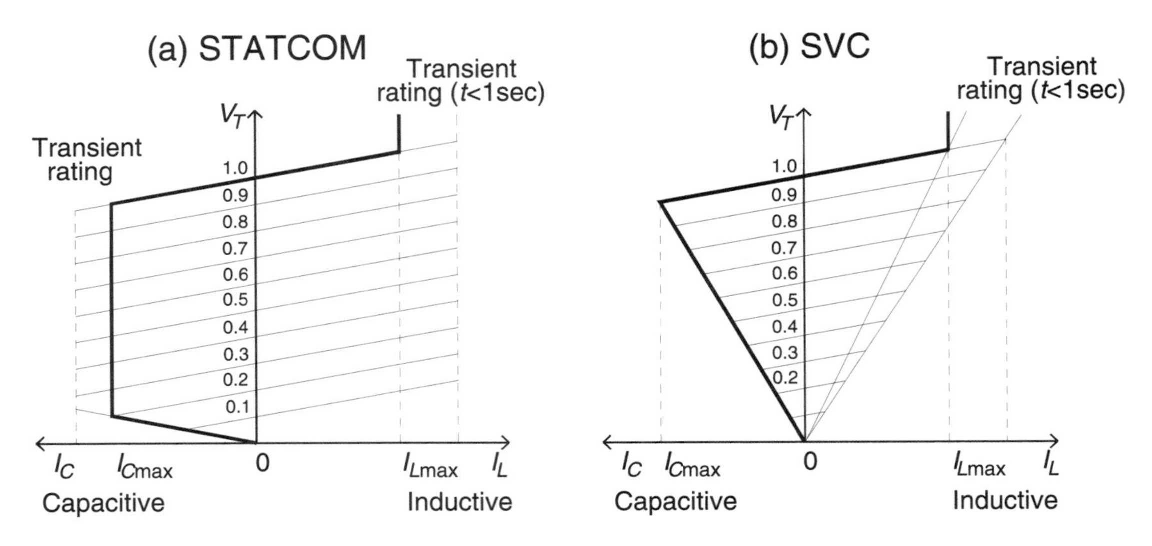

Figure 5.59 *V-I* characteristic of the STATCOM (a) and of the (SVC) (b).

the maximum attainable compensating current of the SVC decreases linearly with ac system voltage, and the maximum var output decreases with the square of this voltage, as shown in Figures 5.59(b) and 5.60(b), respectively. The STATCOM is, therefore, superior to the SVC in providing voltage support under large system disturbances during which the voltage excursions would be well outside of the linear operating range of the compensator. The capability of providing maximum compensating current at reduced system voltage enables the STATCOM to perform in a variety of applications the same dynamic compensation as an SVC of considerably higher rating.

As Figures 5.59(a) and 5.60(a) illustrate, the STATCOM may, depending on the power semiconductors used, have an increased transient rating in both the inductive

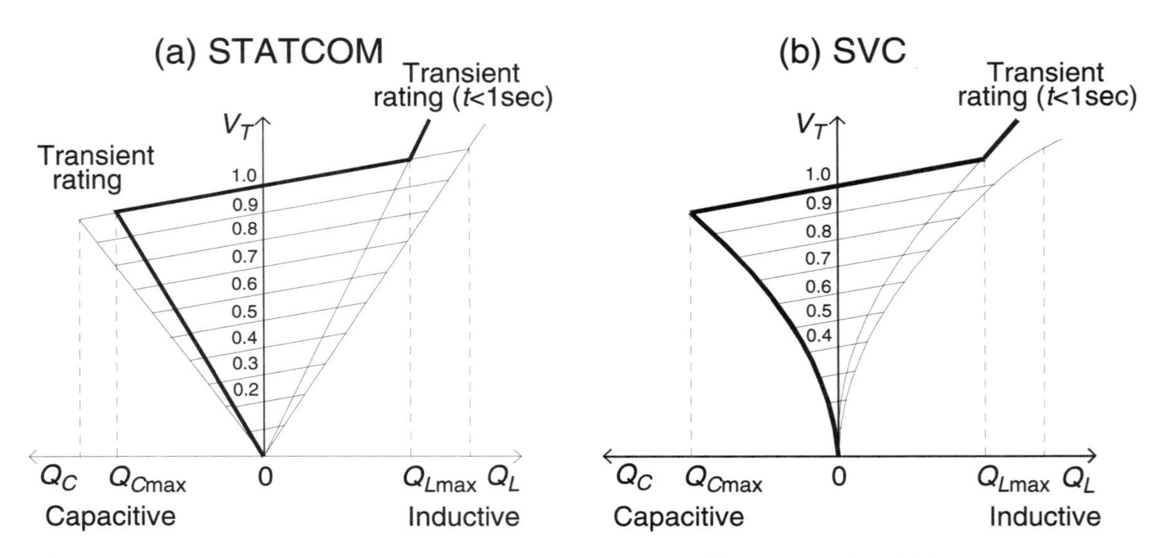

Figure 5.60 *V-Q* characteristic of the STATCOM (a) and of the (SVC) (b).

and capacitive operating regions. (The SVC has no means to increase transiently the var generation since the maximum capacitive current it can draw is strictly determined by the size of the capacitor and the magnitude of the system voltage.) The maximum attainable transient overcurrent of the STATCOM in the capacitive region is determined by the maximum current turn-off capability of the power semiconductors (e.g., GTO thyristors) employed. As shown in Chapter 3, in the inductive operating region the power semiconductors of an elementary converter, switched at the fundamental frequency, are naturally commutated. This means that the transient current rating of the STATCOM in the inductive range is, theoretically, limited only by the maximum permissible GTO junction temperature, which would in principle allow the realization of a higher transient rating in this range than that attainable in the capacitive range. However, it should be pointed out that this possibility would generally not exist if the converter poles were operated to produce a pulse-width modulated waveform, when the current conduction between the upper and lower valves is transferred several times during each fundamental half cycle. Even with non-PWM converters, abnormal operating conditions should be carefully considered in the implementation of transient ratings above the peak turn-off current capability of the semiconductors employed, because if an expected natural commutation would be missed for any reason, converter failure requiring a forced shutdown would likely occur.

5.4.2 Transient Stability

The ability of the STATCOM to maintain full capacitive output current at low system voltage also makes it more effective than the SVC in improving the transient (first swing) stability. The effectiveness of the STATCOM in increasing the transmittable power is illustrated in Figure 5.61(a), where the transmitted power P is shown

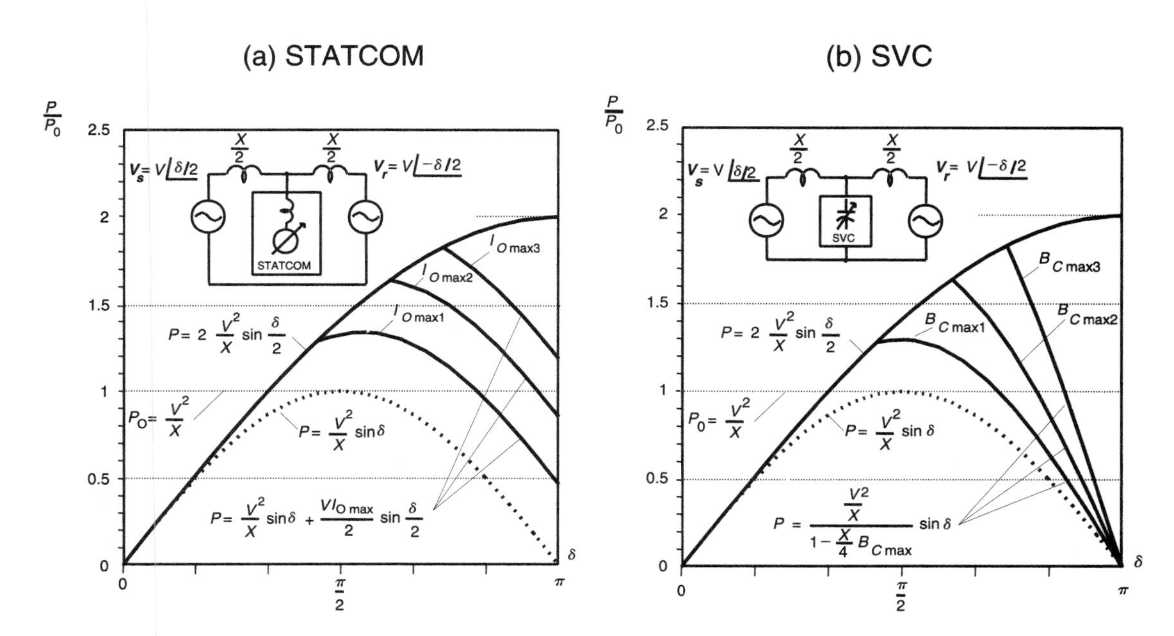

Figure 5.61 Transmitted power versus transmission angle of a two-machine system with a midpoint STATCOM (a) and a midpoint SVC (b) obtained with different var ratings.

against the transmission angle δ for the usual two-machine model at various capacitive ratings defined by the maximum capacitive output current I_{Cmax}. For comparison, an equivalent P versus δ relationship is shown for an SVC in Figure 5.61(b). It can be observed that the STATCOM, just like the SVC, behaves like an ideal midpoint shunt compensator with P versus δ relationship as defined by (5.2b), $P = (2V^2/X)\sin(\delta/2)$ until the maximum capacitive output current I_{Cmax} is reached. From this point, the STATCOM keeps providing this maximum capacitive output current (instead of a fixed capacitive admittance like the SVC), independent of the further increasing angle δ and the consequent variation of the midpoint voltage. As a result, the sharp decrease of transmitted power P in the $\pi/2 < \delta < \pi$ region, characterizing the power transmission of an SVC supported system, is avoided and the obtainable $\int P d\delta$ area representing the improvement in stability margin is significantly increased.

The increase in stability margin obtainable with a STATCOM over a conventional thyristor-controlled SVC of identical rating is clearly illustrated with the use of the previously explained equal-area criterium (Section 5.1.3) in Figures 5.62(a) and (b). The simple two-machine system, discussed at the review of the basic shunt compensation principles [Figure 5.1(a)], is compensated at the midpoint by a STATCOM and an SVC of the same var rating. For the sake of clarity, it is assumed that the system transmitting steady-state electric power P_1, at angle δ_1, is subjected to a fault for a period of time during which P_1 becomes zero. During the fault, the sending-end machine accelerates (due to the constant mechanical input power), absorbing the kinetic energy represented by the shaded area below the constant P_1 line, and increasing δ_1 to δ_c ($\delta_c > \delta_1$). Thus, when the original system is restored after fault clearing, the transmitted power becomes much higher than P_1 due to the larger transmission angle δ_c. As a result, the sending-end machine starts to decelerate, but δ increases further until the machine loses all the kinetic energy it gained during the fault. The recovered kinetic energy is represented by the shaded area between the P versus δ curve and

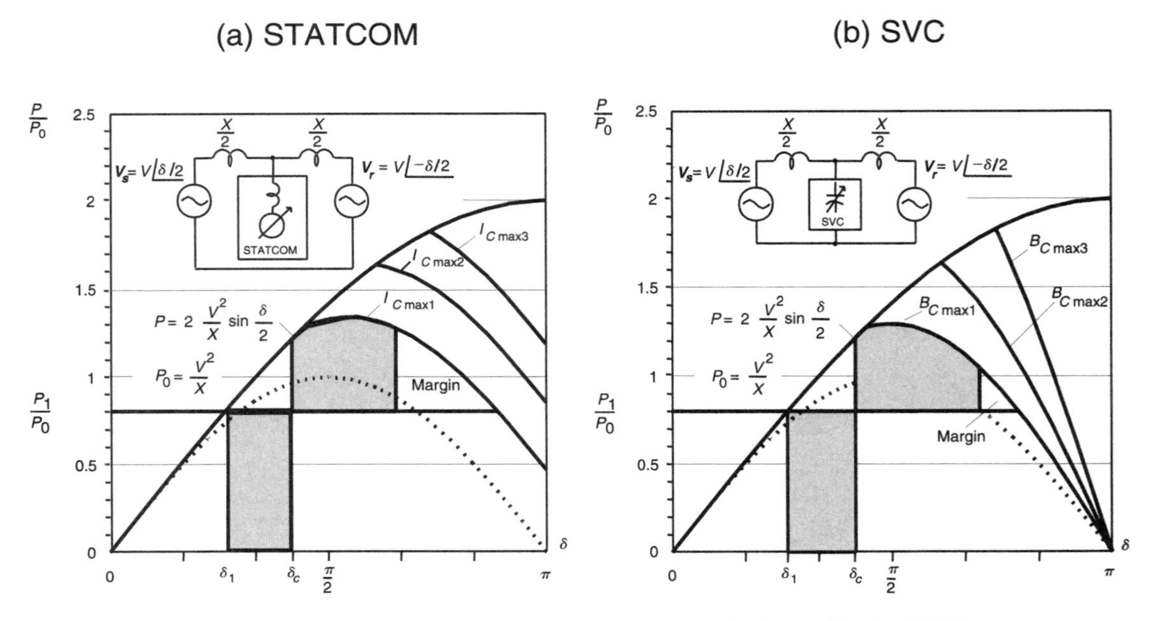

Figure 5.62 Improvement of transient stability obtained with a midpoint STATCOM (a) and a midpoint SVC (b) of a given var rating.

the constant power line P_1. The remaining unshaded area below the P versus δ curve and above the constant power line P_1 provides the transient stability margin. As can be observed, the transient stability margin obtained with the STATCOM, due to the better support of the midpoint voltage, is significantly greater than that attainable with the SVC of identical var rating. This of course means that the transmittable power can be increased if the shunt compensation is provided by a STATCOM rather than by an SVC, or, for the same stability margin, the rating of the STATCOM can be decreased below that of the SVC.

5.4.3 Response Time

As demonstrated in Section 5.3.2, the attainable response time and the bandwidth of the closed voltage regulation loop of the STATCOM are also significantly better than those of the SVC. Although the closed-loop voltage regulation of both compensators can be expressed by the formula given in (5.17), i.e., $\Delta V_T/\Delta V = 1/(1 + G_1G_2HX)$, the time constant T_d in the transfer function G_2 (which characterizes the inherent "transport lag" in the power circuits of the STATCOM and of the SVC) is about an order of magnitude smaller for the STATCOM than it is for the SVC, i.e., it is typically from less than 200 μs to 350 μs for the STATCOM and between 2.5 and 5.0 ms for the SVC. Considering the rapidly changing angle versus frequency characteristic of the e^{-T_ds} term, this improvement is important from the standpoint of attainable frequency bandwidth. The practical importance of wide frequency bandwidth cannot be overstated for applications requiring fast response, but even in typical transmission applications the STATCOM can provide stable operation with respectable response over a much wider variation of the transmission network impedance than is possible with an SVC.

5.4.4 Capability to Exchange Real Power

For applications requiring active (real) power compensation it is clear that the STATCOM, in contrast to the SVC, can interface a suitable energy storage with the ac system for real power exchange. That is, the STATCOM is capable of drawing controlled real power from an energy source (large capacitor, battery, fuel cell, superconducting magnetic storage, etc.) at its dc terminal and deliver it as ac power to the system. It can also control energy absorption from the ac system to keep the storage device charged. This potential capability provides a new tool for enhancing dynamic compensation, improving power system efficiency and, potentially, preventing power outages.

The reactive and real power exchange between the STATCOM and the ac system can be controlled independently of each other and any combination of real power generation and absorption with var generation and absorption is achievable. Thus, by equipping the STATCOM with an energy storage device of suitable capacity, extremely effective control strategies for the modulation of the reactive and real output power can be executed for the improvement of transient stability and the damping of power oscillation. It should be noted that for short-term dynamic disturbances an energy consuming device (e.g., a switched resistor) may be effectively used in place of the more expensive energy storage to absorb power from the ac system via the STATCOM. With this simple scheme, the STATCOM would transfer energy from the ac system to the dc terminals where it would be dissipated by the energy-consuming device that would be switched on whenever surplus energy at those terminals is detected (by, e.g., the increase of dc voltage).

5.4.5 Operation With Unbalanced AC System

The ac system voltages are normally balanced (maximum unbalance does not usually exceed 1%) and therefore compensators normally control all three phases of their output current together. This for the SVC normally means that its control establishes three identical shunt admittances, one for each phase. Consequently, with unbalanced system voltages the compensating currents in each phase would become different. It is possible to control the three compensating admittances individually by adjusting the delay angle of the TCRs so as to make the three compensating currents identical. However, in this case the triple-n harmonic content would be different in each phase and their normal cancellation through delta connection would not take place. This operation mode thus would generally require the installation of the usually unneeded third harmonic filters. For this reason, individual phase control for SVCs in transmission line compensation is rarely employed.

The operation of the STATCOM under unbalanced system conditions is different from that of the SVC, but the consequences of such operation are similar. The STATCOM operation is governed by the fundamental physical law requiring that the net instantaneous power at the ac and dc terminals of the voltage-sourced converter employed must always be equal. This is because the converter has no internal energy storage and thus energy transfer through it is absolutely direct, and consequently the net instantaneous power at its ac and dc terminals must be equal.

Assume that the dc terminal voltage of the STATCOM is supported entirely by an appropriately charged dc capacitor (i.e., there is no source or sink of power attached to this terminal), and that the losses of the converter are zero and its pulse number is infinite (ideal converter). With perfectly balanced sinusoidal ac terminal voltages (provided by the ac power system), the STATCOM will draw a set of balanced, sinusoidal currents in quadrature with the system voltages, but the dc capacitor will experience no charging current because no real power is exchanged with the ac system and, furthermore, because the net instantaneous power remains invariably zero at the ac terminals of the converter. However, if the ac system voltages become unbalanced, then an alternating power component at twice the fundamental frequency will appear at the ac terminals of the STATCOM converter and this will be matched by an alternating second harmonic charging current in the dc terminals, producing in turn an associated alternating voltage component of the same frequency across the capacitor shunting the dc terminals. If the converter control ignores this ac voltage component, that is, if it is operated to produce the ac output voltage as if the dc terminal voltage was constant, then the second harmonic voltage component from the dc terminal will be transformed (by the converter switching operation) as a negative sequence fundamental component and a positive sequence third harmonic component to the ac terminals. As a result, the STATCOM will, in general, draw a negative sequence fundamental current component (in proportion to the difference between its internally generated negative sequence voltage and the negative sequence component of the ac system voltage) as well as a (positive sequence) third harmonic current component.

Out of the two voltage components, generated in the output of the STATCOM as a result of system unbalance, the third harmonic is clearly "unwanted." Whereas the negative sequence fundamental voltage, generated "naturally" by the converter with properly sized dc capacitor, reduces significantly the negative sequence current that would otherwise be forced to flow by the negative sequence system voltage (which could be very large during single-phase faults), the third harmonic has clearly no useful function.

The "natural" behavior of the STATCOM is illustrated in Figure 5.63 where relevant voltage and current waveforms records, representing a TNA simulated power system with precisely scaled 48-pulse STATCOM hardware model that was subjected to a severe line to ground fault. The traces in the figure show (from top to bottom): line to line voltages v_{ab} and v_{cb} (phase a is faulted to ground); the three currents drawn by the STATCOM, i_a, i_b, i_c; the dc capacitor voltage v_{dc}; and the reactive current reference I_{QRef} (limited to 2 p.u.). In steady state, the STATCOM was producing 1.0 p.u. capacitive current, when it was subjected to a line to ground fault lasting for about five cycles. It can be observed that, due to the internally generated negative sequence converter voltage (that largely matched the negative sequence voltage of the unbalanced ac system), the STATCOM was providing during the fault substantially balanced, capacitive compensating currents with the maximum magnitude of 2.0 p.u., but with considerable third harmonic distortion. However, the harmonics, present only during the five cycle fault period are, arguably, of no significant consequence (since, presumably, under this condition significantly more distortion is generated by various static

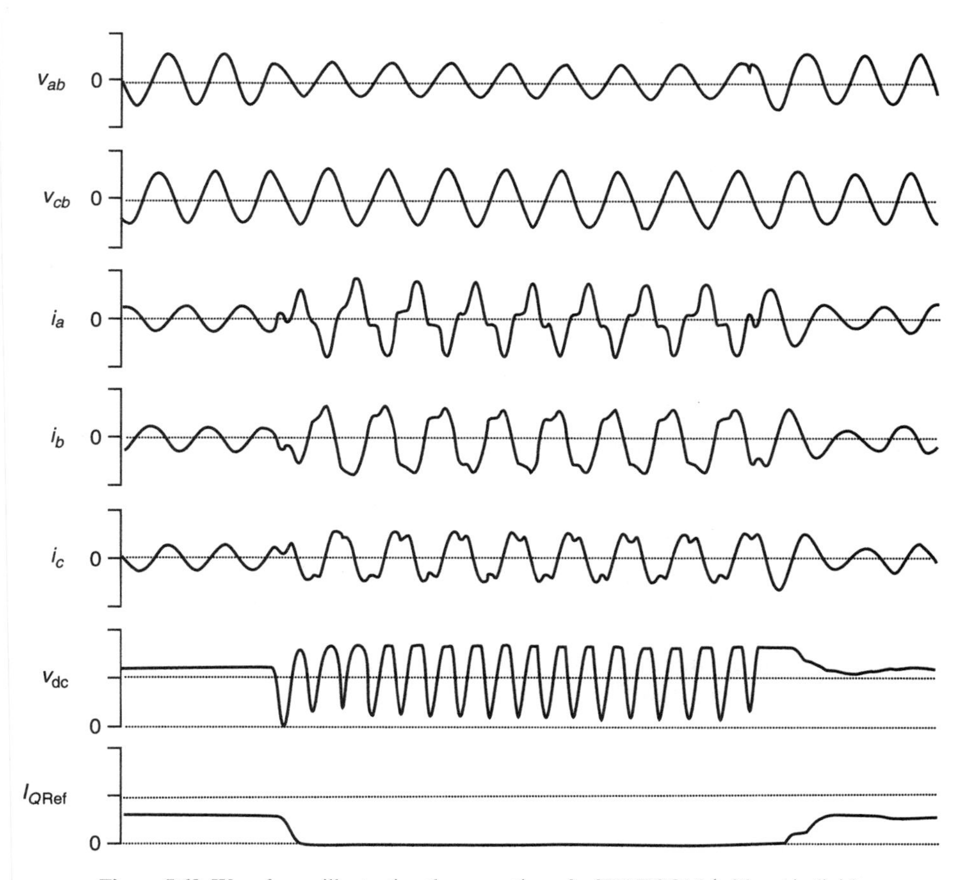

Figure 5.63 Waveforms illustrating the operation of a STATCOM (without individual phase voltage control) during and following a line to ground fault at the regulated bus.

and rotating electromagnetic components present in the system). On the other hand, the STATCOM, as evidenced in the figure by the capacitive compensation maintained, can provide strong system support during the fault. This, with the STATCOM's fast response at fault clearing, could greatly contribute to system stability. (It should be noted that in practice the over-current limit of the STATCOM, for economic reasons, would normally be set lower than 2.0 p.u. used in this illustrative example.)

The natural behavior of the STATCOM characterized above is related to the operating mode in which the three output voltages are controlled together; that is, single-phase voltage control is not applied. This operating mode provides the best VA utilization of the converter and generally the lowest harmonic generation obtainable under normal system conditions with a given method of waveform synthesis employed. However, in special applications where considerable system unbalance exists, or where large unbalanced loads are to be compensated, the STATCOM converter with appropriate pole structure and waveform synthesis method can be operated to control each of the output phases individually, that is, to control the positive and negative sequence compensating currents independently. However, this type of operation is usually associated with some amount of fluctuating ac power and, therefore, it requires a larger dc capacitor than typically used in a transmission system STATCOM to accommodate the consequent second harmonic ripple current at the dc terminals.

5.4.6 Loss Versus Var Output Characteristic

As shown in Figures 5.38 and 5.27 the overall loss versus reactive output characteristic, as well as the actual operating losses, of the STATCOM are comparable to those of the SVC using both thyristor-controlled reactors and thyristor-switched capacitors. Both types of compensator have relatively low losses (about 0.1 to 0.2%) at and in the vicinity of zero var output. On the average, the losses in both cases increase with increasing var output reaching about 1.0% at rated output. This type of loss versus output characteristic is generally considered favorable for transmission applications where the average var output demand is normally low and the compensator is primarily applied to handle dynamic events, system contingencies, and possibly the coordination of the overall area var control.

The loss contribution of power semiconductors and related components to the total compensator losses is higher for the STATCOM than for the SVC. This is because presently available power semiconductor devices with internal turn-off capability have higher conduction losses than conventional thyristors. Also switching losses with forced current interruption tend to involve more losses than natural commutation. However, it is reasonable to expect that the historically rapid semiconductor developments will reduce the device losses in the coming years, whereas the losses of conventional power components, such as reactors, are not likely to change significantly. Thus, the technological advances probably will have help to reduce the overall losses of the STATCOM more than those of the SVC.

5.4.7 Physical Size and Installation

From the standpoint of physical installation, because the STATCOM not only controls but also internally generates the reactive output power (both capacitive and inductive), the large capacitor and reactor banks with their associated switchgear and protection, used in conventional thyristor-controlled SVCs, are not needed. This results in a significant reduction in overall size (about 30 to 40%), as well as in installation labor and cost. The small physical size of the STATCOM makes it eminently suitable for installations in areas where land cost is at a premium, and for applications where anticipated system changes may require the relocation of the installation.

5.4.8 Merits of Hybrid Compensator

From the *V-I* characteristics shown in Figures 5.39 through 5.40 for hybrid var generators employing a switching power converter with thyristor-switched capacitors and thyristor-controlled reactors the obvious deduction can be made that, in general, the operating and performance advantages attainable with the STATCOM will also be exhibited by a hybrid compensator, in direct proportion to the rating of the STATCOM relative to the total controllable var range of the hybrid compensator.

In considering hybrid compensator schemes, a particularly good case can be made for the replacement of the TCR with a converter-based var generator in the presently used TCR-TSC compensator arrangements. This replacement would result in a number of significant operating and performance advantages, including:

1. Faster response, since the converter can immediately provide capacitive output before the TSCs could be switched in. (The TCR can only absorb reactive power.)
2. Reduced harmonic generation and the possible elimination of filters, since the converter can be designed to have very low harmonic generation. (TCR is the harmonic source in the SVC.)
3. Greater flexibility to optimize for loss evaluation criteria, since the converter can generate and absorb reactive power, which makes it possible to switch the capacitors with either a net var output surplus or a net var output shortage that the converter then needs to absorb or generate. (The TCR can only absorb reactive power, making it necessary to switch the capacitors with a net var surplus. For example, even at very small capacitive output demand the first TSC must be switched in, which forces the TCR to absorb a very large surplus capacitive vars, causing significant internal losses at that operating point of the compensator.)

It should be also noted that, in contrast to the TCR and mechanically-switched capacitor (MSC) arrangement, the combination of a STATCOM with MSCs may provide a good and economical solution for many applications. This is because the STATCOM would be able to provide capacitive var output immediately upon demand and only the additional capacitive output would be delayed by the MSC operation. By contrast, in the TCR-MSC arrangement, there would be no compensation before the first MSC would be switched in.

5.5 STATIC VAR SYSTEMS

A static var system is, per CIGRE/IEEE definition, a combination of static compensators and mechanically-switched capacitors and reactors whose operation is coordinated. A static var system is thus not a well-defined compensating arrangement because it does not have a uniform *V-I* characteristic and its overall response time is greatly dependent on the mechanical switching devices used.

The emphasis in a static var system is on coordination. The major objective is usually to ensure that the static compensator, having a well-defined *V-I* characteristic and fast speed of response, is available for dynamic compensation and other elements of the overall var system handle the steady-state var demands. Another reason for coordination is to minimize the steady-state losses in the compensator and the overall power system.

The var output coordination may follow different strategies. In the simplest form, it may be nothing more than the previously described var reserve control, which forces the output of the static compensator to return to a specific (or adjustable) var reference after each significant var demand change in the power system. With this arrangement, the availability of a specific amount of fast compensation capacity is enforced by an automatic control action, but the steady var demand is left for unidentified "other means" in the power system (which may include generator voltage regulators, synchronous compensators (condensers), and mechanically-switched capacitor and reactor banks activated by under- and over-voltage relays) to provide.

An equally simple, but philosophically opposite, policy is to let the static compensator pick up the reactive compensation as required, but provide an alarm signal to the power system dispatcher if a specific var output is exceeded. It is left to the dispatcher to determine whether the compensator should keep providing the compensation or other available means should be brought in operation.

In a more rigorous coordination scheme, the compensator would control a number of dedicated capacitor and reactor banks within the overall static var system. That is, if the capacitive output of the compensator would exceed a preset level for a specified time duration, then the compensator control would activate, in a predetermined sequence, the mechanically-switched capacitor banks until the output of the compensator is reduced below that level. Similarly, an excess in inductive var output would initiate the systematic disconnection of capacitor banks and, if required, actuate an appropriate number of mechanically-switched reactor banks.

In providing automatic coordination, due attention must be paid to the capabilities of the mechanical switches with regard to the frequency of operation and, also, the limitation of possible surge currents. A microprocessor-based control is usually the most convenient for monitoring switch status, storing switching history, and effecting overall coordination according to established priorities and compensation policies.

REFERENCES

Bergmann, K., et al., "Application of GTO-Based SVCs for Improved Use of the Rejsby Hede Windfarm," Ninth National Power System Conference, Kanpur NPSC '96, vol. 2, pp. 391–395.

Edwards, C. W., et al., "Advanced Static Var Generator Employing GTO Thyristors," *IEEE/PES Winter Power Meeting,* Paper No. 38WM109-1, 1988.

Erinmez, I. A., ed., "Static Var Compensators," Working Group 38-01, Task Force No. 2 on SVC, CIGRE, 1986.

Erinmez, I. A., and Foss, A. M., eds., "Static Synchronous Compensator (STATCOM)," Working Group 14.19, CIGRE Study Committee 14, Document No. 144, August 1999.

Gyugyi, L., "Power Electronics in Electric Utilities: Static Var Compensators," *Proceeding of IEEE,* vol. 76, no. 4, April 1988.

Gyugyi, L., "Reactive Power Generation and Control by Thyristor Circuits," Paper SPCC 77-29 presented at Power Electronics Specialist Conference, Cleveland, OH, June 8–10, 1976. Reprinted in the *IEEE Trans. Ind. Appl.,* vol. IA-15, no. 5, September/October 1979.

Gyugyi, L., et al., "Advanced Static Var Compensator Using Gate Turn-off Thyristors for Utility Applications," *CIGRE paper 23-203,* 1990.

Gyugyi, L., "Dynamic Compensation of AC Transmission Lines by Solid-State Synchronous Voltage Sources," *IEEE/PES Summer Meeting,* Paper No. 93SM431-1PWRD, 1993.

Gyugyi, L., and Taylor, E. R., "Characteristics of Static Thyristor. Controlled Shunt Compensators for Power Transmission System Applications," *IEEE Trans. PAS,* vol. PAS-99, no. 5, pp. 1795–1804, September/October 1980.

Gyugyi, L., et al., "Principles and Applications of Static, Thyristor-Controlled Shunt Compensators," *IEEE Trans. PAS,* vol. PAS-97, no. 5, September/October 1978.

Hauth, R. L., and Moran, R. J., "Basics of Applying Static Var Systems on HVAC Power Networks," EPRI Seminar on Transmission Static Var Systems, Duluth, MN, October 24, 1978.

Kimbark, E. W., "How to Improve System Stability without Risking Subsynchronous Resonance," *IEEE Trans. PAS-96,* no. 5, September/October 1977.

Larsen E., et al., "Benefits of GTO-Based Compensation Systems for Electric Utility Applications," *IEEE/PES Summer Power Meeting,* Paper No. 91 SM 397-0 TWRD, 1991.

Mathur, R. M., ed., *Static Compensators for Reactive Power Control,* The Committee on Static Compensation, Canadian Electrical Association (CEA), 1984.

Miller, T. J. E., ed., *Reactive Power Control in Electric Systems,* John Wiley & Sons, New York, 1982.

Mori, S., et al., "Development of Large Static Var Generator Using Self-Commutated Converters for Improving Power System Stability," *IEEE/PES Winter Power Meeting,* Paper No. 92WM165-1, 1992.

Povh, D., and Weinhold, M., "Efficient Computer Simulation of STATCON," International Conference on Power System Transients, Lisbon, pp. 397–402, September 1995.

Schauder, C. D., et al. "Development of a ±100 MVAR Static Condenser for Voltage Control of Transmission Systems," *IEEE/PES Summer Meeting,* Paper No. 94SM479-6 PWRD, 1994.

Schauder, C. D., and Mehta, H., "Vector Analysis and Control of Advanced Static Var Compensators," IEE Fifth International Conference on AC and DC Transmission, *Conference Publication No. 345,* pp. 266–272, 1991.

Schauder, C. D., et al., "Operation of ±100 MVAR TVA STATCON," *IEEE/PES Winter Meeting,* Paper No. PE-509-PWRD-0-01-1997.

Schweickardt, H. W., et al., "Closed Loop Control of Static Var Sources (SVC) on EHV Transmission Lines," *IEEE/PES Winter Meeting,* Paper A 78 135-6, 1978.

Sen, K. K., "STATCOM-STATic synchronous COMpensator: Theory, Modeling, and Applications," 99WM706, *Proceedings of IEEE/PES Winter Meeting,* New York, 1999.

Sumi, Y., et al., "New Static Var Control Using Force-Commutated Converters," *IEEE/PES Winter Power Meeting,* Paper No. 81WM228-6, 1981.

6

Static Series Compensators: GCSC, TSSC, TCSC and SSSC

6.1 OBJECTIVES OF SERIES COMPENSATION

It was shown in Chapter 5 that reactive shunt compensation is highly effective in maintaining the desired voltage profile along the transmission line interconnecting two busses of the ac system and providing support to the end voltage of radial lines in the face of increasing power demand. Thus, reactive shunt compensation, when applied at sufficiently close intervals along the line, could theoretically make it possible to transmit power up to thermal limit of the line, if a large enough angle between the two end voltages could be established. However, shunt compensation is ineffective in controlling the actual transmitted power which, at a defined transmission voltage, is ultimately determined by the series line impedance and the angle between the end voltages of line.

It was always recognized that ac power transmission over long lines was primarily limited by the series reactive impedance of the line. Series capacitive compensation was introduced decades ago to cancel a portion of the reactive line impedance and thereby increase the transmittable power. Subsequently, within the FACTS initiative, it has been demonstrated that variable series compensation is highly effective in both controlling power flow in the line and in improving stability.

Controllable series line compensation is a cornerstone of FACTS technology. It can be applied to achieve full utilization of transmission assets by controlling the power flow in the lines, preventing loop flows and, with the use of fast controls, minimizing the effect of system disturbances, thereby reducing traditional stability margin requirements.

In this section the basic approach of reactive series compensation will be reviewed to provide the necessary foundation for the treatment of power electronics-based compensators. The effect of series compensation on the basic factors, determining attainable maximal power transmission, steady-state power transmission limit, transient stability, voltage stability and power oscillation damping, will be examined.

6.1.1 Concept of Series Capacitive Compensation

The basic idea behind series capacitive compensation is to decrease the overall effective series transmission impedance from the sending end to the receiving end, i.e., X in the $P = (V^2/X)\sin\delta$ relationship characterizing the power transmission over a single line. Consider the simple two-machine model, analogous to that shown for shunt compensation in Figure 5.1, but with a series capacitor compensated line, which, for convenience, is assumed to be composed of two identical segments, as illustrated in Figure 6.1(a). The corresponding voltage and current phasors are shown in Figure 6.1(b). Note that for the same end voltages the magnitude of the total voltage across the series line inductance, $V_x = 2V_{x/2}$ is increased by the magnitude of the opposite voltage, V_C, developed across the series capacitor; this results from an increase in the line current.

The *effective* transmission impedance X_{eff} with the series capacitive compensation is given by

$$X_{eff} = X - X_C \tag{6.1}$$

or

$$X_{eff} = (1 - k)X \tag{6.2}$$

where k is the *degree of series compensation*, i.e.,

$$k = X_C/X \qquad 0 \leq k < 1 \tag{6.3}$$

Assuming $V_s = V_r = V$ in Figure 6.1(b), the current in the compensated line, and the corresponding real power transmitted, can be derived in the following forms:

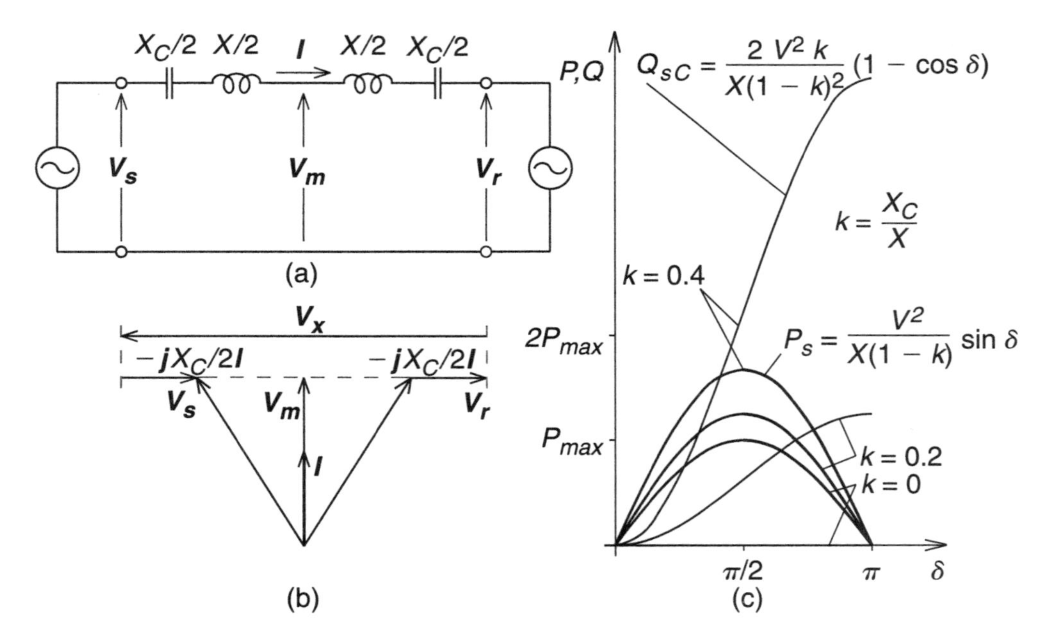

Figure 6.1 Two-machine power system with series capacitive compensation (a), corresponding phasor diagram (b), real power and series capacitor reactive power vs. angle characteristics (c).

$$I = \frac{2V}{(1-k)X}\sin\frac{\delta}{2} \tag{6.4}$$

$$P = V_m I = \frac{V^2}{(1-k)X}\sin\delta \tag{6.5}$$

The reactive power supplied by the series capacitor can be expressed as follows:

$$Q_C = I^2 X_C = \frac{2V^2}{X}\frac{k}{(1-k)^2}(1 - \cos\delta) \tag{6.6}$$

The relationship between the real power P, series capacitor reactive power Q_C, and angle δ is shown plotted at various values of the degree of series compensation k in Figure 6.1(c). It can be observed that, as expected, the transmittable power rapidly increases with the degree of series compensation k. Similarly, the reactive power supplied by the series capacitor also increases sharply with k and varies with angle δ in a similar manner as the line reactive power.

After deriving the simple relationships characterizing series capacitive compensation, the reader should note the duality of the underlying physical explanations. The conventional explanation is that the impedance of the series compensating capacitor cancels a portion of the actual line reactance and thereby the effective transmission impedance, per (6.1), is reduced as if the line was physically shortened. An equally valid physical explanation, which will be helpful to the understanding of converter-based power flow controllers, is that in order to increase the current in the given series impedance of the actual physical line (and thereby the corresponding transmitted power), the voltage across this impedance must be increased. This can be accomplished by an appropriate series connected circuit element, such as a capacitor, the impedance of which produces a voltage opposite to the prevailing voltage across the series line reactance and, as the phasor diagram in Figure 6.1(c) illustrates, thereby causes this latter voltage to increase. It is easy to see that within this second explanation the physical nature of the series circuit element is irrelevant as long as it produces the desired compensating voltage. Thus, an alternate compensating circuit element may be envisioned as an ac voltage source which directly injects the desired compensating voltage in series with the line. As will be seen, the switching power converter used in the shunt-connected STATCOM, applied as a voltage source in series with the line, serves the functional capabilities of series capacitive compensation and also provides additional options for power flow control.

6.1.2 Voltage Stability

Series capacitive compensation can also be used to reduce the series reactive impedance to minimize the receiving-end voltage variation and the possibility of voltage collapse. A simple radial system with feeder line reactance X, series compensating reactance X_C, and load impedance Z is shown in Figure 6.2(a). The corresponding normalized terminal voltage V_r versus power P plots, with unity power factor load at 0, 50, and 75% series capacitive compensation, are shown in Figure 6.2(b). The "nose-point" at each plot given for a specific compensation level represents the corresponding voltage instability. Note that Figure 6.2 is analogous to Figure 5.3 where the same radial system with a reactive shunt compensator, supporting the end voltage, is shown. Clearly, both shunt and series capacitive compensation can effectively increase the voltage stability limit. Shunt compensation does it by supplying the reactive load demand and regulating the terminal voltage. Series capacitive compensation does it

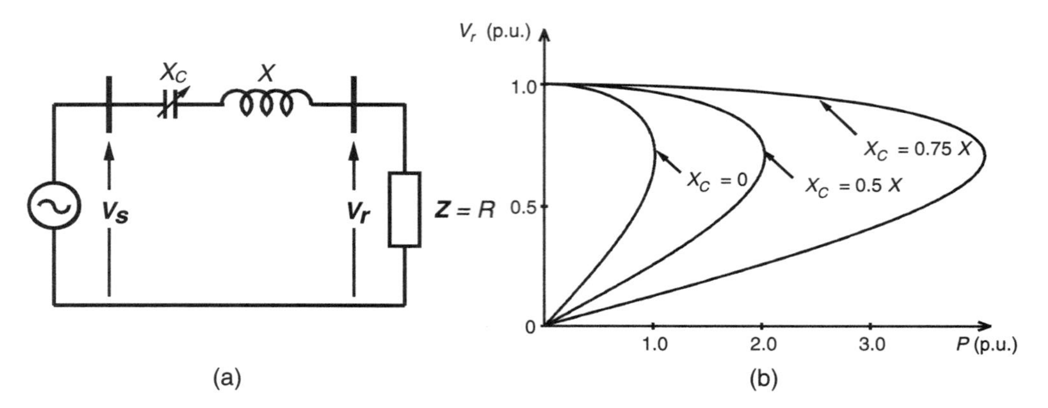

Figure 6.2 Transmittable power and voltage stability limit of a radial transmission line as function of series capacitive compensation.

by canceling a portion of the line reactance and thereby, in effect, providing a "stiff" voltage source for the load. For increasing the voltage stability limit of overhead transmission, series compensation is much more effective than shunt compensation of the same MVA rating.

6.1.3 Improvement of Transient Stability

As shown in Chapter 5, transient stability improvement by controlled shunt compensation is achieved by increasing the power transmission via increasing (or maintaining) the (midpoint) transmission line voltage during the accelerating swing of the disturbed machine(s). The powerful capability of series line compensation to control the transmitted power can be utilized much more effectively to increase the transient stability limit and to provide power oscillation damping. The equal area criterion, introduced in Chapter 5 to investigate the capability of the ideal shunt compensator to improve the transient stability, is used again here to assess the relative increase of the transient stability margin attainable by series capacitive compensation.

Consider the simple system with the series compensated line shown in Figure 6.1(a). As for the shunt compensated system shown in Figure 5.1, it is, for convenience, also assumed for the series compensated case that the pre-fault and post-fault systems remain the same. Suppose that the system of Figure 6.1(a), with and without series capacitive compensation, transmits the same power P_m. Assume that both the uncompensated and the series compensated systems are subjected to the same fault for the same period of time. The dynamic behavior of these systems is illustrated in Figures 6.3(a) and (b). As seen, prior to the fault both of them transmit power P_m at angles δ_1 and δ_{s1}, respectively. During the fault, the transmitted electric power becomes zero while the mechanical input power to the generators remains constant, P_m. Therefore, the sending-end generator accelerates from the steady-state angles δ_1 and δ_{s1} to angles δ_2 and δ_{s2}, respectively, when the fault clears. The accelerating energies are represented by areas A_1 and A_{s1}. After fault clearing, the transmitted electric power exceeds the mechanical input power and therefore the sending-end machine decelerates. However, the accumulated kinetic energy further increases until a balance between the accelerating and decelerating energies, represented by areas A_1, A_{s1}, and A_2, A_{s2}, respectively, is reached at the maximum angular swings, δ_3 and δ_{s3}, respectively. The areas between the P versus δ curve and the constant P_m line over the intervals defined by angles δ_3 and δ_{crit}, and δ_{s3} and δ_{scrit}, respectively, determine the margin of transient stability, represented by areas A_{margin} and $A_{smargin}$.

Comparison of Figures 6.3(a) and (b) clearly shows a substantial increase in the transient stability margin the series capacitive compensation can provide by partial cancellation of the series impedance of the transmission line. The increase of transient stability margin is proportional to the degree of series compensation. Theoretically this increase becomes unlimited for an ideal reactive line as the compensation approaches 100%. However, practical series capacitive compensation does not usually exceed 75% for a number of reasons, including load balancing with parallel paths, high fault current, and the possible difficulties of power flow control. Often the compensation is limited to less than 30% due to subsynchronous concerns.

It is emphasized here again that under practical fault scenarios the pre-fault and post-fault systems are generally different. From the standpoint of transient stability, and of overall system security, the post-fault system is the one that matters. That is, power systems are normally designed to be transiently stable, with defined pre-fault contingency scenarios and post-fault system degradation, when subjected to a major disturbance. For this reason, in most practical systems, the actual capacity of transmission networks is considerably higher than that at which they are normally used. The powerful capability of series compensation, with sufficiently fast controls, to handle dynamic disturbances and increase the transmission capability of post fault or otherwise degraded systems, can be effectively used to reduce the "by-design" underutilization of many power systems.

6.1.4 Power Oscillation Damping

Controlled series compensation can be applied effectively to damp power oscillations. As explained in Chapter 5, for power oscillation damping it is necessary to vary the applied compensation so as to counteract the accelerating and decelerating swings of the disturbed machine(s). That is, when the rotationally oscillating generator accelerates and angle δ increases ($d\delta/dt > 0$), the electric power transmitted must be increased to compensate for the excess mechanical input power. Conversely, when the generator

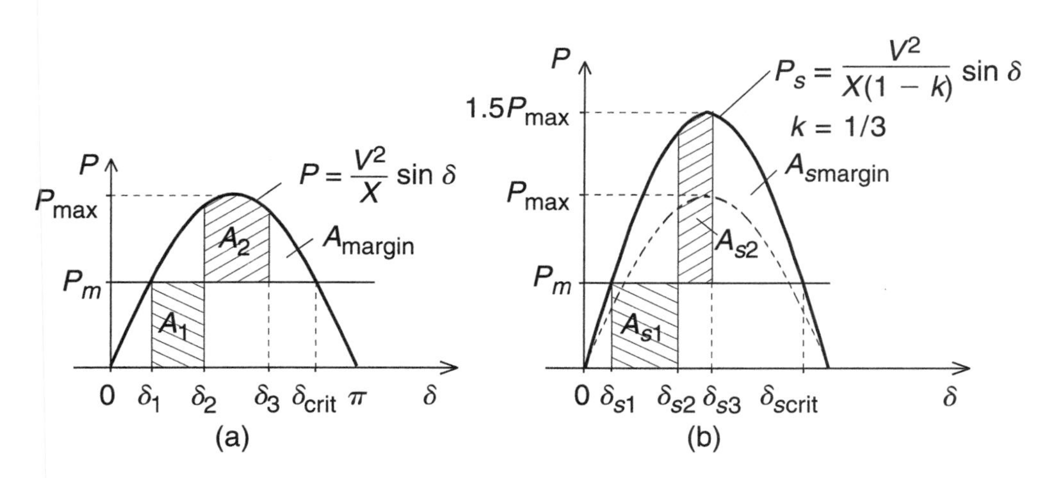

Figure 6.3 Equal area criterion to illustrate the transient stability margin for a simple two-machine system, (a) without compensation, and (b) with a series capacitor.

decelerates and angle δ decreases ($d\delta/dt < 0$), the electric power must be decreased to balance the insufficient mechanical input power.

The required variation of the degree of series compensation, together with the corresponding variation of the transmission angle δ and transmitted power P versus time of an under-damped oscillating system are shown for an illustrative hypothetical case in Figure 6.4. Waveforms in Figure 6.4(a) show the undamped and damped oscillations of angle δ around the steady-state value δ_0. Waveforms in Figure 6.4(b) show the corresponding undamped and damped oscillations of the electric power P around the steady-state value P_0, following an assumed fault (sudden drop in P) that initiated the oscillation. Waveform c shows the applied variation of the degree of series capacitive compensation, k, applied. As seen, k is maximum when $d\delta/dt > 0$, and it is zero when $d\delta/dt < 0$. With maximum k, the effective line impedance is minimum (or, alternatively, the voltage across the actual line impedance is maximum) and consequently, the electric power transmitted over the line is maximum. When k is zero, the effective line impedance is maximum (or, alternatively, the voltage across the actual line impedance is minimum) and the power transmitted is minimum. The illustration shows that k is controlled in a "bang-bang" manner (output of the series compensator is varied between the minimum and maximum values). Indeed, this type of control is the most effective for damping large oscillations. However, damping relatively small power oscillations, particularly with a relatively large series compensator, continuous variation of k, in sympathy with the generator angle or power, may be a better alternative.

6.1.5 Subsynchronous Oscillation Damping

Sustained oscillation below the fundamental system frequency can be caused by series capacitive compensation. The phenomenon, referred to as subsynchronous resonance (SSR), was observed as early as 1937, but it received serious attention only in the 1970s, after two turbine-generator shaft failures occurred at the Mojave Generating Station in southern Nevada. Theoretical investigations showed that interaction between a series capacitor-compensated transmission line, oscillating at the natural

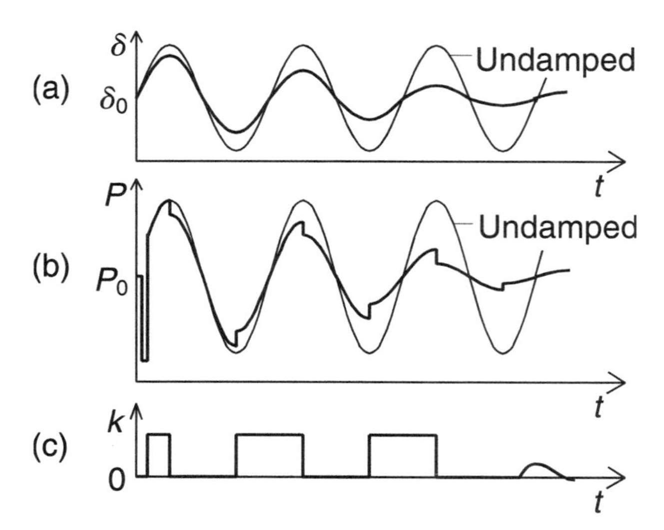

Figure 6.4 Waveforms illustrating power oscillation damping by controllable series compensation: (a) generator angle, (b) transmitted power, and (c) degree of series compensation.

(subharmonic) resonant frequency, and the mechanical system of a turbine-generator set in torsional mechanical oscillation can result in negative damping with the consequent mutual reinforcement of the electrical and mechanical oscillations. The phenomenon of subsynchronous resonance can be briefly described as follows:

A capacitor in series with the total circuit inductance of the transmission line (including the appropriate generator and transformer leakage inductances) forms a series resonant circuit with the natural frequency of $f_e = 1/2\pi\sqrt{LC} = f\sqrt{X_C/X}$, where X_C is the reactance of the series capacitor and X is the total reactance of the line at the fundamental power system frequency f.

Since the degree of series compensation $k = X_C/X$ is usually in the 25 to 75% range, the electrical resonant frequency f_e is less than the power frequency f, i.e., f_e is a subharmonic frequency. (The reader should note that the term subharmonic as it is used here does not mean that there is an integral relationship between the two frequencies; it only means that f_e is less than f.) If the electrical circuit is brought into oscillation (by some network disturbance) then the subharmonic component of the line current results in a corresponding subharmonic field in the machine which, as it rotates backwards relative to the main field (since $f_e < f$), produces an alternating torque on the rotor at the difference frequency of $f - f_e$. If this difference frequency coincides with one of the torsional resonances of the turbine-generator set, mechanical torsional oscillation is excited, which, in turn, further excites the electrical resonance. This condition is defined as subsynchronous resonance. (Of course, this process could also start in the reverse sense: a shock could start a torsional oscillation which, under the condition of subsynchronous resonance, would be reinforced by the response of the electrical network.)

Large generators with multistage steam turbines, which have multiple torsional modes with frequencies below the power frequency, are most susceptible to subsynchronous resonance with series capacitor compensated transmission lines.

In order to be able to fully exploit the functional capabilities of controlled series capacitive compensation for power flow control, transient stability improvement and power oscillation damping, it is imperative that the series compensator, as a minimal requirement, remains passive (nonparticipating) to, or, preferably, actively mitigates subsynchronous resonance. It will be seen later that power electronics-based series compensators can meet this requirement either by their non-capacitive characteristic in the subharmonic frequency range of interest or by active, control-initiated damping action.

6.1.6 Summary of Functional Requirements

The series compensator is primarily applied to solve power flow problems. These problems may be related to the length of the line or the structure of the transmission network. The electric length of the line can be shortened to meet power transmission requirements by fixed (percent) compensation of the line. Network structure related problems, which typically result in power flow unbalance, as well as parallel and loop power flows, may require controlled series compensation, particularly if contingency or planned network changes are anticipated.

Fixed or controlled series capacitive compensation can also be used to minimize the end-voltage variation of radial lines and prevent voltage collapse.

Series compensation, appropriately controlled to counteract prevailing machine swings, can provide significant transient stability improvement for post-fault systems and can be highly effective in power oscillation damping.

Appropriately structured and controlled series compensation can be applied without the danger of subsynchronous resonance to achieve full utilization of transmission lines.

In future Flexible AC Transmission Systems various controlled series compensators will play a key part in maintaining power flow over predefined paths, establishing alternative flow paths under contingency conditions, managing line loading, and generally ensuring the optimal use of the transmission network.

It will be seen that, analogously to shunt compensation, controlled series compensation to meet the above functional requirements can be accomplished by both thyristor-controlled impedance type and converter-based, voltage-source type compensators. However, the operating and performance characteristics of the two types of series compensator are considerably different.

6.1.7 Approaches to Controllable Series Compensation

As described in Chapter 5, there are two basic approaches to modern, power electronics-based shunt compensators: one, which employs thyristor-switched capacitors and thyristor-controlled reactors to realize a variable reactive admittance, and the other, which employs a switching power converter to realize a controllable synchronous voltage source. The series compensator is a reciprocal of the shunt compensator. The shunt compensator is functionally a controlled reactive current source which is connected in parallel with the transmission line to control its voltage. The series compensator is functionally a controlled voltage source which is connected in series with the transmission line to control its current. This reciprocity suggests that both the admittance and voltage source type shunt compensators have a corresponding series compensator. Indeed, as indicated earlier, the series compensator can be implemented either as a variable reactive impedance or as a controlled voltage source in series with the line. Because of this duality between the shunt and series compensators, many of the concepts, and circuit and control approaches discussed in Chapter 5 are applicable, with a complementary view, in the present discussion. That is, in shunt compensation the basic reference parameter is the transmission voltage and in series compensation it is the line current. Therefore, the operation of the shunt compensator is viewed from the perspective of the transmission voltage and that of the series compensator from the perspective of the line current. This complementary relationship will be exploited in this chapter to build upon the principles established for shunt compensation in Chapter 5 and extend them to series compensation so as to develop a general understanding and unified perspective for the role and application of controllable reactive compensation for electric power transmission.

6.2 VARIABLE IMPEDANCE TYPE SERIES COMPENSATORS

Just as in reactive shunt compensation, variable impedance type series compensators are composed of thyristor-switched/controlled-capacitors or thyristor-controlled reactors with fixed capacitors.

6.2.1 GTO Thyristor-Controlled Series Capacitor (GCSC)

An elementary GTO Thyristor-Controlled Series Capacitor, proposed by Karady with others in 1992, is shown in Figure 6.5(a). It consists of a fixed capacitor in parallel

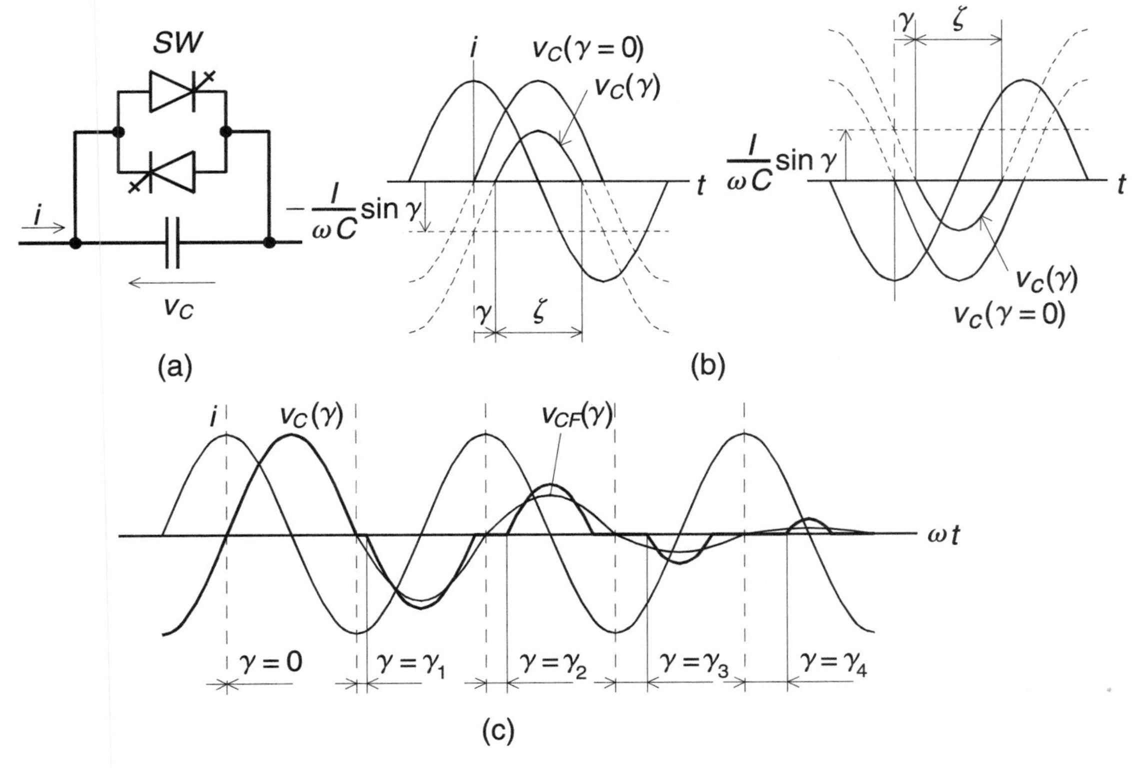

Figure 6.5 Basic GTO-Controlled Series Capacitor (a), principle of turn-off delay angle control (b), and attainable compensating voltage waveform (c).

with a GTO thyristor (or equivalent) valve (or switch) that has the capability to turn on and off upon command.

This compensator scheme is interesting in that it is the perfect combination of the well-established TCR, having the unique capability of directly varying the capacitor voltage by delay angle control. Apart from the theoretical interest, this technique, as will be seen, has some operational merits and it may be incorporated into some series compensator schemes in the future, particularly when larger GTO thyristors become available.

The objective of the GCSC scheme shown in Figure 6.5(a) is to control the ac voltage v_C across the capacitor at a given line current *i*. Evidently, when the GTO valve, *sw*, is closed, the voltage across the capacitor is zero, and when the valve is open, it is maximum. For controlling the capacitor voltage, the closing and opening of the valve is carried out in each half-cycle in synchronism with the ac system frequency. The GTO valve is stipulated to close automatically (through appropriate control action) whenever the capacitor voltage crosses zero. (Recall that the thyristor valve of the TCR opens automatically whenever the current crosses zero.) However, the turn-off instant of the valve in each half-cycle is controlled by a (turn-off) delay angle γ ($0 \le \gamma \le \pi/2$), with respect to the peak of the line current. Refer to Figure 6.5(b), where the line current *i*, and the capacitor voltage $v_C(\gamma)$ are shown at $\gamma = 0$ (valve open) and at an arbitrary turn-off delay angle γ, for a positive and a negative half-cycle. When the valve *sw* is opened at the crest of the (constant) line current

($\gamma = 0$), the resultant capacitor voltage v_C will be the same as that obtained in steady state with a permanently open switch. When the opening of the valve is delayed by the angle γ with respect to the crest of the line current, the capacitor voltage can be expressed with a defined line current, $i(t) = I \cos \omega t$, as follows:

$$v_C(t) = \frac{1}{C}\int_{\gamma}^{\omega t} i(t)\, dt = \frac{I}{\omega C}(\sin \omega t - \sin \gamma) \tag{6.7}$$

Since the valve opens at γ and stipulated to close at the first voltage zero, (6.7) is valid for the interval $\gamma \leq \omega t \leq \pi - \gamma$. For subsequent positive half-cycle intervals the same expression remains valid. For subsequent negative half-cycle intervals, the sign of the terms in (6.7) becomes opposite.

Comparison of (6.7) to (5.4) derived for the current of the TCR indicates that the two equations are formally identical and can be interpreted in the same manner.

In (6.7) the term $(I/\omega C) \sin \gamma$ is simply a γ dependent constant by which the sinusoidal voltage obtained at $\gamma = 0$ is offset, shifted down for positive, and up for negative voltage half-cycles, as illustrated in Figure 6.5(b). Since the GTO valve automatically turns on at the instant of voltage zero crossing (which is symmetrical on the time axis to the instant of turn-off with respect to the peak of the capacitor voltage), this process actually controls the nonconducting (blocking) interval (or angle) of the GTO valve. That is, the turn-off delay angle γ defines the prevailing blocking angle ζ: $\zeta = \pi - 2\gamma$. Thus, as the turn-off delay angle γ increases, the correspondingly increasing offset results in the reduction of the blocking angle ζ of the valve, and the consequent reduction of the capacitor voltage. At the maximum delay of $\gamma = \pi/2$, the offset also reaches its maximum of $I/\omega C$, at which both the blocking angle and the capacitor voltage become zero.

It is evident that the magnitude of the capacitor voltage can be varied continuously by this method of turn-off delay angle control from maximum ($\gamma = 0$) to zero ($\gamma = \pi/2$), as illustrated in Figure 6.5(c), where the capacitor voltage $v_C(\gamma)$, together with its fundamental component $v_{CF}(\gamma)$, are shown at various turn-off delay angles, γ. Note, however, that the adjustment of the capacitor voltage, similar to the adjustment of the TCR current, is discrete and can take place only once in each half-cycle.

Comparison of Figure 6.5 to Figure 5.7 shows that the waveshape obtained for the current of the thyristor-controlled reactor is identical to that derived above for the voltage GTO thyristor-controlled series capacitor and confirms the duality between the GCSC and the TCR. Indeed, the duality between the TCR and the GCSC is quite evident. The TCR is a switch in series with a reactor, the GCSC is a switch in shunt with a capacitor. The TCR is supplied from a voltage source (transmission bus voltage), the GCSC is supplied from a current source (transmission line current). The TCR valve is stipulated to close at current zero, the GCSC at voltage zero. The TCR is controlled by a turn-on delay with respect to the crest of the applied voltage, which defines the conduction interval of the valve. The GCSC is controlled by a turn-off delay with respect to the peak of the line current, which defines the blocking interval of the valve. The TCR controls the current in a fixed inductor from a constant voltage source, thereby presenting a variable reactive admittance as the load to this source. The GCSC controls the voltage developed by a constant current source across a fixed capacitor, thereby presenting a variable reactive impedance to this source.

The duality established above makes it possible to use the analytical results given in (5.5a), (5.5b), and (5.6), and to extend the circuit structural and operational considerations applied to the TCR in Chapter 5 to the GCSC and related series capacitive compensators. Thus, per (5.5a), the amplitude $V_{CF}(\gamma)$ of the fundamental capacitor voltage $v_{CF}(\gamma)$ can be expressed as a function of angle γ:

$$V_{CF}(\gamma) = \frac{I}{\omega C}\left(1 - \frac{2}{\pi}\gamma - \frac{1}{\pi}\sin 2\gamma\right) \tag{6.8a}$$

where I is the amplitude of the line current, C is the capacitance of the GTO thyristor-controlled capacitor, and ω is the angular frequency of the ac system.

The variation of the amplitude $V_{CF}(\gamma)$, normalized to the maximum voltage $V_{CF\max} = I/\omega C$, is shown plotted against delay angle γ in Figure 6.6.

On the basis of Figure 6.6 the GCSC, varying the fundamental capacitor voltage at a fixed line current, could be considered as a variable capacitive impedance. Indeed, an effective capacitive impedance can be found for a given value of angle γ or, in other words, an effective capacitive impedance, X_C, as a function of γ, for the GCSC can be defined. This impedance can be written directly from (6.8a), i.e.,

$$X_C(\gamma) = \frac{1}{\omega C}\left(1 - \frac{2}{\pi}\gamma - \frac{1}{\pi}\sin 2\gamma\right) \tag{6.8b}$$

Evidently, the admittance $X_C(\gamma)$ varies with γ in the same manner as the fundamental capacitor voltage $V_{CF}(\gamma)$.

In a practical application the GCSC can be operated either to control the compensating voltage, $V_{CF}(\gamma)$, or the compensating reactance, $X_C(\gamma)$. In the voltage compensation mode, the GCSC is to maintain the rated compensating voltage in face of decreasing line current over a defined interval $I_{\min} \le I \le I_{\max}$, as illustrated in Figure 6.7(a1). In this compensation mode the capacitive reactance X_C is selected so as to produce the rated compensating voltage with $I = I_{\min}$, i.e., $V_{C\max} = X_C I_{\min}$. As current $I_{\min}$ is increased toward $I_{\max}$, the turn-off delay angle γ is increased to reduce the duration

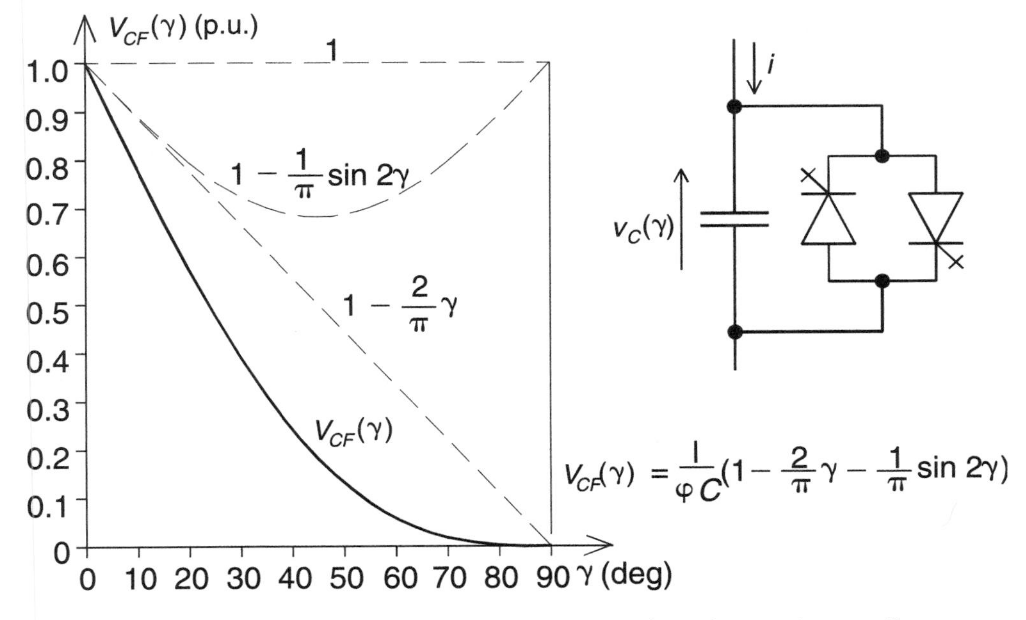

Figure 6.6 Fundamental component of the series capacitor voltage vs. the turn-off delay angle γ.

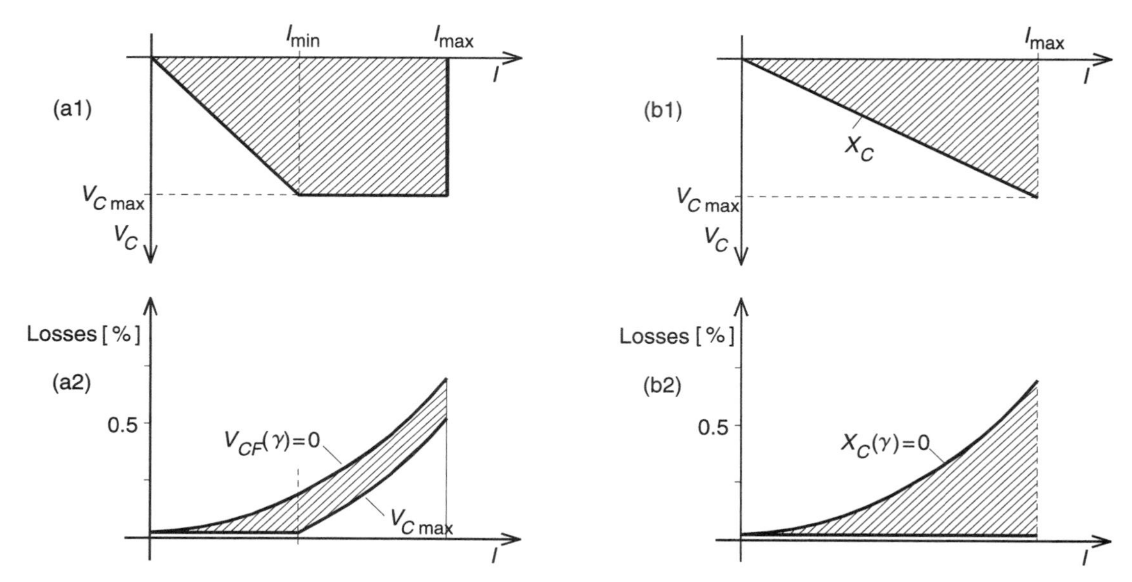

Figure 6.7 Attainable *V-I* (compensating voltage vs. line current) characteristics of the GCSC when operated in voltage control (a1) and reactance control (b1) modes, and the associated loss vs. line current characteristics (a2 and b2, respectively).

of the capacitor injection and thereby maintain the compensating voltage with increasing line current. The loss, as percent of the rated var output, versus line current characteristic of the GCSC operated in the voltage compensation mode is shown in Figure 6.7(a2) for zero voltage injection [the capacitor is bypassed by the GTO valve to yield $V_{CF}(\gamma) = 0$] and for maximum rated voltage injection [$V_{CF}(\gamma) = V_{C\max}$].

In the impedance compensation mode, the GCSC is to maintain the maximum rated compensating reactance at any line current up to the rated maximum, as illustrated in Figure 6.7(b1). In this compensation mode the capacitive impedance is chosen so as to provide the maximum series compensation at rated current, $X_C = V_{C\max}/I_{\max}$, that the GCSC can vary in the $0 \leq X_C(\gamma) \leq X_C$ range by controlling the effective capacitor voltage $V_{CF}(\gamma)$, i.e., $X_C(\gamma) = V_{CF}(\gamma)/I$. The loss versus line current characteristic of the GCSC for this operating mode is shown in Figure 6.7(b2) for zero compensating impedance (capacitor is bypassed by the GTO valve) and for maximum compensating impedance (the GTO valve is open and the capacitor is fully inserted). The impedance and voltage compensating modes are, of course, interchangeable by control action within the rating limitation of the series capacitor controlled.

It should be appreciated that, to realize the theoretically defined operating *V-I* areas shown in Figures 6.7(a1) and 6.7(b1), the power components in the GCSC (the GTO valve and capacitor) must have corresponding voltage and current ratings. In practical applications, series compensators are often required to have increased short duration ratings and they can be exposed to significantly higher than the nominal or short-term rated maximum line current during line faults and other major disturbances. Such fault currents would generally be too high for the GTO valve to conduct. However, if the valve is turned off, as a protection measure, then the voltage across the GCSC capacitor could become excessive both for the capacitor and the parallel connected valve. Thus, in practice it would usually be necessary to protect the GCSC externally, either by an external voltage MOV arrester or other voltage limiting device or by an appropriate bypass switch arrangement (together with a back-up mechanical bypass breaker), which would ensure that the defined voltage and current ratings of the GCSC would not be exceeded under any operating condition. The reader should note that this protection requirement is not unique to the GCSC. Indeed, practically all series connected compensators, including the conventional, uncontrolled series capacitor, require a similar protection arrangement.

The turn-off delay angle control of the GCSC, just like the turn-on delay angle control of the TCR, generates harmonics. For identical positive and negative voltage half-cycles, only odd harmonics are generated. The amplitudes of these are a function of angle γ and, per (5.6), can be expressed in the following form:

$$V_{Cn}(\gamma) = \frac{I}{\omega C}\frac{4}{\pi}\left\{\frac{\sin\gamma\cos(n\gamma) - n\cos\gamma\sin(n\gamma)}{n(n^2-1)}\right\} \tag{6.9}$$

where $n = 2k + 1$, $k = 1, 2, 3, \ldots$

The amplitude variation of the harmonics, expressed as a percentage of the maximum fundamental capacitor voltage, is shown plotted against γ in Figure 6.8.

The elimination of the triple-n and other harmonics families in the capacitor voltage by the usual methods of three-phase operation and multipulse circuit structures are probably not practical for the GCSC, because those approaches would usually require an insertion transformer. The GCSC, like other series compensating capacitors, would normally be inserted directly, without any magnetic coupling, in series with the

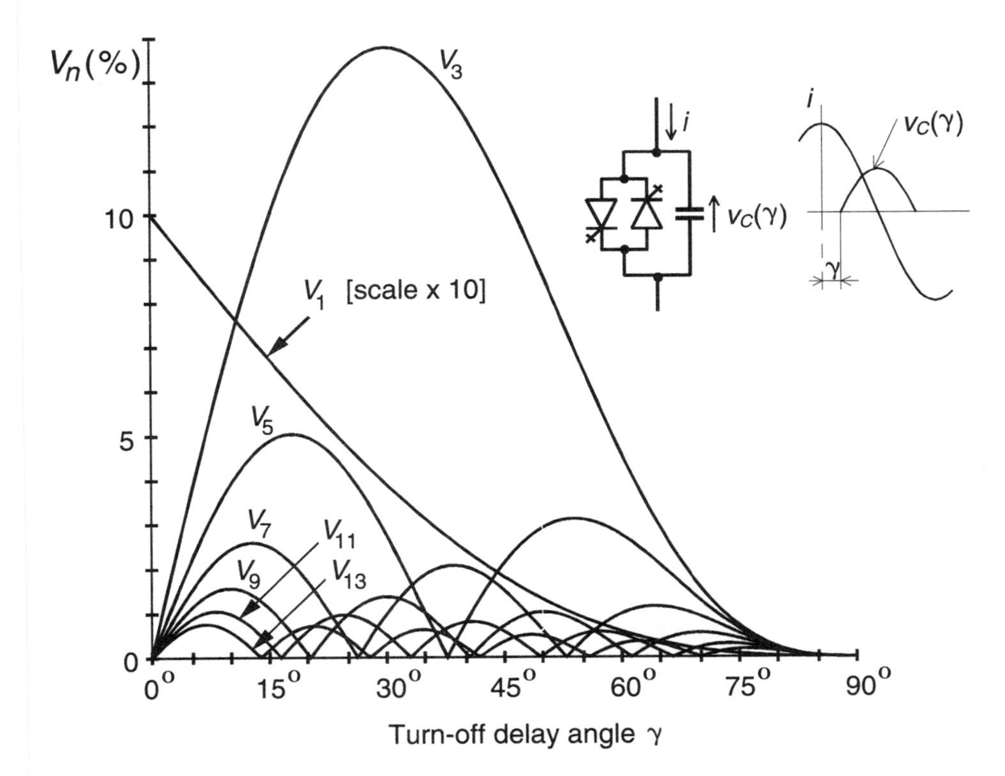

Figure 6.8 The amplitudes of the harmonic voltages, expressed as percents of the maximum fundamental capacitor voltage vs. the turn-off delay angle γ.

line, if operated on a high-voltage platform of sufficient insulation level. The effect of these harmonics may be relatively small, particularly if the transmission line impedance at the harmonic frequencies considered is relatively large. However, if necessary, the magnitudes of the harmonics generated by GCSC can be attenuated effectively by the complementary application of the method of "sequential control" introduced for the reduction of TCR generated harmonics in Section 5.2.1.1.

Recall that this method stipulated the use of m ($m \geq 2$) parallel-connected TCRs, each with $1/m$ of the total rating required. The reactors are "sequentially" controlled; that is, only one of the n reactors is delay angle controlled, and each of the remaining $m - 1$ reactors is either fully "on" or fully "off."

The method of sequential control eminently suits the GCSC. It follows from its duality with the TCR that it requires the use of m ($m \geq 2$) series connected GCSCs, each with $1/m$ of the total (voltage) rating required. As illustrated in Figure 6.9, all but one of m capacitors are "sequentially" controlled to be inserted (valve is off) or bypassed (valve is on). The single capacitor is turn-off delay angle controlled to facilitate continuous voltage control for the whole GCSC over the total operating range. With this arrangement the amplitude of each generated harmonic is evidently reduced by a factor of m in relation to the maximum total fundamental compensating voltage.

Note that, in contrast to the TCR arrangement, where for economic reasons only a relatively small number (usually no more than two) of parallel branches would

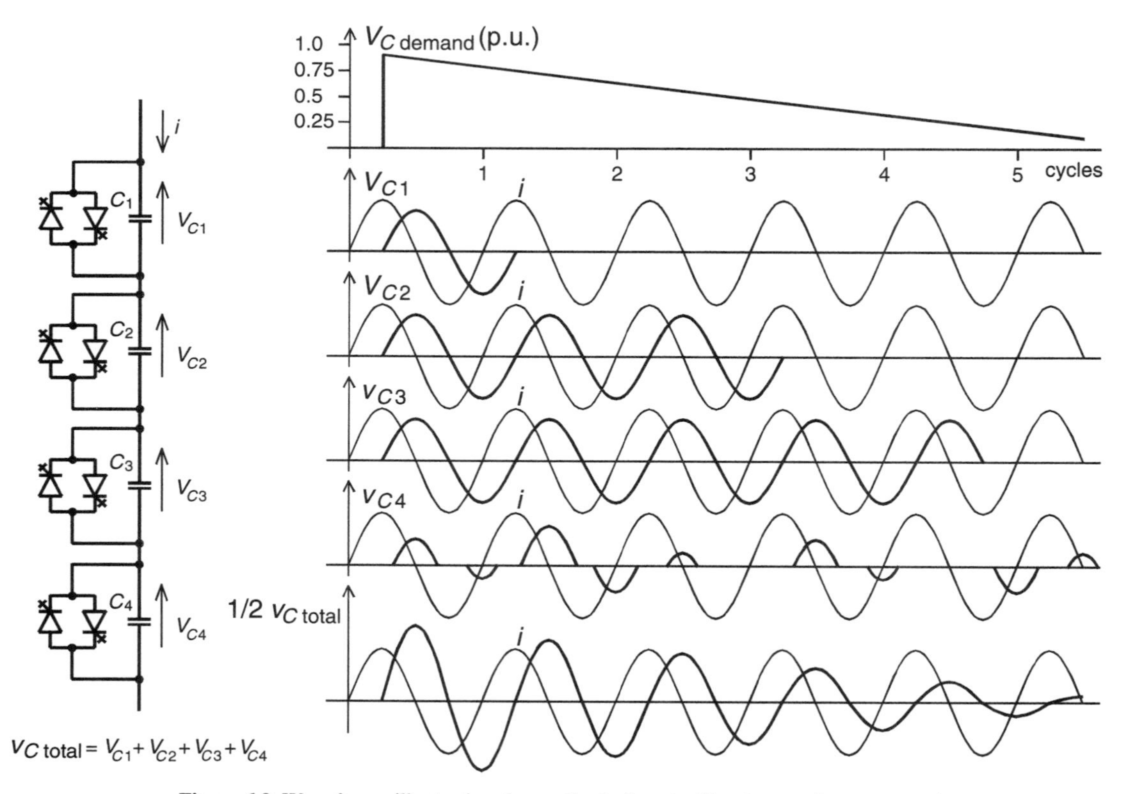

Figure 6.9 Waveforms illustrating the method of controlling four series-connected GCSC banks "sequentially" to achieve harmonic reduction.

be applied, there is no significant economic disadvantage, and may be a technical preference, to break a single high-voltage valve into four or more series connected modules to realize a practical GCSC.

The losses of the sequentially controlled GCSC are inversely proportional to the var output $I_{\text{rated}}^2 \times X_{GCSC}$. The losses are maximum (about 0.7% of the rated var output) when all capacitors of the sequentially controlled GCSC are bypassed (GTO valves are fully on); they are negligible when all capacitors are fully inserted (all GTO valves are off).

In view of the "on and off" operation of all but one of the m GCSC modules, the question naturally arises: why not replace the GTO valves in the $m - 1$ modules with the less expensive conventional thyristor modules? The answer is that with conventional thyristor valves the operation of the total valve would become different. In other words, the conventional thyristor valve cannot imitate GTO valve operation even for full conduction capacitor switching. Refer to Figure 6.5(b). Consider the full capacitor insertion producing the voltage wave $v_C(\gamma = 0)$. Note that in order to obtain this half-cycle wave, the GTO valve must turn on and off when the capacitor voltage is zero, at which instant the line current is at its peak. The conventional thyristor valve could be turned on at the required instant of voltage zero, but it would only turn off at a current zero, which occurs either a quarter cycle before or a quarter cycle after the voltage zero where the proper turn-off should take place. In addition, as shown in the next section, when the conventional thyristor valve turns off at a current zero, it produces a full dc offset for the capacitor voltage, doubling the maximum voltage stress on the valve and the time delay after which the capacitor could again be bypassed.

6.2.2 Thyristor-Switched Series Capacitor (TSSC)

The basic circuit arrangement of the thyristor-switched series capacitor is shown in Figure 6.10. It consists of a number of capacitors, each shunted by an appropriately rated bypass valve composed of a string of reverse parallel connected thyristors, in series. As seen, it is similar to the circuit structure of the sequentially operated GCSC shown in Figure 6.9, but its operation is different due to the imposed switching restrictions of the conventional thyristor valve.

The operating principle of the TSSC is straightforward: the degree of series compensation is controlled in a step-like manner by increasing or decreasing the number of series capacitors inserted. A capacitor is inserted by turning off, and it is bypassed by turning on the corresponding thyristor valve.

A thyristor valve commutates "naturally," that is, it turns off when the current crosses zero. Thus a capacitor can be inserted into the line by the thyristor valve only at the zero crossings of the line current. Since the insertion takes place at line current zero, a full half-cycle of the line current will charge the capacitor from zero to maximum

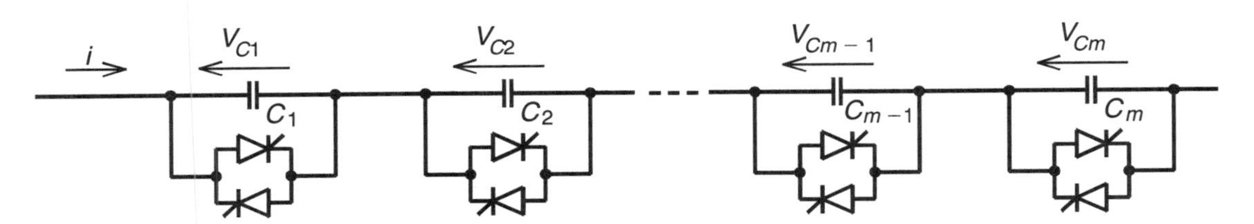

Figure 6.10 Basic Thyristor-Switched Series Capacitor scheme.

and the successive, opposite polarity half-cycle of the line current will discharge it from this maximum to zero, as illustrated in Figure 6.11. As can be seen, the capacitor insertion at line current zero, necessitated by the switching limitation of the thyristor valve, results in a dc offset voltage which is equal to the amplitude of the ac capacitor voltage. In order to minimize the initial surge current in the valve, and the corresponding circuit transient, the thyristor valve should be turned on for bypass only when the capacitor voltage is zero. With the prevailing dc offset, this requirement can cause a delay of up to one full cycle, which would set the theoretical limit for the attainable response time of the TSSC.

The TSSC can control the degree of series compensation by either inserting or bypassing series capacitors but it cannot change the natural characteristic of the classical series capacitor compensated line. This means that a sufficiently high degree of TSSC compensation could cause subsynchronous resonance just as well as an ordinary capacitor. In principle, the TSSC switching could be modulated to counteract subsynchronous oscillations. However, considering the relatively long switching delays encountered, the modulation is likely to be ineffective, if not counterproductive, except for the very low end of the subsynchronous frequency band. Therefore, the pure TSSC scheme of Figure 6.10 would not be used in critical applications where a high degree of compensation is required and the danger of subsynchronous resonance is present. Nevertheless, the TSSC could be applied for power flow control and for damping power oscillation where the required speed of response is moderate.

The basic *V-I* characteristic of the TSSC with four series connected compensator modules operated to control the compensating voltage is shown in Figure 6.12(a1). For this compensating mode the reactance of the capacitor banks is chosen so as to produce, on the average, the rated compensating voltage, $V_{C\max} = 4X_C I_{\min}$, in the face of decreasing line current over a defined interval $I_{\min} \leq I \leq I_{\max}$. As the current $I_{\min}$ is increased toward $I_{\max}$, the capacitor banks are progressively bypassed by the related thyristor valves to reduce the overall capacitive reactance in a step-like manner and thereby maintain the compensating voltage with increasing line current. The loss, as percent of the rated var output, versus line current characteristic of the TSSC operated in the voltage compensating mode is shown in Figure 6.12(a2) for zero voltage injection (all capacitors are bypassed) and for maintaining maximum rated voltage injection (capacitors are progressively bypassed).

In the impedance compensation mode, the TSSC is applied to maintain the maximum rated compensating reactance at any line current up to the rated maximum,

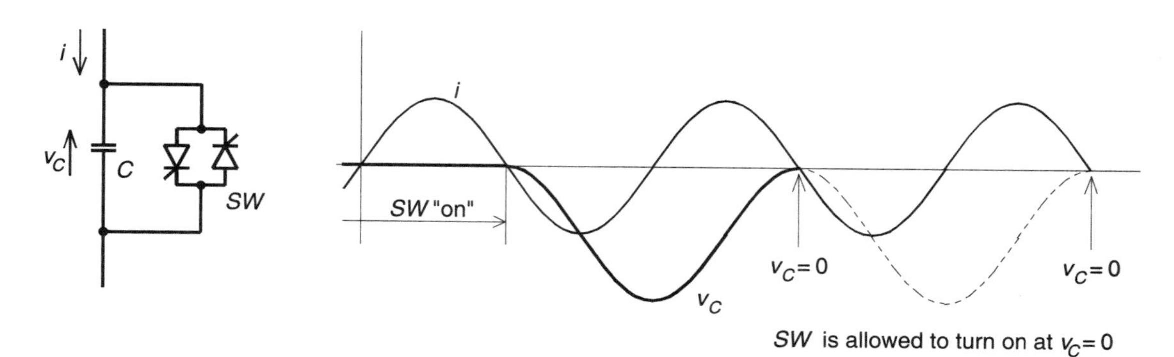

Figure 6.11 Illustration of capacitor offset voltage resulting from the restriction of inserting at zero line current.

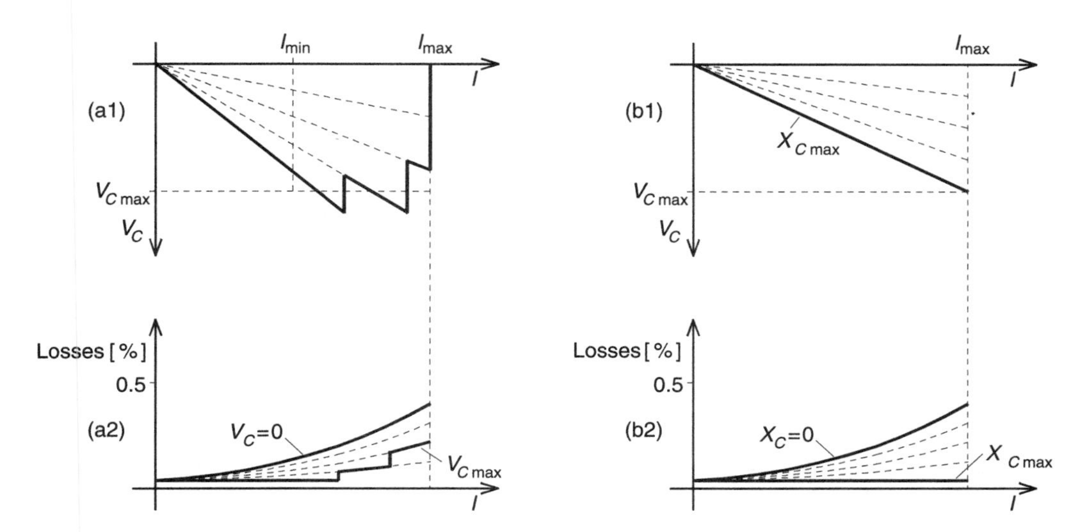

Figure 6.12 Attainable *V-I* (compensating voltage vs. line current) characteristics of the TSSC when operated in voltage control (a1) and reactance control (b1) modes, and the associated loss vs. line current characteristics (a2 and b2, respectively).

as illustrated in Figure 6.12(b1). In this compensation mode the capacitive impedance is chosen so as to provide the maximum series compensation at rated current, $4X_C = V_{C\max}/I_{\max}$, that the TSSC can vary in a step-like manner by bypassing one or more capacitor banks. The loss versus line current characteristic for this compensation mode is shown in Figure 6.12(b2) for zero compensating impedance (all capacitor banks are bypassed by the thyristor valves) and for maximum compensating impedance (all thyristor valves are off and all capacitors are inserted).

The maximum rated line current and corresponding capacitor voltage are design values for which the thyristor valve and the capacitor banks are rated to meet the specific application requirements. The TSSC may also have transient ratings, usually defined as a function of time. Outside the defined ratings the TSSC would be protected against excessive current and voltage surges either by external protection across the capacitor and the parallel valve or, with sufficient rating, by the valve itself in bypass operation.

Constraints imposed by physical device limitation on the turn-on conditions of thyristors (such as *di*/*dt* and surge current magnitude) would necessitate in practice the use of a current limiting reactor in series with the TSSC valve to handle bypass operation, or possible misfirings, which could turn on the valve into a fully charged capacitor of over 2.0 p.u. voltage. However, if a reactor in series with the valve is included in the TSSC structure, then, as is explained in the next section, the resulting power circuit offers new control options which can significantly improve the operating and performance characteristics of the series compensator.

6.2.3 Thyristor-Controlled Series Capacitor (TCSC)

The basic Thyristor-Controlled Series Capacitor scheme, proposed in 1986 by Vithayathil with others as a method of "rapid adjustment of network impedance," is shown in Figure 6.13. It consists of the series compensating capacitor shunted by a

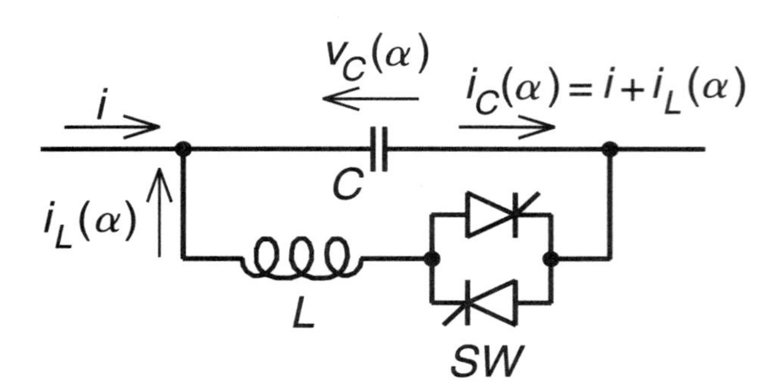

Figure 6.13 Basic Thyristor-Controlled Series Capacitor scheme.

Thyristor-Controlled Reactor. In a practical TCSC implementation, several such basic compensators may be connected in series to obtain the desired voltage rating and operating characteristics. This arrangement is similar in structure to the TSSC and, if the impedance of the reactor, X_L, is sufficiently smaller than that of the capacitor, X_C, it can be operated in an on/off manner like the TSSC. However, the basic idea behind the TCSC scheme is to provide a continuously variable capacitor by means of partially canceling the effective compensating capacitance by the TCR. Since, as shown in Chapter 5, the TCR at the fundamental system frequency is a continuously variable reactive impedance, controllable by delay angle α, the steady-state impedance of the TCSC is that of a parallel LC circuit, consisting of a fixed capacitive impedance, X_C, and a variable inductive impedance, $X_L(\alpha)$, that is,

$$X_{TCSC}(\alpha) = \frac{X_C X_L(\alpha)}{X_L(\alpha) - X_C} \tag{6.10}$$

where, from (5.5b),

$$X_L(\alpha) = X_L \frac{\pi}{\pi - 2\alpha - \sin\alpha}, \; X_L \leq X_L(\alpha) \leq \infty, \tag{6.11}$$

$X_L = \omega L$, and α is the delay angle measured from the crest of the capacitor voltage (or, equivalently, the zero crossing of the line current).

The TCSC thus presents a tunable parallel LC circuit to the line current that is substantially a constant alternating current source. As the impedance of the controlled reactor, $X_L(\alpha)$, is varied from its maximum (infinity) toward its minimum (ωL), the TCSC increases its minimum capacitive impedance, $X_{TCSC,\min} = X_C = 1/\omega C$, (and thereby the degree of series capacitive compensation) until parallel resonance at $X_C = X_L(\alpha)$ is established and $X_{TCSC,\max}$ theoretically becomes infinite. Decreasing $X_L(\alpha)$ further, the impedance of the TCSC, $X_{TCSC}(\alpha)$ becomes inductive, reaching its minimum value of $X_L X_C/(X_L - X_C)$ at $\alpha = 0$, where the capacitor is in effect bypassed by the TCR. Therefore, with the usual TCSC arrangement in which the impedance of the TCR reactor, X_L, is smaller than that of the capacitor, X_C, the TCSC has two operating ranges around its internal circuit resonance: one is the $\alpha_{C\lim} \leq \alpha \leq \pi/2$ range, where $X_{TCSC}(\alpha)$ is capacitive, and the other is the $0 \leq \alpha \leq \alpha_{L\lim}$ range, where $X_{TCSC}(\alpha)$ is inductive, as illustrated in Figure 6.14.

The steady-state model of the TCSC described above is based on the characteristics of the TCR established in an SVC environment, where the TCR is supplied from a constant voltage source. This model is useful to attain a basic understanding of the functional behavior of the TCSC. However, in the TCSC scheme the TCR is connected

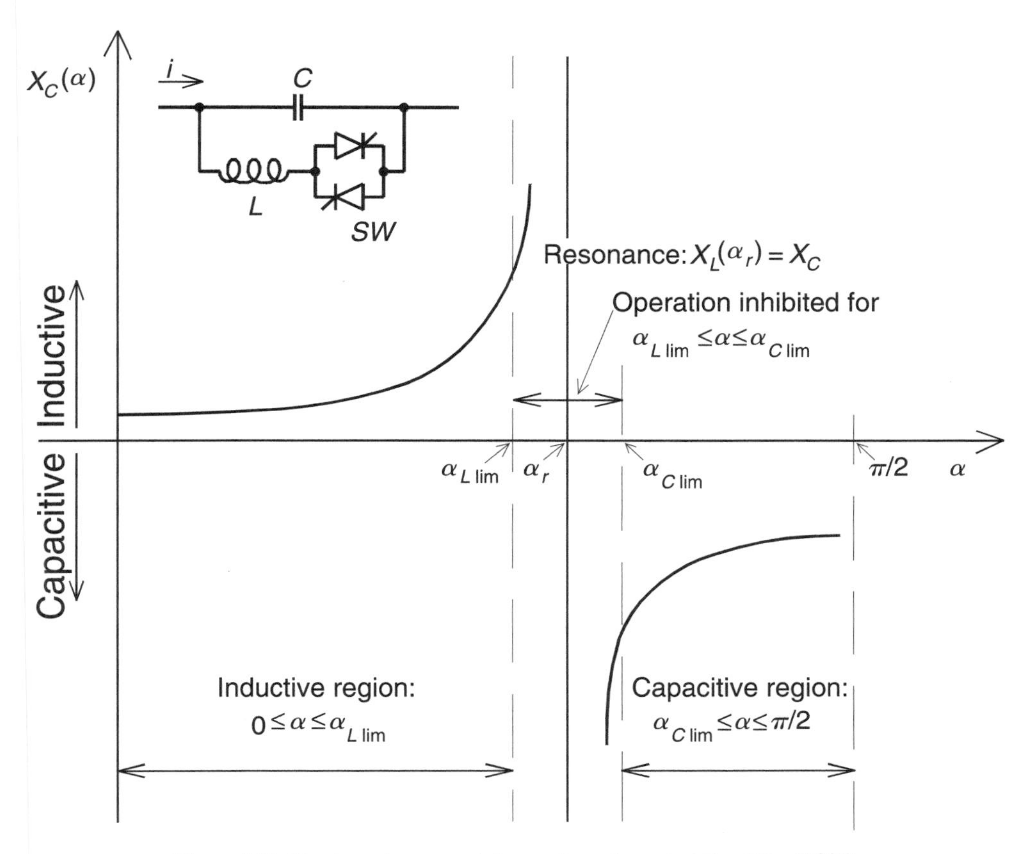

Figure 6.14 The impedance vs. delay angle α characteristic of the TCSC.

in shunt with a capacitor, instead of a fixed voltage source. The dynamic interaction between the capacitor and reactor changes the operating voltage from that of the basic sine wave established by the constant line current. A deeper insight into this interaction is essential to the understanding of the actual physical operation and dynamic behavior of the TCSC, particularly regarding its impedance characteristic at subsynchronous frequencies.

Refer to the basic TCSC circuit shown in Figure 6.13, which, for convenience, is also shown at the top of Figure 6.15(b). Assume that the thyristor valve, *sw*, is initially open and the prevailing line current i produces voltage v_{C0} across the fixed series compensating capacitor, as illustrated in Figure 6.15(a). Suppose that the TCR is to be turned on at α, measured from the negative peak of the capacitor voltage. As seen, at this instant of turn-on, the capacitor voltage is negative, the line current is positive and thus charging the capacitor in the positive direction. During this first half-cycle (and all similar subsequent half-cycles) of TCR operation, the thyristor valve can be viewed as an ideal switch, closing at α, in series with a diode of appropriate polarity to stop the conduction as the current crosses zero, as shown at the bottom of Figure 6.15(b).

At the instant of closing switch *sw*, two substantially independent events will take place: One is that the line current, being a constant current source, continues to (dis)charge the capacitor. The other is that the charge of the capacitor will be reversed

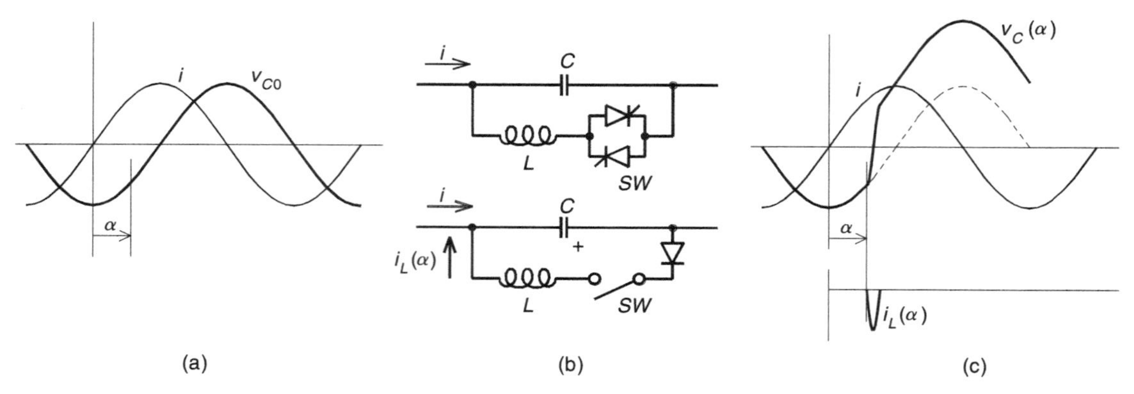

Figure 6.15 Illustration of capacitor voltage reversal by TCR: (a) line current and corresponding capacitor voltage, (b) equivalent circuit of the TCSC at the firing instant α, and (c) the resulting capacitor voltage and related TCR current.

during the resonant half-cycle of the LC circuit formed by the switch closing. (This second event assumes, as stipulated, that $X_L < X_C$.) The resonant charge reversal produces a dc offset for the next (positive) half-cycle of the capacitor voltage, as illustrated in Figure 6.15(c). In the subsequent (negative) half-cycle, this dc offset can be reversed by maintaining the same α, and thus a voltage waveform symmetrical to the zero axis can be produced, as illustrated in Figure 6.16, where the relevant current and voltage waveforms of the TCSC operated in the capacitive region are shown.

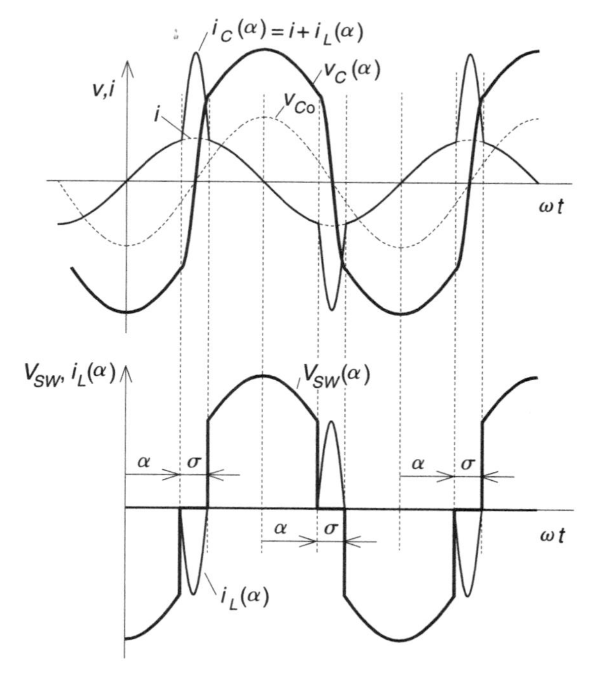

Figure 6.16 Capacitor voltage and current waveforms, together with TCR voltage and current waveforms, characterizing the TCSC in the capacitive region under steady-state operation.

Similar waveforms are shown for the inductive operating range, where the overall impedance of the TCSC is inductive, in Figure 6.17.

The reversal of the capacitor voltage is clearly the key to the control of the TCSC. The time duration of the voltage reversal is dependent primarily on X_L/X_C ratio, but also on the magnitude of the line current. If $X_L \ll X_C$, then the reversal is almost instantaneous, and the periodic voltage reversal produces a square wave across the capacitor that is added to the sine wave produced by the line current. Thus, as illustrated in Figure 6.18, the steady-state compensating voltage across the series capacitor comprises an uncontrolled and a controlled component: The uncontrolled component is v_{C0}, a sine wave whose amplitude is directly proportional to the amplitude of the prevailing line current, and the controlled component is v_{CTCR}, substantially a square wave whose magnitude is controlled through charge reversal by the TCR. For a finite, but still relatively small, X_L, the time duration of the charge reversal is not instantaneous but is quite well defined by the natural resonant frequency, $f = 1/2\ \pi\sqrt{LC}$, of the TCSC circuit, since the TCR conduction time is approximately equal to the half-period corresponding to this frequency: $T/2 = 1/2\ f = \pi\sqrt{LC}$. However, as X_L is increased relative to X_C, the conduction period of the TCR increases and the zero crossings of the capacitor voltage become increasingly dependent on the prevailing line current. (Note that this is in contrast to the TCR operation in the SVC environment, where the voltage zero crossings are singularly determined by the substantially constant applied voltage and are, therefore, independent of the delay and conduction angles of the TCR.)

The reader should appreciate that the impedance of the TCR reactor does not significantly alter the physical operation of the TCSC, provided that it is sufficiently

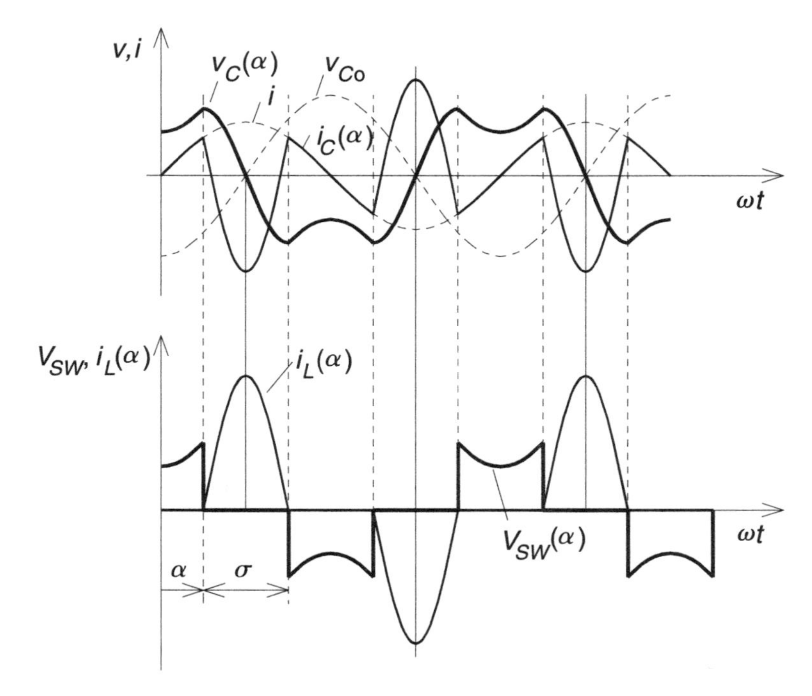

Figure 6.17 Capacitor voltage and current waveforms, together with TCR voltage and current waveforms, characterizing the TCSC in the inductive region under steady-state operation.

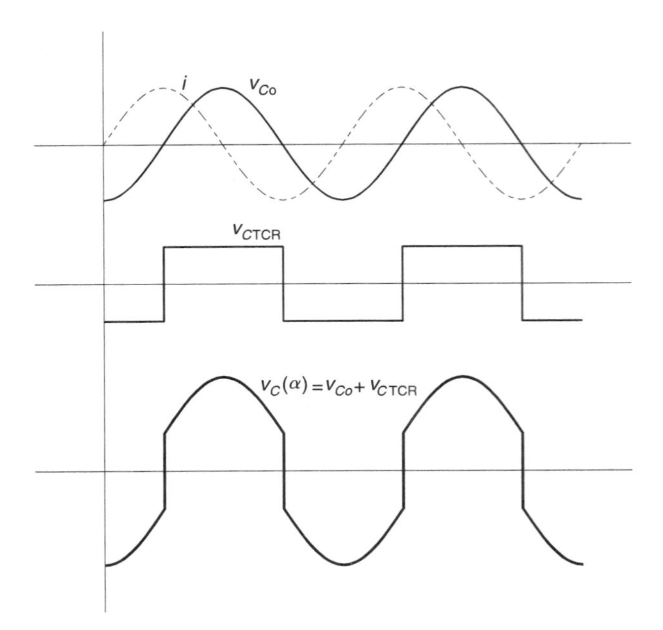

Figure 6.18 Composition of the idealized TCSC compensating voltage waveform from the line current produced (sinusoidal) capacitor voltage and the square wave voltage generated by capacitor voltage reversal.

small in relation to the impedance of the capacitor to facilitate the desired control of the series compensation. However, the design of the reactor for an actual compensator requires careful considerations to reconcile contradictory requirements. On the one hand, small X_L is advantageous in providing well-defined charge reversal and control of the period time of the compensating voltage (important, as will be seen, for handling subsynchronous resonance). A small reactor is also advantageous in facilitating an effective protective bypass for large surge current encountered during system faults. On the other hand, small X_L increases the magnitudes of the current harmonics generated by the TCR and circulated through the series capacitor, and thus also increases the magnitudes of the capacitor voltage harmonics injected into the line. It also decreases the range of actual delay angle control and thus possibly makes the closed-loop parameter regulations more difficult. In addition, small X_L produces large short-duration current pulses in the thyristor valve, necessitating the increase of its current rating and possibly also its voltage rating. In present (prototype) installations the X_L/X_C ratio used is 0.133, and thus the natural resonant frequency of the TCSC circuit is 2.74 times the 60 Hz fundamental frequency. Generally, the X_L/X_C ratio for practical TCSCs would likely be in 0.1 to 0.3 range, depending on the application requirements and constraints. It is important that the natural resonance frequency of the TCSC does not coincide with, or is close to, two and three times the fundamental.

The mechanism of controlling the dc offset by charge reversal is illustrated for the increase and decrease of the capacitor voltage in Figures 6.19(a) and (b), respectively. For the clarity of illustration, the theoretically ideal case of instantaneous voltage reversal is assumed (with an infinitesimal X_L). In Figure 6.19(a), initially the TCR is gated on at $\alpha = \pi/2$, at which the TCR current is zero and the capacitor voltage is entirely due to the line current. To produce a dc offset, the periodically repeated gating in the second cycle is advanced by a small angle ε, i.e., the prevailing half-period is reduced by ε to $\pi - \varepsilon$. This action produces a phase advance for the capacitor voltage with respect to the line current and, as a result, the capacitor absorbs energy from the line, charging it to a higher voltage. (Note that with the phase advance, the

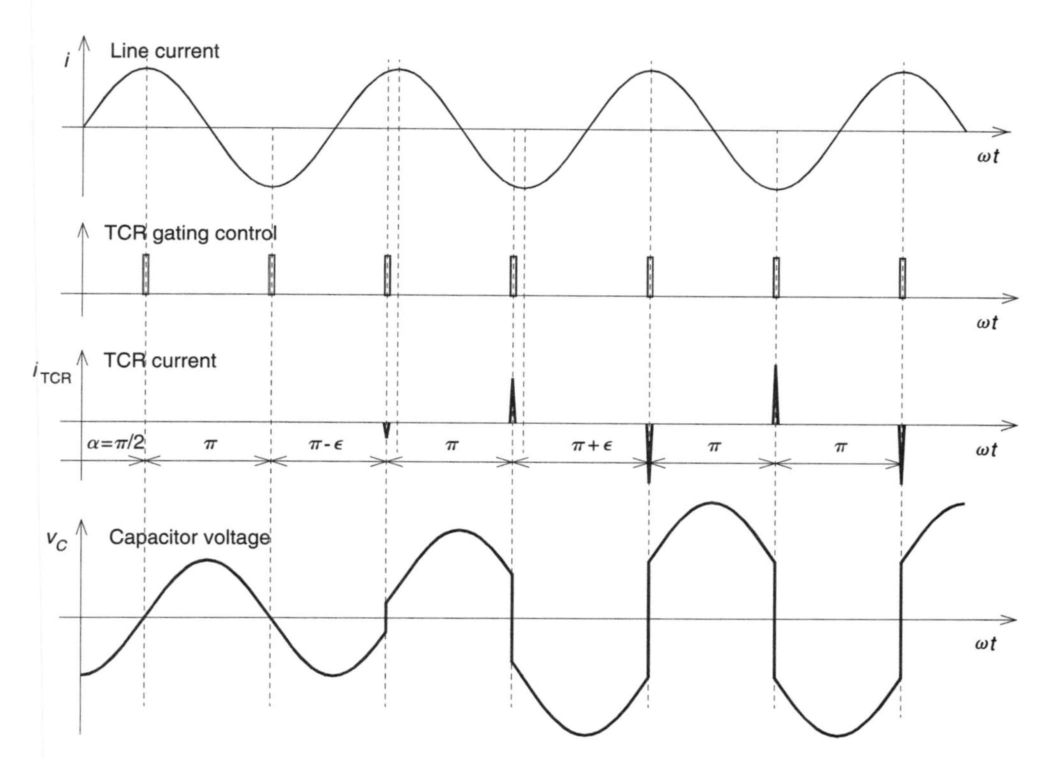

Figure 6.19a Increase of the capacitor voltage by advancing the voltage reversal from $\alpha = \pi$ to $\alpha = \pi - \varepsilon$.

$\int idt$ is greater for the positive current segment than for the negative one in the half-period considered, and the resulting difference between these two integrals provides a net positive offset charge for the capacitor.) Should this phase advance be maintained, then the offset charge of the capacitor would keep increasing its charge at every half-cycle without a theoretical limit, as illustrated for the subsequent half-cycle in the figure. However, if the ε phase advance is negated, when the sufficient offset level of the capacitor voltage is reached, then the capacitor voltage at the desired magnitude can be maintained (for the lossless circuit assumed) by continuing periodic gating at line current zeroes ($\alpha = 0$), as shown for the successive half-cycles in Figure 6.19(a).

In Figure 6.19(b), the opposite process is illustrated, that is, when the magnitude of the capacitor voltage is decreased by retarding the periodic gating from the current zeros until the desired offset voltage level is reached.

With a practical TCR the voltage reversals would take place over a finite conduction period (σ) and this period would vary with the applied phase advancement or retard, and the circuit would behave as if a conventional delay angle control was applied. The process of transition from one capacitor voltage to another therefore would be more complex than the theoretical case illustrated; it would be a function of the prevailing line current, and generally there would be a transitional period before a steady-state condition would be reached. However, the end result would be substantially the same.

It has been shown above that the original objective of providing continuously variable series capacitive compensation can be achieved by the TCSC circuit structure.

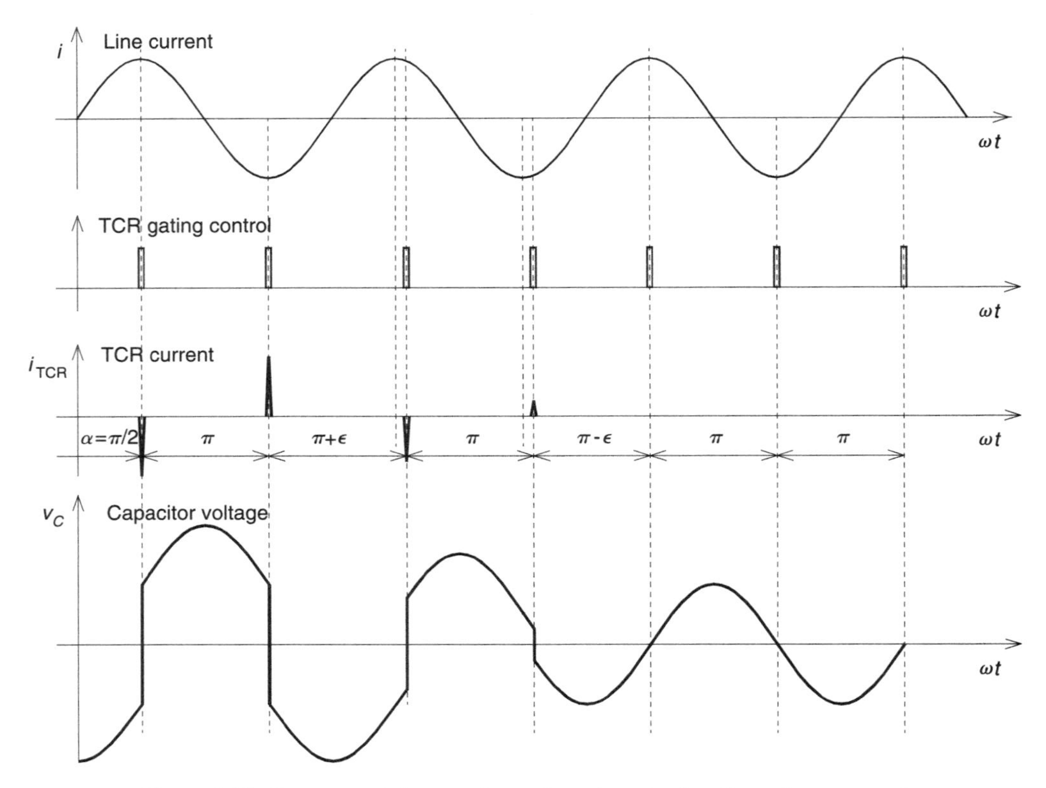

Figure 6.19b Decrease of the capacitor voltage by retarding the voltage reversal from $\alpha = \pi$ to $\alpha = \pi + \varepsilon$.

That is, the effective capacitive impedance of the TCSC can be increased above the actual reactance of the capacitor bank by increasing the conduction angle of the TCR. This increase of the effective impedance is due to the additional voltage the TCR produces across the capacitor by repetitive charge reversals. From the qualitative circuit analysis provided, this process can be summarized in the following way. When the thyristor valve of the TCR is gated on prior to a negative to positive zero crossing of the capacitor voltage ($dv_C/dt > 0$), the prevailing capacitor voltage will drive a current in the TCR that will self-commutate when the integrated volt-seconds over the conduction interval becomes zero (since at the end of the conduction interval the current in the thyristor valve also decreases to zero). The current in the reactor is supplied from the parallel capacitor. Thus, the capacitor voltage will change in proportion to the integral of the reactor current, reversing its polarity (from negative to positive) by the end of the conduction period. This positive voltage offset will remain until reversed again by a similar TCR conduction period around the positive to negative zero crossing of the capacitor voltage ($dv_C/dt < 0$) in the subsequent half-cycle. The periodic reversal of the offset voltage adds a controllable in-phase alternating voltage to the uncontrolled capacitor voltage produced by the line current. In this way, the TCSC can increase the compensating voltage over that produced by the capacitor bank alone at a given line current. In other words, the TCSC can increase the effective impedance of the series compensating capacitor. It is important to appreciate that in contrast to the TCSC the GCSC controls the capacitive impedance from zero to the actual impedance of the capacitor.

The compensating voltage versus line current (V-I) characteristic of a basic TCSC is shown in Figure 6.20(a1). As illustrated, in the capacitive region the minimum delay angle, $\alpha_{C\text{lim}}$, sets the limit for the maximum compensating voltage up to a value of line current (I_{min}) at which the maximum rated voltage, $V_{C\text{max}}$, constrains the operation until the rated maximum current, I_{max}, is reached. In the inductive region, the maximum delay angle, $\alpha_{L\text{lim}}$, limits the voltage at low line currents and the maximum rated thyristor current at high line currents. The loss, as a percent of the rated var output, versus line current for voltage compensation mode in the capacitive operating region is shown in Figure 6.20(a2) for maximum and minimum compensating voltage, as well as for bypass operation (thyristor valve is fully on). The losses are almost entirely due to the TCR, which include the conduction and switching losses of the thyristor valve and the I^2R losses of the reactor. Note that the loss characteristic of the TCSC shown in Figure 6.20(a2) correlates with its voltage compensation characteristic shown in Figure 6.20(a1). That is, the losses increase in proportion with the line current at the fixed maximum TCR conduction angle obtained with the minimum delay angle, $\alpha_{C\text{lim}}$, and then they decline as the conduction angle is continuously decreased with increasing α ($\alpha_{C\text{lim}} < \alpha < \pi/2$) to keep the capacitor voltage constant, below the maximum voltage constraint.

In the impedance compensation mode, the TCSC is applied to maintain the maximum rated compensating reactance at any line current up to the rated maximum. For this operating mode the TCSC capacitor and thyristor-controlled reactor are chosen so that at $\alpha_{C\text{lim}}$ the maximum capacitive reactance can be maintained at and below the maximum rated line current, as illustrated in Figure 6.20(b1). The minimum

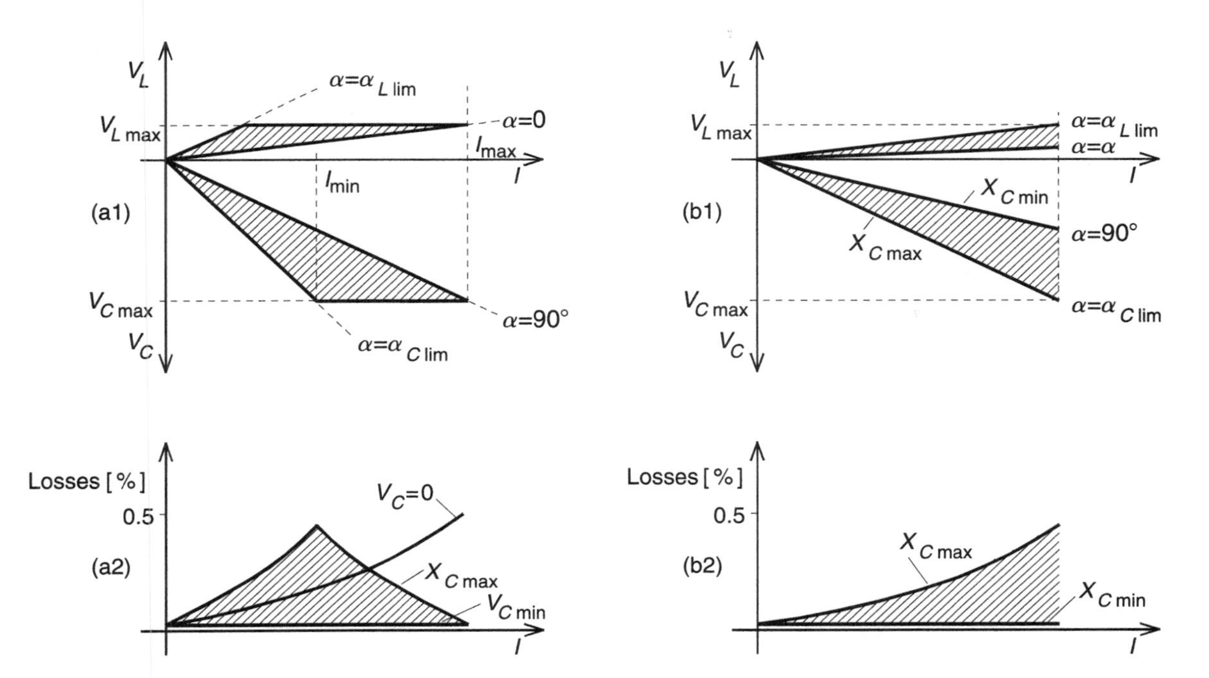

Figure 6.20 Attainable V-I (compensating voltage vs. line current) characteristics of the TCSC when operated in voltage control (a1) and reactance control (b1) modes, and the associated loss vs. line current characteristics (a2 and b2, respectively).

capacitive compensating impedance the TCSC can provide is, of course, the impedance of the capacitor itself, theoretically obtained at $\alpha = 90°$ (with nonconducting thyristor valve). The loss versus line current characteristic for this operating mode is shown in Figure 6.20(b2) for maximum and minimum capacitive compensating reactances. The reader is reminded again that the voltage and impedance compensation modes are interchangeable by control action; however, limitations imposed by component ratings may constrain the attainable range of the desired operating mode. For example, the compensating voltage versus line current characteristic shown in Figure 6.20(a1) can be transformed into the compensating reactance versus line current characteristic shown in Figure 6.21. It can be observed in these figures that constant compensating voltage necessarily results in varying compensating impedance and, vice versa, constant impedance produces varying compensating voltage with changing line current.

The maximum voltage and current limits are design values for which the thyristor valve, the reactor and capacitor banks are rated to meet specific application requirements. The TCSC, like its switched counterpart, the TSSC, is usually required to have transient voltage and current ratings, defined for specific time durations. The TCSC design is complicated by the fact that the internally generated harmonics aggravate the limit conditions. Harmonic currents cause additional losses and corresponding temperature increase in both the thyristor valve and the reactor. The harmonic voltages they produce across the capacitor increase the crest voltage and the stress on the TCSC power components. The effects of the harmonics must be taken into account under the worst case operating conditions to determine the necessary maximum voltage and current ratings of the major TCSC components to satisfy specified operating conditions. Outside these defined operating limits, the TCSC is protected against excessive voltage and current surges either by a shunt connected external protection

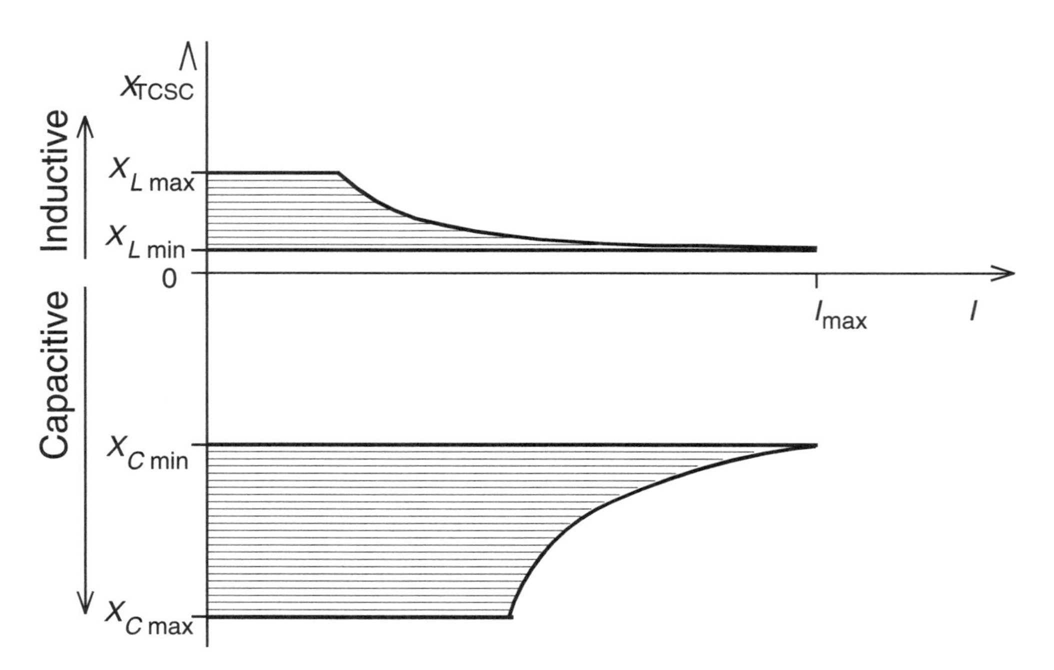

Figure 6.21 The attainable compensating reactance vs. line current characteristic of the TCSC corresponding to the voltage compensation mode *V-I* characteristic shown in Figure 6.20(a1).

(e.g., MOV arresters, triggered spark gap, bypass breaker) or by the TCR itself (with a backup breaker) in bypass operation.

The TCSC, with partial conduction of the TCR, injects harmonic voltages into the line. These harmonic voltages are caused by the TCR harmonic currents which circulate through the series compensating capacitor. The TCR, as established in Chapter 5, generates all odd harmonics, the magnitudes of which are a function of the delay angle α (see Figure 5.10). The harmonic voltages corresponding to these currents in a TCSC circuit are clearly dependent on the impedance ratio of the TCR reactor to the series capacitor, X_L/X_C. For $X_L/X_C = 0.133$ (used in the existing TCSC installations), the most important harmonic voltages, the 3rd, 5th, and 7th, generated in the capacitive operating region, are plotted against the line current I in Figure 6.22, as percents of the fundamental capacitor voltage, V_{Co}, with the TCR off, at rated current. For the plots shown the TCSC is assumed to maintain the rated compensating voltage against varying line current as defined by the V-I characteristic shown in Figure 6.20(a1). It can be observed that the magnitudes of the harmonic voltages rapidly decrease with the frequency, and harmonics above the seventh are totally negligible. The lower order harmonics, although appearing to have relatively high magnitudes, may not contribute significantly to the existing harmonic line currents. This is because these harmonics are substantially voltage sources and the TCSC is usually applied to long, high impedance lines, in which the generated line current harmonics will be relatively low. (In the existing installations, the TCSC-caused line current harmonics was measured lower than the ambient system harmonics.)

Figure 6.21 clearly conveys that the TCSC does not have control below $X_{C\min}$, the system frequency impedance of the capacitor, to $X_{L\min}$, the system frequency impedance of the inductor. If the TCSC is one large single unit, the uncontrolled

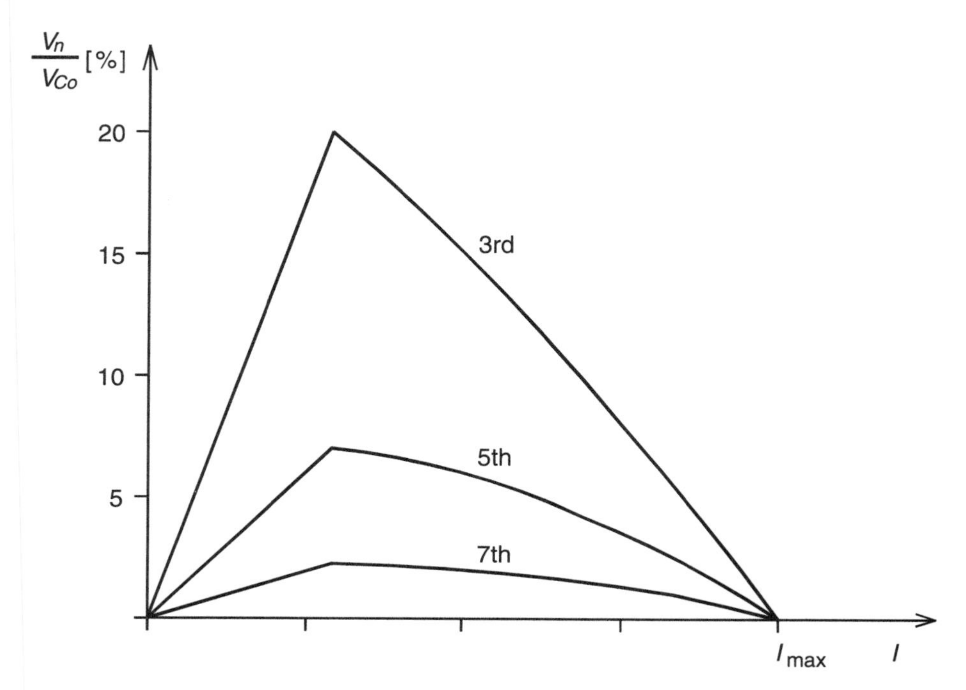

Figure 6.22 Dominant harmonic voltages generated by the TCSC at an X_L/X_C ratio of 0.133.

capacitive band can be quite large. A large uncontrolled band will make it difficult to use the TCSC for control of dynamic stability (power oscillation damping). Dividing TCSC installation into several modules would enable sequential insertion of modules and narrow the uncontrolled band to a small size.

It was established earlier that the GCSC controls the effective capacitor impedance from zero to its system frequency impedance. It follows that in a modular TCSC (or TSSC) arrangement if one module is a GCSC, then the capacitive impedance may, in principle, be continuously controlled over its entire range. However, certain technical problems inherent in this hybrid arrangement, as discussed at the end of Section 6.2.1, may result in some operating constraints.

6.2.4 Subsynchronous Characteristics

As discussed in Section 6.1.5, series capacitive line compensation can cause subsynchronous resonance when the series capacitor resonates with the total circuit inductance of the transmission line at a subsynchronous frequency, f_e, that is equal to the frequency difference between the power frequency, f, and one of the torsional resonant frequencies of the turbine generator set, f_m, i.e., when $f_e = f - f_m$. Variable impedance type series compensators do insert a series capacitor in series with the line, and therefore their behavior in the transmission network at subsynchronous frequencies is critical to their general applicability for unrestricted line compensation and power flow control.

As indicated earlier, the limitations imposed by the subsynchronous resonance (SSR) on the use of series capacitors prompted considerable development effort to find an effective method for the damping of subsynchronous oscillations. In 1981 N. G. Hingorani proposed a thyristor-controlled damping scheme for series capacitors (see Chapter 9 for *NGH Damper*), which has been proven to provide effective SSR mitigation. Although this scheme could be used to complement the series capacitive compensators discussed in this chapter, subsequent research efforts found that the NGH damping principle can be extended to the basic TCSC circuit structure to make it substantially immune to subsynchronous resonance.

The basic principle of the NGH Damper is to force the voltage of the series capacitor to zero at the end of each half-period if it exceeds the value associated with the fundamental voltage component of the synchronous power frequency. Thus, the NGH Damper is basically a thyristor-controlled discharge resistor (in series with a *di*/*dt* limiting reactor), operated synchronously with the power system frequency in the region near the end of the half cycle on the capacitor voltage, as illustrated in Figure 6.23. The NGH Damper clearly interferes with the process of subsynchronous

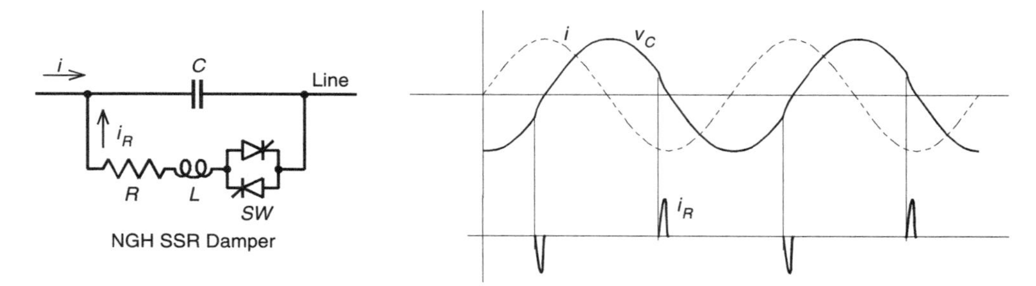

Figure 6.23 Basic NGH SSR Damper.

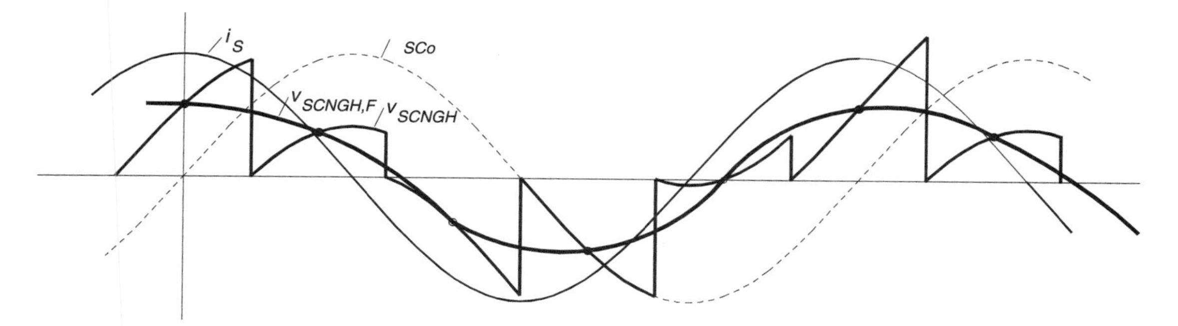

Figure 6.24 The capacitor voltage and its fundamental component produced by a 24 Hz subsynchronous current under the constraints imposed by the NGH Damper.

oscillation build up since the capacitor voltage cannot respond naturally to a subsynchronous line current. The actual effect of the NGH Damper on the capacitor voltage produced by a subsynchronous line current component is illustrated in Figure 6.24. For the illustration, the frequency of the subsynchronous line current is chosen to be 24 Hz and the NGH Damper is operated at 120 Hz to discharge the capacitor at regular half-cycle intervals corresponding to the 60 Hz power frequency. The figure shows the 24 Hz sinusoidal line current, i_S, the corresponding sinusoidal capacitor voltage, v_{SCo}, that would develop without the NGH Damper, the actual capacitor voltage, v_{SCNGH}, and its fundamental component, $v_{SCNGH,F}$ obtained with the activated NGH Damper. Inspection of this figure leads to an interesting observation that the NGH Damper shifts the capacitor voltage at the 24 Hz subsynchronous frequency so as to be in (or almost in) phase with the corresponding subsynchronous line current. In other words, the series capacitor with the NGH Damper exhibits a resistive rather than a capacitive impedance characteristic at the 24 Hz subsynchronous frequency. Although this observation is made on a single illustrative example, and not proven or generalized in a rigorous manner, studies and field measurements indicate conclusively that the actual circuit behavior is in agreement with this resistive impedance characteristic observed.

The reader may already see the striking similarity between the NGH Damper and the TCSC circuit, the former being composed of a thyristor-controlled resistor, and the latter of a thyristor-controlled reactor, both in parallel with the series compensating capacitor. Although the TCSC circuit configuration was originally conceived for the realization of a variable series capacitor, the circuit similarity suggested the possibility to use it also to implement the NGH principles of SSR mitigation.

It has been discussed that the thyristor-controlled reactor, when operated within the TCSC to increase the effective capacitive impedance, reverses the capacitor voltage in the region near to the end of each half cycle corresponding to the power frequency. Thus, it can be expected that this synchronous charge reversal, just like the synchronous discharge of the capacitor in the NGH scheme, will interfere with the normal response of the capacitor to a subsynchronous current excitation so as to hinder or prevent the build up of subsynchronous oscillation. However, it is also obvious that the charge reversal, in contrast to the synchronous discharge action of the NGH Damper, does not result in the dissipation of any energy (ignoring circuit losses) and thus SSR mitigation by actual damping through the extraction and dissipation of energy from the resonant LC circuit cannot take place.

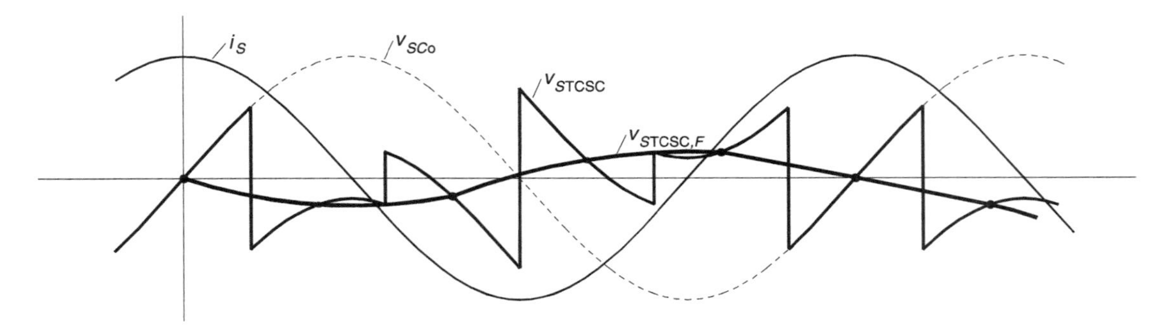

Figure 6.25 The capacitor voltage and its fundamental component produced by a 24 Hz subsynchronous current under the constraints imposed by the TCR executed capacitor voltage reversal.

In order to draw a parallel between the operation of TCSC and the NGH circuit, the voltage waveform obtained across the series capacitor is illustrated in Figure 6.25 for the previously considered 24 Hz subsynchronous current excitation, with the regular half-cycle (60 Hz) capacitor voltage reversal characterizing the operation of the TCSC in the capacitive region. Figure 6.25, similarly to Figure 6.24, shows the 24 Hz line current, i_S, the corresponding capacitor voltage, v_{SCo}, that would develop without the TCR executed charge reversals, the actual capacitor voltage, v_{SCTCSC}, and its fundamental component, $v_{SCTCSC,F}$, obtained with periodically repeated charge reversal. Inspection of this figure reveals that the charge reversal shifts the fundamental 24 Hz voltage component so that it leads the corresponding 24 Hz line current by 90 degrees. In other words, the TCSC circuit exhibits the impedance characteristic of an inductor at subsynchronous frequencies. Thus, whereas the NGH Damper with actual energy dissipation establishes a resistive characteristic for the series capacitor, the TCR executed charge reversal transforms the impedance of the series capacitor into that of an inductor in the subsynchronous frequency band. This observation is evidently important since it would indicate that the TCSC compensated line could not cause or participate in a subsynchronous resonance.

The general validity of the above observation is not proven rigorously to date and applicable relationships for the impedance versus frequency characteristic of TCSC, in terms of the relevant circuit and control parameters (e.g., X_L, X_C and α or σ), are not available in the form of mathematical expressions. However, extensive studies, computer simulations and actual tests in prototype installations seem to indicate that the TCSC is substantially neutral to subsynchronous resonance and would not aggravate subsynchronous oscillations. The single condition for this important circuit property is that the charge reversal must take place at equal half-period intervals corresponding to the fundamental system frequency. With varying delay and conduction angle of the TCR, this condition actually stipulates that the center of different conduction angles remain fixed to the successive half-period intervals, independently from the prevailing delay angle. In other words, the conduction angle in the total operating range is to be symmetrically positioned around the zero crossings of the capacitor voltage. Evidently, it is increasingly difficult to satisfy this condition with decreasing delay angles, since with the correspondingly increasing TCR conduction, the locations of the capacitor voltage zero crossings are, as previously explained, increasingly influenced by the line current.

As shown above, the TCSC circuit arrangement, with proper gating control can be made immune to subsynchronous resonance. This method can, of course, be extended to the TSSC, in which case the TCR conduction would be kept at a minimum and used exclusively to achieve SSR neutrality. (The reader should recall that the thyristor valve must have a reactor in series even in the "classical" TSSC circuit to limit rate of rise of current.) Actually, in a large series capacitive compensator probably several basic TCSC circuits would be connected in series, and most of them would be operated with the TCR fully on (bypass) or fully off (capacitor inserted) to minimize harmonics and operating losses.

The behavior of the NGH scheme and TCSC under subsynchronous system conditions also makes a convincing case for the subsynchronous neutrality of GTO Thyristor-Controlled Series Capacitor (GCSC), presented in Section 6.2.1.1. For this compensator case the capacitor voltage is forcibly kept at zero around the zero crossings of the fundamental capacitor voltage by the normal GTO valve operation (see Figure 6.5), except when the valve is fully off to provide the maximal capacitive compensation. However, there would be no significant loss of compensating voltage if a minimum turn-off delay angle, $\gamma = \gamma_{\min}$, would be maintained which would allow the controlled synchronization of the capacitor voltage to the power system frequency and thereby ensure immunity to subsynchronous resonance.

6.2.5 Basic Operating Control Schemes for GCSC, TSSC, and TCSC

The function of the operating or "internal" control of the variable impedance type compensators is to provide appropriate gate drive for the thyristor valve to produce the compensating voltage or impedance defined by a reference. The internal control operates the power circuit of the series compensator, enabling it to function in a self-sufficient manner as a variable reactive impedance. Thus, the power circuit of the series compensator together with the internal control can be viewed as a "black box" impedance amplifier, the output of which can be varied from the input with a low power reference signal. The reference to the internal control is provided by the "external" or system control, whose function it is to operate the controllable reactive impedance so as to accomplish specified compensation objectives of the transmission line. Thus the external control receives a line impedance, current, power, or angle reference and, within measured system variables, derives the operating reference for the internal control.

As seen, the power circuits of the series compensators operate by rigorously synchronized current conduction and blocking control which not only define their effective impedance at the power frequency but could also determine their impedance characteristic in the critical subsynchronous frequency band. This synchronization function is thus a cornerstone of a viable internal control. Additional functions include the conversion of the input reference into the proper switching instants which result in the desired valve conduction or blocking intervals. The internal control is also responsible for the protection of the main power components (valve, capacitor, reactor) by executing current limitations or initiating bypass or other protective measures.

Structurally the internal controls for the three variable impedance type compensators (GCSC, TCSC, TSSC) could be similar. Succinctly, their function is simply to define the conduction and/or the blocking intervals of the valve in relation to the

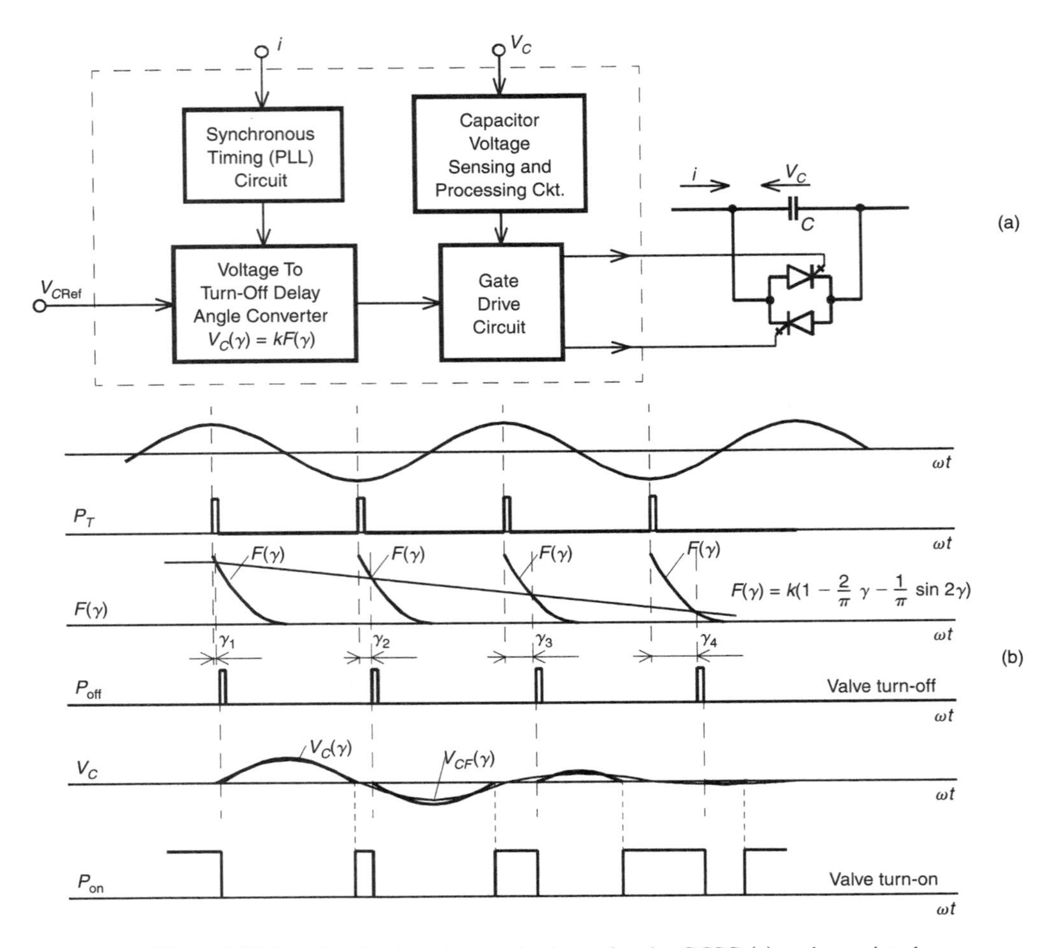

Figure 6.26 Functional internal control scheme for the GCSC (a) and associated waveforms illustrating the basic operating principles (b).

fundamental (power frequency) component of the line current. This requires the execution of three basic functions: synchronization to the line current, turn-on or turn-off delay angle computation, and gate (firing) signal generation. These functions obviously can be implemented by different circuit approaches, with differing advantages and disadvantages. In the following, three possible internal control schemes are functionally discussed: one for the GCSC, and the other two for the TCSC power circuit arrangements. Either of the TCSC schemes could be adapted for the TSSC if subsynchronous resonance would be an application concern.

An internal control scheme for the GTO-Controlled Series Capacitor scheme of Figure 6.5 is shown in Figure 6.26(a). Because of the duality between the shunt-connected GCSC and the series-connected TCR arrangements, this control scheme is analogous to that shown for the TCR in Figure 5.19(a). It has four basic functions.

The *first function* is synchronous timing, provided by a phase-locked loop circuit that runs in synchronism with the line current.

The *second function* is the reactive voltage or impedance to turn-off delay angle conversion according to the relationship given in (6.8a) or (6.8b), respectively.

The *third function* is the determination of the instant of valve turn-on when the capacitor voltage becomes zero. (This function may also include the maintenance of a minimum on time at voltage zero crossings to ensure immunity to subsynchronous resonance.)

The *fourth function* is the generation of suitable turn-off and turn-on pulses for the GTO valve.

The operation of the GCSC power circuit and internal control is illustrated by the waveforms in Figure 6.26(b). Inspection of these waveforms show that, with a "black box" viewpoint, the basic GCSC (power circuit plus internal control) can be considered as a controllable series capacitor which, in response to the transmission line current, will reproduce (within a given frequency band and specified rating) the compensating impedance (or voltage) defined by the reference input. The dynamic performance of the GCSC is similar to that of the TCR, both having a maximum transport lag of one half of a cycle.

The main consideration for the structure of the internal control operating the power circuit of the TCSC is to ensure immunity to subsynchronous resonance. Present approaches follow two basic control philosophies. One is to operate the basic phase-locked-loop (PLL) from the fundamental component of the line current. In order to achieve this, it is necessary to provide substantial filtering to remove the super- and, in particular, the subsynchronous components from the line current and, at the same time, maintain correct phase relationship for proper synchronization. A possible internal control scheme of this type is shown in Figure 6.27. In this arrangement the

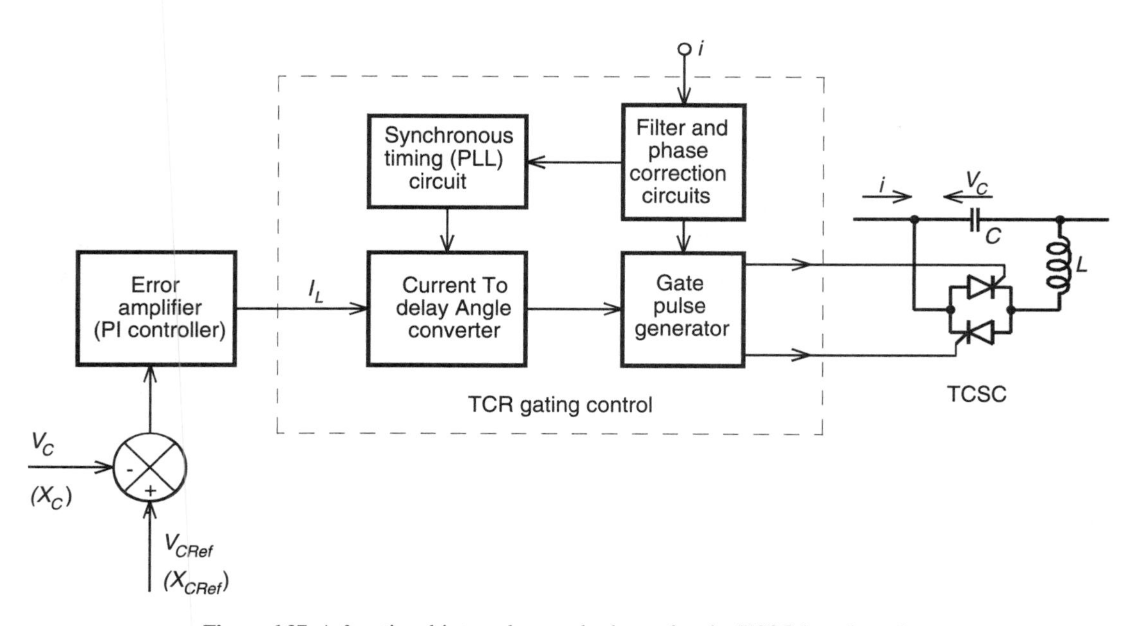

Figure 6.27 A functional internal control scheme for the TCSC based on the synchronization to the fundamental component of the line current.

conventional technique of converting the demanded TCR current into the corresponding delay angle, which is measured from the peak (or, with a fixed 90 degree shift, from the zero crossing) of the fundamental line current, is used. The reference for the demanded TCR current is, as illustrated in Figure 6.27, usually provided by a regulation loop of the external control, which compares the actual capacitive impedance or compensating voltage to the reference given for the desired system operation.

The second approach also employs a PLL, synchronized to the line current, for the generation of the basic timing reference. However, in this method the actual zero crossing of the capacitor voltage is estimated from the prevailing capacitor voltage and line current by an angle correction circuit. The delay angle is then determined from the desired angle and the estimated correction angle so as to make the TCR conduction symmetrical with respect to the expected zero crossing, as illustrated in Figure 6.28. The desired delay angle in this scheme can be adjusted by a closed-loop controlled phase shift of the basic time reference provided by the PLL circuit [refer back to Figures 6.19(a) and (b)]. The delay angle of the TCR, and thus the compensating capacitive voltage, as in the previous case, is controlled overall by a regulation loop of the external control in order to meet system operating requirements. This regulation loop is relatively slow, with a bandwidth just sufficient to meet compensation requirements (power flow adjustment, power oscillation damping, etc.). Thus, from the standpoint of the angle correction circuit, which by comparison is very fast (correction takes place in each half cycle), the output of the phase shifter is almost a steady-state reference.

Although control circuit performances are usually heavily dependent on the actual implementation, the second approach is theoretically more likely to provide faster response for those applications requiring such response.

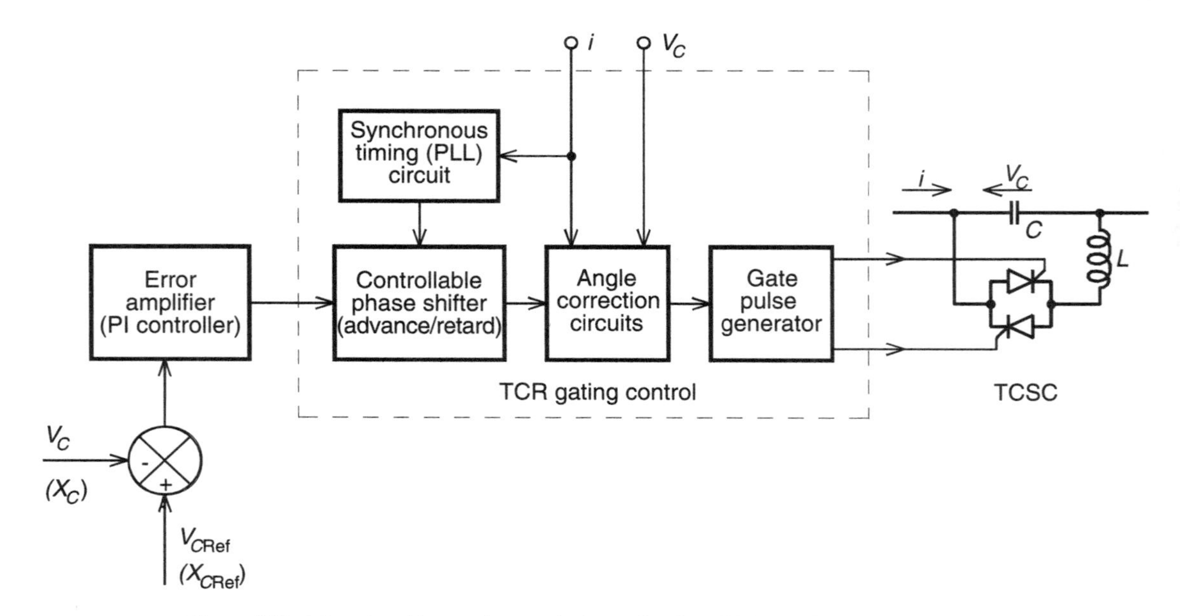

Figure 6.28 A functional internal control scheme for the TCSC based on the prediction of the capacitor voltage zero crossings.

6.3 SWITCHING CONVERTER TYPE SERIES COMPENSATORS

It has been established in Chapter 5 (see Section 5.2.2) that a voltage-sourced converter with its internal control can be considered a synchronous voltage source (SVS) analogous to an ideal electromagnetic generator: it can produce a set of (three) alternating, substantially sinusoidal voltages at the desired fundamental frequency with controllable amplitude and phase angle; generate, or absorb, reactive power; and exchange real (active) power with the ac system when its dc terminals are connected to a suitable electric dc energy source or storage. Thus, the SVS can be operated with a relatively small dc storage capacitor in a self-sufficient manner, like a static var generator, to exchange reactive power with the ac system or, with an external dc power supply like a static generator, or with an energy storage device, to also exchange independently controllable real power. A functional representation of the SVS is shown in Figure 6.29. References Q_{Ref} and P_{Ref} (or other related parameters, such as the desired compensating reactive impedance X_{Ref} and resistance R_{Ref}) define the amplitude V and phase angle ψ of the generated output voltage necessary to exchange the desired reactive and active power at the ac output. If the SVS is operated strictly for reactive power exchange, P_{Ref} (or R_{Ref}) is set to zero.

The concept of using the synchronous voltage source for series compensation is based on the fact that the impedance versus frequency characteristic of the series capacitor, in contrast to filter applications, plays no part in accomplishing the desired compensation of a transmission line. (On the contrary, as discussed in the previous sections, special control techniques are applied in the thyristor-controlled series capacitors to change their impedance versus frequency characteristics in the subsynchronous frequency band to make them immune to subsynchronous resonance.) The function of the series capacitor is simply to produce an appropriate voltage at the fundamental ac system frequency in quadrature with the transmission line current in order to increase the voltage across the inductive line impedance, and thereby increase the

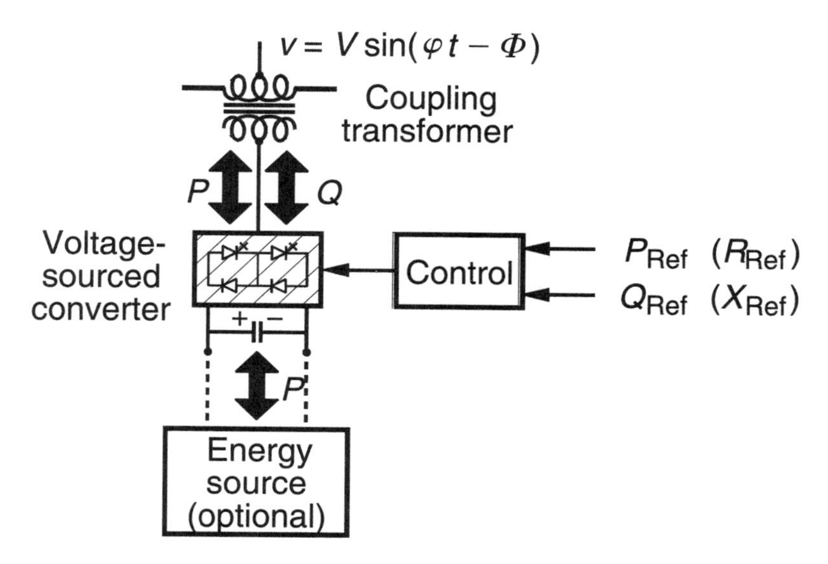

Figure 6.29 Functional representation of the synchronous voltage source based on a voltage-sourced converter.

line current and the transmitted power. The characteristics of the synchronous voltage source clearly suggest a comparable line compensation capability.

6.3.1 The Static Synchronous Series Compensator (SSSC)

The voltage-sourced converter-based series compensator, called Static Synchronous Series Compensator (SSSC), was proposed by Gyugyi in 1989 within the concept of using converter-based technology uniformly for shunt and series compensation, as well as for transmission angle control. The basic operating principles of the SSSC can be explained with reference to the conventional series capacitive compensation of Figure 6.1, shown simplified in Figure 6.30 together with the related voltage phasor diagram. The phasor diagram clearly shows that at a given line current the voltage across the series capacitor forces the opposite polarity voltage across the series line reactance to increase by the magnitude of the capacitor voltage. Thus, the series capacitive compensation works by increasing the voltage across the impedance of the given physical line, which in turn increases the corresponding line current and the transmitted power. While it may be convenient to consider series capacitive compensation as a means of reducing the line impedance, in reality, as explained previously, it is really a means of increasing the voltage across the given impedance of the physical line. It follows therefore that the same steady-state power transmission can be established if the series compensation is provided by a synchronous ac voltage source, as shown in Figure 6.31, whose output precisely matches the voltage of the series capacitor, i.e.,

$$\boldsymbol{V}_q = \boldsymbol{V}_C = -\boldsymbol{j}X_C\boldsymbol{I} = -\boldsymbol{j}kX\boldsymbol{I} \tag{6.12}$$

where, as before, $\boldsymbol{V}_C$ is the injected compensating voltage phasor, $\boldsymbol{I}$ is the line current, X_C is the reactance of the series capacitor, X is the line reactance, $k = X_C/X$ is the degree of series compensation, and $\boldsymbol{j} = \sqrt{-1}$. Thus, by making the output voltage of the synchronous voltage source a function of the line current, as specified by (6.12), the same compensation as provided by the series capacitor is accomplished. However, in contrast to the real series capacitor, the SVS is able to maintain a constant compensating voltage in the presence of variable line current, or control the amplitude of the injected compensating voltage independent of the amplitude of the line current.

For normal capacitive compensation, the output voltage lags the line current by 90 degrees. For SVS, the output voltage can be reversed by simple control action to

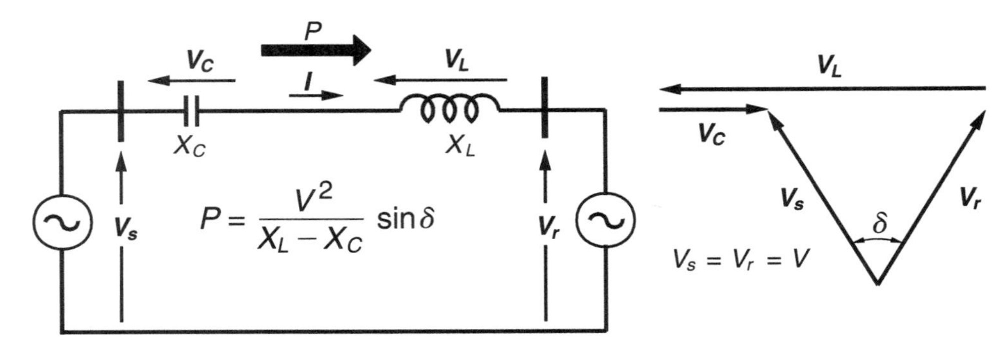

Figure 6.30 Basic two-machine system with a series capacitor compensated line and associated phasor diagram.

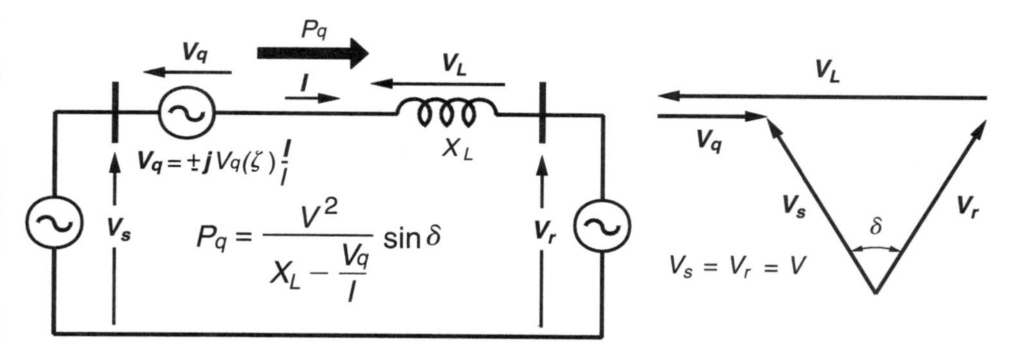

Figure 6.31 Basic two-machine system as in Figure 6.30 but with synchronous voltage source replacing the series capacitor.

make it lead or lag the line current by 90 degrees. In this case, the injected voltage decreases the voltage across the inductive line impedance and thus the series compensation has the same effect as if the reactive line impedance was increased.

With the above observations, a generalized expression for the injected voltage, $\boldsymbol{V}_q$, can simply be written:

$$\boldsymbol{V}_q = \pm \boldsymbol{j} V_q(\zeta) \frac{\boldsymbol{I}}{I} \tag{6.13}$$

where $V_q(\zeta)$ is the magnitude of the injected compensating voltage ($0 \leq V_q(\zeta) \leq V_{q\max}$) and ζ is a chosen control parameter. The series reactive compensation scheme, using a switching power converter (voltage-sourced converter) as a synchronous voltage source to produce a controllable voltage in quadrature with the line current as defined by (6.13) is, per IEEE and CIGRE definition, termed the *Static Synchronous Series Compensator* (*SSSC*).

6.3.2 Transmitted Power Versus Transmission Angle Characteristic

The SSSC injects the compensating voltage in series with the line irrespective of the line current. The transmitted power P_q versus the transmission angle δ relationship therefore becomes a parametric function of the injected voltage, $V_q(\zeta)$, and it can be expressed for a two-machine system as follows:

$$P = \frac{V^2}{X} \sin \delta + \frac{V}{X} V_q \cos \frac{\delta}{2} \tag{6.14}$$

The normalized power P versus angle δ plots as a parametric function of V_q are shown in Figure 6.32 for $V_q = 0$, ± 0.353, and ± 0.707. For comparison, the normalized power P versus angle δ plots of a series capacitor compensated two-machine system are shown in Figure 6.33 as a parametric function of the degree of series compensation k. For this comparison, k is chosen to give the same maximum power as the SSSC with corresponding V_q. That is, at $\delta = 90°$, $k = 1/5$ when $V_q = 0.353$ and $k = 1/3$ when $V_q = 0.707$.

Comparison of the corresponding plots in Figures 6.32 and 6.33 clearly shows that the series capacitor increases the transmitted power by a fixed percentage of that transmitted by the uncompensated line at a given δ and, by contrast, the SSSC can

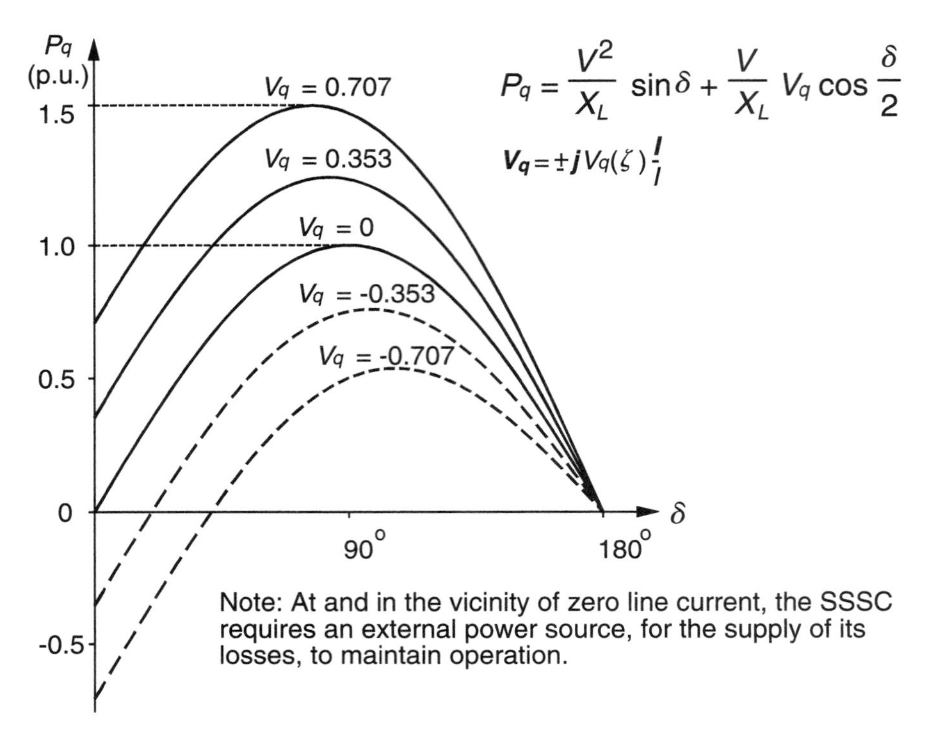

Figure 6.32 Transmitted power vs. transmission angle provided by the SSSC as a parametric function of the series compensating voltage.

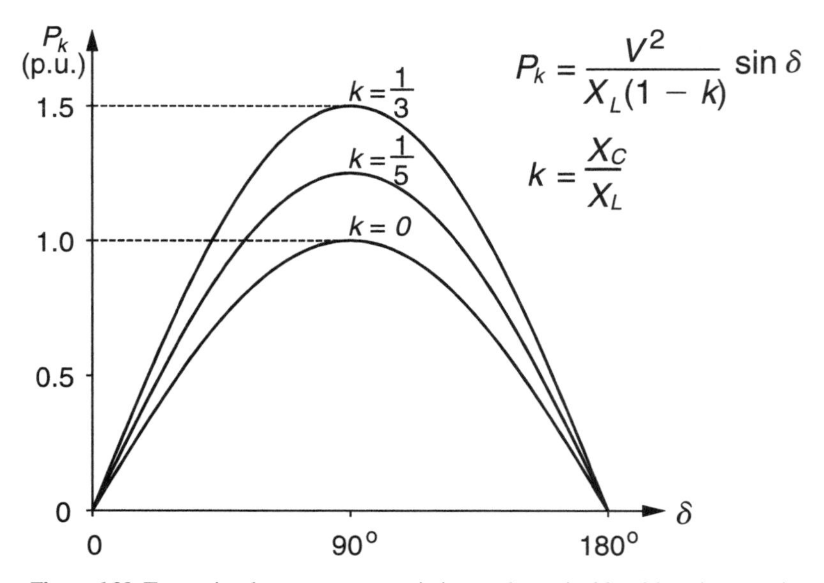

Figure 6.33 Transmitted power vs. transmission angle attainable with series capacitive compensation as a parametric function of the degree of series compensation.

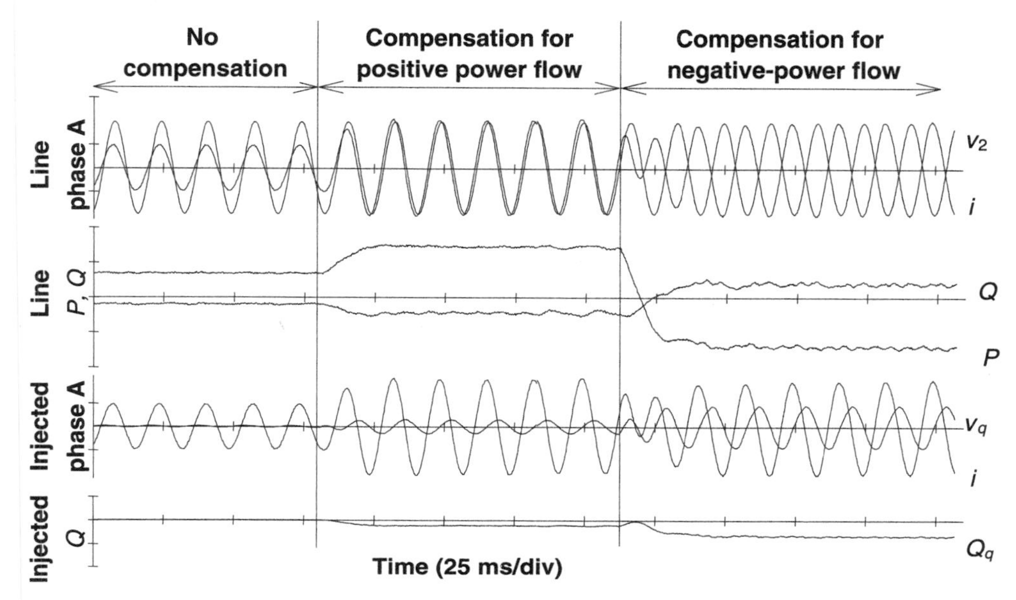

Figure 6.34 Oscillograms from TNA simulation showing the capability of the SSSC to control as well as reverse real power flow.

increase it by a fixed fraction of the maximum power transmittable by the uncompensated line, independent of δ, in the important operating range of $0 \leq \delta \leq \pi/2$.

For applications requiring (steady-state or dynamic) power flow control, the basic P versus δ characteristics shown in Figure 6.32 indicates that the SSSC, similarly to the STATCOM, inherently has twice as wide controlled compensation range as the VA rating of the converter. This means that the SSSC can decrease, as well as increase the power flow to the same degree, simply by reversing the polarity of the injected ac voltage. The reversed (180° phase-shifted) voltage adds directly to the reactive voltage drop of the line as if the reactive line impedance was increased. Furthermore, if this (reverse polarity) injected voltage is made larger than the voltage impressed across the uncompensated line by the sending- and receiving-end systems, that is, if $V_q > |\mathbf{V}_s - \mathbf{V}_r|$, then the power flow will reverse with the line current $I = (V_q - |\mathbf{V}_s - \mathbf{V}_r|)/X$, as indicated in Figure 6.32.

The feasibility of reversing power flow by reactive compensation is demonstrated in Figure 6.34 by the results obtained from the TNA simulation of a simple two-machine system controlled by a precisely detailed SSSC hardware model. The plots in the figure show, at $\delta = 10°$, the line current i together with the receiving-end ($l - n$) voltage $v_r = v_2$ for phase A, the transmitted power P together with the reactive power Q supplied by the receiving end, the same line current i together with the voltage v_q injected by the SSSC in phase A, and the reactive power the SSSC exchanged with the ac system for no compensation ($V_q = 0$), purely reactive compensation for positive power flow ($V_q = IX - |\mathbf{V}_s - \mathbf{V}_r|$), and purely reactive compensation for negative power flow ($V_q = IX + |\mathbf{V}_s - \mathbf{V}_r|$). Apart from the stable operation of the system with both positive and negative power flows, it can also be observed that the SSSC has, as expected, an excellent (subcycle) response time and that the transition from positive to negative power flow through zero voltage injection is perfectly smooth and continuous.

Apart from the bi-directional compensation capability, the basic operating characteristic of the SSSC also suggests a significant difference between the behaviors of SSSC and the series capacitor under the condition of variable line reactance that the reader should note. The gist of this difference is that the SSSC could not be tuned with any finite line inductance to have a classical series resonance (at which the capacitive and inductive voltages would be equal) at the fundamental frequency, because the voltage across the line reactance would, in all practical cases, be greater than, and inherently limited by, the (fixed) compensating voltage produced by the SSSC. This compensating voltage is set by the control and it is independent of network impedance (and, consequently, line current) changes. That is, the voltage V_X across an ideal line of reactance X ($R = 0$) at a fixed δ is the function of only the compensating voltage V_q injected by the SSSC, that is,

$$V_X = IX = V_q + 2V \sin\frac{\delta}{2} \tag{6.15}$$

where again V is the ac system $(l - n)$ voltage, and δ is the transmission angle. As (6.15) shows, V_X can be equal to V_q only if $\delta = 0$, in which case the transmission would be controlled entirely by the SSSC as if it were a generator and the line current would be restricted to the operating range of $0 \leq I \leq V_q/X$. (It should be noted that the SSSC would require an external dc power supply for the replenishment of its internal losses to be able to establish power transmission at zero transmission angle.)

6.3.3 Control Range and VA Rating

The SSSC can provide capacitive or inductive compensating voltage independent of the line current up to its specified current rating. Thus, in voltage compensation mode the SSSC can maintain the rated capacitive or inductive compensating voltage in the face of changing line current theoretically in the total operating range of zero to $I_{q\max}$, as illustrated in Figure 6.35(a1). (The practical minimum line current is that at which the SSSC can still absorb enough real power from the line to replenish its losses.) The corresponding loss, as percent of the (capacitive or inductive) rating of the SSSC, versus line current characteristic is shown in Figure 6.35(a2). The VA rating of the SSSC (solid-state converter and coupling transformer) is simply the product of the maximum line current (at which compensation is still desired) and the maximum series compensating voltage: $\mathrm{VA} = I_{\max} V_{q\max}$.

In impedance compensation mode, the SSSC is established to maintain the maximum rated capacitive or compensating reactance at any line current up to the rated maximum, as illustrated in Figure 6.35(b1). The corresponding loss versus line current characteristic is shown in Figure 6.35(b2).

Note that in practical applications, as indicated previously for variable impedance type compensators, $I_{\max}$ may be separately defined for the rated maximum steady-state line current and for a specified short duration overcurrent. The basic VA rating of the major power components of the SSSC must be rated for these currents and for the relevant maximum voltages.

It is seen in Figure 6.35 that an SSSC of 1.0 p.u. VA rating covers a control range corresponding to 2.0 p.u. compensating vars, that is, the control range is continuous from −1.0 p.u. (capacitive) vars to +1.0 p.u. (inductive) vars. In many practical applications, only capacitive series line compensation is required. In these applications,

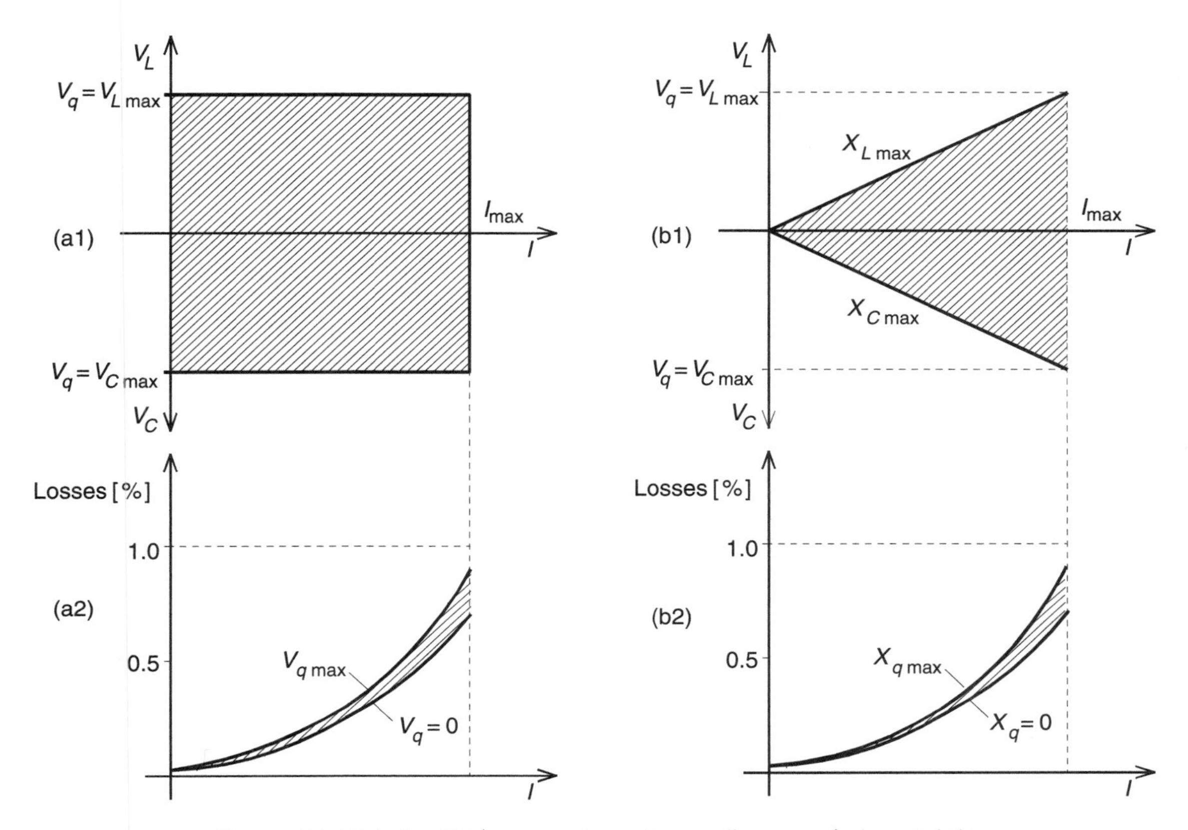

Figure 6.35 Attainable V-I (compensating voltage vs. line current) characteristics of the SSSC when operated in voltage control (a1) and reactance control (b1) modes, and the associated loss vs. line current characteristics (a2 and b2, respectively).

as well as in those which already use or plan to use series capacitors as part of the overall series compensation scheme, the SSSC may be combined cost effectively with a fixed capacitor, as illustrated in Figure 6.36, where an SSSC of 0.5 p.u. VA rating is combined with a fixed capacitor of 0.5 p.u. VAC rating to form a continuously

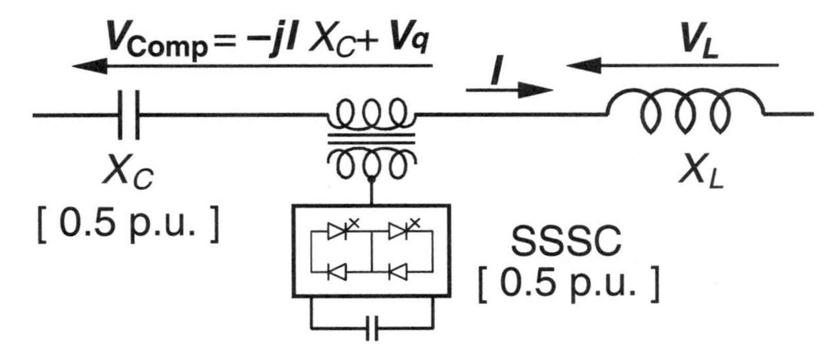

Figure 6.36 Hybrid compensation scheme consisting of a 0.5 p.u. fixed capacitor and a 0.5 p.u. SSSC.

controllable overall series compensator with a maximum compensating range of zero to 1.0 p.u. capacitive. (The reader should note that this is a dual of the hybrid shunt compensator in which a STATCOM of ±0.5 p.u. is used in parallel with a fixed capacitor of 0.5 p.u. rating to provide an overall zero to 1.0 p.u. compensating range.) The injected voltage versus line current characteristic of the SSSC + FC arrangement in the voltage compensation mode is shown in Figure 6.37(a1) and in impedance compensation mode in Figure 6.37(b1). The corresponding loss versus line current characteristics are shown in Figures 6.37(a2) and (b2), respectively. This compensation scheme from the standpoints of major component (converter and fixed capacitor) ratings and operating losses is extremely advantageous, in spite of the fact that the fixed capacitor produces a compensating voltage that is proportional to the line current, and therefore, the controllable compensating voltage range of the overall compensator also becomes, to some degree, a function of the line current.

6.3.4 Capability to Provide Real Power Compensation

In contrast to the series capacitor, which functions in the transmission circuit as a reactive impedance and as such is only able to exchange reactive power, the SSSC can negotiate both reactive and active power with the ac system, simply by controlling the angular position of the injected voltage with respect to the line current. However, as explained previously, the exchange of active power requires that the dc terminal of the SSSC converter be coupled to an energy source/sink, or a suitable energy storage.

The capability of the SSSC to exchange active power has significant application potential. One important application is the simultaneous compensation of both the

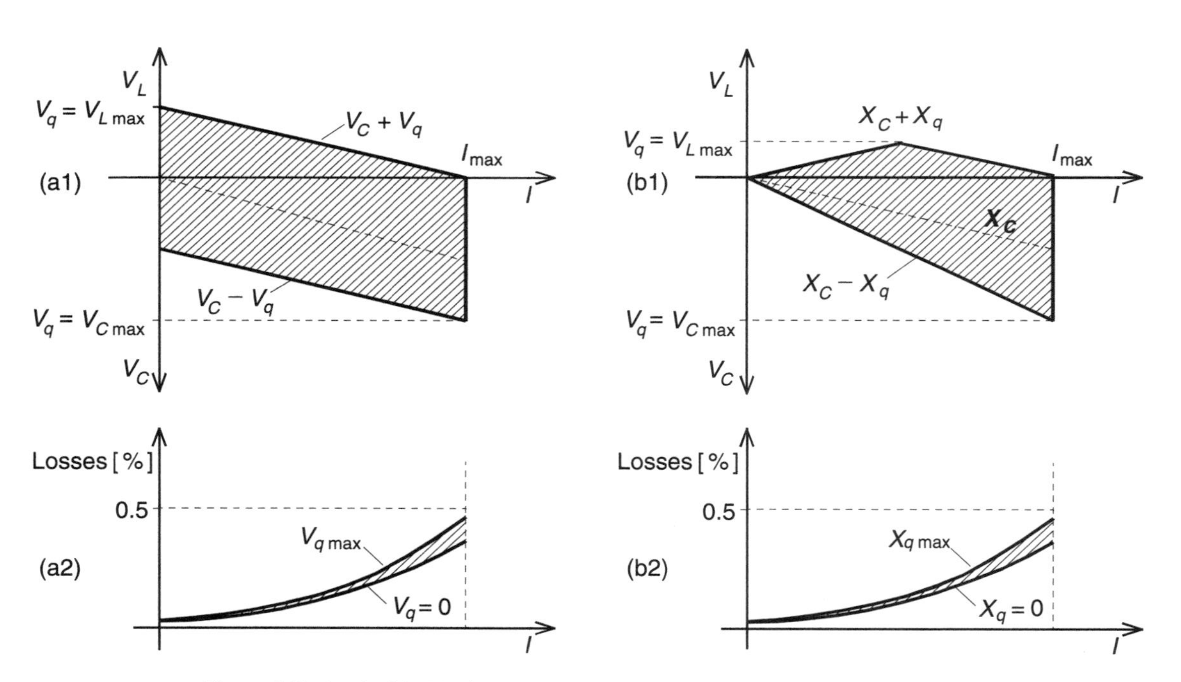

Figure 6.37 Attainable *V-I* (compensating voltage vs. line current) characteristics of the hybrid series compensator of Figure 6.36, when operated in voltage control (a1) and reactance control (b1) modes, and the associated loss vs. line current characteristics (a2 and b2, respectively).

reactive and resistive components of the series line impedance in order to keep the X/R ratio high. In many applications, particularly at transmission voltage levels of 115, 230, and even 340 kV, where the X/R ratio is usually relatively low (in the range of 3 to 10), a high degree of series capacitive compensation could further reduce the effective reactive to resistive line impedance ratio to such low values at which the progressively increasing reactive power demand of the line, and the associated line losses and possible voltage depression, would start to limit the transmittable active power. This situation is illustrated with a phasor diagram in Figure 6.38 for a normal angle-controlled line whose uncompensated X/R ratio is 7.4. As seen, by applying increasing series capacitive compensation (e.g., 50 and 75%), the effective $X_{\text{eff}}/R = (X_L - X_C)/R$ ratio decreases (to 3.7 and 1.85, respectively). As a result, the reactive component of the line current, $I \sin(\delta/2 + \phi)$, supplied by the receiving-end system, progressively increases and the real component, $I \cos(\delta/2 + \phi)$, transmitted to the receiving end, progressively decreases with respect to those which would be obtained with an ideal reactive line ($R = 0$).

The transmittable active power, P, and the reactive power, Q, supplied by the receiving end bus can be expressed for the simple two-machine system, employing angle adjustment for power flow control at constant end voltages ($V_s = V_r = V$), as

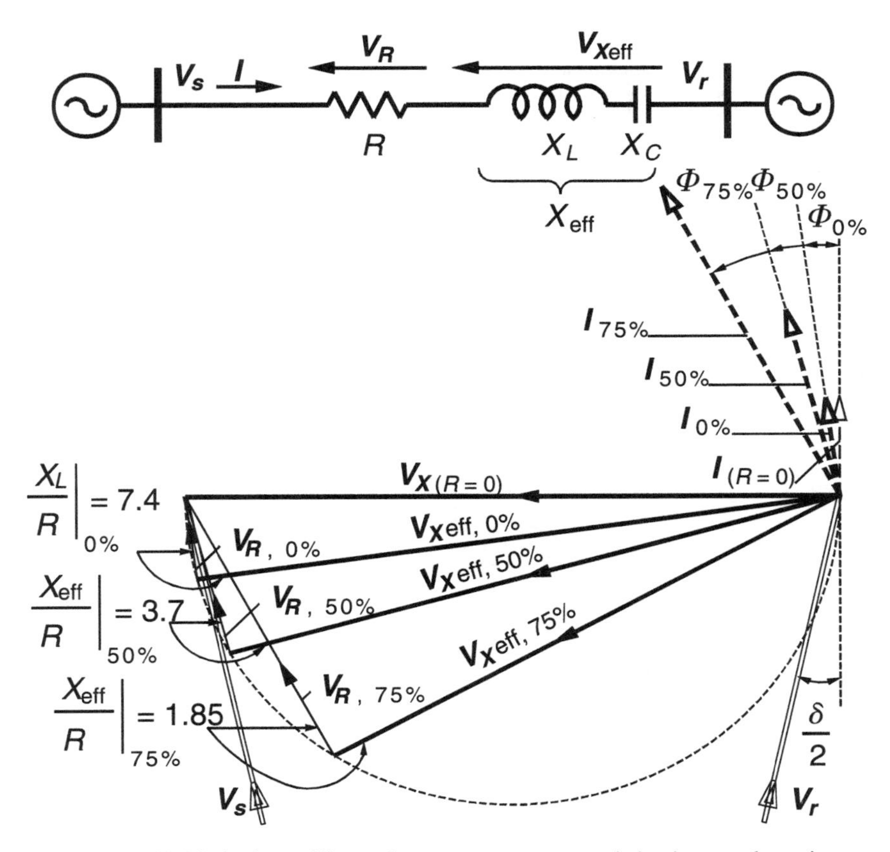

Figure 6.38 Limitations of line resistance on power transmission increase by series capacitive compensation.

functions of the (actual or effective) reactive line impedance, X, the line resistance, R, and transmission angle, δ, as follows:

$$P = \frac{V^2}{X^2 + R^2}[X \sin \delta - R(1 - \cos \delta)] \tag{6.16}$$

$$Q = \frac{V^2}{X^2 + R^2}[R \sin \delta + X(1 - \cos \delta)]. \tag{6.17}$$

The normalized active power P and reactive power Q versus angle δ transmission characteristics described by (6.16) and (6.17) are plotted as a parametric function of the X/R ratio for $X/R = \infty$, 7.4, 3.7, 1.85 in Figure 6.39. These plots clearly show that the maximum transmittable active power decreases, and the ratio of active to reactive power increases, with decreasing X/R ratio.

The SSSC with an appropriate dc power supply (which could be powered simply, for example, from an accessible bus or from the tertiary of a conveniently located transformer) would be able to inject, in addition to the reactive compensating voltage, a component of voltage in antiphase with that developed across the line resistance to counteract the effect of the resistive voltage drop on the power transmission. In this way, by providing simultaneous, and independently controllable compensation

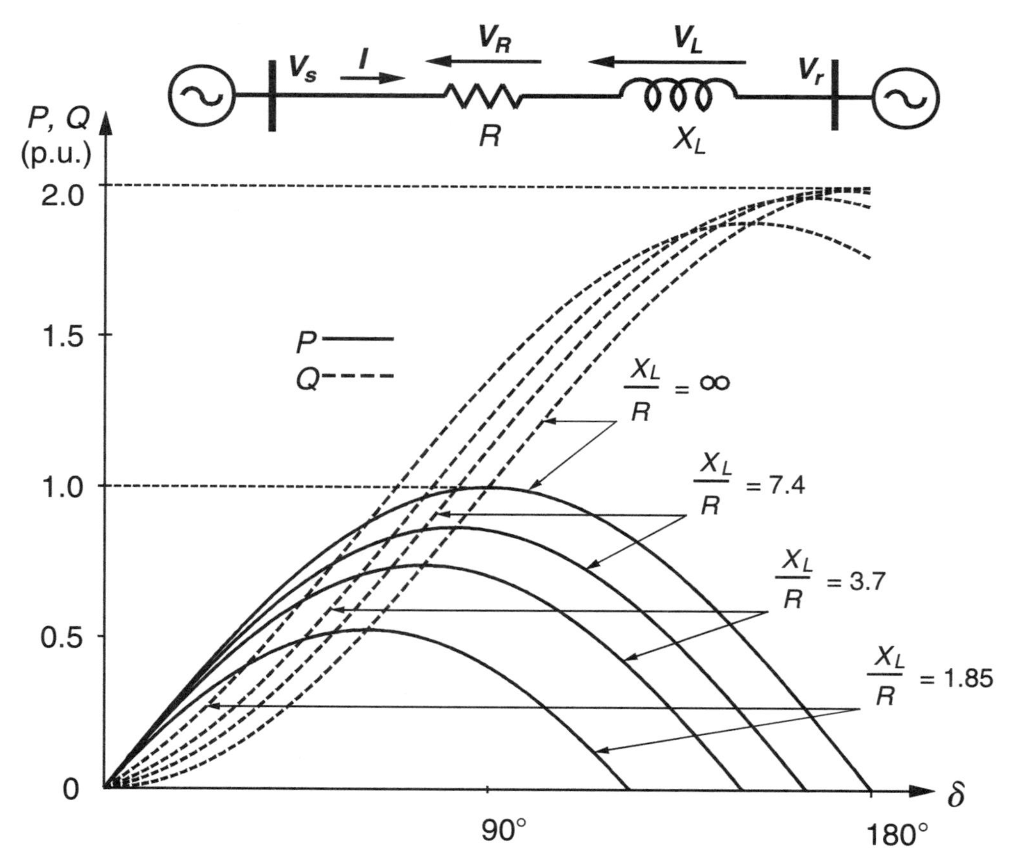

Figure 6.39 Transmitted real power P and reactive power Q vs. transmission angle δ as a parametric function of the line X/R ratio.

of both the reactive and real impedance of the line, in effect an ideal reactive line can be created for maximum power transmission. It should be noted that the power I^2R would, of course, still be dissipated by the physical line. However, this dissipated power would be replenished by the SSSC from the auxiliary power supply. The real power compensation capability could also be used effectively in minimizing loop power flows by balancing both the real and reactive power flows of parallel lines.

The recordings, obtained from the TNA simulation of a two machine system compensated by the SSSC with a dc power supply, illustrate the combined compensation of the line reactance and resistance in Figure 6.40. The plots show the line current i in phase A together with the corresponding receiving-end $(l - n)$ voltage $v_r = v_2$, the transmitted power P together with the reactive power Q supplied by the receiving end, the line current i again in phase A together with the voltage v_q injected by the SSSC, and the active and reactive power the SSSC exchanged with the ac system via the series voltage injection for no compensation ($V_q = 0$), purely reactive compensation, and reactive plus resistive compensation. (The system was operated at $\delta = 20°$ with $X/R = 6$.) It can be observed that the additional resistive compensation increases the transmitted power significantly, while it also decreases the reactive power demand on the receiving end.)

From the standpoint of dynamic system stability, reactive line compensation combined with simultaneous active power exchange can also enhance power oscillation damping. For example, during the periods of angular acceleration, the SSSC with a suitable energy storage can apply maximum capacitive line compensation to increase the transmitted active power and concurrently absorb active power to provide the effect of a damping resistor in series with the line. During the periods of angular deceleration, the SSSC can execute opposite compensating actions, that is, apply

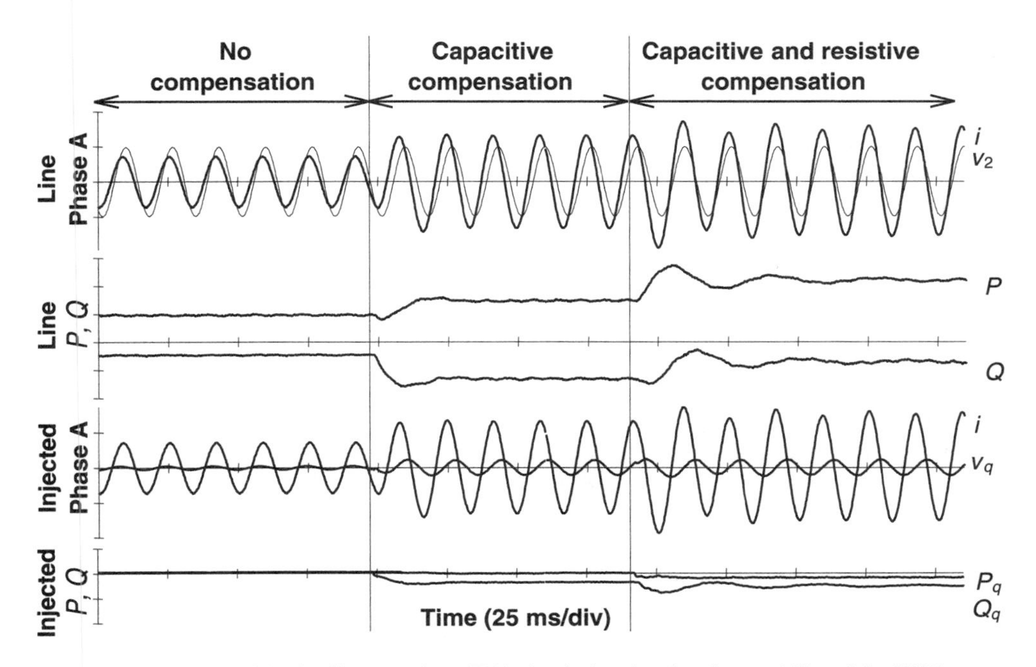

Figure 6.40 Oscillograms from TNA simulation showing the capability of the SSSC to provide both reactive and resistive series line compensation.

maximum inductive compensation to decrease the transmitted active power and concurrently provide the effect of a negative resistance (i.e., a generator) to supply additional active power for the line (negative damping). (Although an energy storage is required for the ideal damping obtained by injecting an alternating positive and negative damping resistor, a more economical and still effective damping can be provided with just an energy sink, e.g., a high energy resistor tied to the dc terminals of the SSSC. With this arrangement the SSSC would be able to absorb active power during the periods of angular accelerations, which then would be dissipated by the energy sink.)

6.3.5 Immunity to Subsynchronous Resonance

The desired function of the series capacitor is to provide a compensating voltage opposite to that which develops across the reactive line impedance at the fundamental system frequency to increase the transmitted power. However, the impedance of the series capacitor is a function of frequency and thus it can cause resonances at various subsynchronous frequencies with other reactive impedances present in the network. As discussed in previous sections, in recent years there has been considerable progress made in modifying the inherent frequency characteristic of the series capacitor in the dominant subsynchronous frequency band by a parallel connected thyristor-controlled reactor, making it immune to subsynchronous resonance with the use of electronic control.

In contrast to a series capacitor and an impedance type series compensator, the voltage-sourced converter-based static synchronous series compensator is essentially an ac voltage source which, with a constant dc voltage and fixed control inputs, would operate only at the selected (fundamental) output frequency, and its output impedance at other frequencies would theoretically be zero. In practice, the SSSC does have a relatively small inductive output impedance provided by the leakage inductance of the series insertion transformer. The voltage drop across this impedance is automatically compensated at the fundamental frequency when the SSSC provides capacitive line compensation. Thus, the effective output impedance versus frequency characteristic of the SSSC, operated at a constant dc voltage, remains that of a small inductor at all but its fundamental operating frequency. Consequently, such an SSSC is unable to form a classical series resonant circuit with the inductive line impedance to initiate subsynchronous system oscillations.

In a practical SSSC, the voltage-sourced converter on the dc side is terminated by a finite (and relatively small) energy storage capacitor to maintain the desired dc operating voltage. (Recall that this dc capacitor is kept charged by the energy absorbed from the system by the converter itself.) Thus the dc capacitor in effect interacts with the ac system via the operating switch (valve) array of the converter. This interaction may conceivably influence the subsynchronous behavior of a practical SSSC.

For the purpose of this discussion let it be assumed (what is a typical practical case) that the SSSC has no source or sink of power attached to its dc terminals, and furthermore let us neglect the power losses of the converter. The dc terminal voltage is thus supported entirely by a dc capacitor bank. With perfectly balanced sinusoidal line currents, the converter (with sufficiently large pulse number) will produce balanced, sinusoidal voltages in quadrature with the line current, and the dc capacitor bank will experience no charging current since no real power is exchanged at the ac terminals of the converter. If the converter is controlled to produce these same voltages while additional sub- and supersynchronous components are introduced into the line current, then alternating power components will appear at the ac terminals of the converter and these will be matched by alternating charging currents in the dc capacitor, producing in turn an associated alternating voltage component on the capacitor. This is because the energy transfer through the converter is absolutely direct, and thus the net instantaneous power at the input terminals must always be equal to the net instantaneous power at the output terminals. For example, if the line current contains sub- and supersynchronous components with frequencies of $f_{sub} = f_o - f_m$ and $f_{sup} = f_o + f_m$, respectively, in addition to the synchronous component of frequency f_o, due to the modulation process taking place at the mechanical system frequency f_m during the subsynchronous resonance stipulated in the example, then an ac current component of frequency f_m will flow through the dc terminals to balance the fluctuating power appearing at the ac terminals of the converter. As a result, the dc terminal voltage of the SSSC converter (which is supported only by a finite capacitor) will have a superimposed ac component with frequency f_m. If the converter control ignores this ac component, that is, if it is operated to produce the ac output voltage as if the dc terminal voltage was constant, then the SSSC will have sub- and supersynchronous voltage components in its output with the same $f_{sub} = f_o - f_m$ and $f_{sup} = f_o + f_m$ frequencies. However, it should be recognized that the variation in dc link voltage does not inherently affect the ability of the converter to produce the designated output voltages unless the dc voltage dips too low, or rises too high for the safe operation of the converter valves. That is, it is possible to produce the designated sinusoidal output voltages at the wanted single synchronous frequency by appropriately controlling the (instantaneous) magnitude of the converter's output voltage. In this way the converter can maintain ideal terminal characteristics: the positive sequence synchronous line currents would flow through the converter valves, bypassing the dc terminals (while the converter would be exchanging reactive power at the synchronous frequency with the line), and the sub- and supersynchronous line current components would circulate through the converter valves and the dc capacitor as if the ac terminals were shunted by ideal, fundamental frequency voltage sources. Thus, in spite of the modulated dc terminal voltage, the power system would "see" the SSSC as a perfect synchronous voltage source, which acts as a short circuit at nonsynchronous frequencies. In other words, from the standpoint of the power system, this SSSC would exhibit ac terminal characteristics identical to those obtained with an ideal converter whose dc terminal voltage is provided by an infinite source of zero impedance.

For converters not having suitable output control capability, an obvious approach to limit the dc bus voltage excursions, and the corresponding effects on the ac output voltage, is to increase the amount of dc capacitance. With sufficient dc bus capacitance the SSSC can produce ideal output voltages irrespective of the composition of the line current and it will clearly remain neutral to SSR.

Without increasing the dc capacitor size, or establishing an internal voltage control capability in the converter, there exists another method of operating the SSSC safely in the presence of subsynchronous line current components. This approach makes use of the intrinsic ability of the converter to change the phase angle of its output voltages very rapidly. The technique is best explained using instantaneous vectors to represent the three-phase voltage and current sets. By this method, each three-phase set (excluding zero sequence components) is entirely represented at each instant of time by a single two-dimensional vector. (Readers not familiar with the instantaneous vector representation should consider it simply an extension of the conventional phasor representation. The major difference is that while a phasor represents a single voltage or current component, usually the fundamental, a vector represents all, except the zero sequence, components, present in the three-phase voltage or current sets. The basic rules for power exchange important in the present discussion are similar: for purely real power exchange the voltage and current vectors, just like the voltage and current phasors, must be in phase. For purely reactive power exchange, these vectors, just like the corresponding phasors, must be in quadrature.) The vector representing the converter terminal voltages can be controlled in magnitude (constant, for example) and, as explained previously, can be assigned any angle virtually instantaneously. By the simple strategy of keeping the converter voltage vector instantaneously and continuously in quadrature with the line current vector, the ac terminal power of the SSSC can be maintained precisely at zero, absolutely irrespective of the nature of the line current. The dc capacitor thus sees no charging currents during abnormal line conditions, and the dc voltage stays constant.

In view of above general conclusions, consider now the consequence of the prescribed control technique in the presence of SSR on the line. In this case too, the magnitude of the voltage vector, representing the output voltages of the SSSC, is freely controlled according to the compensation requirements of the line, but its angle is kept precisely at 90 degrees with respect to the vector representing the three line currents, which are now modulated by the torsional machine oscillations. Thus, by definition, the instantaneous total power at the terminals of the SSSC will be maintained at zero. The SSSC therefore appears in series with the line as an "energy-neutral" device, since at no instant in time does it ever deliver any energy to the network nor absorb any energy from it. It is evident that an energy-neutral device cannot contribute in any way to the occurrence of an SSR which is, after all, a power oscillation between a generator and various energy storage components.

From the above argument, one can conclude that, due to fundamental physical laws, a sufficient condition for the SSSC to remain neutral to subsynchronous oscillations, independently of system conditions, is to keep its instantaneous output voltage vector (representing the output voltages of the converter) in quadrature with the instantaneous line current vector.

While the generality of the above postulated sufficient condition is supported by basic physical laws, it should be recognized that other methods (e.g., the previously established voltage control method) may also be used to exclude the SSSC's active participation in subsynchronous oscillation. Also, if the SSSC is equipped with a dc power supply, for example, to facilitate real power exchange for resistive line compensation, then this power supply may also be used to regulate the dc terminal voltage in the presence of subsynchronous (or other) line components. With this, the converter would naturally generate only the desired fundamental compensating voltage and be neutral to subsynchronous resonance. Although not yet proven, it is also very likely that the SSSC, due to the fast, almost instantaneous response of the voltage-sourced converter, can be controlled to be highly effective in the active damping of prevailing subsynchronous oscillations brought about by conventional series capacitive compensation.

In discussing dynamic interactions, it is also a consideration that the SSSC, like actively controlled equipment, could under abnormal conditions exhibit instability or oscillatory interaction with the ac system if, for example, its closed-loop gains, providing automatic power flow control or other regulative functions, are improperly set, or if the electronic control itself malfunctions. However, these considerations are generic to all actively controlled systems and are soluble by the well-established techniques of control robustness, control redundancy, and protection, which are not the subject of this book.

6.3.6 Internal Control

The discussion on subsynchronous resonance in the preceding section indicates that the implementation of some SSR immunity strategies requires the full (magnitude and angle) controllability of the compensating voltage the SSSC generates. As explained, the SSSC is based on the synchronous voltage source concept, which is implemented by a voltage-sourced converter. Basic approaches to this implementation, from the standpoints of converter power circuit and related internal controls, are discussed in detail in Chapter 5, Section 5.2.2.2. These approaches are summarized here with appropriate changes in the corresponding internal controls necessary for series compensation.

From the standpoint of output voltage control, converters may be categorized as "directly" and "indirectly" controlled. For directly controlled converters both the angular position and the magnitude of the output voltage are controllable by appropriate valve (on and off) gating. Figure 5.34 is an example of this converter. For indirectly controlled converters only the angular position of the output voltage is controllable by valve gating; the magnitude remains proportional to the dc terminal voltage. (Recall, however, that when the dc terminal voltage is provided by a charged capacitor, the magnitude of the output voltage can be varied by angle control which establishes momentary real power exchange at the ac terminals to charge or discharge the dc capacitor.) An example for such an indirectly controlled converter is shown in Figure 5.33.

The control method of maintaining a quadrature relationship between the instantaneous converter voltage and line current vectors, to provide reactive series compensation and handle SSR, can be implemented with an indirectly controlled converter. The method of maintaining a single-frequency synchronous (i.e., fundamental) output independent of dc terminal voltage variation, requires a directly controlled converter. Although high-power directly controlled converters are more difficult and costly to implement than indirectly controlled converters (because their greater control flexibility is usually associated with some penalty in terms of increased losses, greater circuit complexity, and/or increased harmonic content in the output), nevertheless they can be realized to meet practical utility requirements and, indeed, they are used, along with indirectly controlled converters, in existing installations for transmission compensation.

A possible internal control scheme for the indirectly controlled SSSC converter is shown in Figure 6.41. The inputs to the internal control are: the line current i, the injected compensating voltage v_q, and the reference V_q. The control is synchronized to the line current by a phase-locked loop which, after a $+\pi/2$ or $-\pi/2$ phase shift, provides the basic synchronizing signal θ. (The reader should note that the PLL itself could also be operated from the ac system voltage, in which case the phase shift of θ would also take into account the relative phase angle between the system voltage and line current.) The phase shifter is operated from the output of a polarity detector which determines whether reference V_q is positive (capacitive) or negative (inductive). The compensating voltage v_q is controlled by a simple closed loop: the absolute value of reference V_q is compared to the measured magnitude of the injected voltage v_q and the amplified difference (error) is added, as a correction angle $\Delta\alpha$, to the synchronizing signal $\theta(=\omega t)$. Depending on the polarity of $\Delta\alpha$, angle θ, and consequently the con-

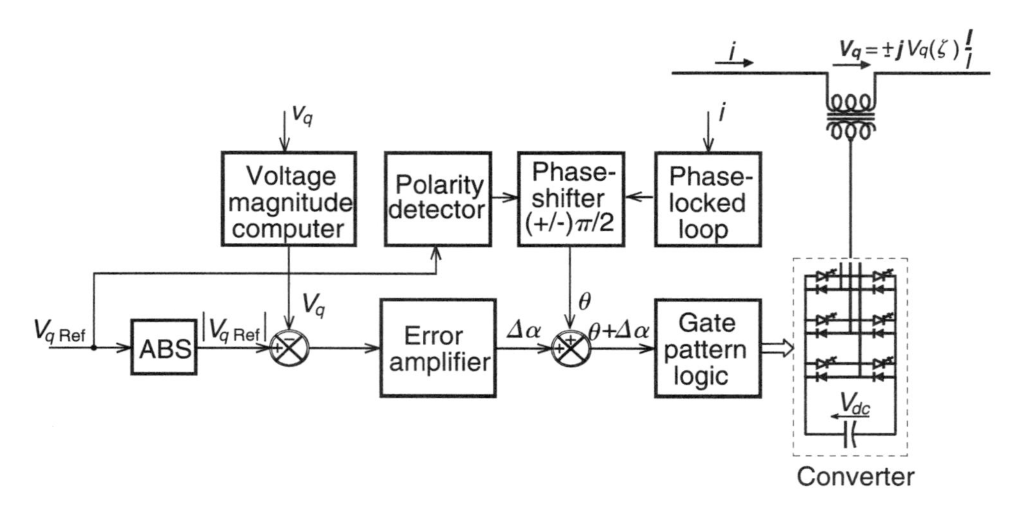

Figure 6.41 Functional internal control scheme for the SSSC employing an indirectly controlled converter.

verter gate drive signals, will be advanced or retarded and, thereby, the compensating voltage v_q will be shifted with respect to the prevailing line current from its original $+\pi/2$ or $-\pi/2$ phase position. This phase shift will cause the converter to absorb real power from the ac system for the dc capacitor or, vice versa, supply that to the ac system from the dc capacitor. As a result, the voltage of the dc capacitor will increase or decrease, causing a corresponding change in the magnitude of the compensating voltage. (Recall that the magnitude of the output voltage generated by the indirectly controlled converter is proportional to the dc voltage.) Once the desired magnitude of v_q is reached (normally within 1 or 2 milliseconds), the substantially quadrature relationship between the line current and compensating voltage gets reestablished with only a remaining small, steady angular difference necessary to absorb power from the ac system to replenish the operating losses of the converter.

A possible control scheme for the directly controlled SSSC converter is shown in Figure 6.42. This scheme can be used to eliminate the unwanted output voltage components due to the modulation of the dc capacitor voltage by subsynchronous or other line current components. It is also suitable to provide both reactive and real (resistive) line compensation if the converter is equipped with a suitable dc power supply (and/or sink).

As shown in Figure 6.42, the synchronization to the line current is again accomplished by a phase-locked loop in a manner discussed above. Overall, the control structure is similar to that discussed in connection with the indirectly controlled converter, except for the continuous and independent control of both the magnitude and angle of the compensating voltage.

In its general form, the control is operated from three reference signals: $V_{qQ\text{Ref}}$, defining the desired magnitude of the series reactive compensating voltage; the optional $V_{qR\text{Ref}}$, defining the desired magnitude of the series real compensating voltage; and $V_{dc\text{Ref}}$, defining the operating voltage of the dc capacitor. The reactive voltage reference $V_{qQ\text{Ref}}$ (which with the line current determines the reactive power exchange for series compensation) and the overall real voltage reference $V_{qR\text{Ref}} + V_{dc\text{Ref}}$ (which with the line current determines the real power exchange for the optional real power compensa-

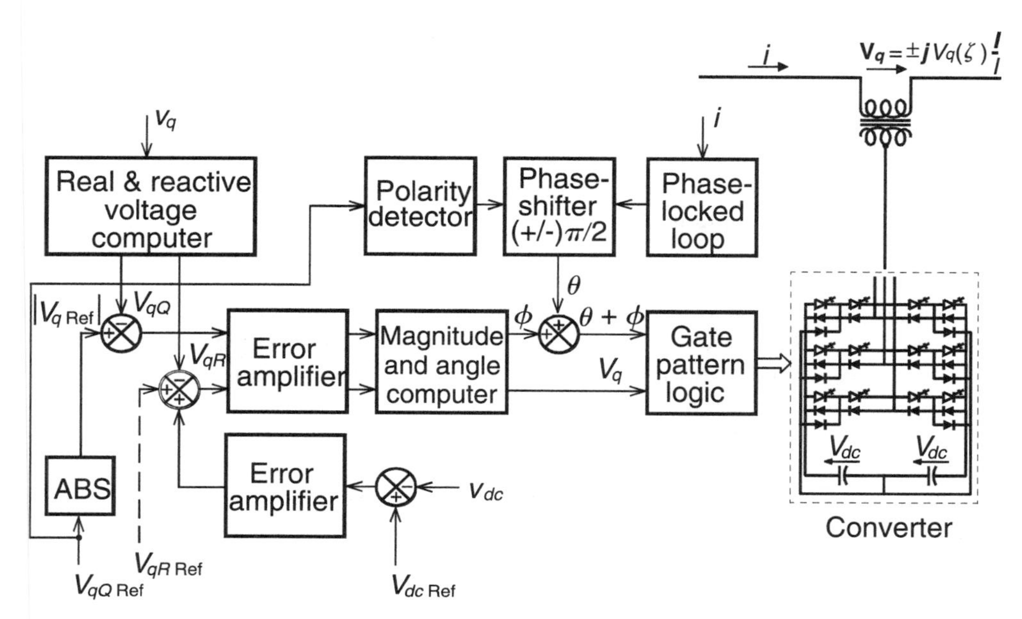

Figure 6.42 Functional internal control scheme for the SSSC employing a directly controlled converter.

tion of the line and for keeping the dc capacitor charged to its operating voltage level) are compared to the corresponding components of the measured compensating voltage v_q. From the resulting signals the magnitude of v_q and its angle ϕ with respect to θ (identifying the crest of the line current) are derived. Magnitude V_q and the $\theta + \phi$ are then used to generate the gate drive signals for the converter. With sufficient closed-loop bandwidth and suitable instantaneous vector-based signal processing, this type of control is able to maintain sinusoidal compensating voltage at the power system frequency in the presence of subsynchronous line current components and the consequent modulation of the dc capacitor voltage. (The reader should note that if resistive line compensation is not applied, i.e., $V_{qRRef} = 0$, angle ϕ in steady state would provide the same function as $\Delta\alpha$, a small, substantially constant angle to ensure energy absorption from the ac system to replenish the converter losses.)

6.4 EXTERNAL (SYSTEM) CONTROL FOR SERIES REACTIVE COMPENSATORS

In the previous sections, variable impedance type and synchronous voltage source type compensators, comprising the pertinent power circuit and related internal control, have been described. As seen, these compensators have different operating principles and exhibit differing compensation and performance characteristics. However, they all can provide series reactive compensation and be operated self-sufficiently from an input reference defining the desired compensating voltage or reactance. This means that the external control that defines the functional operation of the compensator and derives the reference input for it can basically be the same for all types of series compensator discussed. (The reader should recall that the same conclusion was also reached for the external control of all shunt connected static compensators in Chapter 5.)

The principal function of a series reactive compensator is power flow control. This can be accomplished either by the direct control (regulation) of the line current I or the transmitted power P, or, alternatively, by the indirect control of either the compensating impedance $X_q = X_C$ ($X_q = X_L$) or the compensating voltage $V_q = V_C$ ($V_q = V_L$). The direct power flow control has the advantage of maintaining the transmitted power in a closed-loop manner at the value defined by the given reference. However, under some network contingency the maintenance of constant power flow may not be possible or desirable. For this reason, in some applications the impedance (or voltage) control that maintains the impedance characteristic of the line may, from the operating standpoint, be preferred. Independent of the variable controlled, power flow control loops are usually operated with a rather slow (several seconds) response in order to avoid rapid power changes and the corresponding, often oscillatory, response or other stability impact on the ac system.

Additional functions for the improvement of transient (first swing) and dynamic stability (power oscillation damping) and, in some cases, for the damping of subsynchronous oscillation may be included in the external control of the series compensator.

Transient stability enhancement may be applied in response to a major disturbance. The reference to the internal control is a signal demanding maximum capacitive series compensation (to achieve maximal power transmission) for the measured or anticipated duration of the first swing over which the associated generators are accelerating (about 0.2–2.0 sec) after a major disturbance.

Power oscillations typically take place in the 0.2–2 Hz frequency range. For effective damping, the external control has to provide a variable reference that will modulate the output of the series compensator so as to increase the transmitted power over the line when the associated generators are accelerating and to decrease it when the generators are decelerating.

Subsynchronous oscillation damping, in addition to the SSR neutrality of the controlled series compensator, may be required if the line has a significant amount of uncontrolled series capacitive compensation. For SSR damping, the external control needs to produce a variable reference that corresponds to the torsional speed variation of the affected generator(s) and modulates the output of the series compensator so as to oppose this torsional variation. This variable reference may be derived directly from torsional speed variation, if accessible, or from system frequency or line current and voltage variations. In order to be effective for SSR damping, the series compensator needs a relatively wide frequency band, depending on the particular turbine generator set, up to 45 Hz. It is not yet proven that the series compensators discussed can be effective in damping subsynchronous oscillations over the possible total range of SSR. However, on the basis of the inherent transport lag characterizing the thyristor-controlled and converter-based compensators (as discussed in detail in Chapter 5), the latter type of compensator appears to have the necessary bandwidth for this requirement.

A possible structure of the external control is illustrated in Figure 6.43. The main power flow control is executed by a (slow) closed loop, which is operated from one of the selectable references, X_{qRef}, or V_{qRef}, or I_{Ref}, or P_{Ref}. The corresponding network variable (X_q, or V_q, or I, or P) is derived by the voltage and current processor and compared to the selected reference. The amplified error at the output of the PI controller provides the reference, X_q, or V_q, for the internal control.

The auxiliary control signals to improve transient and dynamic stability, and to damp subsynchronous oscillations, are derived from the relevant system variables, such as system frequency variation, power flow variation, or, for subsynchronous

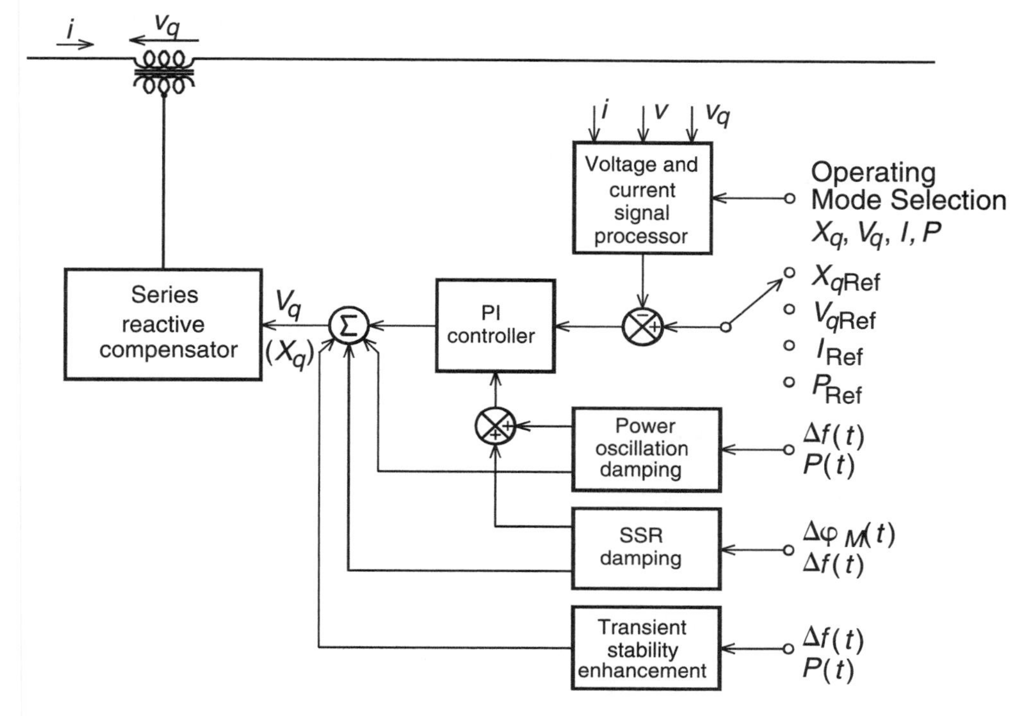

Figure 6.43 Functional external (system) control scheme for the SSSC.

oscillation damping, from torsional speed variation, when available, or synthesized from locally measurable voltages and currents, by the power oscillation damping, transient stability enhancement and subsynchronous oscillation damping auxiliary control circuits and fed directly, as references, to the internal control. These auxiliary control circuits may also control or inhibit the operation of the main power flow regulation loop in order to avoid contradictory reference requirements and maintain an operational set point appropriate under the particular system contingency.

6.5 SUMMARY OF CHARACTERISTICS AND FEATURES

The SSSC is a voltage source type and the TSSC, TCSC, and GCSC are variable impedance type series compensators. Although both types of compensator can provide highly effective power flow control, their operating characteristics and compensation features, are, or can be made, different. The possible differences are related to the inherent attributes of the different power circuits associated with the type of the compensator. The voltage source character of the SSSC offers some inherent capabilities and functional features for series line compensation which may not be attainable with variable impedance type compensators. On the other hand, the circuit structure and power semiconductors used in the thyristor-controlled variable impedance type compensators (TSSC, TCSC) offers easier accommodation of necessary protection features required to handle line fault conditions. The essential differences in characteristics and features of the two types of compensator can be summarized as follows:

1. The SSSC is capable of internally generating a controllable compensating

voltage over an identical capacitive and inductive range independently of the magnitude of the line current. The compensating voltage of the GCSC and TSSC over a given control range is proportional to the line current. The TCSC can maintain maximum compensating voltage with decreasing line current over a control range determined by the current boosting capability of the thyristor-controlled reactor.

2. The SSSC has the inherent ability to interface with an external dc power supply to provide compensation for the line resistance by the injection of real power, as well as for the line reactance by the injection of reactive power, for the purpose of keeping the effective X/R ratio high, independently of the degree of series compensation. The variable impedance type series compensators cannot exchange real power (except for circuit losses) with the transmission line and can only provide reactive compensation.
3. The SSSC with an energy storage (or sink) increases the effectiveness of power oscillation damping by modulating the series reactive compensation to increase and decrease the transmitted power, and by concurrently injecting an alternating virtual positive and negative real impedance to absorb and supply real power from the line in sympathy with the prevalent machine swings. The variable impedance type compensator can damp power oscillation only by modulated reactive compensation affecting the transmitted power.
4. The TSSC and TCSC employ conventional thyristors (with no internal turn-off capability). These thyristors are the most rugged power semiconductors, available with the highest current and voltage ratings, and they also have the highest surge current capability. For short-term, they are suitable to provide bypass operation to protect the associated capacitors during line faults. The GCSC and the SSSC use GTO thyristors. These devices presently have lower voltage and current ratings, and considerably lower short-term surge current rating. They are suitable for short-term bypass operation only if the anticipated line fault current is relatively low. Therefore, they may need external fast protection during severe line faults by an auxiliary conventional thyristor bypass switch, or a MOV arrester type voltage limiter, or by some other means of suitable speed. (Note that all series compensators would also typically have a mechanical bypass switch for backup.)
5. The variable impedance type compensators, TSSC, TCSC, and GCSC, are coupled directly to the transmission line and therefore are installed on a high-voltage platform. The cooling system and control are located on the ground with high voltage insulation requirements and control interface. The SSSC requires a coupling transformer, rated for 0.5 p.u. of the total series var compensating range, and a dc storage capacitor. However, it is installed in a building at ground potential and operated at a relatively low voltage (typically below 20 kV). Thus, this installation needs only relatively low voltage insulation for the cooling system and a control interface.
6. The voltage source and the different type of variable impedance type compensators also exhibit different loss characteristics. At zero compensation, the line current would flow through the semiconductor valves in all compensators and the losses would be proportional to the valve losses plus the reactor or transformer losses. At rated line current these losses would be about 0.5% of the rated var output for the TSSC and TCSC, about 0.7 to 0.9% for the GCSC and SSSC. At full and uncontrolled capacitive output the TSSC, TCSC, and

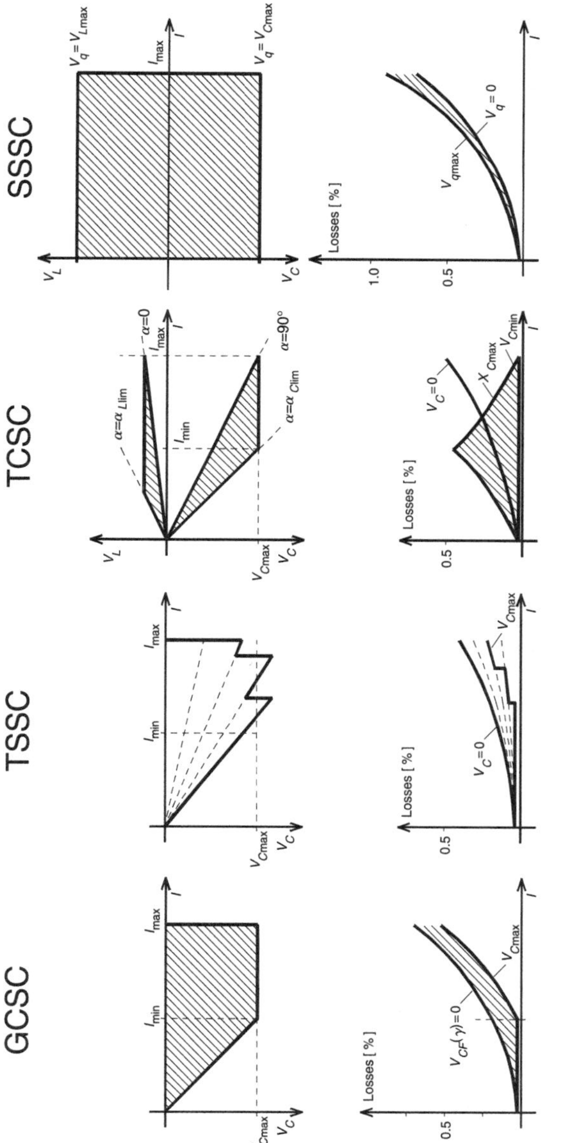

Figure 6.44 Summary of attainable compensating voltage vs. line current, and corresponding loss vs. line current, characteristics of basic series compensating schemes.

GCSC losses would be very low, independent of the line current, since the semiconductor valves would be nonconductive. With controlled compensating voltage, the losses would not be proportional to, but would be a function of the line current, and for the TSSC they could reach a maximum value of about 0.3%, for the TCSC about 0.4%, and for the GCSC about 0.6%. With this operating mode, the SSSC losses would be proportional to the line current and would reach a maximum of about 0.9% at rated line current.

The *V-I* characteristics of the major series compensators, operated in voltage compensation mode, with the corresponding loss versus line current characteristics, are summarized for the reader's convenience in Figure 6.44.

REFERENCES

Agrawal, B. L., et al., "Advanced Series Compensation (ASC) Steady-State, Transient Stability, and Subsynchronous Resonance Studies," *Proceedings of Flexible AC Transmission Systems (FACTS) Conference,* Boston, MA, May 1992.

Ängquist, L., et al., "Dynamic Performance of TCSC Schemes," *CIGRE Paper No. 14-302,* 1996.

Ängquist, L., et al., "Synchronous Voltage Reversal (SVR) Scheme—A New Control Method for Thyristor-Controlled Series Capacitors," *EPRI Conference on Flexible AC Transmission Systems (FACTS 3): The Future of High Voltage Transmission,* Baltimore, MD, October 5–7, 1994.

Bowler, C. E. J., et al., "FACTS and SSSR—Focus on TCSC Application and Mitigation of SSR Problems," *Proceedings of Flexible AC Transmission Systems (FACTS) Conference,* Boston, MA, May 1992.

Christl, N., et al., "Advanced Series Compensation with Variable Impedance," *EPRI Conference on Flexible AC Transmission Systems (FACTS): The Future of High Voltage Transmission,* Cincinnati, OH, November 14–16, 1990.

Christl, N., et al., "Advanced Series Compensation (ASC) with Thyristor-Controlled Impedance," *CIGRE Paper 14/37/38-05,* 1992.

De Souza, L. F. W., et al., "A Gate-Controlled Series Capacitor for Distribution Lines," *CIGRE Paper 14-201,* 1998.

Gribel, J., et al., "Brazilian North-South Interconnection—Application of Thyristor-Controlled Series Compensation (TCSC) to Damp Inter Area Oscillation Mode," *CIGRE Paper No. 14-101,* 1998.

Gyugyi, L., "Dynamic Compensation of AC Transmission Lines by Solid-State Synchronous Voltage Sources," *IEEE/PES Summer Meeting,* Paper No. 93SM431-1PWRD, 1993.

Gyugyi, L., "Solid-State Control of Electric Power in AC Transmission Systems," *International Symposium on 'Electric Energy Conversion in Power Systems',* Paper No. T-IP. 4, Capri, Italy, 1989.

Gyugyi, L., "Solid-State Control of AC Power Transmission," *EPRI Conference on Flexible AC Transmission Systems (FACTS): The Future of High Voltage Transmission,* Cincinnati, OH, November 14–16, 1990.

Gyugyi, L., et al., "Static Synchronous Series Compensator: A Solid-State Approach to the Series Compensation of Transmission Lines," *IEEE Trans. on Power Delivery,* vol. 12, no. 1, January 1997.

Hedin, R. A., et al., "SSSR Characteristics of Alternative Types of Series Compensation Schemes," *IEEE/PES Summer Meeting,* Paper No. 94 SM 534-8 PWRS, 1994.

Hingorani, N. G., et al., "Prototype NGH Subsynchronous Resonance Damping Scheme. Part 1, Field Installation and Operating Experience," *IEEE/PES Winter Meeting,* New Orleans, 1987.

Hingorani, N. G., "A New Scheme for Subsynchronous Resonance Damping of Torsional Oscillations and Transient Torque," Part I, *IEEE Trans. PAS-100,* no. 4, April 1981.

Jalali, S. G., et al., "A Stability Model for Advanced Series Compensator (ASC)," *IEEE/PES Summer Meeting,* Paper No. 95 SM 404-4 PWRD, 1995.

Karady, G. G., et al., "Continuously Regulated Series Capacitor," *IEEE/PES Summer Meeting,* Paper No. 92 SM 492-9 PWRD, 1992.

Larsen, E. V., et al., "Characteristics and Rating Considerations of Thyristor Controlled Series Compensation," *IEEE/PES Summer Meeting,* Paper No. 93 SM 433–3 PWRD, 1993.

Nyati, S., et al., "Effectiveness of Thyristor-Controlled Series Capacitor in Enhancing Power System Dynamics: An Analog Simulation Study," *IEEE/PES Summer Meeting,* Paper No. 93 SM 432-5 PWRD, 1993.

Paserba, J. J., et al., "Thyristor-Controlled Series Compensation Prototype Installation at the Slatt 500 kV Substation," *IEEE/PES Summer Meeting,* Paper No. 94 SM 476-2 PWRD, 1994.

Piwko, R. J., et al., "The Slatt Thyristor-Controlled Capacitor Project—Design, Installation, Commissioning and System Testing," *CIGRE Paper 14-104,* 1994.

Sen, K. K., "SSSC—Static Synchronous Series Compensator: Theory, Modeling, and Applications," *IEEE Trans. on Power Delivery,* vol. 13, no. 1, January 1998.

Torgerson, D. R. (Convenor), "Thyristor-Controlled Series Capacitor," *Working Group 14.18, CIGRE,* 1997.

Urbanek, J., et al., "Thyristor-Controlled Series Compensation Prototype Installation at the Slatt 500 kV Substation," *IEEE/PES Summer Meeting,* Paper No. 92 SM 467-1 PWRD, 1992.

Vithayathil, J. J., "Case Studies of Conventional and Novel Methods of Reactive Power Control on an AC Transmission System," *CIGRE Paper 38-02, 1986.*

7

Static Voltage and Phase Angle Regulators: TCVR and TCPAR

7.1 OBJECTIVES OF VOLTAGE AND PHASE ANGLE REGULATORS

The elementary equations (5.3) and (5.4) derived in Chapter 5 for the determination of transmitted real power, P, and reactive line power, Q, indicate that both are a function of the transmission line impedance, the magnitude of the sending- and receiving-end voltages, and the phase angle between these voltages. The discussion shows that increased real power transmission will inevitably result in increased reactive power demand on the end-voltage bus systems (generators) and increased voltage variation along the transmission line.

Chapter 5 focuses on the control of transmission voltage by attacking the root cause of its variation, the changing reactive line power. It was established that controlled reactive shunt compensation is highly effective in maintaining the desired voltage profile along the transmission line in spite of changing real power demand. However, controlled reactive shunt compensation of bulk transmission systems to maintain specified voltage levels for loads, sometimes at the end of a subtransmission or distribution system, is often not an issue. For example, the interconnection of a high voltage line with a lower voltage line for increased power transmission is usually accomplished with a mechanical on-load tap changer to isolate the lower voltage system from the large voltage variation of the high-voltage line caused by seasonal or daily load changes. Similarly, voltage regulators employing on-load tap changers have been used since the early days of ac transmission to maintain the desired user voltage in the face of changing transmission voltage and loads. In addition to voltage regulation, tap changers can, in general, be used to control reactive power flow in the line. Since transmission line impedances are predominantly reactive, an in-phase voltage component introduced into the transmission circuit causes a substantially quadrature (reactive) current flow that, with appropriate polarity and magnitude control, can be used to improve prevailing reactive power flows.

Although reactive compensation and voltage regulation by on-load tap changers appear to provide the same transmission control function, there is an important operating difference to note between them. Whereas a reactive compensator supplies reactive power to, or absorbs that from the ac system to change the prevailing reactive power flow and thereby indirectly control the transmission line voltage, the tap changer-based voltage regulator cannot supply or absorb reactive power. It directly manages the transmission voltage on one side and leaves it to the power system to provide the necessary reactive power to maintain that voltage. Should the power system be unable to provide the reactive power demand, overall voltage collapse in the system could occur. On-load tap changers contribution to voltage collapse under certain conditions is well recognized. For example, when the tap changers increase the transformation ratio in order to minimize the voltage drop for predominantly motor loads in the face of an overloaded transmission system, the transmission line needs to supply increasing load current at decreasing load power factor. This of course further reduces the transmission voltage and in turn further increases the current until the voltage collapses and the protection relays remove the load. Nevertheless, the on-load tap changer and its electronic counterparts to be discussed in this, and the subsequent chapter, play a significant role in transmission power flow control by providing the important functions of voltage regulation and reactive power management.

Chapter 6 concentrates on the control of transmitted power by series reactive compensation which can be a highly effective means to control power flow in the line as well as improving the dynamic behavior of the power system. However, whereas series reactive compensation is generally highly effective for power flow control, its application to certain transmission problems can be impractical, cumbersome, or economically not viable. These problems are related to the transmission angle. For example, the prevailing transmission angle may not be compatible with the transmission requirements of a given line or it may vary with daily or seasonal system loads over too large a range to maintain acceptable power flow in some affected lines. Other problems involve the control of real and reactive loop flows in a meshed network. The solution to these types of problems usually requires the control of the effective angle [the third variable of the basic transmission relationships given in (5.3) and (5.4)] to which the relevant transmission line or network is exposed.

Mechanical phase angle regulators (PARs) or phase shifting transformers (PSTs), using on-load tap changers with quadrature voltage injection, were introduced in the 1930s to solve power flow problems and increase the utilization of transmission lines. Whereas on-load tap changers with in-phase voltage injection control reactive power via voltage magnitude adjustment, those with quadrature voltage injection control real power via phase adjustment. Their combined use makes both the reactive and real power flow control possible. As a result, PARs have historically been used to redirect current flows and alleviate inherent loop flows in interconnected systems and thereby improve and balance the loading of interconnected transmission links.

Apart from steady-state voltage and power flow control, the role of modern voltage phase angle regulators with fast electronic control can also be extended to handle dynamic system events. Potential areas of application include: transient stability improvement, power oscillation damping, and the minimization of post-disturbance overloads and the corresponding voltage dips. As compared to reactive compensators, voltage and phase angle regulators bring a new element to the control of dynamic events, the capability to exchange real power. At the same time, the voltage and angle regulators, based on the classical arrangement of tap-changing transformers, lack the ability of reactive compensators to supply or absorb reactive power and thus this burden is left to the power system to handle.

In the following section, the basic techniques of voltage and angle regulation are reviewed to provide the necessary foundation for the treatment of power electronics-based approaches. The effect of phase angle control on the basic factors determining attainable maximal power transmission (steady-state power transmission limit, transient stability, and power oscillation damping) is also examined.

7.1.1 Voltage and Phase Angle Regulation

The basic concept of voltage and phase angle regulation is the addition of an appropriate in-phase or a quadrature component to the prevailing terminal (bus) voltage in order to change (increase or decrease) its magnitude or angle to the value specified (or desired). Thus, voltage regulation could, theoretically, be achieved by a synchronous, in-phase voltage source with controllable amplitude, $\pm\Delta V$, in series with the ac system and the regulated terminal, as illustrated in Figure 7.1(a). A commonly used implementation of this concept is shown schematically in Figure 7.1(b). An adjustable voltage is provided by means of a tap changer from a three-phase (auto) transformer (usually referred to as regulating or excitation transformer) for the primary of a series insertion transformer which injects it to achieve the required voltage regulation. From the arrangement shown, it is evident that injected voltages $\pm\Delta v_a$, $\pm\Delta v_b$, and $\pm\Delta v_c$, are in phase with the line to neutral voltages v_a, v_b, and v_c, respectively, as illustrated by the phasor diagram in Figure 7.1(c). In two winding transformers, on load tap changers are provided on the neutral side of the windings.

In a similar manner, the arrangement of Figure 7.1(a) can be used for phase-angle control simply by stipulating the injected voltage, Δv, to have a phase of $\pm 90°$ relative to the system voltage, v, as illustrated in Figure 7.2(a). With this stipulation, the injected voltage will change the prevailing phase-angle of the system voltage. A possible arrangement for phase angle control is shown schematically in Figure 7.2(b) with the corresponding phasor diagram in Figure 7.2(c). For relatively small angular adjustments, the resultant angular change is approximately proportional to the injected voltage, while the voltage magnitude remains almost constant. However, for large angular adjustments, the magnitude of the system voltage will appreciably increase and, for this reason, is often referred to as a quadrature booster transformer (QBT). The voltage magnitude could be maintained independent of the angular adjustment by a more complex winding arrangement. Nevertheless, because of its relative simplic-

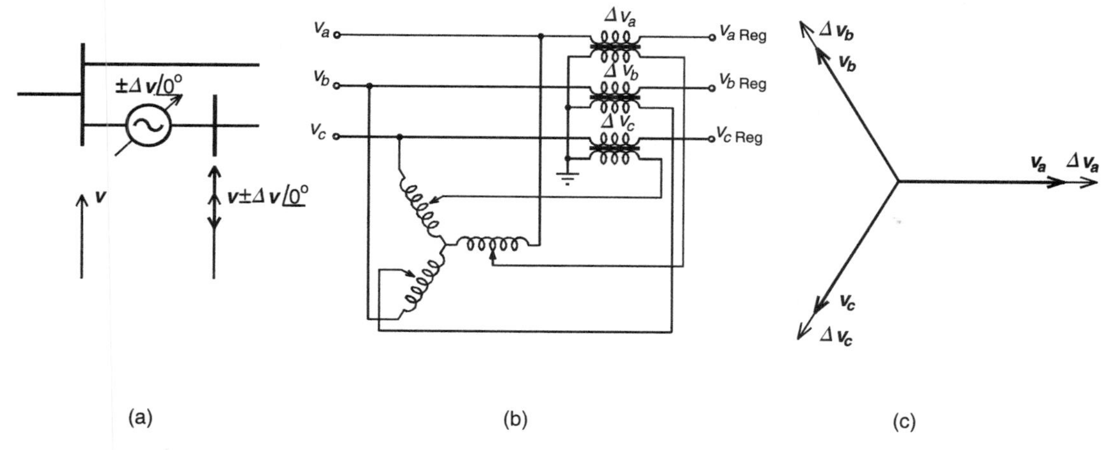

Figure 7.1 Concept and basic implementation of a Voltage Regulator.

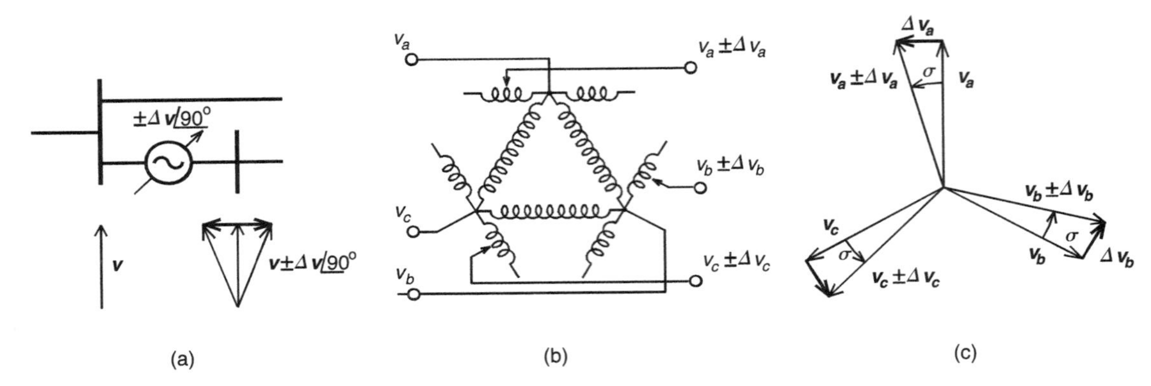

Figure 7.2 Concept and basic implementation of a Phase Angle Regulator.

ity, the QBT arrangement has typically been used in conventional phase shifting applications.

7.1.2 Power Flow Control by Phase Angle Regulators

As indicated above, the optimal loading of transmission lines in practical power systems cannot always be achieved at the prevailing transmission angle. Such cases would occur, for example, when power between two buses is transmitted over parallel lines of different electrical length or when two buses are inserted whose prevailing angle difference is insufficient to establish the desired power flow. In these cases a Phase Angle Regulator (PAR) is frequently applied.

The basic concept of power flow control by angle regulation, illustrated in Figure 7.3(a), is represented in terms of the usual two-machine model in which a Phase Angle Regulator is inserted between the sending-end generator (bus) and the transmission line. Theoretically, the Phase Angle Regulator can be considered a sinusoidal (fundamental frequency) ac voltage source with controllable amplitude and phase angle. Thus, the effective sending-end voltage $\mathbf{V}_{seff}$ becomes the sum of the prevailing sending-end bus voltage $\mathbf{V}_s$ and the voltage $\mathbf{V}_\sigma$ provided by the PAR, as the phasor diagram shown in Figure 7.3(b) illustrates. For an ideal Phase Angle Regulator, the angle of phasor $\mathbf{V}_\sigma$ relative to phasor $\mathbf{V}_s$ is stipulated to vary with σ so that the angular change does not result in a magnitude change, that is,

$$\mathbf{V}_{seff} = \mathbf{V}_s + \mathbf{V}_\sigma \text{ and } |\mathbf{V}_{seff}| = |\mathbf{V}_s| = V_{seff} = V_s = V \tag{7.1}$$

The basic idea behind the independent angle regulation is to keep the transmitted power at the desired level, independent of the prevailing transmission angle δ, in a predetermined operating range. Thus, for example, the power can be kept at its peak value after angle δ exceeds $\pi/2$ (the peak power angle) by controlling the amplitude of quadrature voltage $\mathbf{V}_\sigma$ so that the effective phase angle (δ-σ) between the sending- and receiving-end voltages stays at $\pi/2$. In this way, the actual transmitted power may be increased significantly, even though the Phase Angle Regulator per se does not increase the steady-state power transmission limit.

With the phase-angle control arrangement stipulated by (7.1) the effective phase angle between the sending- and receiving-end voltages becomes (δ-σ), and with this the transmitted power P and the reactive power demands at the ends of the line can simply be expressed as follows:

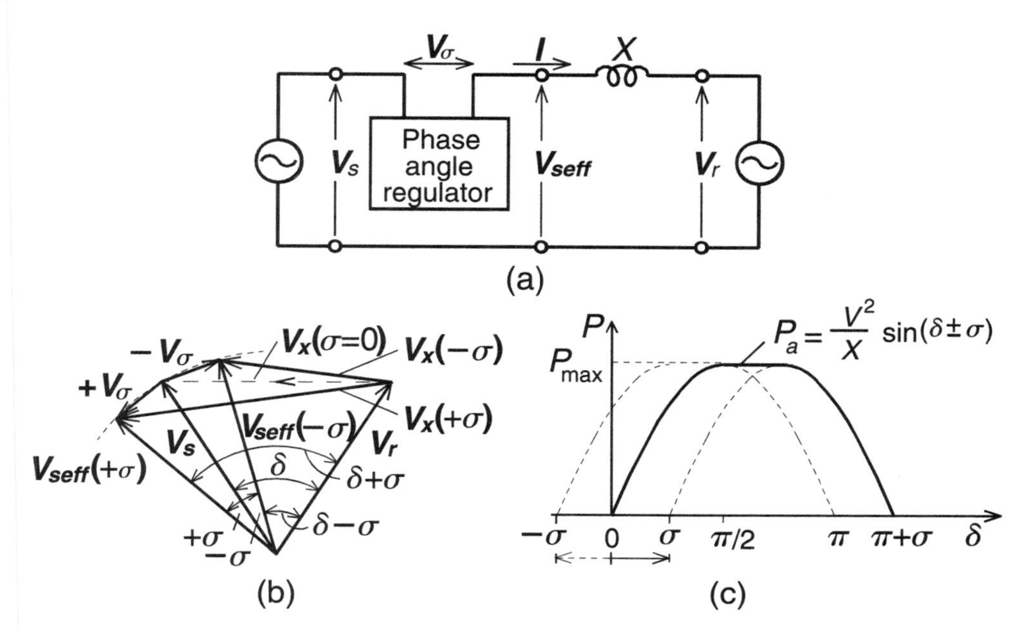

Figure 7.3 Two-machine power system with a Phase Angle Regulator (a), corresponding phasor diagram (b), and transmitted power vs. angle characteristics (c).

$$P = \frac{V^2}{X}\sin(\delta - \sigma) \tag{7.2}$$

and

$$Q = \frac{V^2}{X}\{1 - \cos(\delta - \sigma)\} \tag{7.3}$$

The relationships between real power P, reactive power Q, and angles δ and σ are shown plotted in Figure 7.3(c). It can be observed that, although the Phase Angle Regulator does not increase the transmittable power of the uncompensated line, it makes it theoretically possible to keep the power at its maximum value at any angle δ in the range $\pi/2 < \delta < \pi/2 + \sigma$ by, in effect, shifting the P versus δ curve to the right. It should be noted that the P versus δ curve can also be shifted to the left by inserting the voltage of the angle regulator with an opposite polarity. In this way, the power transfer can be increased and the maximum power reached at a generator angle less than $\pi/2$ (that is, at $\delta = \pi/2 - \sigma$).

It can also be observed that the relationship between the real power P and reactive power Q remains the same as that for the uncompensated system with equivalent transmission angle.

If the angle of phasor $\mathbf{V}_\sigma$ relative to phasor $\mathbf{V}_s$ is stipulated to be fixed at $\pm 90°$, the Phase Angle Regulator becomes a Quadrature Booster (QB), with the following relationships:

$$\mathbf{V}_{seff} = \mathbf{V}_s + \mathbf{V}_\sigma \text{ and } |\mathbf{V}_{seff}| = V_{seff} = \sqrt{V^2 + V_\sigma^2} \tag{7.4}$$

For a Quadrature Booster type angle regulator the transmitted power P can be expressed in the following form:

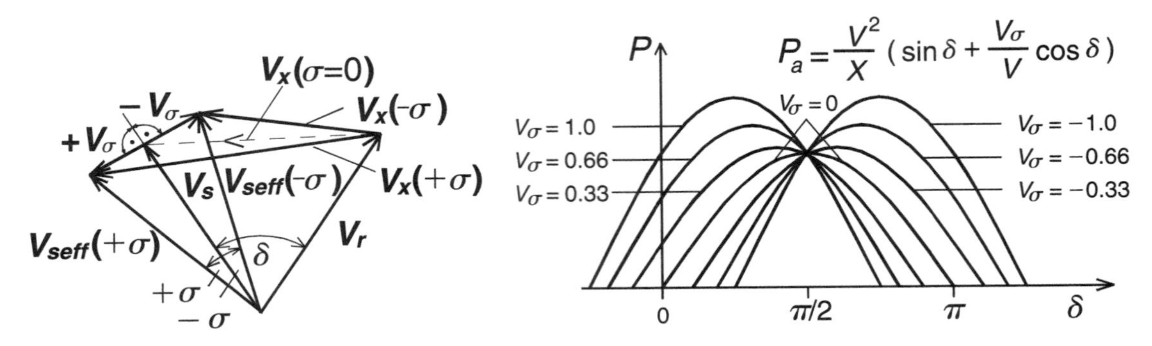

Figure 7.4 Phasor diagram and transmitted power vs. angle characteristics of a Quadrature Booster.

$$P = \frac{V^2}{X}\left(\sin\delta + \frac{V_\sigma}{V}\cos\delta\right) \tag{7.5}$$

The transmitted power P versus angle δ as a parametric function of the injected quadrature voltage V_σ is shown, together with the characteristic phasor diagram of the Quadrature Booster, in Figure 7.4. It can be observed that the maximum transmittable power increases with the injected voltage V_σ since, in contrast to the proper Phase Angle Regulator, the Quadrature Booster increases the magnitude of the effective sending-end voltage.

In contrast to the previously investigated reactive shunt and series compensation schemes, the phase angle regulators generally have to handle both real and reactive powers. The total VA throughput of the angle regulator (viewed as a voltage source) is

$$VA = |\boldsymbol{V}_{seff} - \boldsymbol{V}_s|\,|\boldsymbol{I}| = |\boldsymbol{V}_\sigma|\,|\boldsymbol{I}| = V_\sigma I \tag{7.6}$$

The rating of the angle regulator is thus determined by the product of the maximum injected voltage and the maximum continuous line current.

7.1.3 Real and Reactive Loop Power Flow Control

Consider two power systems "s" and "r" in Figure 7.5(a) connected by a single transmission with reactance X and resistance R. The transmission of power P from "s" to "r" results in a difference in magnitude between the terminal voltages, $\boldsymbol{V}_s$ and $\boldsymbol{V}_r$, and also a shift in phase angle, as illustrated in Figure 7.5(b). Vectorial voltage difference, $\boldsymbol{V}_l = \boldsymbol{V}_s - \boldsymbol{V}_r$, appears across the transmission line impedance $\boldsymbol{Z} = R + jX$, resulting in the line current $\boldsymbol{I}$. Phasor $\boldsymbol{V}_l$ is normally considered to be composed of the resistive and inductive voltage drops $\boldsymbol{I}R$ and $j\boldsymbol{I}X$, respectively, as indicated. However, for the present consideration of loop power flow, it is more meaningful to decompose $\boldsymbol{V}_l$ into two components, one in phase and the other in quadrature with the sending-end voltage phasor $\boldsymbol{V}_s$, as shown in Figure 7.5(b). These voltage components, as the reader should recall, determine the reactive and real power supplied by the sending-end system.

In practice, power systems are normally connected by two or more parallel transmission paths, resulting in one or more circuit loops with the potential for circulating current flow. Consider the above defined system with two parallel transmission lines, as shown in Figure 7.6. Basic circuit considerations indicate that if the X/R ratio

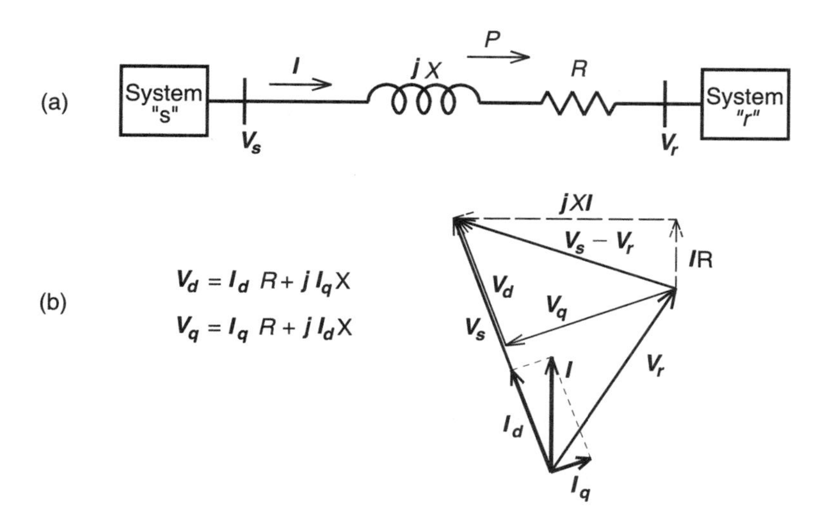

Figure 7.5 Two systems with a single line inter-tie and the corresponding phasor diagram.

for the two lines are not equal, that is, if $X_1/R_1 \neq X_2/R_2$, then a circulating current will flow through the two lines. Assuming such an inequality for the two X/R ratios and decomposing both line currents, $\boldsymbol{I}_1$ and $\boldsymbol{I}_2$, into an in-phase and a quadrature component with respect to the sending-end voltage $\boldsymbol{V}_s$, as illustrated previously in Figure 7.5(b), then the corresponding in-phase and quadrature voltage components for the lines, $\boldsymbol{V}_{1d}$, $\boldsymbol{V}_{1q}$ and $\boldsymbol{V}_{2d}$, $\boldsymbol{V}_{2q}$ can be expressed with the circuit resistances and reactances, R_1, R_2, and X_1, X_2, the line current components $\boldsymbol{I}_{1d}$, $\boldsymbol{I}_{1q}$ and $\boldsymbol{I}_{2d}$, $\boldsymbol{I}_{2q}$, and with the similar components of the assumed circulating current, $\boldsymbol{I}_{cd}$ and $\boldsymbol{I}_{cq}$, as follows:

$$\boldsymbol{V}_{1d} = (\boldsymbol{I}_{1d} + \boldsymbol{I}_{cd})R_1 + \boldsymbol{j}(\boldsymbol{I}_{1q} + \boldsymbol{I}_{cq})X_1 \tag{7.7}$$

$$\boldsymbol{V}_{1q} = (\boldsymbol{I}_{1q} + \boldsymbol{I}_{cq})R_1 + \boldsymbol{j}(\boldsymbol{I}_{1d} + \boldsymbol{I}_{cd})X_1 \tag{7.8}$$

and

$$\boldsymbol{V}_{2d} = (\boldsymbol{I}_{2d} - \boldsymbol{I}_{cd})R_2 + \boldsymbol{j}(\boldsymbol{I}_{2q} - \boldsymbol{I}_{cq})X_2 \tag{7.9}$$

$$\boldsymbol{V}_{2q} = (\boldsymbol{I}_{2q} - \boldsymbol{I}_{cq})R_2 + \boldsymbol{j}(\boldsymbol{I}_{2d} - \boldsymbol{I}_{cd})X_2 \tag{7.10}$$

Inspection of (7.7) through (7.10) shows that if $X_1/R_1 = X_2/R_2$ then I_{cd} and I_{cq}

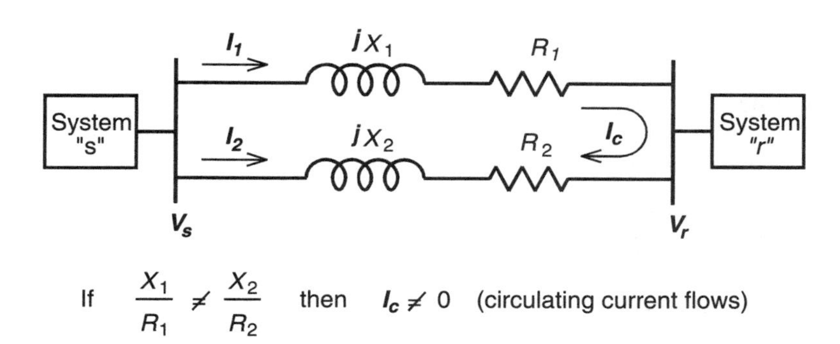

Figure 7.6 Two systems with a double line inter-tie.

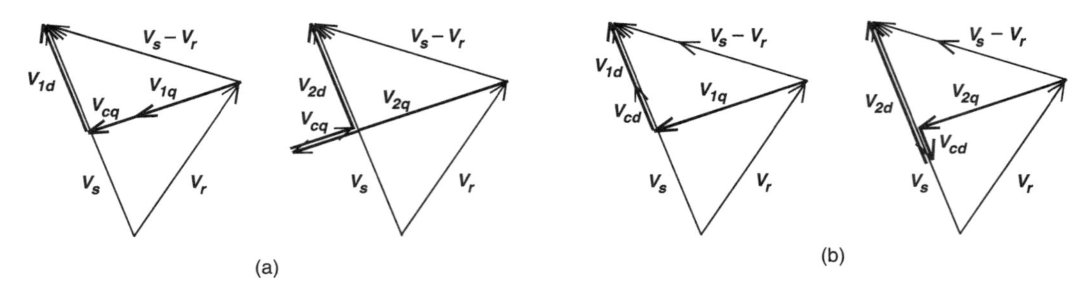

Figure 7.7 Phasor diagrams illustrating voltage unbalances resulting in real (a) and reactive (b) loop power flows.

must be equal to zero. It also indicates that a difference in either or both of the in-phase and quadrature voltage components may exist. For clarity, these possibilities are considered individually.

Figure 7.7(a) illustrates the case when there is a difference in the quadrature voltage components, $V_{1q} - V_{2q}$. Making the generally valid practical assumption that $R_1 \ll X_1$ and $R_2 \ll X_2$ this difference will primarily maintain the in-phase circulating current component $\boldsymbol{I}_{cd}$, thus increasing the real power in one line (line 1) and decreasing it in the other (line 2).

Figure 7.7(b) illustrates the case when a difference in the in-phase voltage components, $V_{1d} - V_{2d}$ exists. This difference, under the above assumption, will maintain primarily the quadrature circulating current component, $\boldsymbol{I}_{cq}$, and thus change the reactive power flow balance between the lines. In a general case, of course, differences may exist between both the in-phase and quadrature voltage components which will maintain both the in-phase and quadrature components of the circulating current, changing both the real and reactive power flow balance between the lines.

Generally, the distribution of real power flow over interconnections forming loop circuits can be controlled by Phase Angle Regulators. The flow of reactive power can be controlled by Voltage Regulators. These statements follow from the fact that the transmission circuit impedances are predominantly reactive. The Phase Angle Regulator injects a quadrature voltage in series with the circuit loop resulting in the flow of in-phase circulating current. The Voltage Regulator introduces a series in-phase voltage into the loop, and quadrature current is circulated through the loop since the impedances are substantially reactive. Thus, the insertion of a PAR in either tie-line 1 or tie-line 2 in the system of Figure 7.6 can correct the difference in quadrature voltage drops and can generally control the real power distribution between the lines. An added Voltage Regulator (separate or embedded in the PAR) can cancel the in-phase voltage difference and generally control the reactive power flow balance.

Both voltage and angle regulators can clearly provide major benefits for multiline and meshed systems: the full utilization of transmission assets (decrease of reactive power flow, control and balance of real power flow) and reduction of overall system losses through the elimination of circulating loop currents.

7.1.4 Improvement of Transient Stability with Phase Angle Regulators

As shown in Chapter 5, controlled shunt compensation increases transient stability by increasing (or maintaining) the (midpoint) transmission line voltage during the accelerating swing of the disturbed machine(s). In Chapter 6, controlled series reactive

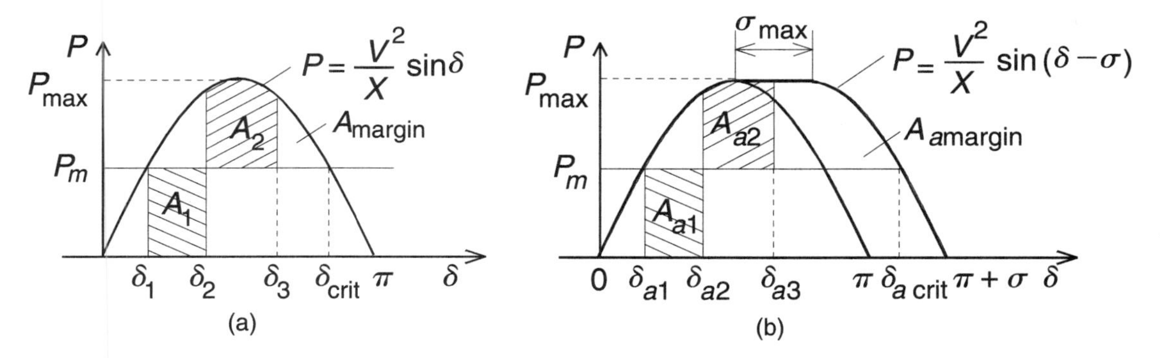

Figure 7.8 Equal area criterion illustrating the transient stability margin for a simple two-machine system, (a) without phase-angle control and, (b) with the use of an Phase Angle Regulator.

compensation is shown to improve transient stability by maximizing the power transmission during the first swing through the reduction of the effective transmission impedance. The capability of the Phase Angle Regulator to maintain the maximum effective transmission angle during the first swing can also be utilized effectively to increase the transient stability limit.

The equal area criterion, used in Chapters 5 and 6 to investigate the capability of the shunt series compensator to improve the transient stability, is used again here to assess the relative increase of the transient stability margin attainable by transmission angle control. In this and the subsequent section, an ideal Phase Angle regulator with characteristics as defined by (7.1) through (7.3) is assumed for uniformity and clarity.

Consider the simple system with a Phase Angle Regulator as shown in Figure 7.3(a). As for the previously investigated shunt and series compensated systems, it is, for convenience, also assumed for the case of transmission angle control that the pre-fault and post-fault systems remain the same. Suppose that the system of Figure 7.3(a), with and without the angle regulator, transmits the same power P_m. Assume that both the uncompensated and the series compensated systems are subjected to the same fault for the same period of time. The dynamic behavior of these systems is illustrated in Figures 7.8(a) and (b). As seen, prior to the fault both systems transmit power P_m at angles δ_1 and δ_{a1}, respectively. During the fault, the transmitted electric power becomes zero while the mechanical input power to the generators remains constant, P_m. Therefore, the sending-end generator accelerates from the steady-state angles δ_1 and δ_{a1} to angles δ_2 and δ_{a2}, respectively, when the fault clears. The accelerating energies are represented by areas A_1 and A_{a1}. After fault clearing, the transmitted electric power exceeds the mechanical input power and therefore the sending-end machine decelerates. However, the accumulated kinetic energy further increases until a balance between the accelerating and decelerating energies, represented by areas A_1, A_{a1} and A_2, A_{a2}, respectively, is reached at the maximum angular swings, δ_3 and δ_{a3}, respectively. The areas between the P versus δ curve and the constant P_m line over the intervals defined by angles δ_3 and δ_{crit}, and δ_{a3} and δ_{acrit}, respectively, determine the margin of transient stability, represented by areas A_{margin} and $A_{amargin}$.

Comparison of Figures 7.8(a) and (b) clearly shows a substantial increase in the transient stability margin which the Phase Angle Regulator can provide. It is noteworthy that in contrast to the shunt and series compensators, which provided the transient stability improvement via their inherent capability to increase the steady-state transmission limit of the uncompensated line, the angle regulator achieved that by maintaining the power transmission at the steady-state limit of the uncompensated line during the first swing as if the sending-end and receiving-end system were asynchronously connected. The increase of transient stability margin is proportional to the angular range, and thereby ultimately the *VA* rating of the Phase Angle Regulator.

It is pointed out again that under practical fault scenarios the pre-fault and post-fault systems are generally different and that the transient stability requirements would be established on the basis of the post-fault system.

7.1.5 Power Oscillation Damping with Phase Angle Regulators

Transmission angle control can also be applied to damp power oscillations. As explained in the two preceding chapters, power oscillation damping is achieved by varying the active power flow in the line(s) so as to counteract the accelerating and decelerating swings of the disturbed machine(s). That is, when the rotationally oscillating generator accelerates and angle δ increases ($d\delta/dt > 0$), the electric power transmitted must be increased to compensate for the excess mechanical input power. Conversely, when the generator decelerates and angle δ decreases ($d\delta/dt < 0$), the electric power must be decreased to balance the insufficient mechanical input power.

The requirements of output control and the process of power oscillation damping by transmission angle control are illustrated in Figure 7.9. Waveforms at (a) show the undamped and damped oscillations of angle δ around the steady-state value δ_0. Waveforms at (b) show the undamped and damped oscillations of the electric power P around the steady-state value P_0. (The momentary drop in power shown in the figure represents an assumed disturbance that initiated the oscillation.)

Waveform (c) shows the variation of angle σ produced by the phase shifter. (For the illustration it is assumed that α has an operating range of $-\sigma_{max} \leq \sigma \leq \sigma_{max}$, and δ is in the range of $0 < \delta < \pi/2$.) When $d\delta/dt > 0$, angle σ is negative, making the power versus δ curve (refer to Figure 7.3) shift to the left, which increases both the

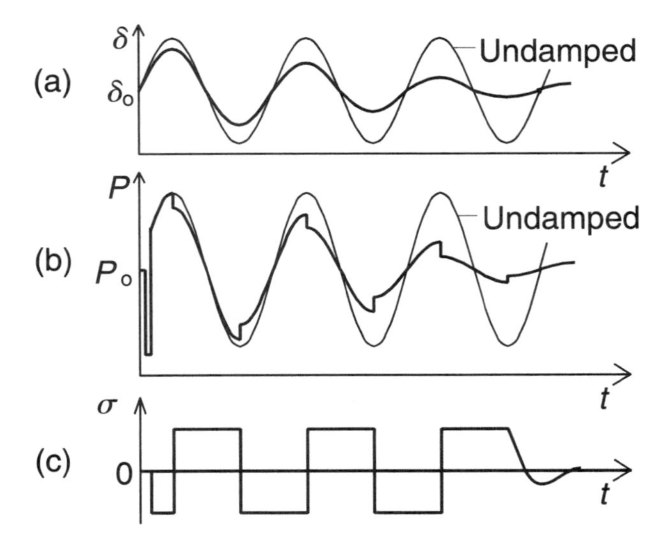

Figure 7.9 Waveforms illustrating power oscillation damping by phase angle control: (a) generator angle, (b) transmitted power, and (c) phase shift provided by the Phase Angle Regulator.

angle between the end terminals of the line and the real power transmitted. When $d\delta/dt < 0$, angle σ is made positive, which shifts the power versus angle curve to the right and thus decreases the overall transmission angle and transmitted power.

As the illustrations show, a "bang-bang" type control (output is varied between minimum and maximum values) is the most effective if large oscillations are encountered. However, for damping relatively small power oscillations, continuous variation of the angle may be preferred.

7.1.6 Summary of Functional Requirements

Phase angle regulators are primarily applied for power flow management, i.e., to control line loading and mitigate loop flows. Voltage regulators are used for reactive power flow and terminal voltage control. Their functional capability is vital for controlling real and reactive loop power flows. Voltage regulators also play an important role in subtransmission and distribution systems in maintaining operating voltage levels.

Phase angle regulators, with appropriate control capability to counteract prevailing machine swings, can improve transient stability and provide power oscillation damping.

In future flexible ac transmission systems, the functional capabilities of conventional voltage and phase angle regulators, with modern solid-state implementations, will play an important role in the optimal utilization of the transmission network by real and reactive power flow management and voltage control.

It will be seen that, analogous to controlled shunt and series compensation, the above functional requirements can be accomplished by adapting the conventional on load tap changer concept for fast and unrestricted thyristor control, or by using a new approach in which switching converters, as voltage sources, are configured to provide the desired voltage injection for voltage and phase angle regulation.

7.2 APPROACHES TO THYRISTOR-CONTROLLED VOLTAGE AND PHASE ANGLE REGULATORS (TCVRs AND TCPARs)

It is shown in Chapters 5 and 6 that there are two basic approaches to modern, power electronics-based reactive compensators: one that uses conventional thyristors (which commutate "naturally" at current zeros) to control current in reactive impedances, and the other that employs turn-off (GTO) thyristors (or similar devices) in switching power converters to realize controllable synchronous voltage sources. This dual approach, in a different form, extends also to voltage and phase angle regulators.

As indicated at the beginning of this chapter, voltage and angle regulation are generally accomplished by in-phase and, respectively, quadrature voltage injection. Thus, both the conventional thyristor- and GTO-controlled converter-based approaches insert a controlled voltage between the given bus and the controlled terminal or line. The major difference between the two is that whereas the thyristor-based approach obtains the insertion voltage from appropriate taps of the regulating ("excitation") transformer, the GTO-based approach generates this voltage from a dc power supply. Consequently, the function of the thyristor-based controller in the first approach is that of an on-load tap-changer: selecting the proper tap of the regulating transformer and injecting the thus obtained voltage, usually by an insertion transformer, in series with the line. The function of the GTO-based voltage source is to generate the required voltage and inject it, also by an insertion transformer, in series with the line. At first sight, the difference between the two approaches appears to be minor. However, it will be seen that the two approaches result in important operational and performance differences, the most significant of which is the capability of the voltage-source-based approach to be self-sufficient in supplying or absorbing the reactive power the voltage or angle regulation demands. The thyristor-based approach must be supplied externally, usually by the system, both the real and reactive power required to accomplish the desired regulation. Another major difference is in the unrestricted controllability of the injected voltage which, as is explained in the next chapter, leads to broad, new functional possibilities in Flexible AC Transmission Systems.

In this section the Thyristor-Controlled Voltage and Phase Angle Regulators (TCVRs and TCPARs) are discussed and in the following section the converter-based Voltage and Phase Angle Regulators are introduced as a preparation for the more general treatment given in Chapter 8.

There are two main reasons for the application of Thyristor-Controlled Voltage and Phase Angle Regulators instead of mechanical on-load tap changers: One is the elimination of the expensive regular maintenance and the other is to provide the high speed of response necessary for dynamic system control.

From the standpoint of the thyristor controller, voltage and angle regulators operate in the same manner, that is, they provide a voltage with a variable magnitude from a fixed voltage source. Of course, this is functionally the same as producing a voltage with a fixed magnitude from a source of variable magnitude. In other words, the thyristor-controller does not intentionally change the angle of the voltage that it controls; generally, the voltage at its output is in phase with the voltage applied at its input. However, it will be seen that voltage regulation based on delay angle control does result in some angular shift. Thus, whether the regulator arrangement is a voltage or angle regulator is determined entirely by the transformer winding configuration, i.e., whether it provides an in-phase or a quadrature voltage input for the thyristor controller. For this reason, in the following discussions voltage and angle regulators will not be distinguished and reference will be made only collectively to "regulators." The thyristor controller used in (voltage and angle) regulators will be referred to as the thyristor tap changer, analogous to its mechanical counterpart, the on-load tap changer.

Thyristor tap changers may be configured to provide continuous or discrete level control. Continuous control is based on delay angle control, similar to that introduced in Chapter 5 for the TCR. Delay angle control, as seen, inevitably generates harmonics. To achieve little or no harmonic generation, thyristor tap changer configurations must provide discrete level control.

The main constituent parts of a thyristor tap changer offering discrete level control are transformers and the thyristor valves, together with their heat sinks, snubbers, and gate-drive control. There are a number of possible configurations capable of discrete level control with tap step sizes comparable to those generated by conventional electromechanical units. Some of these arrangements can lead to a reduction in the number of transformer taps required which usually is advantageous. Figure 7.10 shows a single line diagram of two basic concepts. One concept, shown in Figure 7.10(a), is based on n identical windings and bi-directional thyristor bridge circuits to provide from zero to n voltage steps in either direction, since each bridge circuit can connect the related transformer winding with either polarity, or can bypass it. The other concept, shown in Figure 7.10(b), is based on ternary progression: the transformer windings and the voltage ratings of the thyristor bridges are proportioned in the ratio of 1 : 3 : 9 : . . . and the number of steps, n, in one direction is given with the number

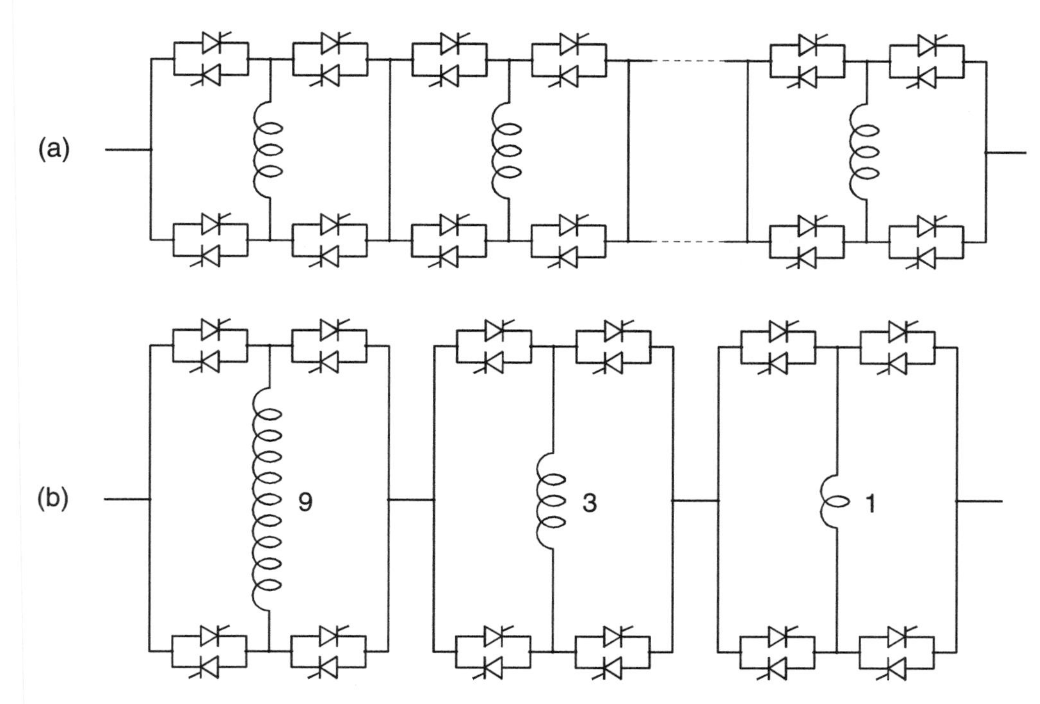

Figure 7.10 Basic thyristor tap-changer configurations: (a) identical windings and thyristor valve ratings, (b) windings and thyristor valve voltage ratings in ternary progression.

of windings, l, by the expression $n = (3^l - 1)/2$. Thus, the three-winding arrangement illustrated in Figure 7.10(b) has 13 steps in each direction.

Structurally, the thyristor tap changer can be further simplified if continuous voltage control is implemented by using phase-delay angle control. Figure 7.11 shows a single line diagram of the basic approach. The two thyristor valves employed need only be rated according to maximum percentage voltage regulation required. However, as indicated earlier, such schemes giving continuous control suffer from two drawbacks: they create harmonics of the supply frequency in their terminal voltage and implementation of their control for all load power factors is relatively complex. The output of a continuously controlled regulator contains, in a single-phase case, all odd order harmonics of the system frequency. In the three-phase case, with identical loads and control settings on all three phases, it contains the usual six-pulse type harmonics of the order of $(6m \pm 1)$, where m is any integer from 1 to ∞. If the loads or control settings are not identical on all three phases, then the harmonic content for the three-phase case becomes identical, relative to frequency components present, to that for the single-phase case.

In all the above approaches where thyristors are to be used in utility applications, the devices must tolerate the fault currents and transient voltages endemic to utility systems. Thus, the number of thyristors required for a voltage regulator application is heavily influenced by the actual transient voltages and currents occurring during surges and faults. Consequently, specification requirements and protection arrangements can significantly impact the equipment cost.

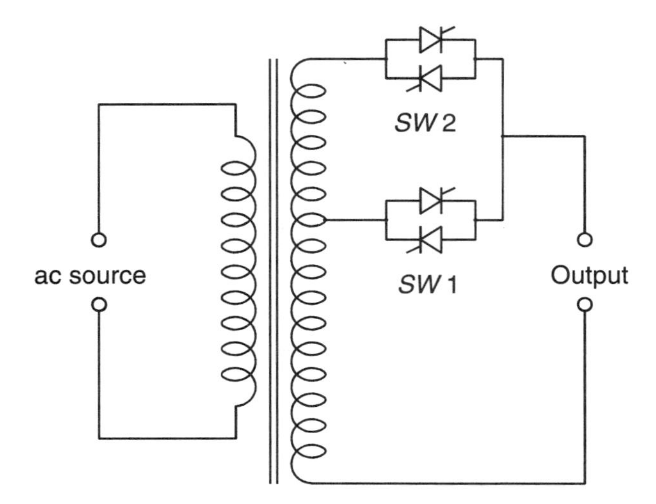

Figure 7.11 Basic thyristor tap-changer circuit configuration for continuous (delay angle) control of the output voltage.

7.2.1 Continuously Controllable Thyristor Tap Changers

There are a number of possible thyristor tap changer configurations which can give continuous control with varying degrees of circuit complexity. The basic power circuit scheme of a thyristor tap changer considered here is shown in Figure 7.11 and repeated in Figure 7.12 for convenience. This arrangement can give continuous voltage magnitude control by initiating the onset of thyristor valve conduction. Consider Figure 7.12 and assume that a resistive load is connected to the output terminals of the thyristor tap changer. This load of course could be the line current in phase with the terminal voltage. The two voltages obtainable at the upper and lower taps, v_2 and v_1, respectively, are shown in Figure 7.12(a). The gating of the thyristor valves is controlled by the delay angle α with respect to the voltage zero crossing of these voltages. For example, Figure 7.12(b) shows that at $\alpha = 0$, at which, in the present case of a resistive load, the current crosses zero and thus the previously conducting valve turns off, valve sw_1 turns on to switch the load to the lower tap. At $\alpha = \alpha_1$, valve sw_2 is gated on, which commutates the current from the conducting thyristor valve sw_1 by forcing a negative anode to cathode voltage across it and connecting the output to the upper tap with voltage v_2. Valve sw_2 continues conducting until the next current zero is reached (in the present case, the next current zero coincides with the voltage zero crossing, $\alpha = 0$), whereupon the previous gating sequence continues, as shown by the load voltage waveform in Figure 7.12(b). Inspection of this waveform indicates that, by delaying the turn-on of sw_2 from zero to π, any output voltage between v_2 and v_1 can be attained.

Fourier analysis of the output voltage waveform for an idealized continuously controlled thyristor tap changer, operating between voltages v_1 and v_2 with resistive load and delay angle α with respect to zero crossing of the voltage, can be easily carried out, yielding the following expressions for the fundamental component:

$$V_{of} = \sqrt{a_1^2 + b_1^2} \tag{7.11}$$

$$\psi_{of} = \tan^{-1}\left(\frac{a_1}{b_1}\right) \tag{7.12}$$

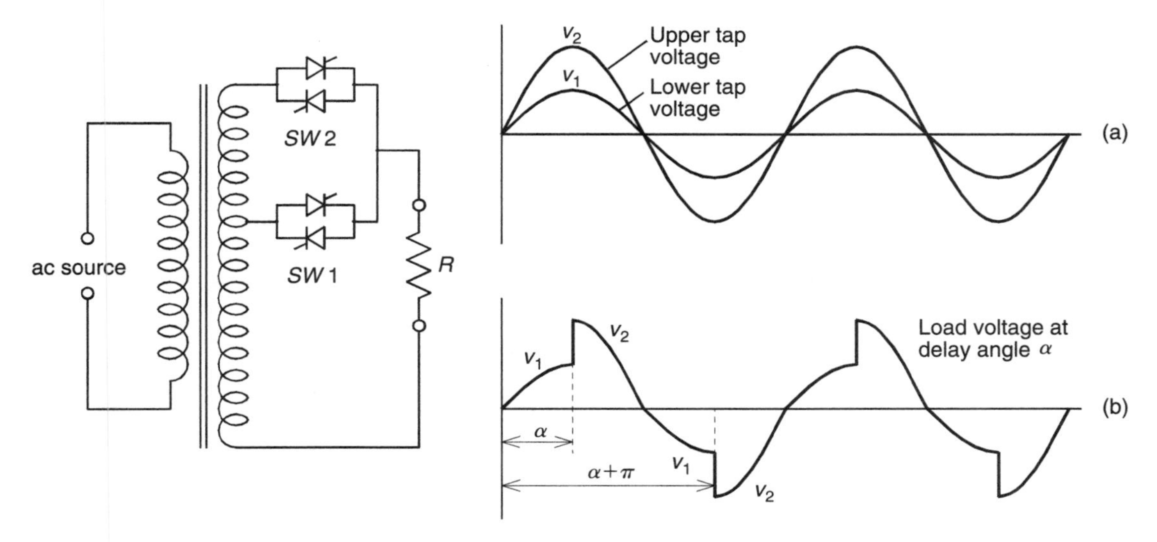

Figure 7.12 Output voltage waveform of the delay angle controlled thyristor tap changer supplying a resistive load.

where V_{of} is the amplitude of the fundamental and ψ_{of} is the phase angle of the fundamental with respect to the unregulated voltage and

$$a_1 = \left(\frac{V_2 - V_1}{2\pi}\right)(\cos 2\alpha - 1) \tag{7.13}$$

$$b_1 = V_1 + \left(\frac{V_2 - V_1}{\pi}\right)\left(\pi - \alpha + \frac{\sin 2\alpha}{2}\right) \tag{7.14}$$

The variation of amplitude V_{of} and angle ψ_{of} of the fundamental voltage V_{of} with delay angle α for an assumed $\pm 10\%$ regulation range ($V_1 = 0.9$ and $V_2 = 1.1$ p.u.) is shown in Figure 7.13. It is to be noted that, as Figure 7.13 and (7.11) through (7.14) indicate,

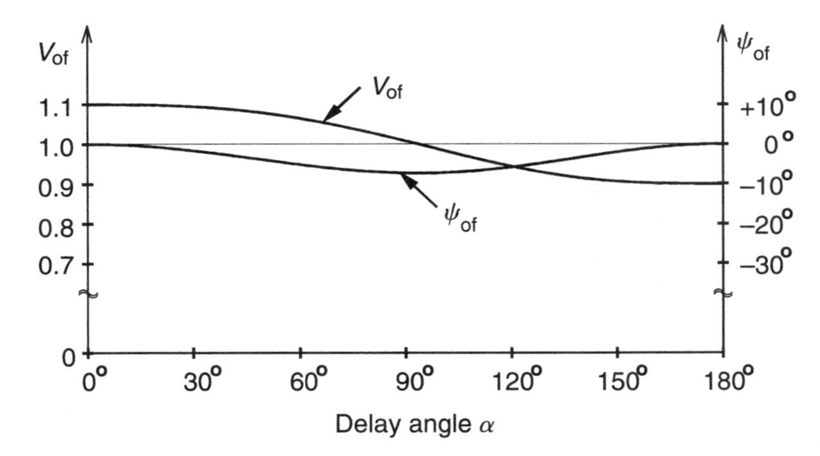

Figure 7.13 Variation of the amplitude and phase-angle of the fundamental output voltage obtained with a delay angle controlled thyristor tap changer supplying a resistive load.

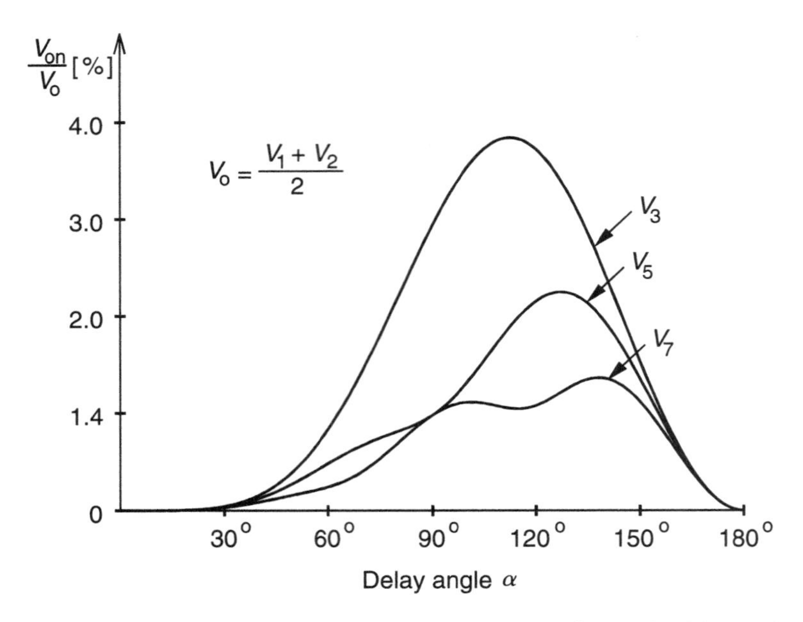

Figure 7.14 Amplitude variation of the dominant harmonics vs. the delay angle present in the output voltage of a controlled thyristor tap changer supplying a resistive load.

the fundamental component of the regulated output (load) voltage generally lags the unregulated supply voltage in the range of $0 \leq \alpha \leq \pi$, even with the assumed unity load power factor (resistive load). The maximum phase shift is in the vicinity of $\alpha =$ 90° and it is proportional to the regulation range. (In the present case, $\psi_{ofmax} = 7.3°$.)

The harmonics in the output voltage can be expressed in the following way:

$$V_{on} = \sqrt{a_n^2 + b_n^2} \tag{7.15}$$

$$a_n = \left(\frac{V_2 - V_1}{\pi}\right)\left(\frac{1}{n-1} - \frac{1}{n+1} + \frac{\cos(n+1)\alpha}{n+1} - \frac{\cos(n-1)\alpha}{n-1}\right) \tag{7.16}$$

$$b_n = \left(\frac{V_2 - V_1}{\pi}\right)\left(\frac{\sin(n+1)\alpha}{n+1} - \frac{\sin(n-1)\alpha}{n-1}\right) \tag{7.17}$$

where $n = 2k + 1$ and $k = 1, 2, 3, \ldots$

Figure 7.14 shows the dominant harmonic components as a percentage of the nominal fundamental output voltage $(V_1 + V_2)/2$ for the stipulated tap voltage difference $(V_2 - V_1) = 0.2$ p.u. which, as mentioned, corresponds to a ±10% regulation range. The amplitudes of the harmonics at any given α are of course proportional to the regulation range; for example, with a ±20% regulation range, the amplitudes plotted in Figure 7.14 would double. As the amplitude plots and the waveshape of the output voltage in Figure 7.12(b) show, the maximum third harmonic distortion occurs at a delay angle of 90 degrees, and the maximum fifth harmonic at a delay angle of 120 degrees.

In order to operate the continuously controllable thyristor tap changer properly with reactive (inductive or capacitive) loads, certain rules regarding the conduction and commutation of the thyristor valves need to be established. Consider the basic

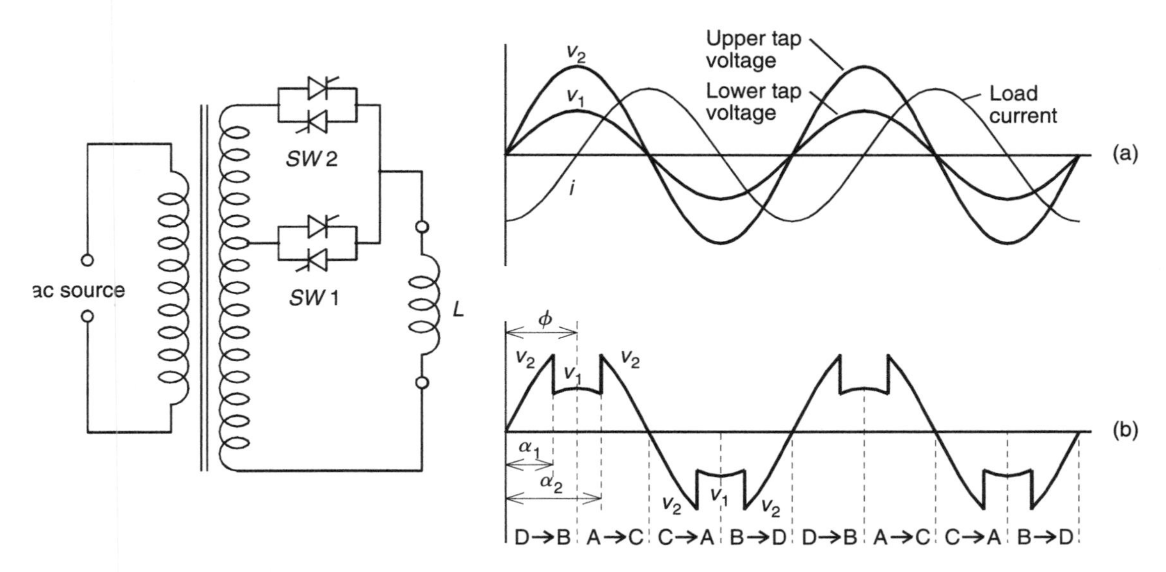

Figure 7.15 Output voltage waveform of the delay angle controlled thyristor tap changer supplying a purely inductive load.

power circuit of the thyristor tap changer with fully inductive load, as shown in Figure 7.15. In this figure the unidirectional thyristor valves capable of conducting only positive or only negative current are labeled individually. That is, the "positive" valve at the lower tap labeled "A," the negative one labeled "B" and, similarly, at the upper tap the "positive" valve is labeled "C" and the "negative" labeled "D." The voltages at the upper and lower taps, together with an inductive load current of arbitrary magnitude for reference, are shown at Figure 7.15(a).

Assume a sequence of operation similar to that shown for resistive load, i.e., changing first from the upper to the lower tap, and then from the lower to the upper tap, to be carried out during the positive half-cycle of the supply voltage. It can be observed that at $\alpha = 0$ the inductive load current is negative and, consequently, thyristor valve D of the upper tap must be conducting. Thus, in the interval of zero voltage ($\alpha = 0$) to zero current, which is, without harmonics, at $\alpha = \pi/2$, commutation to the lower tap is possible by turning on thyristor valve B of the lower tap. This action will impose a negative anode to cathode voltage on D, forcing it to turn off. In the subsequent interval of the positive voltage half-cycle, i.e., from current zero crossing to the next voltage zero crossing, the current is positive so either thyristor valve A or C could be gated on, but commutation would be possible only from valve A to C. Turning on valve C would impose a negative anode to cathode voltage on A to turn it off, but C would stay in conduction until the natural current zero crossing is reached because C is connected to the highest tap voltage available. These observations are summarized and extended also to the negative half-cycle, of the supply voltage in Tables 7.1 and 7.2.

From the above information it can be seen that in positive time progression from voltage zero to current zero in any half-cycle, when the thyristor tap changer is supplying a resistive load, only one tap change, from lower to the upper tap, can take place. This is because the voltage and current zeros for resistive load coincide, allowing

TABLE 7.1 Positive Half-Cycle of the Input Voltage

Time Period	Valve(s) Conducting	Permissible Tap Change
Voltage zero to current zero	D	D to B
Current zero to voltage zero	A, C	A to C

only one commutation from lower to higher voltage in each half-cycle. However, during this same time period, there is, for an inductive load, also a permissible and achievable tap change from upper to lower tap, that is, from thyristor D to B and from thyristor C to A in the positive and the negative half-cycles, respectively. This can be explained as follows. During the positive half-cycle of the voltage waveform, from the voltage zero to the current zero, the load current flows against the input voltage. If during this time thyristor valve B is turned on, the positive inter-tap voltage will commutate valve D and load voltage will change from v_2 to v_1. Similar considerations apply to thyristor valves C and A during the negative half-cycle of the input voltage. It is clear that in each cycle two tap-changing operations can be performed. Between voltage zero and current zero, a tap change can occur from upper to lower tap and between current zero to succeeding voltage zero a tap change can occur from lower to upper tap. Thus, by using two tap changes per half-cycle, control over the entire half-cycle can be achieved as shown in Figure 7.15(b). Tap change D to B or C to A, controlled by delay angle α_1, can occur at any angle from the voltage zero to the current zero, i.e., in the interval $0 \leq \alpha_1 \leq \phi$ (ϕ is the load angle), and then tap change A to C or B to D, controlled by delay angle α_2 can occur at any angle from current zero to voltage zero, i.e., in the interval $\phi \leq \alpha_2 \leq \pi$, in positive time progression. Since there now are two tap changes per half cycle and appropriate valves are required to be gated as the current passes through zero, the control process becomes more complex than that required for a purely resistive load. Note that as the load power factor approaches unity, the interval for α_1, $0 \leq \alpha_1 \leq \phi$, diminishes and the interval for α_2, $\phi \leq \alpha_2 \leq \pi$, stretches over the whole half-cycle, from zero to π. In other words, at a unity power factor (resistive) load, $\alpha_1 = 0$ and $\alpha_2 = \alpha$.

The above considerations can be extended to the case where the continuously controllable thyristor tap changer has a purely capacitive load as shown in Figure 7.16. The capacitive load here is assumed to be capacitive at supply frequency only, and highly inductive to all harmonics, in order not to terminate the thyristor controller with a voltage source at both terminals. With a control strategy similar to that established for inductive load, that is, when the load current flow is opposite to the supply voltage polarity [see Figure 7.16(a)], gating of appropriate lower tap thyristor valve takes place to achieve commutation from the upper tap to the lower tap, then 180 degrees continuous control can be obtained over each half-cycle. Figure 7.16(b) shows the load voltage waveform with such a control employed. The tap changes can occur anywhere in the time periods indicated by arrows.

TABLE 7.2 Negative Half-Cycle of the Input Voltage

Time Period	Valve(s) Conducting	Permissible Tap Change
Voltage zero to current zero	C	C to A
Current zero to voltage zero	B, D	B to D

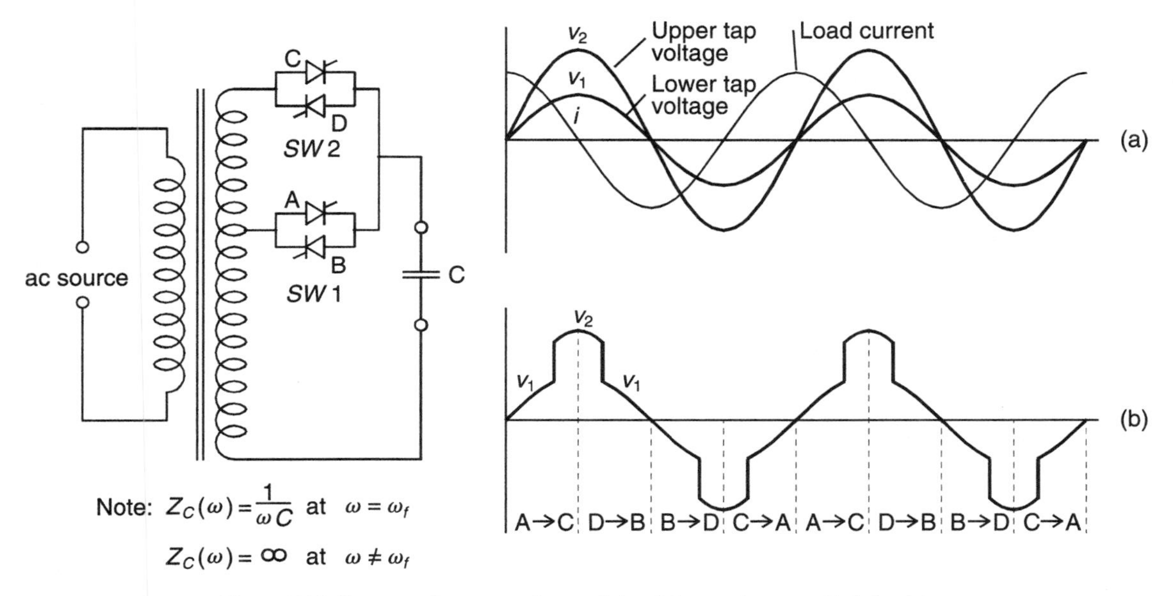

Figure 7.16 Output voltage waveform of the delay angle controlled thyristor tap changer supplying a purely capacitive load.

It is clear that, with an appropriate control providing correct thyristor firing sequence and correct timing of tap change, it is possible to operate a continuously controllable thyristor tap changer into loads of any phase angle and maintain 180 degree control. A possible basic structure for the internal control of the continuously controllable thyristor tap changer is shown in Figure 7.17. As shown, this control is based on the detection of the voltage and current zero crossing points, which determine the achievable thyristor commutation and define the control intervals for the delay angles α_1 and α_2 for all possible load power factors. The delay angle is usually controlled directly by a closed regulation loop to maintain the regulated voltage at a given reference value. This schematic block diagram intends to provide only the conceptual control functions, the implementation of which may require sophisticated circuitry with auxiliary functions not shown in the simple block diagram. For example, precise and reliable sensing of voltage and current zeros in the presence of superimposed harmonics, switching transients and other nonfundamental voltage components resulting from disturbances may require the implementation of sophisticated filters and estimation routines in the control.

From the preceding analysis the implications of the two major disadvantages resulting from the use of a delay angle controlled thyristor tap changer to achieve continuous control, become clear. First, the fundamental component of the terminal voltage is phase shifted from the source (bus) voltage by an amount depending on the gating angle α or on α_1, α_2, and ϕ, the load phase angle, and the direction of this phase shift depends on the load phase angle. This can lead to significant problems in transmission applications where even small arbitrary phase shifts can have serious consequences in an interconnected ac network. Thus, the thyristor tap changer with delay angle control is more likely to be applied as a voltage regulator in distribution systems than as a more general, voltage and angle regulator in transmission systems.

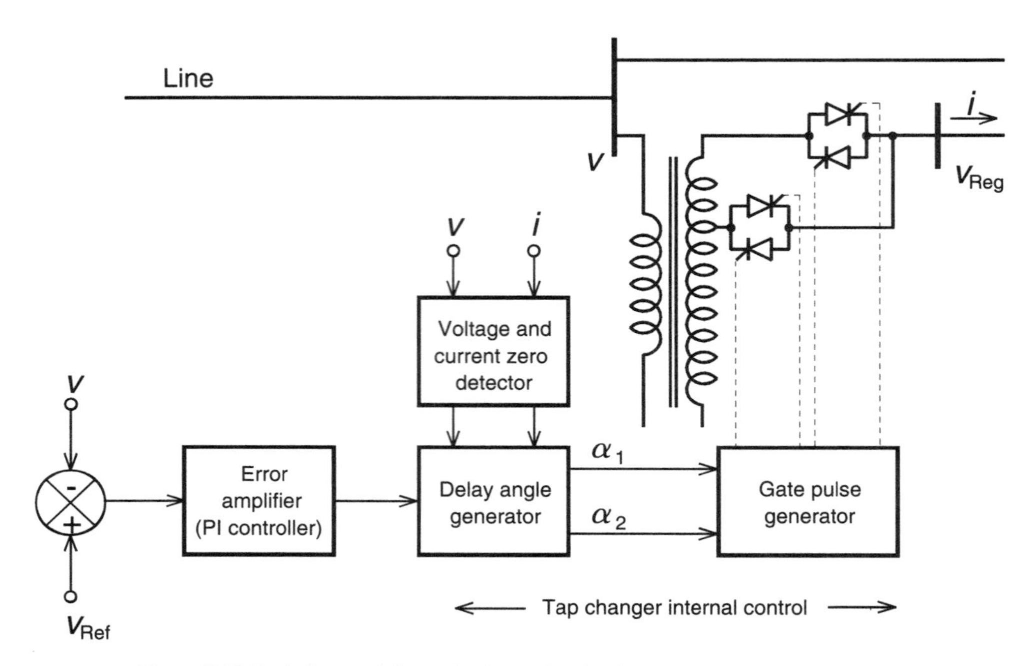

Figure 7.17 Basic Internal Control scheme for the delay angle controlled thyristor tap changer.

Second, the low-order harmonics can have magnitudes, even for small values of inter-tap voltages, which may be unacceptable in many utility applications. Thus, an output harmonic filter is almost certainly required for a continuously controlled thyristor tap changer applied in a utility environment. The output filters add to the loading of the thyristor tap changer, increase the cost, and may introduce undesirable resonances to the ac system. These problems can be avoided by using discrete level control.

7.2.2 Thyristor Tap Changer with Discrete Level Control

The problems associated with a continuously controllable thyristor tap changer can be avoided with the application of discrete level voltage control employed in conventional electromechanical tap changers. With discrete level control the tap-changing function can be achieved without introducing harmonic distortion or undesirable phase shifts and without the control complexities associated with continuous control. The choice of power circuit is decided by performance requirements and cost. Two possible circuit configurations are discussed next.

A conceptually simple tap-changing configuration is shown in Figure 7.18. In this scheme each winding section is bridged by four bi-directional thyristor valves, and thus may be inserted in the outside (transmission line) circuit in either polarity or bypassed, giving 0 to $\pm$ V volts availability in V/n steps. If 16 equal sections are used to give 33 steps capability over V volts with current rating I, then the total required thyristor valve rating is $(V/16 \times 64)I = 4\ VI$ (or some multiple thereof to accommodate over-voltages and surge currents). It should be noted that with this rating the actual voltage control range is 2 V and thus the ratio of controlled VA

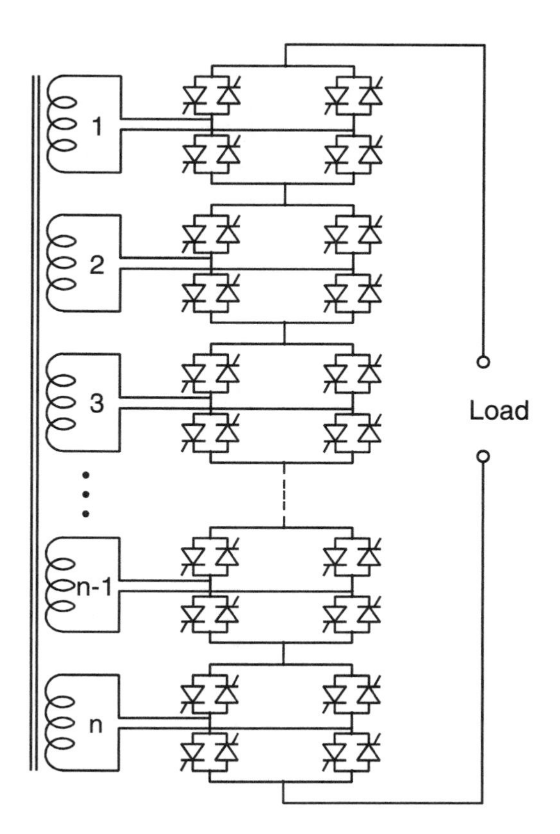

Figure 7.18 Thyristor tap changer using equal winding sections for discrete level voltage control.

range to valve VA rating is only two. However, it is also important to recognize that the thyristor tap changer, in contrast to the previously considered shunt and reactive compensators, can neither generate nor absorb reactive power. The reactive power supplied to or absorbed from a line when it injects in phase or quadrature voltage must be absorbed by or supplied to it by the ac system. For this reason, both the series insertion and shunt regulating transformers must be fully rated for the total VI product (injected voltage times line current). Furthermore, a tap changer, particularly that with a significant quadrature voltage injection for angle regulation, cannot readily be applied without external var support at connection points remote from generation, since the reactive voltage drop across the line impedance due to the increased line current will oppose the increase of power transfer intended.

The circuit configuration shown in Figure 7.18 also has some practical disadvantages. The winding must be broken into n equal sections for $2n + 1$ total number of required steps, and $4 \times n$ thyristor valves (of which only $2n$ are operating at any given time) are used. A major problem with this is the difficulty of producing a transformer with n small and isolated winding sections, with $2n$ leads coming from the winding structure. Another disadvantage of this configuration is that at lower system voltages, i.e., smaller controlled voltage V, the voltage per winding section becomes much lower than the minimum economic voltage application point of power thyristors currently available. Consider, for example, control of 10% of a nominal 115 kV system. The total controlled voltage is 11500 V, and if $n = 16$ for 33 steps, then voltage per section

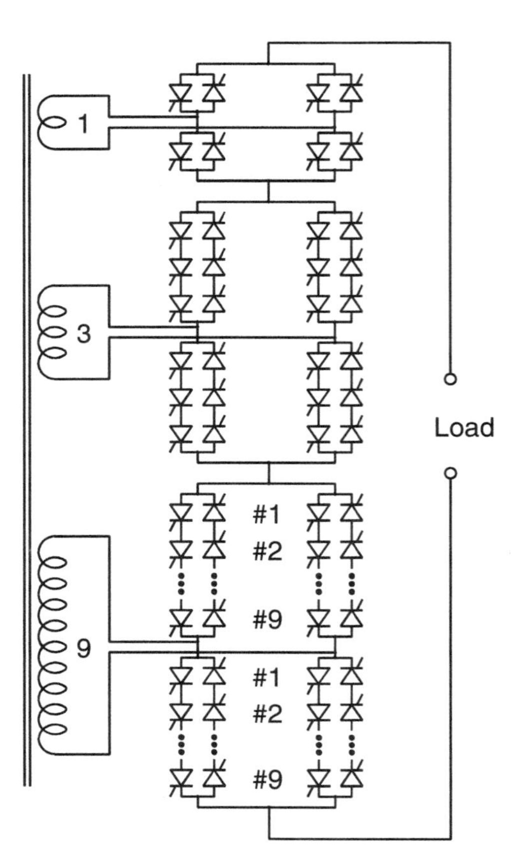

Figure 7.19 Thyristor tap changer using ternary proportioned winding sections for discrete level voltage control.

is only 715 V r.m.s. or 1011 V peak—considerably below the voltage rating of presently available high-power thyristors (typically 6 to 8 kV).

A possible approach to solving the above practical problems, proposed by Güth with others in 1981, is to use nonidentical winding sections with winding turns increasing in geometrical progression, as illustrated in Figure 7.19. With the previously introduced ternary windings, proportioned 1:3:9, a total of 27 steps can be obtained with only 12 bi-directional thyristor valves (of different voltage ratings). The total rating of the valves is the same as that of the previous scheme, 4 *VI* (which, as indicated above, is equal to twice the controlled VA range), and also half of the total number of valves, i.e., 6 out of 12, operating at any given time. The thyristors for higher voltage sections are fully utilized, while those for the smallest voltage section may still be operating below the minimum economic voltage level. The practical problems of transformer winding are also much reduced. However, the structure of a thyristor controller with progressively larger (higher voltage) valves increases both complexity and cost. The number of obtainable voltage steps is also reduced to 27. (The number of steps obtainable can, of course, be increased to 81 by extending the winding arrangement to four winding sections proportioned 1:3:9:27. This will increase the number of valves to 16 but the total volt-ampere rating of the valves will remain 4 *VI*.) Of course, the ternary progression type arrangement does not alter the above cited shortcoming

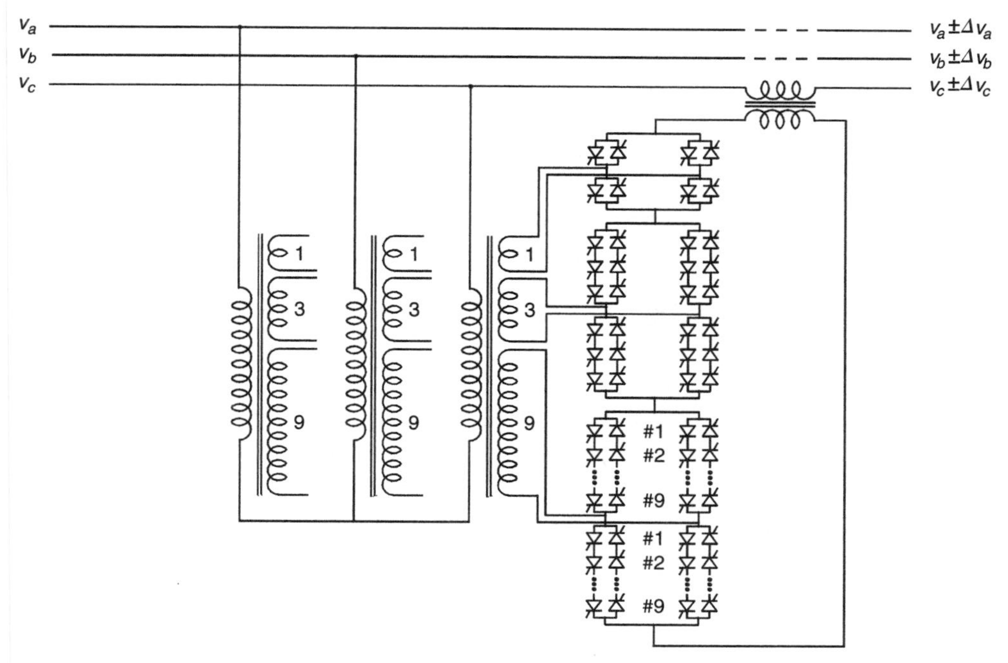

Figure 7.20 (a) Voltage regulator employing a thyristor tap changer using ternary proportioned winding sections for discrete level voltage control.

of all thyristor tap changers: their inherent inability to supply or absorb reactive power.

From the review of the equal and ternary progression type discrete level tap changers, it appears that the arrangement using equal, isolated, winding sections will probably be uneconomical except for very high voltage application; even in that case, the increased complexity of transformer connections and winding construction will likely make it less attractive than the 1:3:9 (:27) configuration. This leads to the conclusion that the most practical configuration for discrete level voltage control in utility applications is probably the ternary progression arrangement using 1:3:9 transformer winding sections that gives 27 steps with 12 thyristor valves.

A thyristor-controlled voltage regulator and a thyristor-controlled phase angle regulator scheme are shown in Figures 7.20(a) and (b) respectively, for the 1:3:9 transformer winding arrangement to provide discrete level control.

7.2.3 Thyristor Tap Changer Valve Rating Considerations

The number of thyristors required in a tap changer for a given application is determined by the operating voltage and consequent transient voltage exposure of the valves, and the operating current and consequent fault current exposure.

Transient fault current exposure is the governing factor in determining the maximum permissible steady-state current rating of thyristors used in a thyristor tap changer because the highest junction temperatures seen by the devices occur under fault conditions. The surge current rating of a thyristor exposed to only a single conduction period of fault current and then required to block full voltage is less than that of a

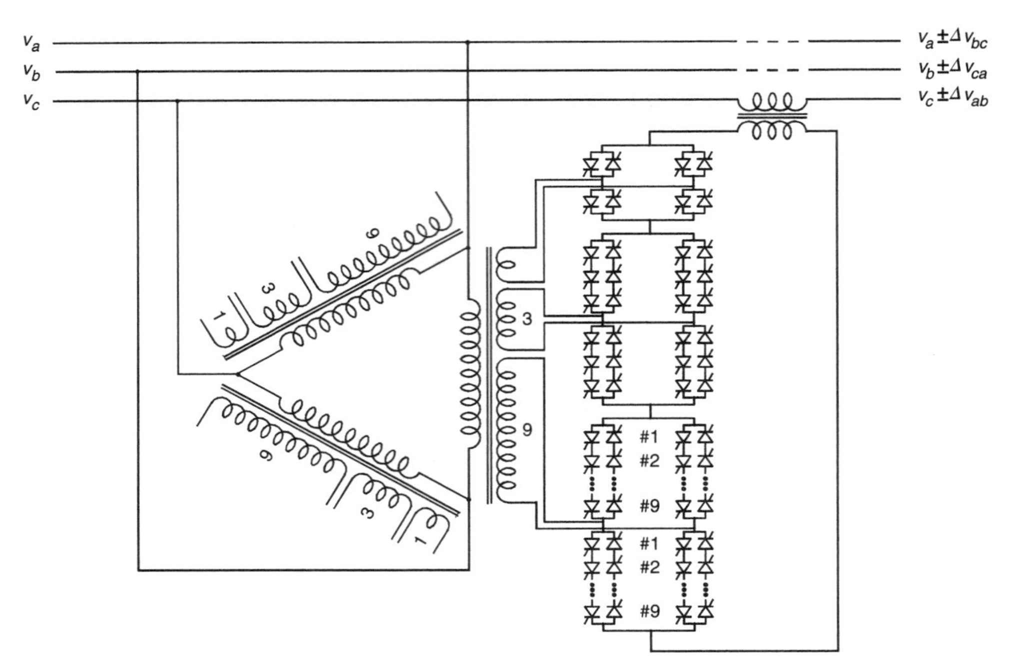

Figure 7.20 (b) Phase Angle Regulator employing a thyristor tap changer using ternary proportioned winding sections for discrete level voltage control.

thyristor exposed to several conduction periods, but not required to block voltage thereafter. In addition, if the thyristor valves are to act as fault interrupting devices, they will need to be designed for full ac system voltage, and not the inter-tap voltage, thus resulting in a considerable increase in the number of thyristors needed. A much better alternative is to have thyristor valves designed only for tap voltage and use a conventional electromechanical breaker for fault clearing. The thyristors must, in this case, withstand the fault current until the system breaker clears the fault (typically 4 to 8 cycles). However, if a breaker is used to clear the fault, then the thyristors will not be required to block any voltage following the final extinction of the fault current. Under all circumstances, thyristor surge current capability can have a significant impact on the final cost of a thyristor tap changer, the main factor being the fault current to steady-state current ratio the application requires.

The required voltage rating of thyristor valves is determined not by steady-state operating conditions, but by the transient voltages occurring during surges. The transient voltage ratios (ratio of peak transient voltage to peak normal operating voltage) that can be obtained with the MVO arresters is about 2.5. Thus the thyristor valves would have to be rated nominally for about 2.5 times the peak steady-state voltage.

7.3 SWITCHING CONVERTER-BASED VOLTAGE AND PHASE ANGLE REGULATORS

The concept of a synchronous voltage source (SVS) has been introduced in Chapters 5 and 6 and shown that a voltage-sourced converter exhibits the characteristics of a

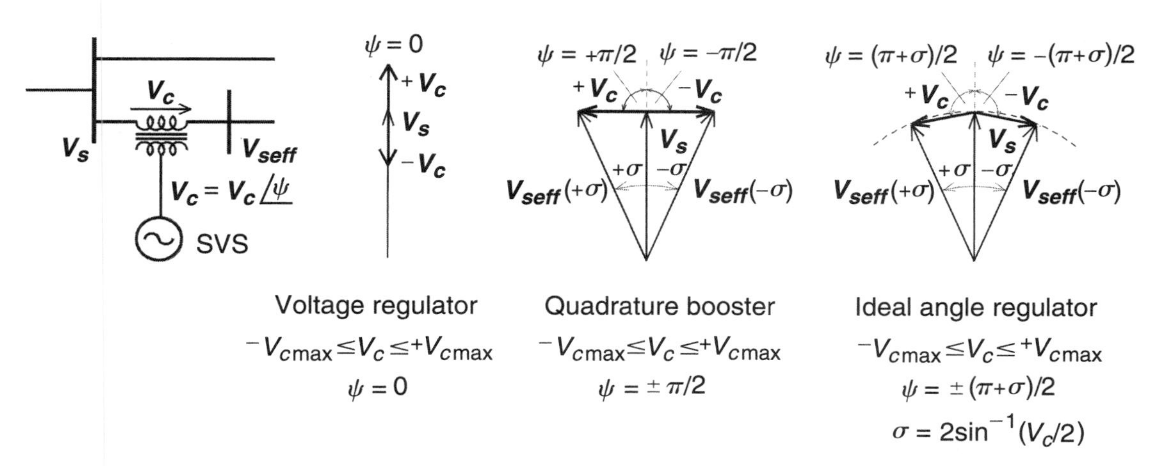

Figure 7.21 Synchronous Voltage Source used for voltage regulation, "quadrature boosting" and phase angle regulation.

theoretical SVS: it can produce a substantially sinusoidal output voltage at the fundamental frequency with controllable amplitude and phase angle and can self-sufficiently generate reactive power for and exchange real (active) power with the ac system (if the real power is supplied to it at its dc terminals). In Chapter 6 such a synchronous voltage source is applied as a series reactive compensator to inject a controllable voltage in quadrature with the line current. It is shown that such a compensator, when appropriately supplied with dc power, can also provide compensation for the resistive voltage drop across the line by injecting a voltage component that is in phase with the line current. At the beginning of this chapter the conceptual possibility of voltage regulation and phase angle control with a series-connected in-phase and quadrature voltage source (see Figures 7.1 and 7.2) is explained. Therefore it should be plausible that a converter-based SVS with controllable amplitude V_c and phase angle ψ can be used for voltage and phase angle regulation, as illustrated in Figure 7.21, by setting $\psi = 0$ and $\psi = \pm\,\pi/2$ (quadrature booster) or $\psi = \pm(\pi + \sigma)/2$, where σ is the desired angular phase shift produced by the injection of voltage phasor $V_c e^{\pm j\psi}$.

The synchronous voltage source used as a voltage or angle regulator will generally exchange both real and reactive power as illustrated in Figure 7.22 where the in-phase and quadrature components of an assumed load current with respect to the voltage inserted for voltage regulation, quadrature boosting and ideal phase angle control are shown together with the corresponding expressions for the real and reactive power the SVS exchanged. It can be observed that, particularly for phase angle control, the reactive power demand on the SVS is generally a significant, and often the dominant portion, of the total VA $= V_c I$ throughput. The SVS, in contrast to a thyristor tap changer, has the inherent capability to generate or absorb the reactive power it exchanges. However, it must be supplied at its dc terminals with the real power portion of the total VA demand resulting from the voltage or angle regulation.

Depending on the application, the voltage regulation or phase angle control may require either unidirectional or bi-directional real power flow. In the case of unidirectional real power flow (the voltage injection at the ac terminals only supplies real power), the real power could be supplied from the ac system by a relatively simple

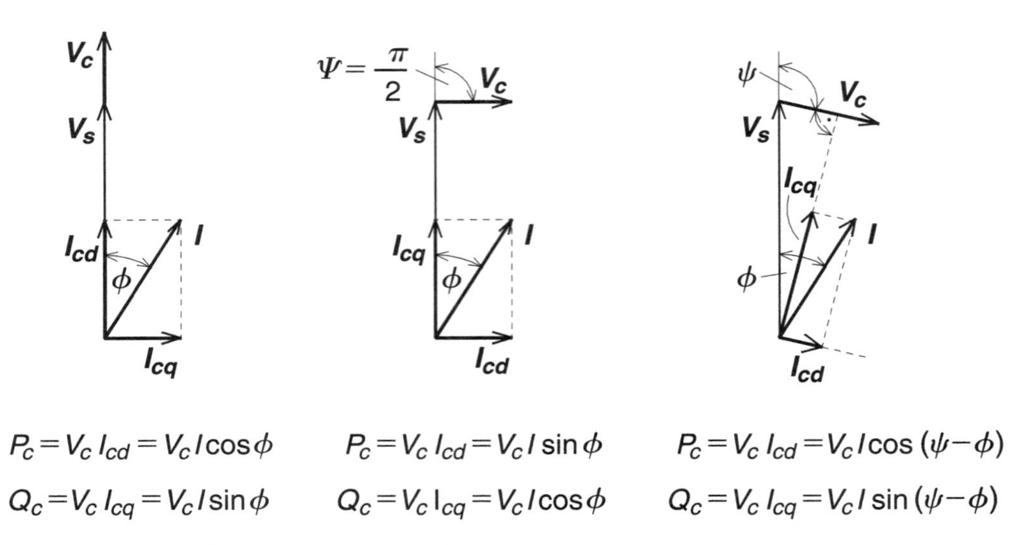

Figure 7.22 Phasor diagrams showing the real and reactive power the Synchronous Voltage Source exchanges when operated as a Voltage Regulator, Quadrature Booster, and Phase Angle Regulator.

line-commutated ac-to-dc thyristor converter (the type discussed in Chapter 4), as shown in Figure 7.23(a). If the application requires bi-directional power flow (the ac voltage injection supplies and absorbs real power under different operating conditions), the power supply must be "regenerative," capable of controlling the flow of current in and out of the dc terminal of the injection converter. Although there are various ac-to-dc converter arrangements to do this, the back-to-back voltage-sourced converter arrangement, shown in Figure 7.23(b), may be the best from the standpoint of uniformity, flexibility, and performance. The reader should note that the dc power supply for the voltage source type implementation of the voltage and angle regulator fulfills the same function as the excitation transformer for its more conventional counterparts employing a thyristor tap changer. It should also be noted that the rating of the dc power supply (shunt transformer and ac-to-dc converter) is, particularly for angle

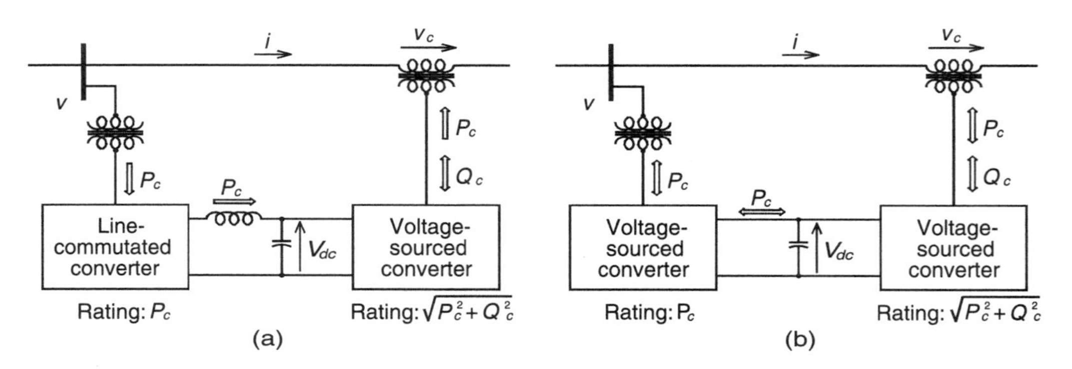

Figure 7.23 Realization of a Synchronous Voltage source (a) for supplying active power and (b) for supplying and absorbing real power.

regulators, appreciably lower than that of the ac excitation transformer since it is rated only for the real power portion of the total VA exchanged through the series voltage injection.

The internal capability of the voltage-sourced converter to generate reactive power is a significant advantage in both voltage and phase angle regulation applications. This is because the ac system has to supply only the real power demand of the regulation and consequently it is not burdened by the transmission of reactive power, the corresponding voltage drops, and resultant line losses if the regulator is remotely located. The self-sufficiency of the regulator to supply reactive power is also important to avoid voltage collapse. The action of the converter-based voltage regulator to maintain load voltages in the face of decreasing system voltage does not result in increased reactive line current and the corresponding voltage drop which may result in degenerative voltage collapse.

More generally, the arrangement of the back-to-back voltage-sourced converter has broad possibilities for the implementation of extremely powerful FACTS Controllers with multiple and convertible functional capabilities. These include voltage regulation and phase-angle control in addition to combined real and reactive series and shunt compensation of transmission lines. These generalized Controllers are the subjects of Chapter 8 and the converter-based voltage and angle regulators are, within that broader scope, further discussed there with respect to functional operation, power circuit structures and control.

7.4 HYBRID PHASE ANGLE REGULATORS

The hybrid phase angle regulator is a combination of two (or more) different types of phase angle regulators to achieve specific objectives at a minimum cost. For example, a mechanical tap-changer type phase angle regulator may be combined with a continuously controllable, fast voltage source type angle regulator, as illustrated in Figure 7.24. In this arrangement the mechanical tap changer would provide the quadrature voltage injection as needed to maintain the required steady-state power flow. The voltage source converter would provide superimposed dynamic phase angle control during and following system disturbances. This hybrid arrangement can be highly cost effective if the steady-state flow control requires only relatively large, infrequent (daily or seasonal) angular changes for an inter-area tie or other line with inadequate dynamic stability for which a converter with relatively small rating could provide effective oscillation damping.

The above concept can be extended to the combination of thyristor tap-changer type and voltage source converter type angle regulator to achieve a different objective. For example, the discrete level voltage regulator with n identical transformer windings and thyristor valve arrangements (Figure 7.18) can be made simple and economically attractive by reducing n to a manageable number (e.g., four). Of course, with that small number of winding sections, the size of the injected voltage step, and the corresponding angular change, would become too large. However, adding to this arrangement a continuously controllable converter type angle regulator with only $\pm 12.5\%$ control range (i.e., one-half of the control range of a single winding section) in a similar manner as shown for the mechanical tap changer in Figure 7.24, the output of the thus obtained hybrid regulator could be controlled continuously, and rapidly, in the total range of -112.5 to $+112.5\%$. The converter should operate in a somewhat wider

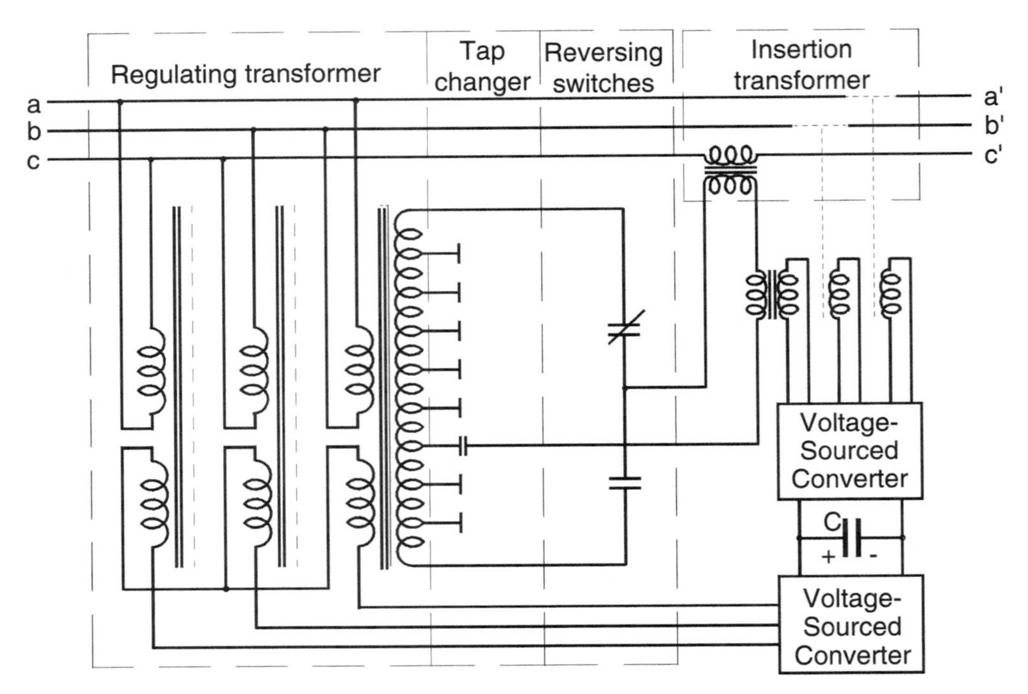

Figure 7.24 Hybrid Phase Angle regulator scheme using a mechanical tap changer with a voltage-sourced converter-based Phase Angle Regulator.

range than ±12.5% in order to provide some hysteresis to prevent undefined switching at the midpoint voltage of a winding segment.

REFERENCES

Arrillaga, J., et al., "Thyristor-Controlled Regulating Transformer for Variable Voltage Boosting," *Proceedings of IEE,* vol. 123, no. 10, October 1976.

Arrillaga, J., et al., "Thyristor-Controlled Quadrature Boosting," *Proceedings of IEE,* vol. 126, no. 6, June 1979.

Arrillaga, J., and Duke, R. M., "A Static Alternative to Transformer on Load Tap Changers," *IEEE PAS,* vol. 3, no. 1, January–February 1980.

Baker, R., et al., "Control Algorithm for a Static Phase Shifting Transformer to Enhance Transient and Dynamic Stability," *IEEE PAS,* vol. 101, no. 9, September 1982.

Etis, A., "Enhancement of First Swing Stability Using a High Speed Phase Shifter," *IEEE Trans. on Power Systems,* vol. 6, no. 3, August 1991.

Güth, G., et al., "Static Thyristor-Controlled Regulating Transformer for AC Transmission," *IEE International Conference on Thyristor and Variable Static Equipment for AC and DC Transmission,* no. 205, London, November 30–December 3, 1981.

Gyugyi, L., "Solid-State Control of Electric Power in AC Transmission Systems," *International Symposium on 'Electric Energy Conversion in Power Systems',* Invited Paper, no. T-IP. 4, Capri, Italy, 1989.

Gyugyi, L., "Solid-State Control of AC Power Transmission," *EPRI Conference on Flexible AC Transmission Systems (FACTS): The Future of High Voltage Transmission,* Cincinnati, OH, November 14–16, 1990.

Iravani, M. R., and Maratukulam, D., "Review of Semiconductor-Controlled (Static) Phase Shifters for Power System Applications," *IEEE/PES Winter Meeting,* Paper no. 94 WM 182-6 PWRS, 1994.

Mathur, R. M., and Basati, R. S., "A Thyristor-Controlled Static Phase Shifter for AC Power Transmission," *IEEE PAS,* vol. 100, no. 5, May 1981.

Ooi, B. T., et al., "A Solid-State PWM Phase-Shifter," *IEEE Trans.,* PWRD-8, no. 2, April 1993.

Wood, P., et al., "Study of Improved Load Tap-Changing for Transformers and Phase-Angle Regulators," *EPRI Report EL-6079,* Project 2763-1, November 1988.

8

Combined Compensators: Unified Power Flow Controller (UPFC) and Interline Power Flow Controller (IPFC)

8.1 INTRODUCTION

In the previous three chapters, Controllers acting individually on one of the three transmission line operating parameters which determine the transmitted power (voltage, impedance and angle) have been discussed. It has been shown that there are two distinctly different technical approaches to the realization of these controllers, both resulting in a comprehensive group of controllers able to address targeted transmission system compensation and control problems.

The first group employs reactive impedances and tap-changing transformers with conventional thyristor valves (switches) as controlled elements. This group, the Static Var Compensator (SVC), Thyristor-Controlled Series Capacitor, and Thyristor-Controlled Voltage and Phase Angle Regulators (TCVR and TCPAR) as illustrated in Figure 8.1, is similar in circuit configuration to the breaker-switched capacitors, reactors and mechanical tap-changing transformers, but have a much faster response and are operated by sophisticated controls.

The second group employs self-commutated, voltage-sourced switching converters to realize rapidly controllable, static synchronous voltage sources. The operation and performance of this group, as illustrated in Figure 8.2, include the Static Synchronous Compensator (STATCOM), Static Synchronous Series Compensator (SSSC), and the Static Synchronous Voltage and Angle Regulators, are analogous to those characterizing ideal synchronous machines, providing almost instantaneous speed of response and control characteristics which are independent of system voltage (shunt operation) and current (series operation). The reader should note that the terms "synchronous voltage regulator" and "synchronous angle regulator" are not commonly used since these functions are usually combined with others within the same controller.

The most significant difference between the approaches is related to the capability to generate reactive power and exchange real power. In the first group these capabilities are separated: the controllers are either reactive compensators (i.e., SVC and TCSC),

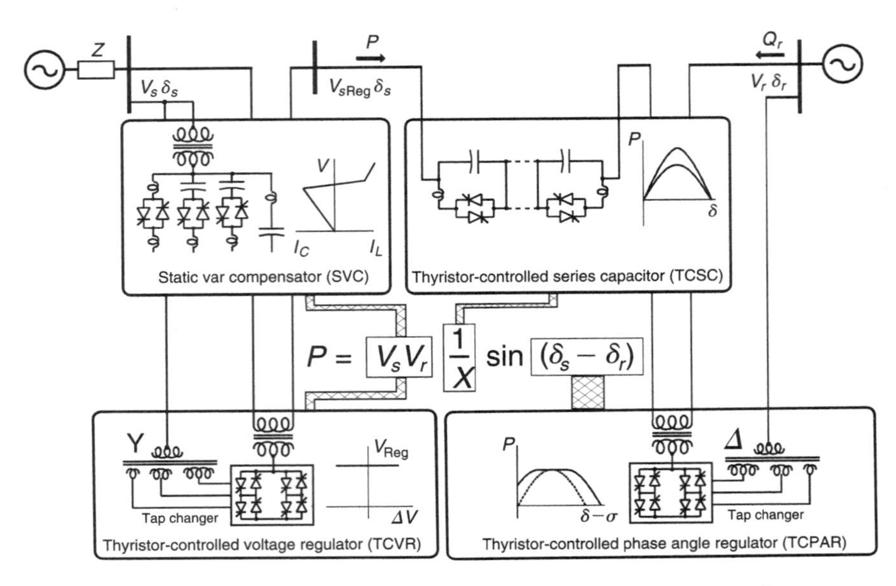

Figure 8.1 Group of conventional thyristor-controlled FACTS Controllers.

which are unable to exchange real power with the ac system (ignoring losses) or regulators (i.e., TCVR and TCPAR), which can exchange real (and reactive) power, but are unable to generate reactive power and thus cannot provide reactive compensation. The second group of controllers has the inherent capability, like a synchronous machine, to exchange both real and reactive power with the ac system. Furthermore, this group automatically generates or absorbs the reactive power exchanged and thus

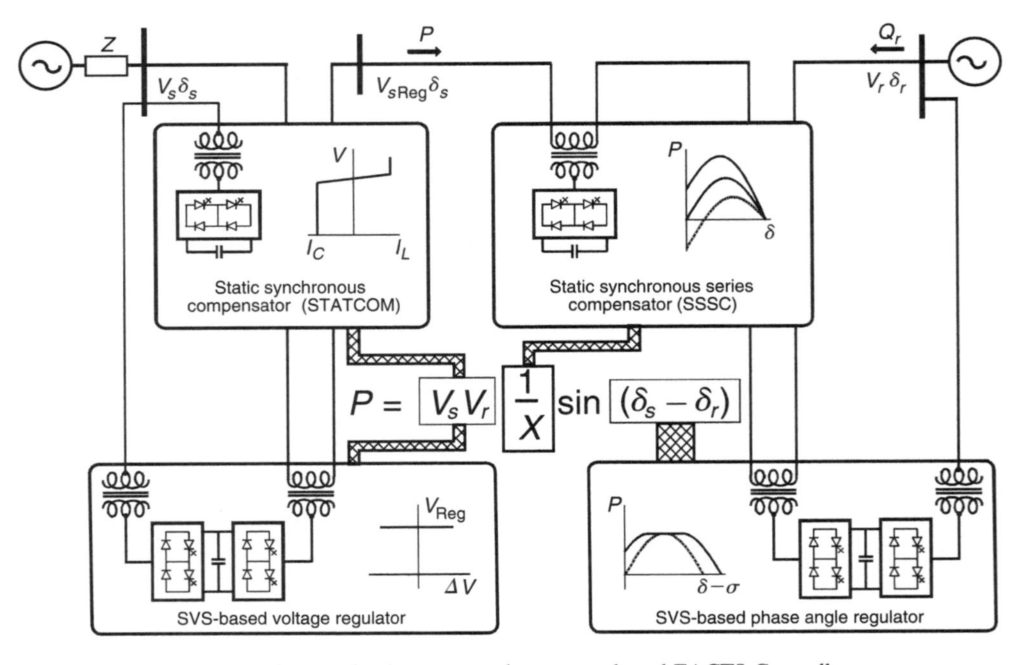

Figure 8.2 Group of voltage-sourced converter-based FACTS Controllers.

provides reactive compensation without ac capacitors or reactors. However, the real power exchanged must be supplied to them, or absorbed from them, by the ac system (or an independent energy source or storage).

In Chapters 5 and 6 the STATCOM and SSSC, each implemented by a voltage-sourced converter operated as a synchronous voltage source, is shown to provide effective voltage and power flow control, respectively, in a self-sufficient manner by internally generated shunt and series reactive compensation. However, it is seen in Chapter 7 that voltage and phase angle control generally involves the exchange of both reactive and real power with the ac system. Thus, in general, unrestricted Voltage and Phase Angle Regulators need to operate in all four quadrants of the complex $P + jQ$ plane. As shown, this requirement can be realized with two voltage-sourced converters, one operated in series and the other in shunt with the transmission line from a common dc capacitor in back-to-back connection. The series converter providing the regulation exchanges the total VA corresponding to the injected series voltage and the prevailing line current. It also internally generates the reactive power component of the total VA but transfers the separated real power component to the common dc link. It is the shunt converter's function to supply to or absorb from the dc link this real power component and transfer it back to the ac system bus that feeds the transmission line. Thus, with the real power support of the shunt converter, the series converter can be operated, within its VA rating, as an ideal ac generator capable of supplying or absorbing both real and reactive power. It is easy to visualize that this arrangement, beyond voltage regulation and angle control, represents a potentially powerful and highly effective FACTS Controller with not only a full four quadrant operating capability in the complex $P + jQ$ plane, but also with an internal var generation capability to support self-sufficiently the reactive power demand in any quadrant. Indeed, as will be shown, the back-to-back voltage-sourced converter arrangement, representing a general ac-to-ac converter with full four-quadrant operation at both ac terminals, provides the basic implementation tool for unique, multifunctional FACTS Controllers.

The objective of this chapter is to investigate the application potential of multifunctional FACTS Controllers based on the back-to-back voltage-sourced converter arrangement. This arrangement offers three basic possibilities. In the first Controller one converter of the back-to-back arrangement is in series and the other is in shunt with the transmission line. This arrangement in transmission applications is, for reasons explained later, generally referred to as the *Unified Power Flow Controller* (*UPFC*). This UPFC identification is quite universal for this arrangement, although the UPFC, like the STATCOM and SSSC, could be implemented by a variety of different switching converters. In the second Controller both converters of the back-to-back arrangement are connected in series with, usually, a different line. This arrangement is referred to as the *Interline Power Flow Controller* (*IPFC*). In the third Controller both converters are connected in shunt, each with a different power system. This arrangement functions as an asynchronous tie, sometimes referred to as a back-to-back STATCOM tie. In the following two major sections, the UPFC and IPFC are discussed in detail. The subject of the STATCOM tie is analogous to the back-to-back HVDC and is not treated in this book.

8.2 THE UNIFIED POWER FLOW CONTROLLER

The Unified Power Flow Controller (UPFC) concept was proposed by Gyugyi in 1991. The UPFC was devised for the real-time control and dynamic compensation of ac transmission systems, providing multifunctional flexibility required to solve many of the problems facing the power delivery industry. Within the framework of traditional power transmission concepts, the UPFC is able to control, simultaneously or selectively, all the parameters affecting power flow in the transmission line (i.e., voltage, impedance, and phase angle), and this unique capability is signified by the adjective "unified" in its name. Alternatively, it can independently control both the real and reactive power flow in the line. The reader should recall that, for all the Controllers discussed in the previous chapters, the control of real power is associated with similar change in reactive power, i.e., increased real power flow also resulted in increased reactive line power.

8.2.1 Basic Operating Principles

From the conceptual viewpoint, the UPFC is a generalized synchronous voltage source (SVS), represented at the fundamental (power system) frequency by voltage phasor $\boldsymbol{V}_{pq}$ with controllable magnitude V_{pq} ($0 \leq V_{pq} \leq V_{pq\max}$) and angle ρ ($0 \leq \rho \leq 2\pi$), in series with the transmission line, as illustrated for the usual elementary two-machine system (or for two independent systems with a transmission link intertie) in Figure 8.3. In this functionally unrestricted operation, which clearly includes voltage and angle regulation, the SVS generally exchanges both reactive and real power with the transmission system. Since, as established previously, an SVS is able to generate only the reactive power exchanged, the real power must be supplied to it, or absorbed from it, by a suitable power supply or sink. In the UPFC arrangement the real power exchanged is provided by one of the end buses (e.g., the sending-end bus), as indicated in Figure 8.3.

In the presently used practical implementation, the UPFC consists of two voltage-sourced converters, as illustrated in Figure 8.4. These back-to-back converters, labeled "Converter 1" and "Converter 2" in the figure, are operated from a common dc link provided by a dc storage capacitor. As indicated before, this arrangement functions as an ideal ac-to-ac power converter in which the real power can freely flow in either direction between the ac terminals of the two converters, and each converter can independently generate (or absorb) reactive power at its own ac output terminal.

Converter 2 provides the main function of the UPFC by injecting a voltage $\boldsymbol{V}_{pq}$ with controllable magnitude V_{pq} and phase angle ρ in series with the line via an insertion transformer. This injected voltage acts essentially as a synchronous ac voltage

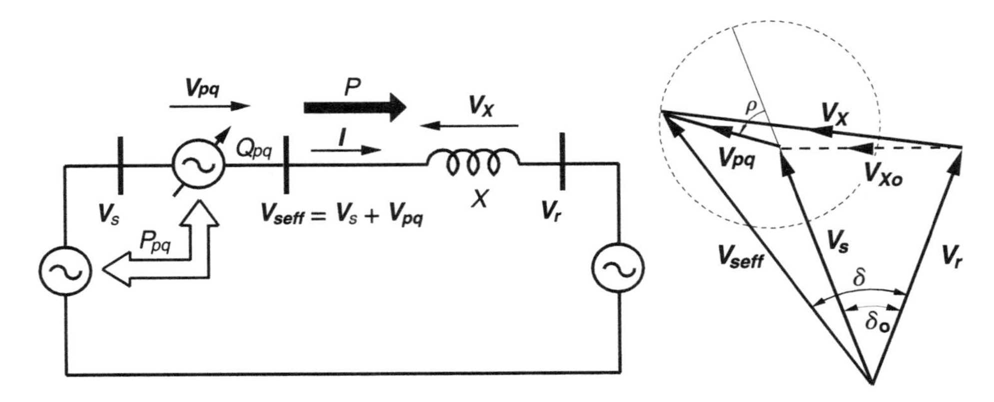

Figure 8.3 Conceptual representation of the UPFC in a two-machine power system.

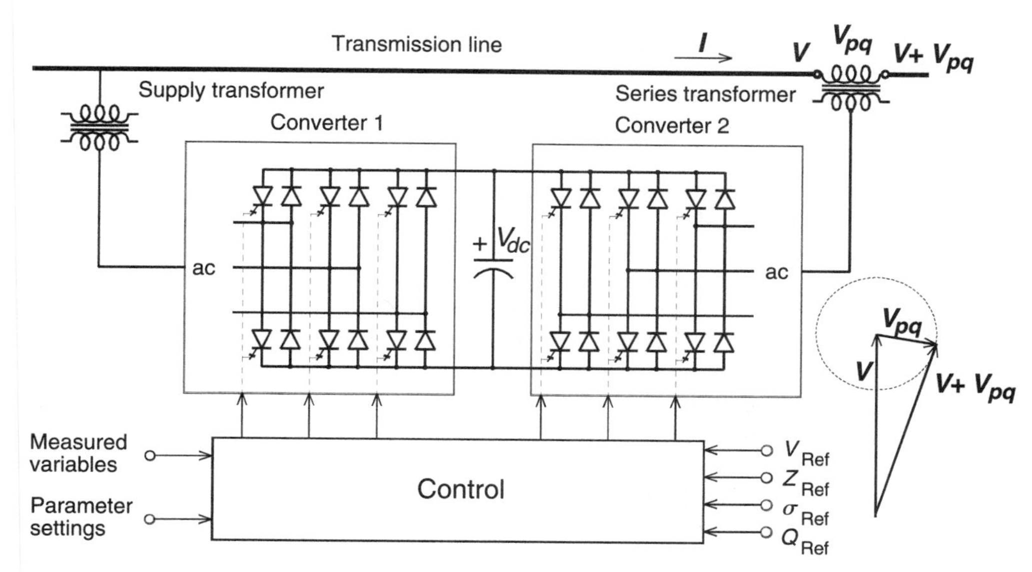

Figure 8.4 Implementation of the UPFC by two back-to-back voltage-sourced converters.

source. The transmission line current flows through this voltage source resulting in reactive and real power exchange between it and the ac system. The reactive power exchanged at the ac terminal (i.e., at the terminal of the series insertion transformer) is generated internally by the converter. The real power exchanged at the ac terminal is converted into dc power which appears at the dc link as a positive or negative real power demand.

The basic function of Converter 1 is to supply or absorb the real power demanded by Converter 2 at the common dc link to support the real power exchange resulting from the series voltage injection. This dc link power demand of Converter 2 is converted back to ac by Converter 1 and coupled to the transmission line bus via a shunt-connected transformer. In addition to the real power need of Converter 2, Converter 1 can also generate or absorb controllable reactive power, if it is desired, and thereby provide independent shunt reactive compensation for the line. It is important to note that whereas there is a closed direct path for the real power negotiated by the action of series voltage injection through Converters 1 and 2 back to the line, the corresponding reactive power exchanged is supplied or absorbed locally by Converter 2 and therefore does not have to be transmitted by the line. Thus, Converter 1 can be operated at a unity power factor or be controlled to have a reactive power exchange with the line independent of the reactive power exchanged by Converter 2. Obviously, there can be no reactive power flow through the UPFC dc link.

8.2.2 Conventional Transmission Control Capabilities

Viewing the operation of the Unified Power Flow Controller from the standpoint of traditional power transmission based on reactive shunt compensation, series compensation, and phase angle regulation, the UPFC can fulfill all these functions and thereby meet multiple control objectives by adding the injected voltage V_{pq}, with appropriate amplitude and phase angle, to the (sending-end) terminal voltage V_s. Using phasor representation, the basic UPFC power flow control functions are illustrated in Figure 8.5.

Voltage regulation with continuously variable in-phase/anti-phase voltage injection, is shown in Figure 8.5(a) for voltage increments $\boldsymbol{V}_{pq} = \pm\Delta\boldsymbol{V}$ ($\rho = 0$). As shown in Chapter 7, this is functionally similar to that obtainable with a transformer tap changer having infinitely small steps.

Series reactive compensation is shown in Figure 8.5(b) where $\boldsymbol{V}_{pq} = \boldsymbol{V}_q$ is injected in quadrature with the line current $\boldsymbol{I}$. Functionally this is similar to series capacitive and inductive line compensation attained by the SSSC in Chapter 6: the injected series compensating voltage can be kept constant, if desired, independent of line current variation, or can be varied in proportion with the line current to imitate the compensation obtained with a series capacitor or reactor.

Phase angle regulation (phase shift) is shown in Figure 8.5(c) where $\boldsymbol{V}_{pq} = \boldsymbol{V}_\sigma$ is injected with an angular relationship with respect to $\boldsymbol{V}_s$ that achieves the desired σ phase shift (advance or retard) without any change in magnitude. Thus the UPFC can function as a perfect Phase Angle Regulator which, as discussed in Chapter 7, can also supply the reactive power involved with the transmission angle control by internal var generation.

Multifunction power flow control, executed by simultaneous terminal voltage regulation, series capacitive line compensation, and phase shifting, is shown in Figure 8.5(d) where $\boldsymbol{V}_{pq} = \Delta\boldsymbol{V} + \boldsymbol{V}_q + \boldsymbol{V}_\sigma$. This functional capability is unique to the UPFC. No single conventional equipment has similar multifunctional capability.

The general power flow control capability of the UPFC, from the viewpoint of conventional transmission control, can be illustrated best by the real and reactive power transmission versus transmission angle characteristics of the simple two-machine system shown in Figure 8.3. With reference to this figure, the transmitted power P and the reactive power $-jQ_r$, supplied by the receiving end, can be expressed as follows:

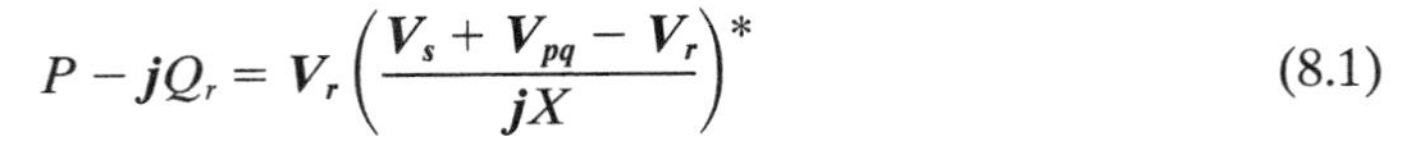

$$P - jQ_r = \boldsymbol{V}_r\left(\frac{\boldsymbol{V}_s + \boldsymbol{V}_{pq} - \boldsymbol{V}_r}{jX}\right)^* \tag{8.1}$$

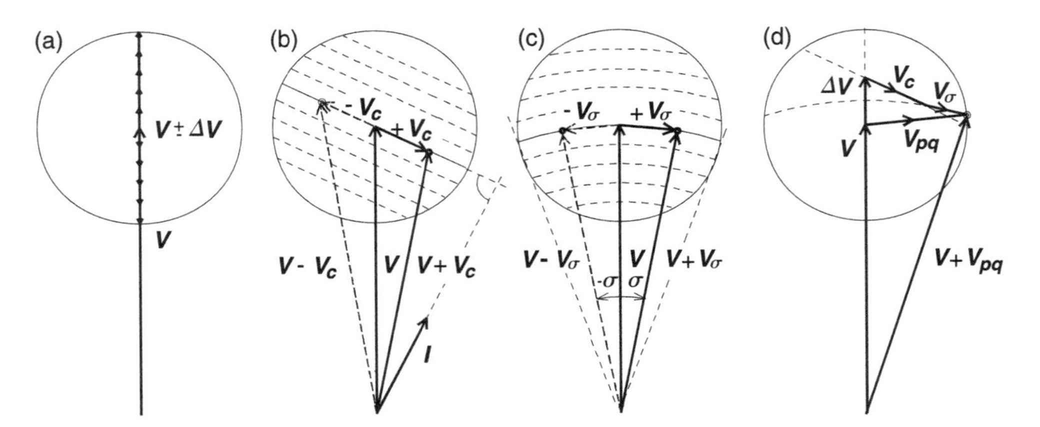

(a) Voltage regulation
(b) Line impedance compensation
(c) Phase shifting
(d) Simultaneous control of voltage, impedance, and angle

Figure 8.5 Phasor diagrams illustrating the conventional transmission control capabilities of the UPFC.

where symbol * means the conjugate of a complex number and $\boldsymbol{j} = e^{j\pi/2} = \sqrt{-1}$. If $\boldsymbol{V}_{pq} = 0$, then (8.1) describes the uncompensated system, that is,

$$P - \boldsymbol{j}Q_r = \boldsymbol{V}_r\left(\frac{\boldsymbol{V}_s - \boldsymbol{V}_r}{\boldsymbol{j}X}\right)^* \tag{8.2}$$

Thus, with $\boldsymbol{V}_{pq} \neq 0$, the total real and reactive power can be written in the form

$$P - \boldsymbol{j}Q_r = \boldsymbol{V}_r\left(\frac{\boldsymbol{V}_s - \boldsymbol{V}_r}{\boldsymbol{j}X}\right)^* + \frac{\boldsymbol{V}_r\boldsymbol{V}^*_{pq}}{-\boldsymbol{j}X} \tag{8.3}$$

Substituting

$$\boldsymbol{V}_s = Ve^{j\delta/2} = V\left(\cos\frac{\delta}{2} + \boldsymbol{j}\sin\frac{\delta}{2}\right) \tag{8.4}$$

$$\boldsymbol{V}_r = Ve^{-j\delta/2} = V\left(\cos\frac{\delta}{2} - \boldsymbol{j}\sin\frac{\delta}{2}\right) \tag{8.5}$$

and

$$\boldsymbol{V}_{pq} = V_{pq}e^{j(\delta/2+\rho)} = V_{pq}\left\{\cos\left(\frac{\delta}{2} + \rho\right) + \boldsymbol{j}\sin\left(\frac{\delta}{2} + \rho\right)\right\} \tag{8.6}$$

the following expressions are obtained for P and Q_r:

$$P(\delta, \rho) = P_0(\delta) + P_{pq}(\rho) = \frac{V^2}{X}\sin\delta - \frac{VV_{pq}}{X}\cos\left(\frac{\delta}{2} + \rho\right) \tag{8.7}$$

and

$$Q_r(\delta, \rho) = Q_{0r}(\delta) + Q_{pq}(\rho) = \frac{V^2}{X}(1 - \cos\delta) - \frac{VV_{pq}}{X}\sin\left(\frac{\delta}{2} + \rho\right) \tag{8.8}$$

where

$$P_0(\delta) = \frac{V^2}{X}\sin\delta \tag{8.9}$$

and

$$Q_{0r}(\delta) = -\frac{V^2}{X}(1 - \cos\delta) \tag{8.10}$$

are the real and reactive power characterizing the power transmission of the uncompensated system at a given angle δ. Since angle ρ is freely variable between 0 and 2π at any given transmission angle δ ($0 \leq \delta \leq \pi$), it follows that $P_{pq}(\rho)$ and $Q_{pq}(\rho)$ are controllable between $-VV_{pq}/X$ and $+\ VV_{pq}/X$ independent of angle δ. Therefore, the transmittable real power P is controllable between

$$P_0(\delta) - \frac{VV_{pq\max}}{X} \leq P_0(\delta) \leq P_0(\delta) + \frac{VV_{pq\max}}{X} \tag{8.11}$$

and the reactive power Q_r is controllable between

$$Q_{0r}(\delta) - \frac{VV_{pq\max}}{X} \leq Q_{0r}(\delta) \leq Q_{0r}(\delta) + \frac{VV_{pq\max}}{X} \tag{8.12}$$

at any transmission angle δ, as illustrated in Figure 8.6. The wide range of control for the transmitted power that is independent of the transmission angle δ, observable in the figure, indicates not only superior capability of the UPFC in power flow applications, but it also suggests powerful capacity for transient stability improvement and power oscillation damping.

To illustrate further the meaning of (8.7) and (8.8), consider again Figure 8.3 which is, for convenience, also shown in a simplified manner in Figure 8.7(a).

A phasor diagram, defining the relationship between $\mathbf{V}_s$, $\mathbf{V}_r$, $\mathbf{V}_X$ (the voltage phasor across X) and the inserted voltage phasor $\mathbf{V}_{pq}$, with controllable magnitude ($0 \leq V_{pq} \leq V_{pq\max}$) and angle ($0 \leq \rho_{pq} \leq 360°$), is shown in Figure 8.7(a). (For the illustrations, $\delta = 30°$ and $V_s = V_r = 1$, $X = 0.5$, $V_{pq\max} = 0.25$ p.u. values were assumed.) As illustrated, the inserted voltage phasor $\mathbf{V}_{pq}$ is added to the fixed sending-end voltage phasor $\mathbf{V}_s$ to produce the effective sending-end voltage $\mathbf{V}_{seff} = \mathbf{V}_s + \mathbf{V}_{pq}$. The difference, $\mathbf{V}_{seff} - \mathbf{V}_r$, provides the compensated voltage phasor, $\mathbf{V}_X$, across X. As angle ρ_{pq} is varied over its full 360 degree range, the end of phasor $\mathbf{V}_{pq}$ moves along a circle with its center located at the end of phasor $\mathbf{V}_s$. The area within this circle, obtained with $\mathbf{V}_{pq\max}$, defines the operating range of phasor $\mathbf{V}_{pq}$ and thereby the achievable compensation of the line.

The rotation of phasor $\mathbf{V}_{pq}$ with angle ρ_{pq} modulates both the magnitude and angle of phasor $\mathbf{V}_X$ and, therefore, both the transmitted real power, P, and the reactive

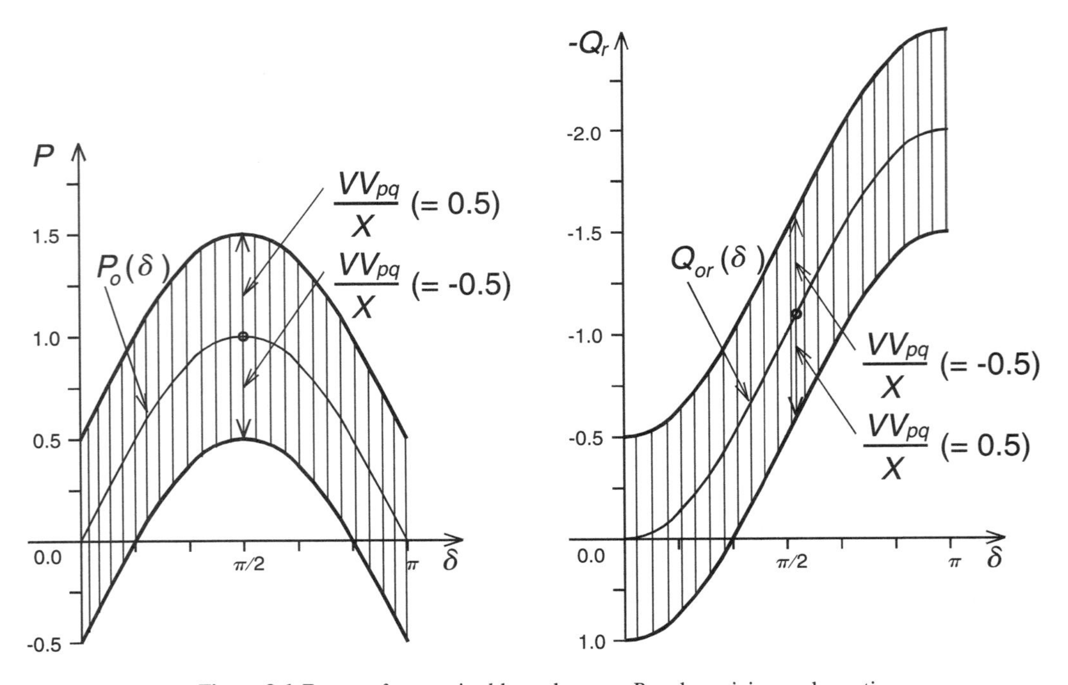

Figure 8.6 Range of transmittable real power P and receiving-end reactive power demand Q vs. transmission angle δ of a UPFC controlled transmission line.

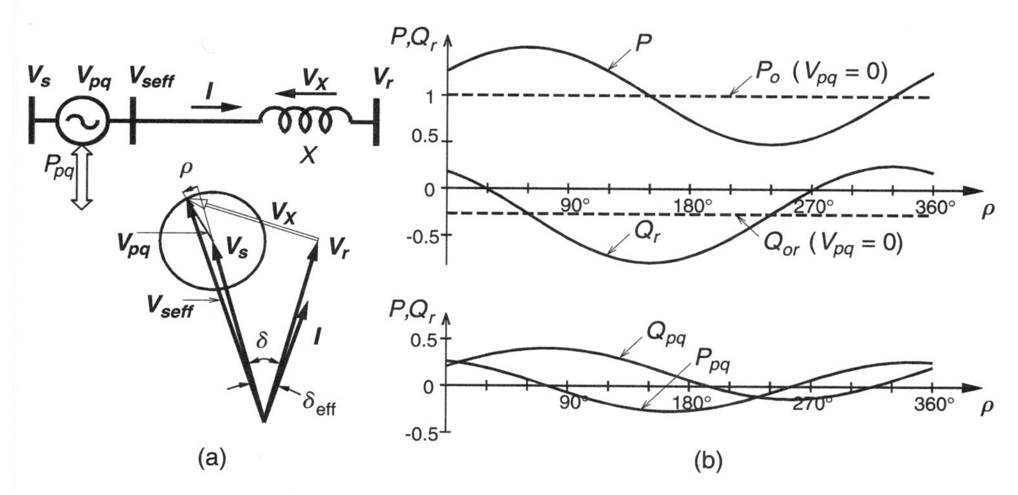

Figure 8.7 Phasor diagram representation of the UPFC (a) and variation of the receiving-end real and reactive power, and the real and reactive power supplied by the UPFC, with the angular rotation of the injected voltage phasor (b).

power, Q_r, vary with ρ_{pq} in a sinusoidal manner, as illustrated in Figure 8.7(b). This process, of course, requires the voltage source (V_{pq}) to supply and absorb both reactive and real power, Q_{pq} and P_{pq}, which are also sinusoidal functions of angle ρ_{pq}, as shown in the figure.

The powerful, previously unattainable, capabilities of the UPFC summarized above in terms of conventional transmission control concepts, can be integrated into a generalized power flow controller that is able to maintain prescribed, and independently controllable, real power P and reactive power Q in the line. Within this concept, the conventional terms of series compensation, phase shifting, etc., become irrelevant; the UPFC simply controls the magnitude and angular position of the injected voltage in real time so as to maintain or vary the real and reactive power flow in the line to satisfy load demand and system operating conditions.

8.2.3 Independent Real and Reactive Power Flow Control

In order to investigate the capability of the UPFC to control real and reactive power flow in the transmission line, refer to Figure 8.7(a). Let it first be assumed that the injected compensating voltage, V_{pq}, is zero. Then the original elementary two-machine (or two-bus ac intertie) system with sending-end voltage V_s, receiving-end voltage V_r, transmission angle δ, and line impedance X is restored. With these, the normalized transmitted power, $P_0(\delta) = \{V^2/X\} \sin \delta = \sin \delta$, and the normalized reactive power, $Q_0(\delta) = Q_{0s}(\delta) = -Q_{0r}(\delta) = \{V^2/X\}\{1-\cos \delta\} = 1-\cos \delta$, supplied at the ends of the line, are shown plotted against angle (δ) in Figure 8.8(a). The relationship between real power $P_0(\delta)$ and reactive power $Q_{0r}(\delta)$ can readily be expressed with $V^2/X = 1$ in the following form:

$$Q_{0r}(\delta) = -1 - \sqrt{1 - \{P_0(\delta)\}^2} \tag{8.13}$$

or

$$\{Q_{0r}(\delta) + 1\}^2 + \{P_0(\delta)\}^2 = 1 \tag{8.14}$$

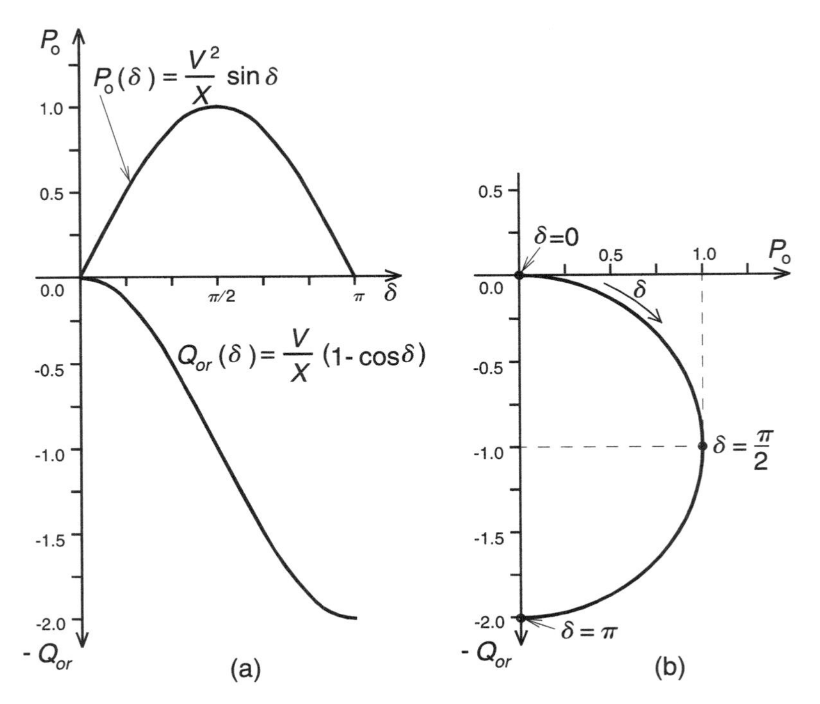

Figure 8.8 Transmittable real power P_0 and receiving-end reactive power demand Q_{or} vs. transmission angle δ of a two-machine system (a) and the corresponding Q_{or} vs. P_0 loci (b).

Equation (8.14) describes a circle with a radius of 1.0 around the center defined by coordinates $P = 0$ and $Q_r = -1$ in a $\{Q_r, P\}$ plane, as illustrated for positive values of P in Figure 8.8(*b*). Each point of this circle gives the corresponding P_0 and Q_{0r} values of the uncompensated system at a specific transmission angle δ. For example, at $\delta = 0$, $P_0 = 0$ and $Q_{0r} = 0$; at $\delta = 30°$, $P_0 = 0.5$ and $Q_{0r} = -0.134$; at $\delta = 90°$, $P_0 = 1.0$ and $Q_{0r} = -1.0$; etc.

Refer again to Figure 8.7(a) and assume now that $\boldsymbol{V_{pq}} \neq 0$. It follows from (8.3), or (8.7) and (8.8), and from Figure 8.7(b), that the real and reactive power change from their uncompensated values, $P_0(\delta)$ and $Q_{0r}(\delta)$, as functions of the magnitude V_{pq} and angle ρ of the injected voltage phasor $\boldsymbol{V_{pq}}$. Since angle ρ is an unrestricted variable ($0 \leq \rho \leq 2\pi$), the boundary of the attainable control region for $P(\delta, \rho)$ and $Q_r(\delta, \rho)$ is obtained from a complete rotation of phasor $\boldsymbol{V_{pq}}$ with its maximum magnitude $V_{pq\max}$. It follows from the above equations that this control region is a circle with a center defined by coordinates $P_0(\delta)$ and $Q_{0r}(\delta)$ and a radius of $V_r V_{pq}/X$. With $V_s = V_r = V$, the boundary circle can be described by the following equation:

$$\{P(\delta, \rho) - P_0(\delta)\}^2 + \{Q_r(\delta, \rho) - Q_{0r}(\delta)\}^2 = \left\{\frac{VV_{pq\max}}{X}\right\}^2 \tag{8.15}$$

The circular control regions defined by (8.15) are shown in Figures 8.9(a) through (d) for $V = 1.0$, $V_{pq\max} = 0.5$, and $X = 1.0$ (per unit or p.u. values) with their centers on the circular arc characterizing the uncompensated system (8.14) at transmission angles $\delta = 0°$, 30°, 60°, and 90°. In other words, the centers of the control regions are defined

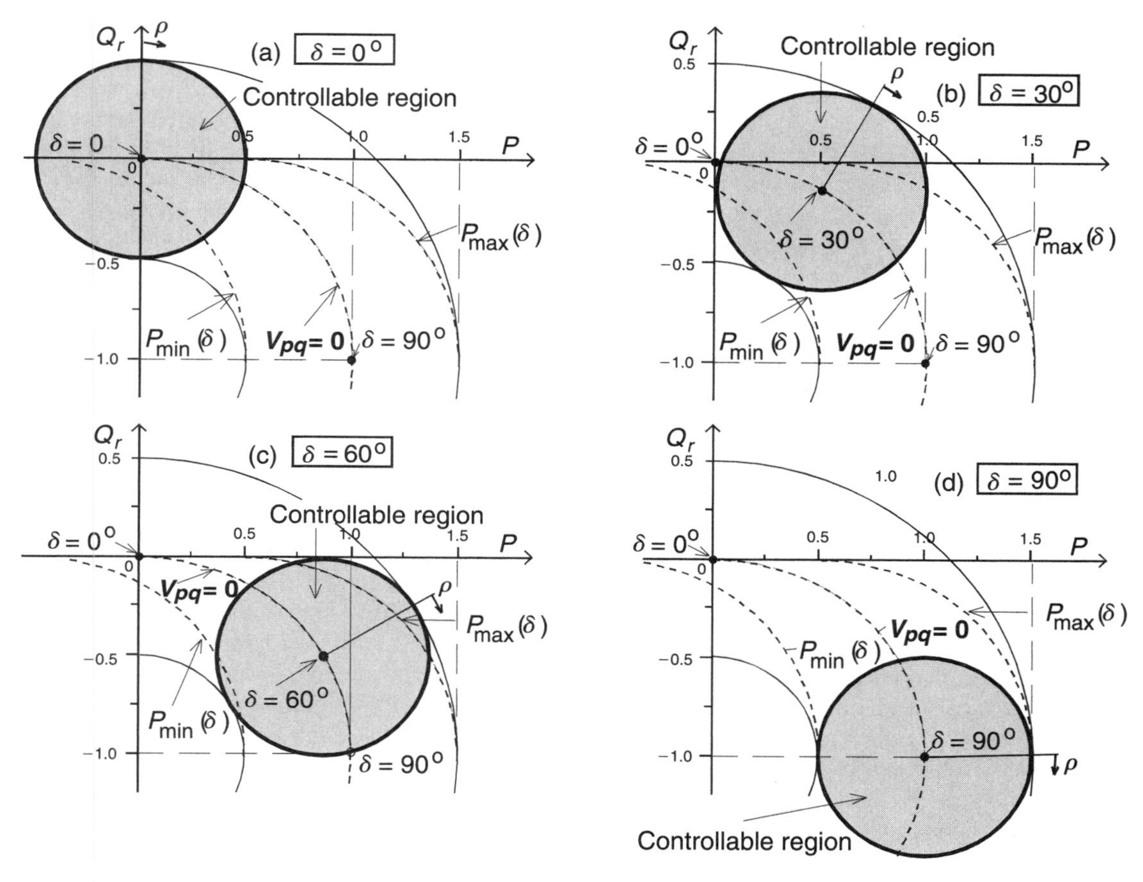

Figure 8.9 Control region of the attainable real power P and receiving-end reactive power demand Q_r with a UPFC-controlled transmission line at $\delta = 0°$ (a), $\delta = 30°$ (b), $\delta = 60°$ (c), and $\delta = 90°$ (d).

by the corresponding $P_0(\delta)$, $Q_{0r}(\delta)$ coordinates at angles $\delta = 0$, 30°, 60°, and 90° in the $\{Q_r, P\}$ plane.

Consider first Figure 8.9(a), which illustrates the case when the transmission angle is zero ($\delta = 0$). With $V_{pq} = 0$, P, Q_r, (and Q_s) are all zero, i.e., the system is at standstill at the origin of the Q_r, P coordinates. The circle around the origin of the $\{Q_r, P\}$ plane is the loci of the corresponding Q_r and P values, obtained as the voltage phasor $\mathbf{V}_{pq}$ is rotated a full revolution ($0 \leq \rho \leq 360°$) with its maximum magnitude $V_{pq\max}$. The area within this circle defines all P and Q_r values obtainable by controlling the magnitude V_{pq} and angle ρ of phasor $\mathbf{V}_{pq}$. In other words, the circle in the $\{Q_r, P\}$ plane defines all P and Q_r values attainable with the UPFC of a given rating. It can be observed, for example, that the UPFC with the stipulated voltage rating of 0.5 p.u. is able to establish 0.5 p.u. power flow, in either direction, without imposing any reactive power demand on either the sending-end or the receiving-end generator. (This statement tacitly assumes that the sending-end and receiving-end voltages are provided by independent power systems which are able to supply and absorb real power without any internal angular change.) Of course, the UPFC, as illustrated, can force the system at one end to supply reactive power for, or absorb that from, the system at the other end. Similar control characteristics for real power P and the reactive power Q_r can be observed at angles $\delta = 30°$, $60°$, and $90°$ in Figures 8.9(b), (c), and (d).

In general, at any given transmission angle δ, the transmitted real power P, as well as the reactive power demand at the receiving end Q_r, can be controlled freely by the UPFC within the boundary circle obtained in the $\{Q_r, P\}$ plane by rotating the injected voltage phasor V_{pq} with its maximum magnitude a full revolution. Furthermore, it should be noted that, although the above presentation focuses on the receiving-end reactive power, Q_r, the reactive component of the line current, and the corresponding reactive power can actually be controlled with respect to the voltage selected at any point of the line.

Figures 8.9(a) through (d) clearly demonstrate that the UPFC, with its unique capability to control independently the real and reactive power flow at any transmission angle, provides a powerful, hitherto unattainable, new tool for transmission system control.

In order to put the capabilities of the UPFC into proper perspective in relation to other related power flow controllers, such as the Thyristor-Switched, Thyristor-Controlled and GTO-Controlled Series Capacitor (TSSC, TCSC and GCSC), the Static Synchronous Series Compensator (SSSC), and the Thyristor-Controlled Phase Angle Regulator (TCPAR), a basic comparison between the respective power flow control characteristics are presented in the next section. The basis chosen for this comparison is the capability of each type of power flow controller to vary the transmitted real power and the reactive power demand at the receiving end. The receiving end var demand is usually an important factor because it significantly influences the variation of the line voltage with load demand, the overvoltage at load rejection, and the steady-state system losses. Similar comparison, of course, could easily be made at the sending end or at other points of the transmission line, but the results would be quite similar for all practical transmission angles.

8.2.4 Comparison of the UPFC to Series Compensators and Phase Angle Regulators

The objective of this section is to compare the character of compensation attainable with controlled series compensators, phase angle regulators and the UPFC for power flow control. Consider once more the simple two-machine system shown in Figure 8.3. This model can be used to establish the basic transmission characteristics of the Controlled Series Compensators (TSSC, GCSC, TCSC, and SSSC) and the Controlled Phase Angle Regulator (TCPAR and SVS-based) by using them in place of the UPFC, as symbolically illustrated in Figure 8.10 for controlled capacitive compensation and in Figure 8.12 for phase shift, to establish their effect on the Q versus P characteristic of the system at different transmission angles.

8.2.4.1 Comparison of the UPFC to Controlled Series Compensators. As shown in Chapter 6, Thyristor-Switched Series Capacitor (TSSC) and Gate-Controlled Series Capacitor (GCSC) schemes employ a number of capacitor banks in series with the line, each with a thyristor or, respectively, a GTO bypass switch. This arrangement in effect is equivalent to a series capacitor whose capacitance is adjustable in a step-like or continuous manner. The controllable series capacitive impedance provided by the TSSC or GCSC cancels part of the reactive line impedance resulting in a reduced overall transmission impedance (i.e., in an electrically shorter line) and correspondingly increased transmittable power. For the purpose of the present investigation, both the

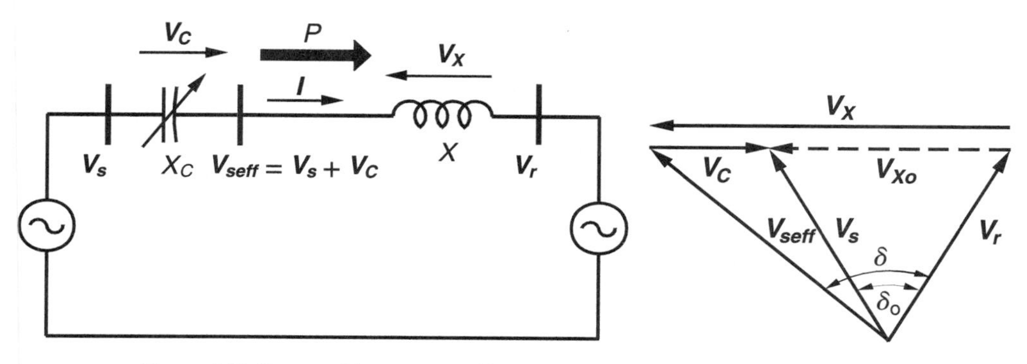

Figure 8.10 Two-machine system with controlled series capacitive compensation.

TSSC and GCSC can be considered simply as a continuously variable capacitor whose impedance is controllable in the range of $0 \leq X_0 \leq X_{C\max}$.

The Static Synchronous Series Compensator (SSSC) injects a continuously variable series compensating voltage in quadrature with the line current. In contrast to the TSSC and GCSC, the compensating voltage of the SSSC can be controlled totally independent of the line current (i.e., independent of the transmission angle δ) over the theoretical operating range of zero to rated line current.

The Thyristor-Controlled Series Capacitor (TCSC) schemes typically use one or more modules in series, each consisting of a Thyristor-Controlled Reactor in parallel with a capacitor, to vary the effective series compensating impedance and thereby the compensating voltage. The TCSC with sufficient number of constituent modules is able to control and maintain the compensating voltage with decreasing line current in its operating range. It can also provide, with a suitably designed thyristor-controlled reactor, inductive compensation. The inductive compensation range, for practical reasons (e.g., losses, thyristor valve current and voltage ratings, reactor rating) is usually less than the capacitive range. For the purpose of the intended comparison, an ideal TCSC with a sufficient number of constituent modules is assumed. Such a TCSC is capable of providing equal capacitive and inductive compensation and is able to continuously control and maintain the compensating voltage in the face of decreasing line current. With this assumption the TCSC can be considered as a SSSC until the transmission angle becomes too small to maintain sufficient line current (about 0.25 p.u.) needed for normal operation. This ideal TCSC is probably beyond economic practicality, but this stipulation does not significantly affect the present comparison because the idealization of the TCSC alters only the control range and not the character of the compensation it provides.

Controlled series compensators provide a series compensating voltage that, by definition, is in quadrature with the line current. Consequently, they can affect only the magnitude of the current flowing through the transmission line. At any given setting of the capacitive impedance of the TSSC and GCSC, or of the effective capacitive/inductive impedance of the TCSC and SSSC, a particular overall transmission impedance is defined at which the transmitted power is strictly determined by the transmission angle (assuming a constant amplitude for the end voltages). Therefore, the reactive power demand at the endpoints of the line is determined by the transmitted real power in the same way as if the line was uncompensated but had a lower line impedance. Consequently, the relationship between the transmitted power, P, and the reactive power demand at the receiving end, Q_r, can be represented by a single Q_r –

P circular locus, similar to that shown for the uncompensated system in Figure 8.8(b), at a given output (compensating impedance or voltage) setting of the series compensator. This means that for a continuously controllable compensator an infinite number of $Q_r - P$ circular loci can be established by using the basic transmission relationships, i.e., $P = \{V^2/(X - X_q)\}\sin\delta$ and $Q_r = \{V^2/(X - X_q)\}\{1 - \cos\delta\})$ with X_q varying between 0 and $X_{q\max}$ or 0 and $\pm X_{q\max}$ (where $X_q = X_C$ for the TSSC and GCSC and $X_q = |\boldsymbol{V_q}/\boldsymbol{I}| = V_q/I$ for the SSSC and, with limitations, for the TCSC). Evidently, a given transmission angle defines a single point on each $Q_r - P$ locus obtained with a specific value of X_q. Thus, the progressive increase of X_q from zero to $X_{q\max}$ could be viewed as if the point defining the corresponding P and Q_r values at the given transmission angle on the first $Q_r - P$ locus (uncompensated line) moves through an infinite number of $Q_r - P$ loci representing progressively increasing series compensation, until it finally reaches the last $Q_r - P$ locus that represents the system with maximum series compensation. The first $Q_r - P$ locus, representing the uncompensated power transmission, is the lower boundary curve for the TSSC and GCSC and identified by $(Q_r - P)_{Xq=0}$. The last $Q_r - P$ locus, representing the power transmission with maximum capacitive impedance compensation, is the upper boundary curve for the TSSC and GCSC and identified by $(Q_r - P)_{Xc\max}$. The lower and upper boundary curves for the SSSC and the TCSC considered are different from that of TSSC and GCSC and therefore they are identified by $(Q_r - P)_{+Vq\max}$ and $(Q_r - P)_{-Vq\max}$. The difference is partially due to the capability of the SSSC and TCSC to inject the compensating voltage both with 90 degree lagging (capacitive) or 90 degree leading (inductive) relationship with respect to the line current. Thus, they can both increase and decrease the transmitted power. Also, the SSSC and TCSC can maintain maximum compensating voltage with decreasing line current (the SSSC, theoretically, even with zero line current). For these reasons, the SSSC and TCSC have a considerably wider control range at low transmission angles than the TSSC and GCSC. Note that all $Q_r - P$ circular curves are considered only for the normal operating range of the transmission angle $(0 \leq \delta \leq 90°)$.

The plots in Figures 8.11(a) through (d) present, at the four transmission angles ($\delta = 0°$, $30°$, $60°$, and $90°$), the range of $Q_r - P$ control for the series reactive compensators, TSSC, GCSC, TCSC, and SSSC; with the same 0.5 p.u. maximum voltage rating stipulated for the UPFC. The previously derived circular control region of the UPFC is also shown in Figure 8.11 (by heavy broken lines) for comparison. In each figure the upper and lower boundary curves identified above for the TSSC/GCSC and SSSC/TCSC are shown by dotted and broken lines, respectively, for reference. The upper and lower boundary curves for an ideal TCSC, as explained above, are the same as those of the SSSC, as indicated in the figure. The boundary of a practical TCSC in its capacitive operating range (in which it is able to maintain rated compensating voltage) would be the same as that of the SSSC; however, in the inductive range, depending on the design, it would likely be lower (typically 30 to 50% of rated capacitive voltage). Outside of its normal operating range (i.e., at small transmission angles and relatively low line currents), the boundary curves of the practical TCSC would converge to those of the TSSC/GCSC.

Consider Figure 8.11(a) which illustrates the case when the transmission angle, δ, is zero. (For this special case, the two-machine system is again assumed to represent two independent power systems with an ac line intertie.) Since the TSSC, GCSC, and TCSC are an actively controlled, but functionally passive impedance, the current through the line with the compensated line impedance $(X - X_C)$ remains invariably zero, regardless of the actual value of X_C. Thus both P and Q_r are zero, the system

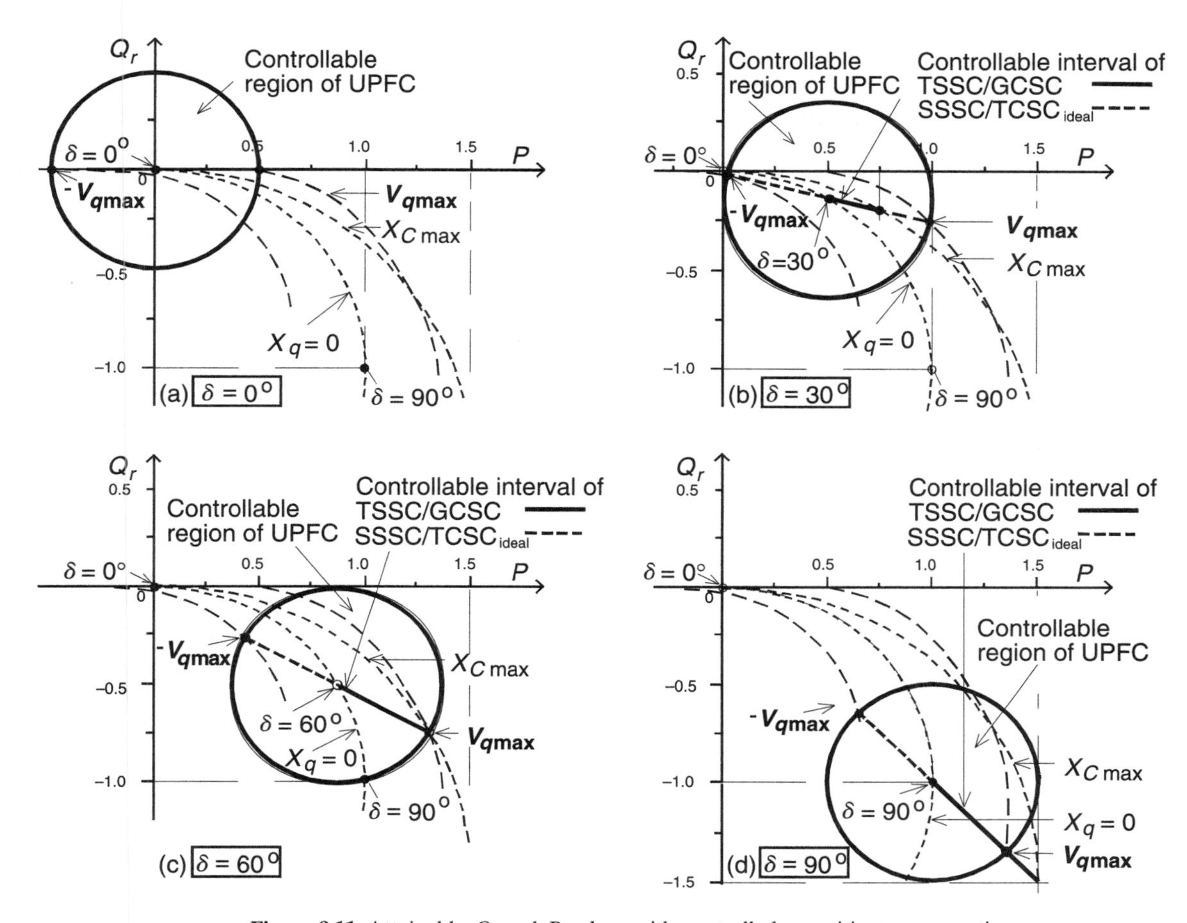

Figure 8.11 Attainable Q_r and P values with controlled capacitive compensation (points on the heavy straight line inside the circle) and those with the UPFC (any point inside the circle) at $\delta = 0°$ (a), $\delta = 30°$ (b), $\delta = 60°$ (c), and $\delta = 90°$ (d).

is at standstill, and therefore cannot be changed by reactive impedance compensation. An ideal SSSC, represented by a theoretical ac voltage source could establish power flow between the two sources. However, a practical SSSC with internal losses could only do that if an external power supply (connected to the dc terminals of the converter) replenished these losses and kept the SSSC in operation at zero (and at small) line current. By contrast, since the UPFC is a self-sufficient voltage source (whose losses are supplied by the shunt converter), it can force up to 0.5 p.u. real power flow in either direction and also control reactive power exchange between the sending-end and receiving-end buses within the circular control region shown in the figure.

The $Q_r - P$ characteristic of the TSSC, GCSC, TCSC, and SSSC at $\delta = 30°$ is shown in Figure 8.11(b). The relationship between Q_r and P, for the total range of series compensation, $0 \leq \delta \leq 90°$, is defined by a straight line connecting the two related points on the lower and upper boundary curves, $(Q_r - P)_{Xq=0}$ and $(Q_r - P)_{Xcmax}$, which represent the power transmission at $\delta = 30°$ with zero and, respectively, maximum series compensation. It can be observed that, as expected, the TSSC and GCSC control the real power by changing the degree of series compensation (and thereby the magnitude of the line current). However, the reactive power demand of the line cannot be controlled independently; it remains a direct function of the transmitted real power obtained at $\delta = 30°$ with varying X_C ($0 \leq X_C \leq X_{cmax}$). It is also to be noted that the achievable maximum increase in transmittable power attainable with the TSSC and GCSC is a constant percentage (defined by the maximum degree of series compensation) of the power transmitted with the uncompensated line at the given transmission angle. In other words, the maximum increase attainable in actual transmitted power is much less at small transmission angles than at large ones. This is due to the fact that the TSSC and GSSC are a series impedance and thus the compensating voltage they produce is proportional to the line current, which is a function of angle δ. (However, in steady-state, the angle is a function of the surplus power generated by the sending end. Thus, often, when the angle is small, the required power flow is also small.) The SSSC, being a reactive voltage source (with respect to the line current) can provide maximum compensating voltage, theoretically down to zero line current. (The limit, as previously mentioned, is the ability of the line to replenish the losses of the SSSC.) The TCSC can also provide constant compensating voltage over the operating range it is designed for. They both can also reverse the polarity of the compensating voltage. These characteristics mean that the SSSC and the TCSC (in its operating range) can provide compensation over a wider range, between the boundaries of $(Q_r - P)_{+Vq\max}$ and $(Q_r - P)_{-Vq\max}$, than can the TSSC and GCSC. However, due to its strictly reactive compensation capability, the SSSC and TCSC cannot control the reactive line power and thus the reactive power, Q_r, remains proportional to the real power, P. In other words, the SSSC and TCSC simply lengthens the control range of the TSSC and GCSC, without changing their Q_r versus P characteristic. By contrast, since the UPFC is a self-sufficient voltage source with the capability of exchanging both reactive and real power, the magnitude and angle of the compensating voltage it produces is independent of the line current (and of angle δ). Therefore the maximum change (increase or decrease) in transmittable power, as well as in receiving-end reactive power, achievable by the UPFC is not a function of angle δ and is determined solely by the maximum voltage the UPFC is rated to inject in series with the line (assumed 0.5 p.u. in this comparison). This characteristic can be observed in the figures where the radius of the circular loci defining the control region of the UPFC remains the same for all four transmission angles considered.

Figures 8.11(c) and (d), showing the $Q_r - P$ characteristics of the TSSC, GCSC, TCSC, SSSC, and the UPFC at $\delta = 60°$ and $\delta = 90°$, respectively, further confirms the above observations. The range of the TSSC and GCSC for real power control remains a constant percentage of the power transmitted by the uncompensated line at all transmission angles ($0 \leq \delta \leq 90°$), but the maximum actual change in transmitted real power progressively increases with increasing δ and reaches that of the TCSC, SSSC, and UPFC at $\delta = 90°$. However, it should be noted that whereas the maximum transmitted power of 1.5 p.u., obtained with all the reactive compensators, TSSC, GCSC, TCSC and SSSC, at full compensation is associated with 1.5 p.u. reactive power demand at the receiving end, the same 1.5 p.u. power transmission is achieved only with 1.0 p.u. reactive power demand when the line is compensated by a UPFC.

From Figures 8.11(a) through (d), it can be concluded that the UPFC has superior power flow control characteristics compared to the TSSC, GCSC, TCSC, and SSSC: it can control independently both real and reactive power over a broad range, its control range in terms of actual real and reactive power is independent of the transmission angle, and it can control both real and reactive power flow in either direction at zero (or at a small) transmission angle.

8.2.4.2 Comparison of the UPFC to Controlled Phase Angle Regulators. As discussed in Chapter 7, ideal Phase Angle Regulators provide voltage injection in series with the line so that the voltage applied at their input terminals appears with the same magnitude but with $\pm\sigma$ phase difference at their output terminals. Mechanical and thyristor-controlled Phase Angle Regulators and Quadrature Boosters provide controllable quadrature voltage injection in series with the line. These employ a shunt-connected regulation (excitation) transformer with appropriate taps on the secondary windings, a mechanical or thyristor-based tap changer, and a series insertion transformer. The excitation transformer has wye to delta (or delta to wye) windings and thus the phase-to-phase secondary voltages are in quadrature with respect to the corresponding primary phase to neutral voltages. Synchronous Voltage Source-Based Phase Angle Regulators have a power circuit structure similar to the UPFC and, as shown, can provide angle control without changing the magnitude of the system voltage.

There is a fundamental, and, from the application standpoint, important, difference between the thyristor-controlled and SVS-Based Phase Angle Regulators (or the UPFC). In the case of the former, the total VA (i.e., both the real power and the vars), exchanged by the series insertion transformer appears at the primary of the regulation transformer as a load demand on the power system. However, in the latter case only the real power exchanged is supplied by the shunt transformer from the power system.

The purpose of the present investigation is to compare the power flow control capability of the UPFC to that of the Phase Angle Regulator. For this comparison, the practical implementation of the Phase Angle Regulator and its ability to generate reactive power is unimportant. For simplicity, both the thyristor-controlled and SVS-based PARs are considered as ideal phase angle regulators. That is, both are assumed to be able to vary the phase angle between the voltages at the two ends of the insertion transformer in the control range of $-\sigma_{\max} \leq \sigma \leq \sigma_{\max}$, without changing the magnitude of the phase shifted voltage from that of the original line voltage. (Recall that the magnitude of the phase shifted voltage would increase by quadrature voltage injection usually employed in thyristor-controlled PAR implementations.)

Consider Figure 8.12 in which the previously considered two-machine system is shown again with a PAR assumed to function as an ideal phase angle regulator. The transmitted power and the reactive power demands at the sending end and receiving end can be described by relationships analogous to those characterizing the uncompensated system: $P = \{V^2/X\} \sin \delta'$ and $Q_r = Q_s = \{V^2/X\}\{1 - \cos \delta\}$, where $\delta' = \delta - \sigma$.

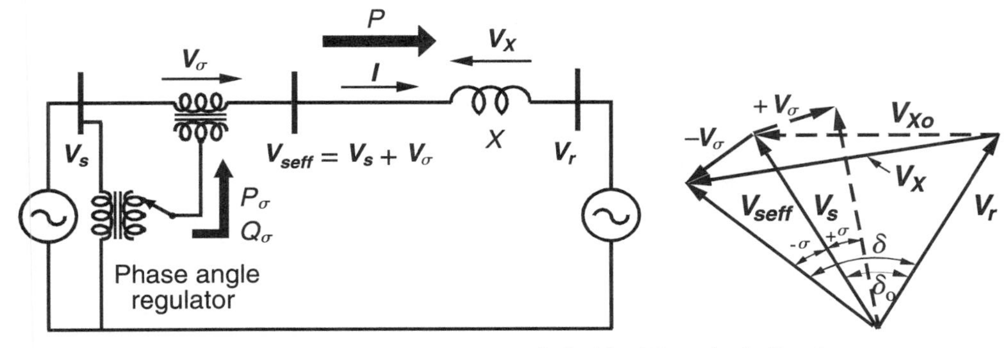

Figure 8.12 Two-machine system controlled with a Phase Angle Regulator.

Thus, as established in Chapter 7, the PAR cannot increase the maximum transmittable power, $P = V^2/X$, or change Q_r at a fixed P. Consequently, the $Q_r - P$ relationship, with transmission angle $\delta'(0 \leq \delta' \leq 90°)$ controlling the actual power transmission, is identical to that of the uncompensated system shown in Figure 8.8. The function of the PAR is simply to establish the actual transmission angle, δ', required for the transmission of the desired power P, by adjusting the phase shift angle σ so as to satisfy the equation $\delta' = \delta - \sigma$ at a given δ, the prevailing angle between the sending-end and receiving-end voltages. In other words, the PAR can vary the transmitted power at a fixed δ, or maintain the actual transmission angle δ' constant in the face of a varying δ, but it cannot increase the maximum transmittable power or control the reactive power flow independently of the real power.

The plots in Figures 8.13(a) through (d) present the $Q_r - P$ relationship for the ideal PAR in comparison to that of the UPFC at $\delta = 0°$, 30°, 60°, and 90°. It can be observed in these figures that the control range centers around the point defined by the given value of angle $\delta = \delta_0$ (= 0°, 30°, 60°, 90°) on the $Q_r - P$ locus characterizing the uncompensated power transmission ($V_0 = 0$ for the PAR and $V_{pq} = 0$ for the

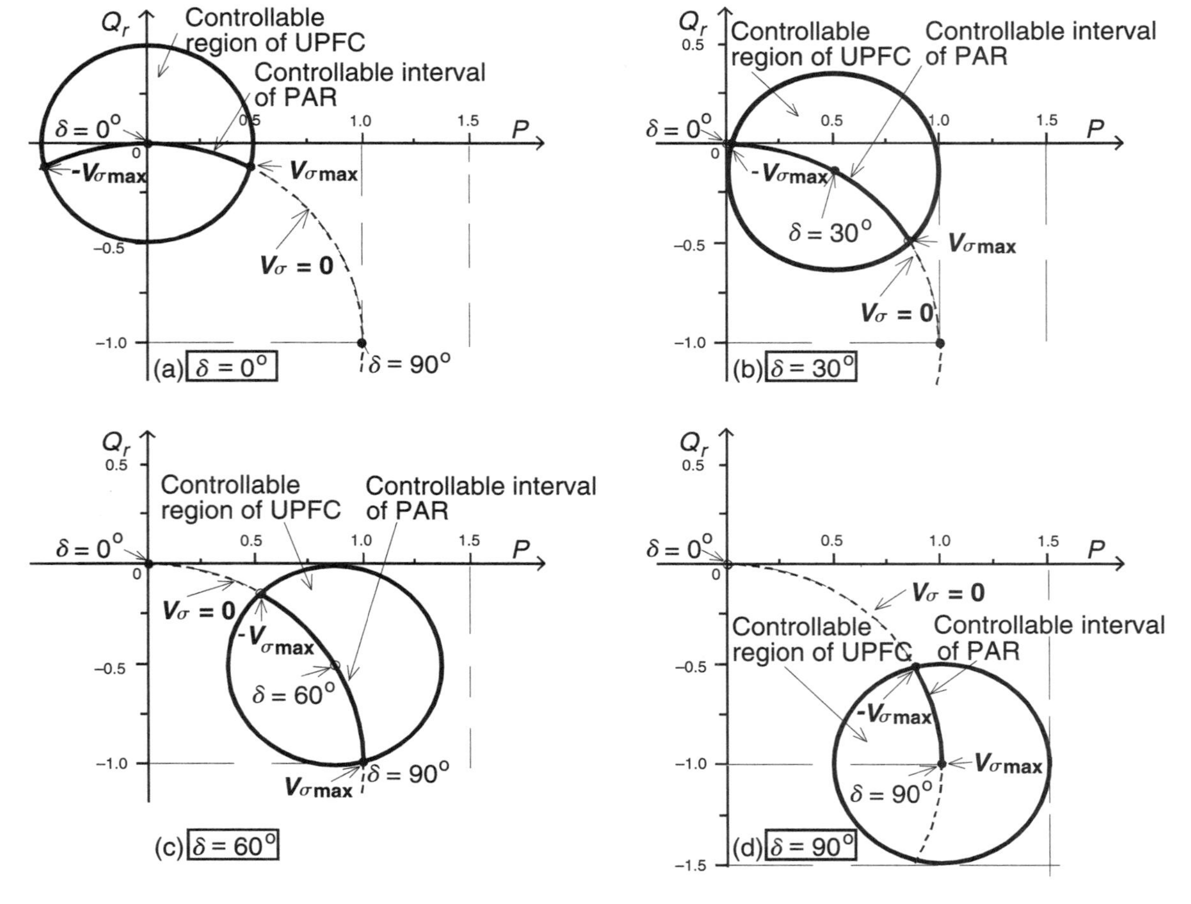

Figure 8.13 Attainable Q_r and P values with a PAR (points on the heavy arc inside the circle) and those with the UPFC (any point inside the circle) at $\delta = 0°$ (a), $\delta = 30°$ (b), $\delta = 60°$ (c), and $\delta = 90°$ (d).

UPFC). As the phase-shift angle σ is varied in the range of $-30° \leq \sigma \leq 30°$ (which corresponds to the maximum inserted voltage of $V_0 = 0.5$ p.u.), this point moves on the uncompensated $Q_r - P$ locus in the same way as if the fixed angle δ_0 was varied in the range of $\delta_0 - 30° \leq \delta_0 \leq \delta_0 + 30°$. Consequently, the PAR, in contrast to the series reactive compensators TSSC, TCSC, and SSSC, can effectively control the real power flow, as illustrated in Figure 8.13(a), when the angle between the sending-end and receiving-end voltages is zero ($\delta = 0$).

The Q_r versus P plots in Figure 8.13 clearly show the superiority of the UPFC over the PAR for power flow control: The UPFC has a wider range for real power control and facilitates the independent control of the receiving-end reactive power demand over a broad range. For example, it is seen that the UPFC can facilitate up to 1.0 p.u. real power transmission with unity power factor at the receiving end ($Q_r = 0$), whereas the reactive power demand at the receiving end would increase with increasing real power (Q_r reaching 1.0 p.u. at $P = 1.0$ p.u.), in the manner of an uncompensated line, when the power flow is controlled by the PAR.

8.2.5 Control Structure

The superior operating characteristics of the UPFC are due to its unique ability to inject an ac compensating voltage *vector* with arbitrary magnitude and angle in series with the line upon command, subject only to equipment rating limits. With suitable electronic controls, the UPFC can cause the series-injected voltage vector to vary rapidly and continuously in magnitude and/or angle as desired. Thus, it is not only able to establish an operating point within a wide range of possible P, Q conditions on the line, but also has the inherent capability to transition rapidly from one such achievable operating point to any other.

The control of the UPFC is based upon the vector-control approach proposed by Schauder and Mehta for "advanced static var compensators" (i.e., for STATCOMs) in 1991. The term *vector,* instead of phasor, is used in this section to represent a set of three instantaneous phase variables, voltages, or currents that sum to zero. The symbols $\tilde{v}$ and $\tilde{i}$ are used for voltage and current vectors. The reader will recall that these vectors are not stationary, but move around a fixed point in the plane as the values of the phase variables change, describing various trajectories, which become circles, identical to those obtained with phasors, when the phase variables represent a balanced, steady-state condition. For the purpose of power control it is useful to view these vectors in an orthogonal coordinate system with p and q axes such that the p axis is always coincident with the instantaneous voltage vector $\tilde{v}$ and the q axis is in quadrature with it. In this coordinate system the p-axis current component, i_p, accounts for the instantaneous real power and the q-axis current component, i_q, for the reactive power. Under balanced steady-state conditions, the p-axis and q-axis components of the voltage and current vector are constant quantities. This characteristic of the described vector representation makes it highly suitable for the control of the UPFC by facilitating the decoupled control of the real and reactive current components.

The UPFC control system may, in the previously established manner, be divided functionally into internal (or converter) control and functional operation control. The internal controls operate the two converters so as to produce the commanded series injected voltage and, simultaneously, draw the desired shunt reactive current. The internal controls provide gating signals to the converter valves so that the converter output voltages will properly respond to the internal reference variables, $i_{p\text{Ref}}$, $i_{q\text{Ref}}$, and $\tilde{v}_{pq\text{Ref}}$, in accordance with the basic control structure shown in Figure 8.14. As can

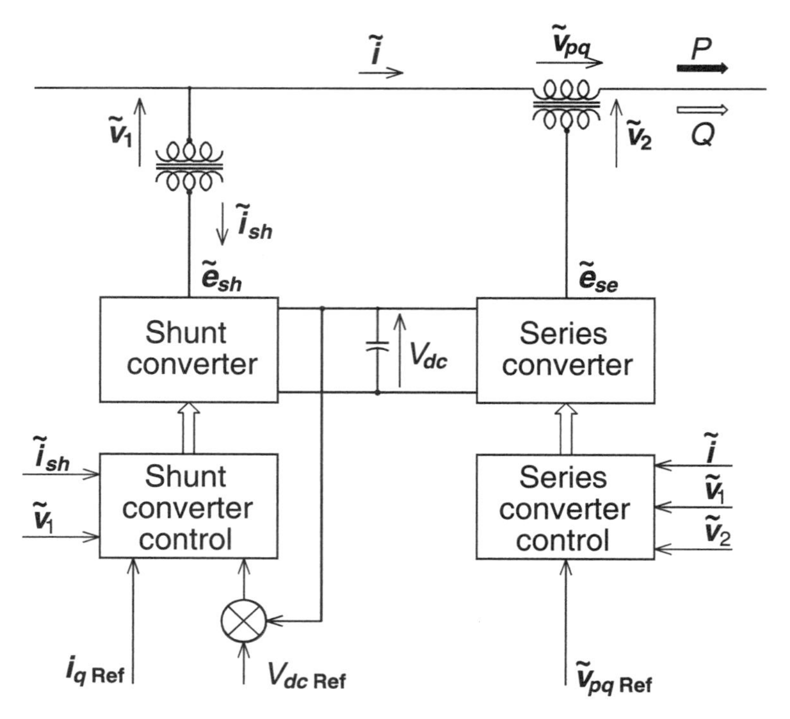

Figure 8.14 Basic UPFC control scheme.

be observed, the series converter responds directly and independently to the demand for series voltage vector injection. Changes in series voltage vector, $\tilde{v}_{pq}$, can therefore be affected virtually instantaneously. In contrast, the shunt converter operates under a closed-loop current control structure whereby the shunt real and reactive power components are independently controlled. The shunt reactive power (if this option is used, for example, for terminal voltage control) responds directly to an input demand. However, the shunt real power is dictated by another control loop that acts to maintain a preset voltage level on the dc link, thereby providing the real power supply or sink needed for the support of the series voltage injection. In other words, the control loop for the shunt real power ensures the required real power balance between the two converters. As mentioned previously, the converters do not (and could not) exchange reactive power through the link.

The external or functional operation control defines the functional operating mode of the UPFC and is responsible for generating the *internal* references, $\tilde{v}_{pq\text{Ref}}$ and $i_{q\text{Ref}}$, for the series and shunt compensation to meet the prevailing demands of the transmission system. The functional operating modes and compensation demands, represented by external (or system) reference inputs, can be set manually (via a computer keyboard) by the operator or dictated by an automatic system optimization control to meet specific operating and contingency requirements. An overall control structure, showing the *internal*, the *functional operation*, and *system optimization* controls with the internal and external references is presented in Figure 8.15.

As shown in Figure 8.15 the capability of unrestricted series voltage injection together with independently controllable reactive power exchange offered by the circuit structure of two back-to-back converters, facilitate several operating and control

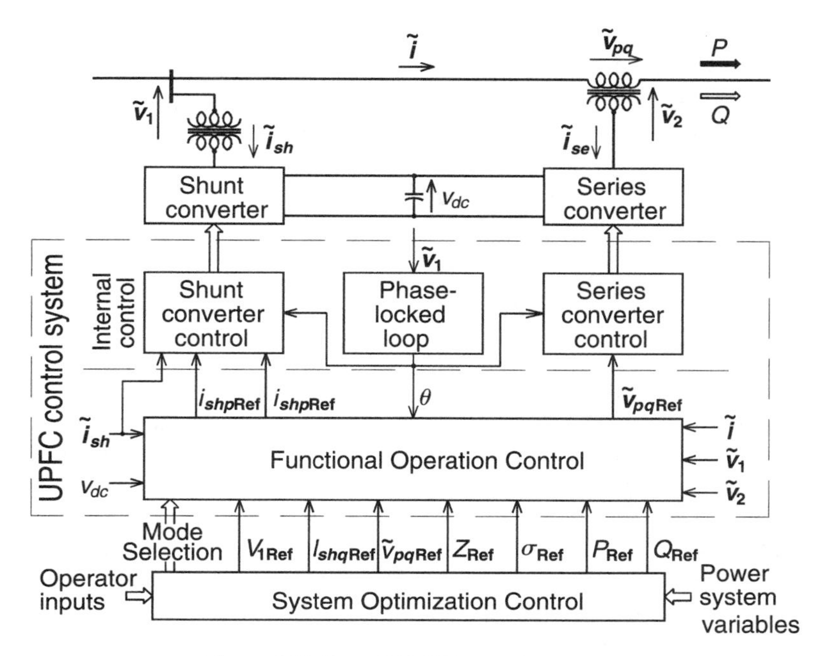

Figure 8.15 Overall UPFC control structure.

modes for the UPFC. These include the option of reactive shunt compensation and the free control of series voltage injection according to a prescribed functional approach selected for power flow control. The UPFC circuit structure also allows the total decoupling of the two converters (i.e., separating the dc terminals of the two converters) to provide independent reactive shunt compensation (STATCOM) and reactive series compensation (SSSC) without any real power exchange.

8.2.5.1 Functional Control of the Shunt Converter. The shunt converter is operated so as to draw a controlled current, $\tilde{i}_{sh}$, from the line. One component of this current, i_{shp}, is automatically determined by the requirement to balance the real power of the series converter. The other current component, i_{shq}, is reactive and can be set to any desired reference level (inductive or capacitive) within the capability of the converter. The reactive compensation control modes of the shunt converter are, of course, very similar to those commonly employed for the STATCOM and conventional static var compensator.

Reactive Power (VAR) Control Mode. In *reactive power control mode* the reference input is an inductive or capacitive var request. The shunt converter control translates the var reference into a corresponding shunt current request and adjusts the gating of the converter to establish the desired current. The control in a closed-loop arrangement uses current feedback signals obtained from the output current of the shunt converter to enforce the current reference. A feedback signal representing the dc bus voltage, v_{dc}, is also used to ensure the necessary dc link voltage.

Automatic Voltage Control Mode. In voltage control mode (which is normally used in practical applications), the shunt converter reactive current is automatically regulated to maintain the transmission line voltage to a reference value at the point of connection, with a defined droop characteristic. The droop factor defines the per unit voltage error per unit of converter reactive current within the current range of the converter. The automatic voltage control uses voltage feedback signals, usually representing the magnitude of the positive sequence component of bus voltage $\tilde{v}_1$.

8.2.5.2 Functional Control of the Series Converter. The series converter controls the magnitude and angle of the voltage vector $\tilde{v}_{pq}$ injected in series with the line. This voltage injection is, directly or indirectly, always intended to influence the flow of power on the line. However, $\tilde{v}_{pq}$ is dependent on the operating mode selected for the UPFC to control power flow. The principal operating modes are as follows in the next subsections.

DIRECT VOLTAGE INJECTION MODE. The series converter simply generates the voltage vector, $\tilde{v}_{pq}$, with the magnitude and phase angle requested by the reference input. This operating mode may be advantageous when a separate system optimization control coordinates the operation of the UPFC and other FACTS controllers employed in the transmission system. Special functional cases of direct voltage injection include those having dedicated control objectives, for example, when the injected voltage vector, $\tilde{v}_{pq}$, is kept in phase with the system voltage for voltage magnitude control, or in quadrature with it for controlled "quadrature boosting," or in quadrature with the line current vector, $\tilde{i}$, to provide controllable reactive series compensation.

BUS VOLTAGE REGULATION AND CONTROL MODE. The injected voltage vector, $\tilde{v}_{pq}$, is kept in phase with the "input" bus voltage vector $\tilde{v}_1$ and its magnitude is controlled to maintain the magnitude "output" bus voltage vector $\tilde{v}_2$ at the given reference value.

LINE IMPEDANCE COMPENSATION MODE. The magnitude of the injected voltage vector, $\tilde{v}_{pq}$, is controlled in proportion to the magnitude of the line current, $\tilde{i}$, so that the series insertion emulates an impedance when viewed from the line. The desired impedance is specified by reference input and in general it may be a complex impedance with resistive and reactive components of either polarity. A special case of impedance compensation occurs when the injected voltage is kept in quadrature with respect to the line current to emulate purely reactive (capacitive or inductive) compensation. This operating mode may be selected to match existing series capacitive line compensation in the system.

PHASE ANGLE REGULATION MODE. The injected voltage vector $\tilde{v}_{pq}$ is controlled with respect to the "input" bus voltage vector $\tilde{v}_1$ so that the "output" bus voltage vector $\tilde{v}_2$ is phase shifted, without any magnitude change, relative to $\tilde{v}_1$ by an angle specified by the reference input. One special case of phase shifting occurs when $\tilde{v}_{pq}$ is kept in quadrature with $\tilde{v}_1$ to emulate the "quadrature booster."

AUTOMATIC POWER FLOW CONTROL MODE. The magnitude and angle of the injected voltage vector, $\tilde{v}_{pq}$, is controlled so as to force such a line current vector, $\tilde{i}$, that results in the desired real and reactive power flow in the line. In automatic power flow control mode, the series injected voltage is determined automatically and continuously by a closed-loop control system to ensure that the desired P and Q are maintained despite power system changes. The transmission line containing the UPFC thus appears to the rest of the power system as a high impedance power source or sink. This operating mode, which is not achievable with conventional line compensating equipment, has far reaching possibilities for power flow scheduling and management. It can also be applied effectively to handle dynamic system disturbances (e.g., to damp power oscillations).

8.2.5.3 Stand-Alone Shunt and Series Compensation. The UPFC circuit structure offers the possibility of operating the shunt and series converters independently of each other by disconnecting their common dc terminals and splitting the capacitor bank. In this case, the shunt converter operates as a stand-alone STATCOM, and the series converter as a stand-alone SSSC. This feature may be included in the UPFC structure in order to handle contingencies (e.g., one converter failure) and be more adaptable to future system changes (e.g., the use of both converters for shunt only or series only compensation). In the stand-alone mode, of course, neither converter is capable of absorbing or generating real power so that operation only in the reactive power domain is possible. In the case of the series converter this means considerable limitation in the available control modes. Since the injected voltage must be in quadrature with the line current, only controlled reactive voltage compensation or reactive impedance emulation is possible for power flow control.

8.2.6 Basic Control System for *P* and *Q* Control

As illustrated in Figure 8.15, the UPFC has many possible operating modes. However, in order to keep focused on the subject of this chapter, only the automatic power flow control mode, providing independent control for real and reactive power flow in the line, will be considered in further detail. This control mode utilizes most of the unique capabilities of the UPFC and it is expected to be used as the basic mode in the majority of practical applications, just as the shunt compensation is used normally for automatic voltage control. Accordingly, block diagrams giving greater details of the control schemes are shown for the series converter in Figure 8.16(a) and for the shunt converter in Figures 8.16(b) and (c) for operation in these modes. For clarity

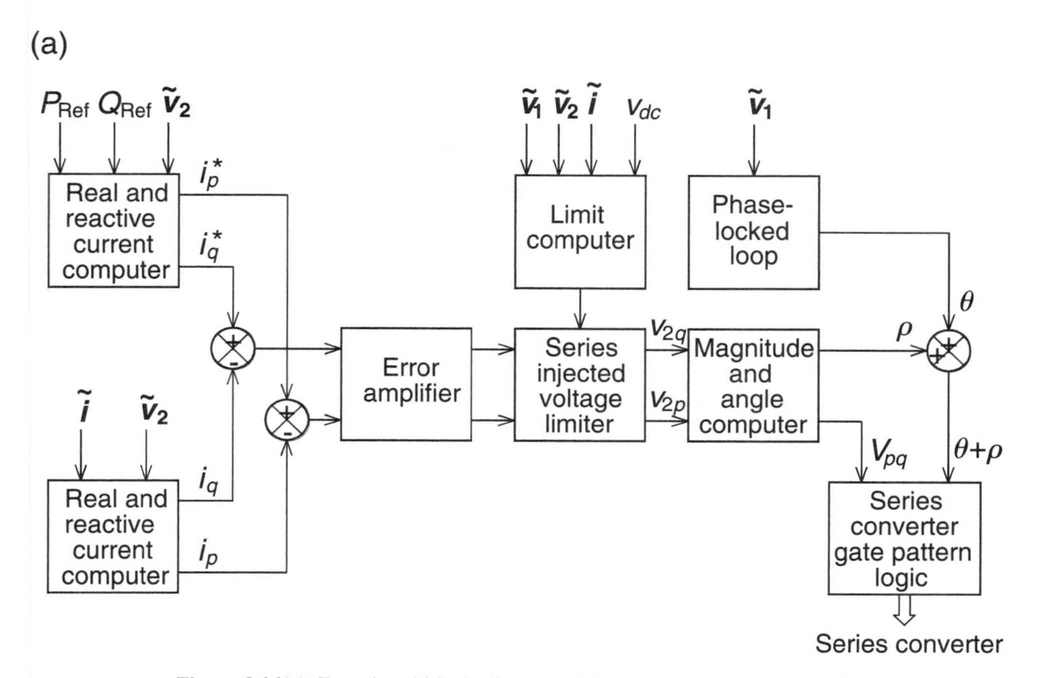

Figure 8.16(a) Functional block diagram of the series converter control.

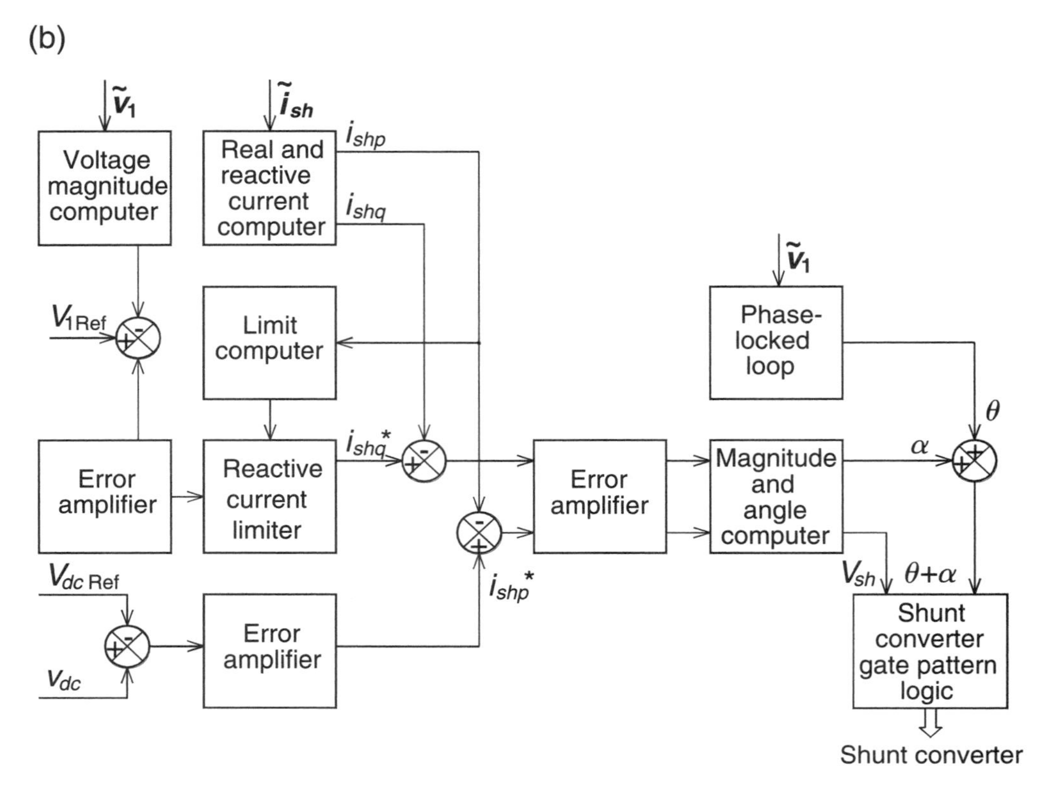

Figure 8.16(b) Functional block diagram of the shunt converter control for operation with constant dc link voltage.

and conciseness, only the most significant features are shown in these figures while less important signal processing and limiting functions have been omitted.

The control scheme shown in Figure 8.16(a) assumes that the series converter can generate output voltage with controllable magnitude and angle at a given dc bus voltage. The control scheme for the shunt converter shown in Figure 8.16(b) also assumes that the converter can generate output voltage with controllable magnitude and angle. However, this may not always be the case, since the converter losses and harmonics can be reduced by allowing the dc voltage to vary according to the prevailing shunt compensation demand. Although the variation of the dc voltage inevitably reduces the attainable magnitude of the injected series voltage when the shunt converter is operated with high reactive power absorption, in many applications this may be an acceptable trade-off. In the control scheme for the shunt converter shown in Figure 8.16(c) the magnitude of the output voltage is directly proportional to the dc voltage and only its angle is controllable. With this control scheme the dc capacitor voltage is changed (typically up to the $\pm 12\%$ range) by momentary angle adjustment that forces the converter to exchange real power with the ac system to meet the shunt reactive compensation requirements.

As shown in Figure 8.16(a) the automatic power flow control for the series converter is achieved by means of a vector control scheme that regulates the transmission line current using a synchronous reference frame (established with an appropriate

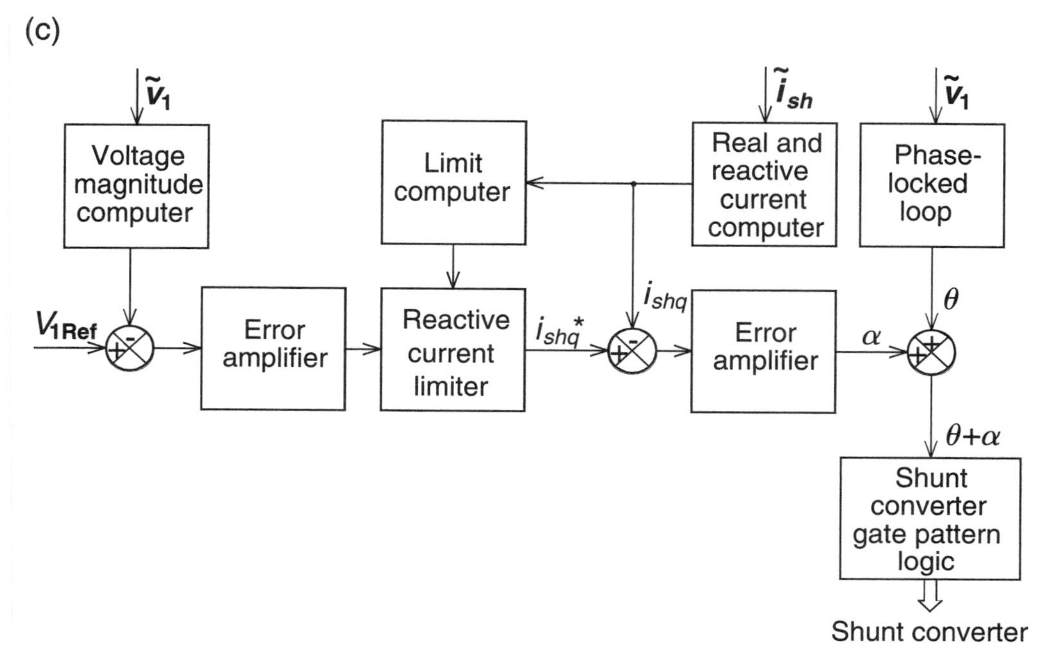

Figure 8.16(c) Functional block diagram of the shunt converter control for operation with varying dc link voltage.

phase-locked loop producing reference angle θ) in which the control quantities appear as dc signals in the steady state. The appropriate reactive and real current components, i_q^* and i_p^*, are determined for a desired P_{Ref} and Q_{Ref}. These are compared with the measured line currents, i_q and i_p, and used to drive the magnitude and angle of the series converter voltage, V_{pq} and ρ, respectively. Note that a voltage limiter in the forward path is employed to enforce practical limits, resulting from system restrictions (e.g., voltage and current limits) or equipment and component ratings, on the series voltage injected.

A vector control scheme is also used for the shunt converter, as illustrated by the block diagrams of Figures 8.16(b) and (c). In this case the controlled quantity is the current $\tilde{\boldsymbol{i}}_{sh}$ drawn from the line by the shunt converter. The real and reactive components of this current, however, have a different significance. For the scheme of Figure 8.16(b), the reference for the reactive current, i^*_{shq}, is generated by an outer voltage control loop, responsible for regulating the ac bus voltage, and the reference for the real power bearing current, i^*_{shp}, is generated by a second voltage control loop that regulates the dc bus voltage. In particular, the real power negotiated by the shunt converter is regulated to balance the dc power from the series converter and maintain a desired bus voltage. The dc voltage reference, $V_{dc\text{Ref}}$, may be kept substantially constant. In the scheme shown in Figure 8.16(c), the outer voltage loop regulates the ac bus voltage and also controls the dc capacitor voltage. This outer loop changes the angle α of the converter voltage with respect to the ac bus voltage until the dc capacitor voltage reaches the value necessary to achieve the reactive compensation demanded. The closed-loop controlling the output of the series converter is responsible for maintaining the magnitude of the injected voltage, $\tilde{\boldsymbol{v}}_{pq}$, in spite of the variable dc voltage.

The most important limit imposed on the shunt converter is the shunt reactive current which is a function of the real power being passed through the dc bus to support the real power demand of the series converter. This prevents the shunt converter current reference from exceeding its maximum rated value.

The control block diagrams shown in Figure 8.16 represent only a selected part of the numerous control algorithms needed if additional operating modes of the UPFC are also to be implemented. The block diagrams omit control functions related to converter protection, as well as sequencing routines during operating mode changes and start-up and shutdown procedures.

8.2.7 Dynamic Performance

The dynamic performance of the UPFC is illustrated by real-time voltage and current waveforms obtained in a representative TNA (Transient Network Analyzer) hardware model shown schematically by a simplified single-line diagram in Figure 8.17. The simple, two-bus power system modeled includes the sending-end and receiving-end generators with two parallel transmission lines which are represented by lumped reactive impedances. One of the lines is controlled by a model UPFC. The converters and the magnetic structure of the UPFC model accurately represent a 48-pulse structure used in an actual transmission application (refer to Chapter 10). The UPFC power circuit model is operated by the actual control used in the full scale system. The performance of the UPFC is demonstrated for power flow control with operation under power system oscillation and transmission line faults.

8.2.7.1 Power Flow Control. The performance of the UPFC for real and reactive power flow control is demonstrated with the objectives of keeping the sending- and receiving-end bus voltages constant (the same 1.0 p.u. magnitude, and a fixed

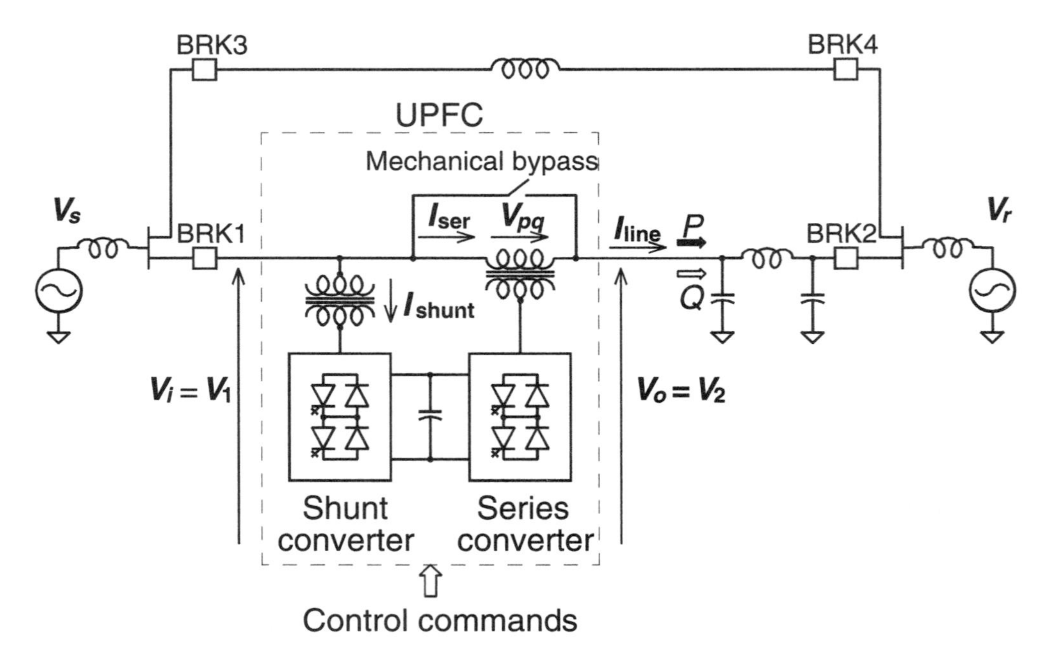

Figure 8.17 Simplified schematic of the power system and UPFC model used for TNA evaluation.

transmission angle) and operating the UPFC in the automatic power flow control mode. As established previously, in this operating mode the UPFC regulates the real and reactive line power to given reference values. As illustrated in Figure 8.18, the UPFC, via appropriate P_{Ref} and Q_{Ref} inputs to its control, is instructed to perform a series of step changes in rapid succession. First, P is increased, then Q, followed by a series of decreases, ending with a negative value of Q. The waveforms, showing the injected voltage $v_{pq}(t)$, the system voltage at the two ends of the insertion transformer, $v_i(t) = v_1(t)$ and $v_0(t) = v_2(t) = v_1(t) + v_{pq}(t)$, and the line current, illustrate clearly the operation of the UPFC. It can be observed that the closed-loop controlled UPFC easily follows the P_{Ref} and Q_{Ref} references, changing the power flow in approximately a quarter cycle. (Note that in a real system the power flow would typically be changed much more gradually in order to avoid possible dynamic disturbances.)

8.2.7.2 Operation Under Power System Oscillation. The unique versatility of the UPFC for power flow control can well be demonstrated when the power system is subjected to dynamic disturbances resulting in power oscillations. Since the UPFC actually controls the effective sending-end voltage, it is capable of forcing a desired power flow on the transmission line under dynamic system conditions as well as in

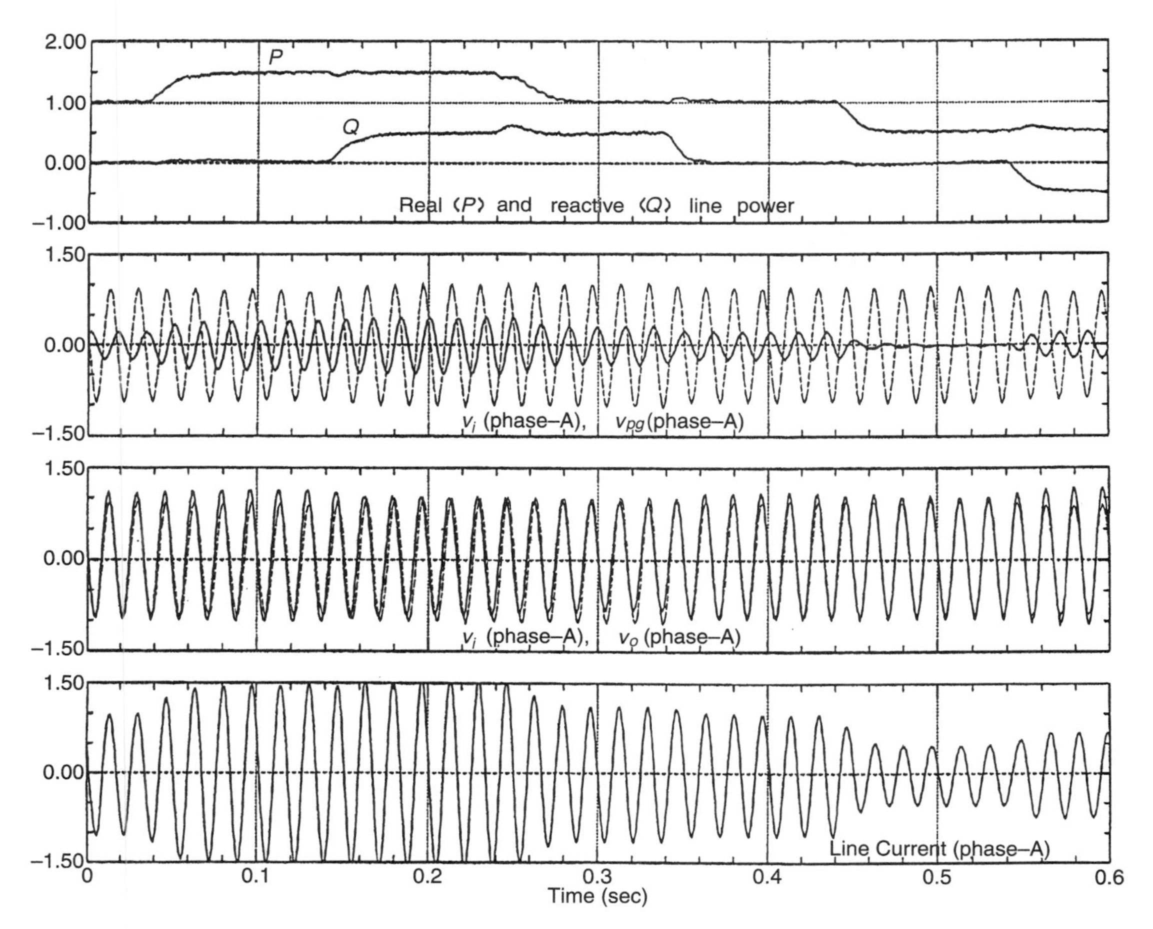

Figure 8.18 TNA simulation results for step-like changes in P and Q demands.

the steady state. This capability can be used in several different ways to meet system requirements. If constant power flow is to be maintained, the UPFC will act to provide it, even if the conditions on the sending- and receiving-end buses are varying. In essence, the UPFC will dynamically decouple the two (sending- and receiving-end) buses. If preferred, the UPFC can be commanded to force an appropriately varying power level on the line that will effectively damp the prevailing power oscillation.

To demonstrate the potential of the UPFC under dynamic (oscillatory) conditions, the TNA system model was provided with a receiving-end (V_r) bus programmed to have a damped second-order phase angle response characteristic of a generator with a large inertia. The simple algorithm governing this mechanism assumes a defined source of mechanical power to the generator supporting the V_r bus and matches this against the electrical power being delivered from the bus to the two transmission lines. Excess mechanical power causes an acceleration with a resultant phase angle advance and, vice versa, insufficient mechanical power causes deceleration with phase angle retardation. In the steady state, the bus angle assumes a value in which the electrical and mechanical powers are exactly equal. The modified system model is shown in Figure 8.19.

To initiate a power oscillation in this simple system model, a fault was applied for a duration of several cycles through an impedance to ground at the V_r bus as shown in Figure 8.19, simulating a distant fault condition. Three cases are presented in Figures 8.20 through 8.22 to illustrate dynamic response of the UPFC under various operating modes. The initial conditions for all three cases are identical, with the mechanical power request for generator V_r programmed to produce a 1.0 p.u. real power flow from V_s to V_r. Line impedances are such that with zero compensation ($v_{pq} = 0$), the UPFC line carries 0.75 p.u. real power while the remaining 0.25 p.u. real power flows through the parallel line. The UPFC is then operated to obtain 1.0 p.u. real power flow on its line, so that no power is transferred through the parallel

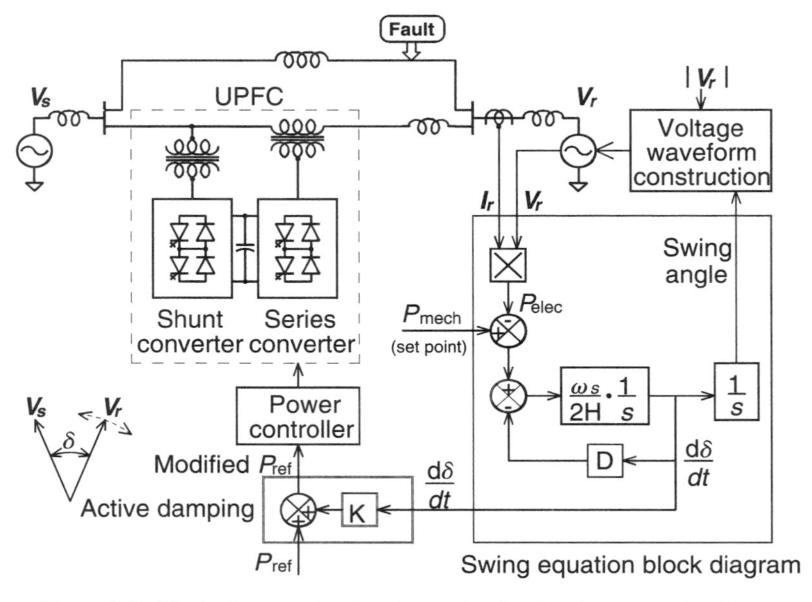

Figure 8.19 Block diagram showing the "swing bus" and control algorithms for power oscillation damping.

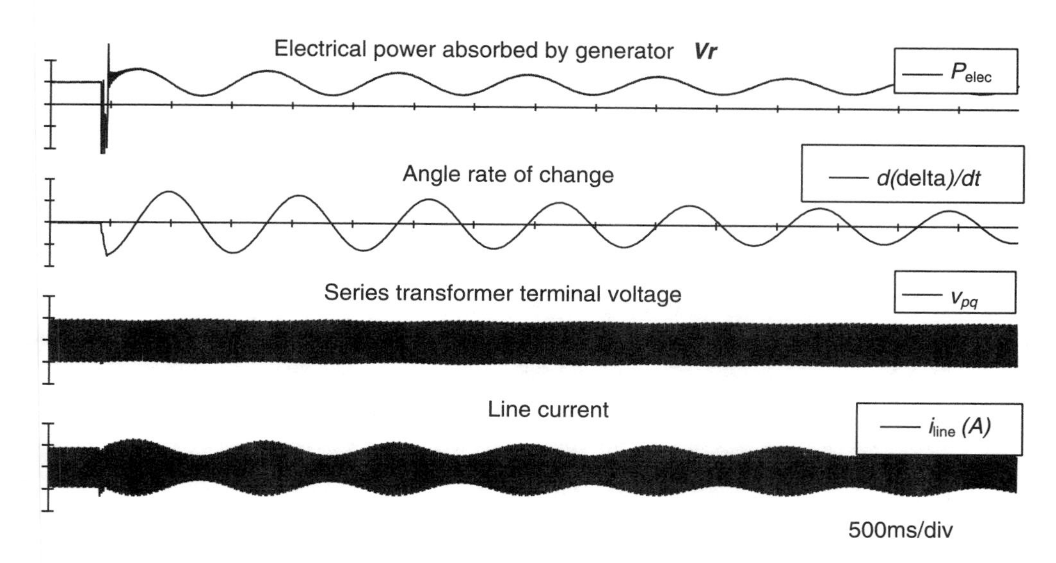

Figure 8.20 Power flow control during power oscillation with the UPFC in *direct voltage injection mode* (UPFC remains neutral to damping).

line. In each of the three cases, the UPFC achieves this initial power flow using a different operating mode.

The results of Figure 8.20 show the UPFC in *direct voltage injection mode* where the magnitude and angle of the injected voltage are adjusted by the operator (or by the *system optimization control*) until the desired power flow is achieved. In this operating mode, the UPFC has no effect on the dynamic performance of the system due to the applied fault, appearing only as a voltage source of fixed magnitude and angle added to the sending-end voltage.

The results of Figure 8.21 show the UPFC in *automatic power flow control* mode with a constant reference. Here the UPFC holds the power flow on its line constant while the oscillating power required to synchronize the generators is carried entirely by the parallel line. Note the change in swing frequency since the impedance of the UPFC's line no longer dynamically couples the two generators.

In the final case, shown in Figure 8.22, the UPFC is again in *automatic power flow control mode*, but with active damping control. The oscillation damping control algorithm is shown in Figure 8.19, where the rate of change of the differential phase angle between the sending-end and receiving-end buses ($d\delta/dt$) is sensed and fed into the real power command, P_{Ref}, for the UPFC with the correct polarity and an appropriate gain. Clearly, it may be difficult to obtain the feedback for this algorithm in a real system, but the purpose here is to show the powerful capability of the UPFC to damp system oscillations. However, it seems plausible that any of those system variables available at, or transmittable to a given location of the power system, which were found in practice to be effective inputs to other types of power flow controllers (e.g., TCSC) for power oscillation damping, would be applicable to the UPFC as well.

8.2.7.3 Operation Under Line Faults. The current of the compensated line flows through the series converter of the UPFC. Depending on the impedance of the line and the location of the system fault, the line current during faults may reach a magnitude which would far exceed the converter rating. Under this condition the

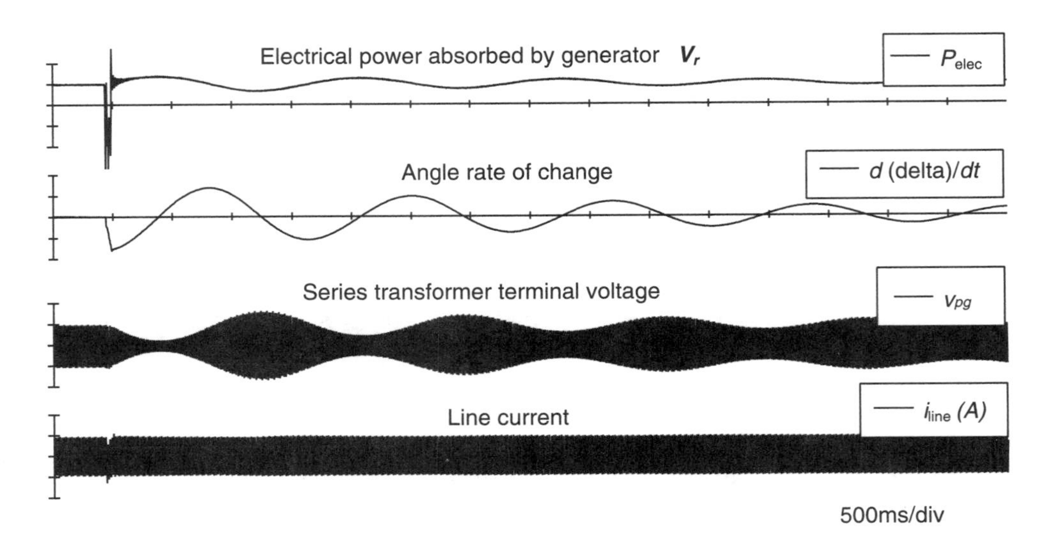

Figure 8.21 Power flow control during power oscillation with the UPFC in *automatic power flow control* with constant reference (UPFC maintains constant power flow in the controlled line).

UPFC would typically assume a bypass operating mode. In this mode the injected voltage would be reduced to zero and the line current, depending on its magnitude, would be bypassed through either the converter valves, electronically reconfigured for terminal shorting, or through a separate high current thyristor valve. For the contingency of delayed fault clearing, a mechanical bypass breaker would also be normally employed.

Two cases of line faults are considered in this section. The first is an *external*

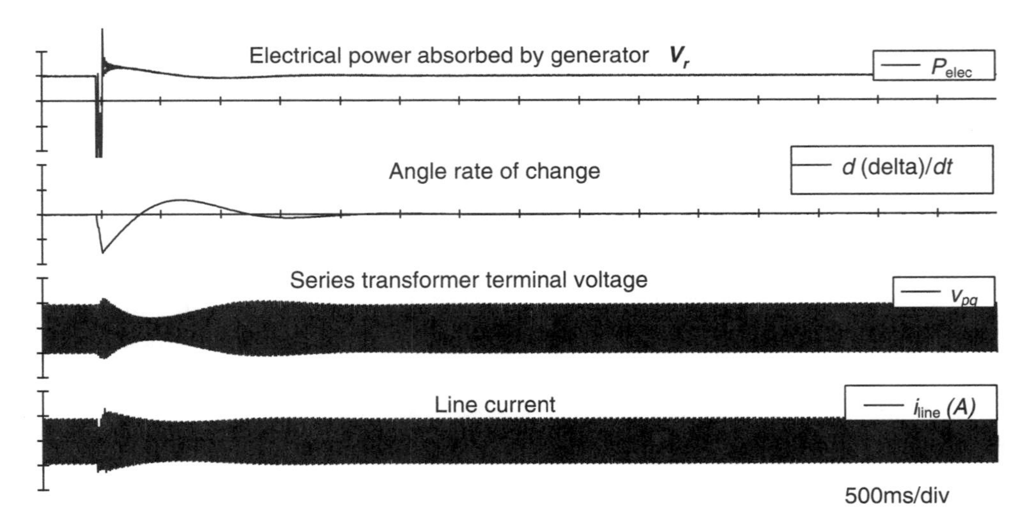

Figure 8.22 Power flow control during power oscillation with an active *damping control* added to the *automatic power flow control* of the UPFC (UPFC acts to damp the power oscillation).

fault (fault is on the non-UPFC line) with normal clearing time. As illustrated in Figure 8.23, phase A of the parallel line is faulted to ground at point **Fault #1**, effectively at the stiff sending-end bus. The fault path has zero impedance, and prior to the onset, the UPFC is in automatic power flow control mode controlling the power on its line at $P = 1.0$ p.u. and $Q = 0.02$ p.u. Six cycles after the fault, breakers BRK3 and BRK4 open, clearing the fault and restoring voltage to the UPFC. The breakers reclose nine cycles after opening.

Resulting waveforms for this case are given in Figure 8.24. When the UPFC senses the overcurrent on faulted phase A, it immediately (at point 1 in the figure) activates the electronic bypass to protect the series converter. During the fault, the shunt converter may, if desired, remain operational to supply reactive compensation. However, the gross voltage unbalance caused by the single line-to-ground fault, may cause considerable distortion on the compensating currents. These waveforms show normal fault clearing conditions, where the fault current is conducted by the electronic bypass switch. Should the fault clearing be delayed beyond the thermal capacity of the electronic switch, a mechanical bypass would be initiated. If the series transformer is mechanically bypassed, a specific reinsertion sequence for the UPFC would be required, as shown in the next fault case.

Six cycles after initiation the fault is cleared (at point 2) when BRK3 and BRK4 open, restoring balanced voltage to the line with the UPFC. Line conditions quickly return to normal and the UPFC responds by removing the electronic bypass and immediately returning to the pre-fault power flow control mode (point 3). Breakers BRK3 and BRK4 reclose with no noticeable effect (point 4).

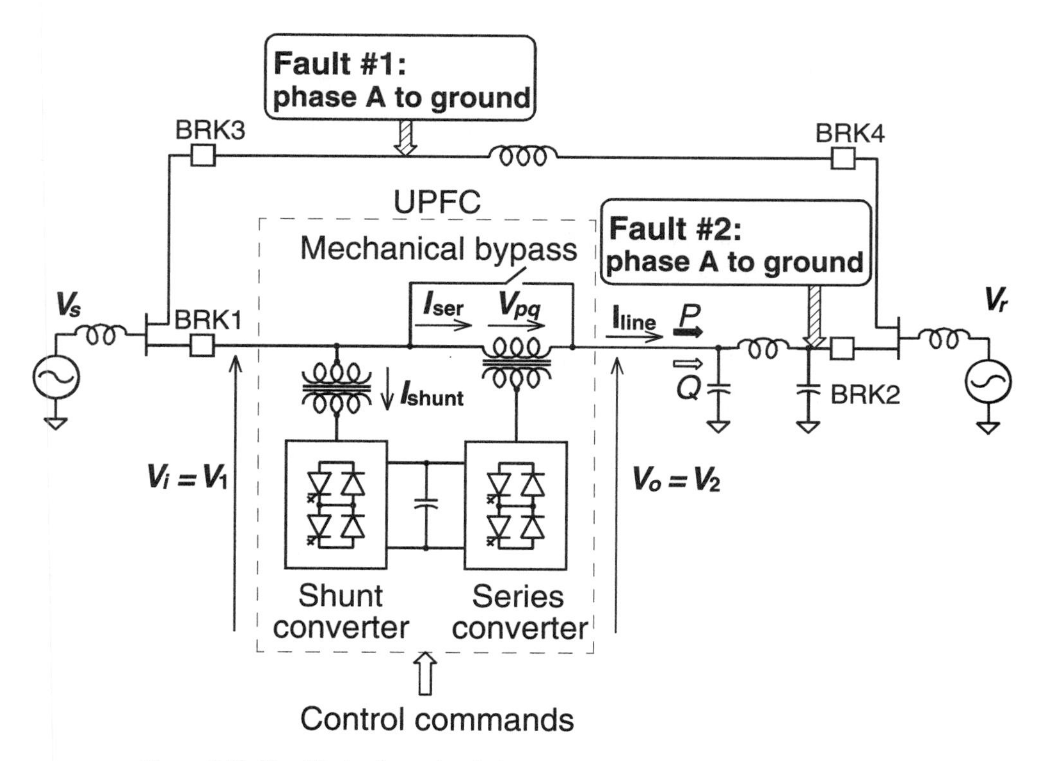

Figure 8.23 Simplified schematic of the power system and UPFC TNA model showing the locations of *external* and *internal* system faults.

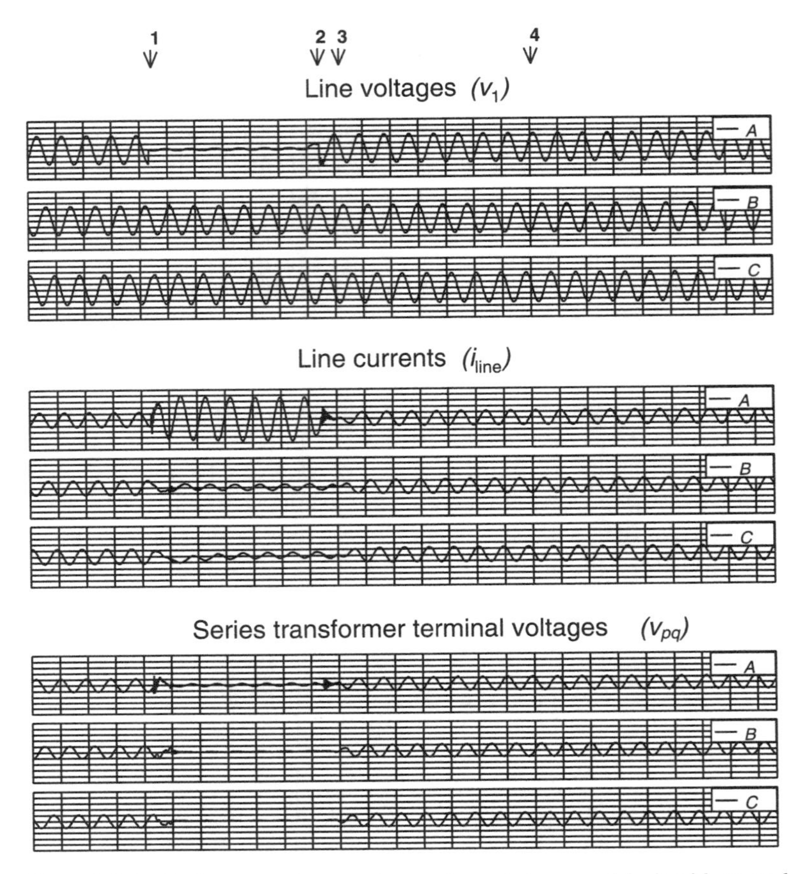

Figure 8.24 UPFC response to an external phase A to ground fault with normal clearing.

The second fault case represents an internal fault (i.e., fault is on the UPFC line) with delayed fault clearing. This case consists of a single line-to-ground internal fault, with phase A faulted to ground at point **Fault #2**, the receiving-end bus in the single-line diagram of Figure 8.23. The fault path has zero impedance and, as before, at the onset the UPFC is operating in automatic power flow control mode with P = 1.0 p.u. and Q = 0.02 p.u. Fault clearing by opening breakers BRK1 and BRK2 is delayed until 12 cycles after fault initiation. The breakers reclose nine cycles after opening.

Resulting waveforms for this case are given in Figure 8.25. As indicated previously, when the UPFC senses the overcurrent on faulted phase A, it immediately (at point 1 in the figure) activates the electronic bypass to protect the series converter. However, these waveforms illustrate the case in which the fault clearing is assumed delayed beyond the thermal capacity of the electronic switch and a mechanical bypass is initiated (point 2). Approximately four cycles later the mechanical bypass across the primary of the insertion transformer closes and carries the line current, thus relieving the electronic bypass switch (point 3).

A short time later (point 4), breakers BRK1 and BRK2 open to clear the fault. This action leaves the UPFC in its mechanically bypassed state on an isolated line. Once the fault has been cleared, the breakers reclose, restoring voltage to the UPFC

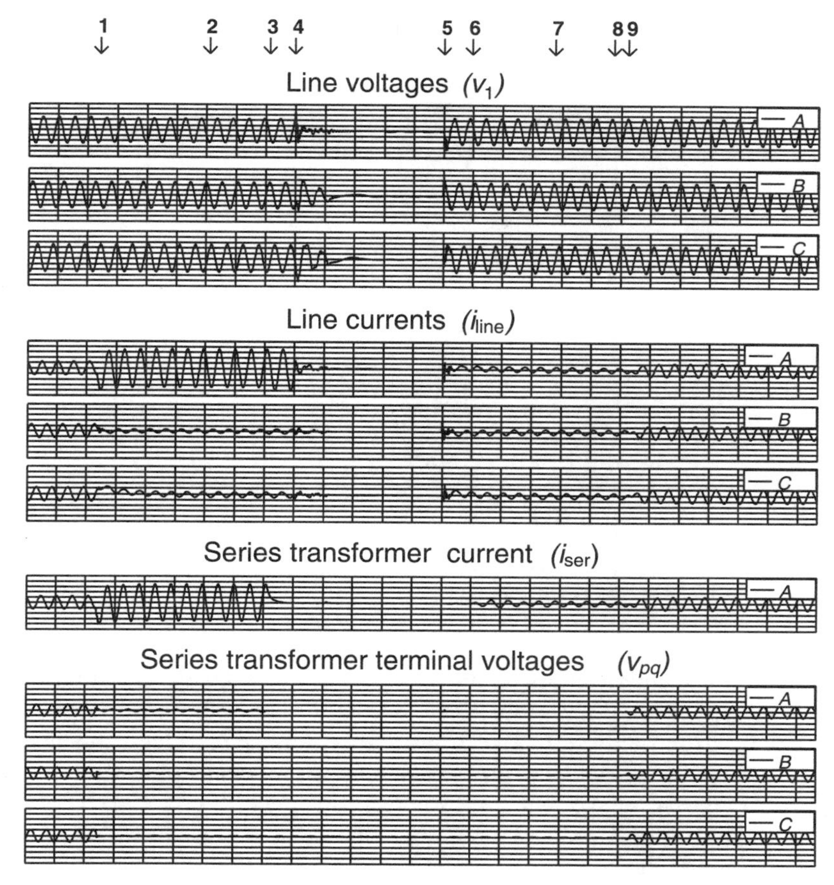

Figure 8.25 UPFC response to an internal phase A to ground fault with delayed clearing.

(point 5). Line conditions quickly return to normal and the UPFC is able to initiate reinsertion of the series transformer (point 6). During the reinsertion procedure the series converter forces current (i_{ser}) through the series transformer to match to the line current (i_{line}), driving the current through the mechanical bypass to zero in order to achieve a transient-free reinsertion. At this point the value of the series-injected voltage is fixed and a signal is sent to open the mechanical bypass (point 7). Several cycles later the mechanical bypass, carrying zero current, opens (point 8). The "bumpless" reinsertion of the series transformer into the line is complete.

Once the series transformer is in the line again, the UPFC can remain in voltage insertion mode or transition to any other desired post-fault operating mode. To illustrate this capability, this case shows the UPFC returning to the pre-fault operating mode (automatic power flow control) and reference values (point 9).

8.2.8 Hybrid Arrangement: UPFC with a Phase Shifting Transformer

As shown in Figures 8.3 and 8.5, the voltage injection of the UPFC is in a circular region around the end of the (sending-end) transmission voltage phasor. This means that the UPFC voltage injection in general results in an advance or a retard of the

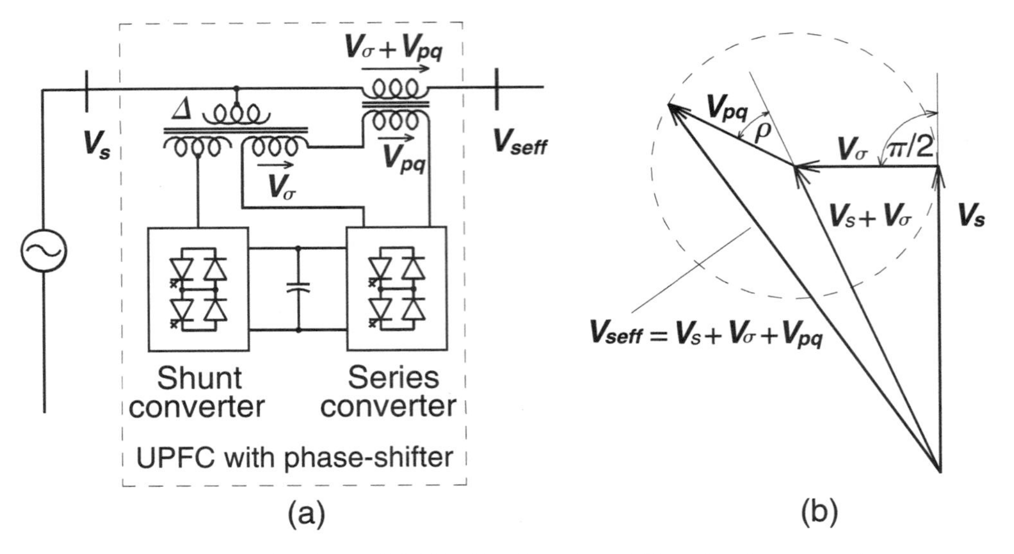

Figure 8.26 Basic scheme of UPFC with a fixed phase shifter (a) and corresponding phasor diagram (b).

prevailing transmission angle. In other words, the UPFC has two equal half-circle segments as operating regions, one characterized by the advancement, the other by the retardation of the effective transmission angle. Although this general capability of the UPFC may well be utilized in many practical applications, there are other applications in which voltage injection resulting in unidirectional change, advancement, or retardation of the transmission angle is satisfactory or in which unequal operating regions for advancement and retardation are preferred. For example, if the attainable transmission angle is too small for the desired power transmission, the UPFC would have to provide first an appropriate phase angle advancement to establish the correct steady-state operating point around which the control of real and reactive power would be executed under the prevailing system conditions. In these applications the converters of the UPFC are not utilized effectively because with a given MVA rating they could control, for example, the transmission angle over the range of $-\sigma_{max} \leq \sigma \leq +\sigma_{max}$, but they are actually used to control the transmission angle over either the positive range of $0 \leq \sigma_{max}$ or the negative range of $-\sigma_{max} \leq 0$, i.e., they are used to control up to a maximum angular change of only σ_{max} of the available maximum angular change of $2\sigma_{max}$. Since the required MVA rating is proportional to the control range, it is easy to see that the UPFC converters are at best 50% utilized. Another consideration in this application is that a significant portion of the UPFC rating may be used up for steady-state angle control, which could be provided by more economical means.

The operating requirements for unequal operating range and steady-state angular shift can be satisfied if the UPFC is combined with a phase-shifting transformer providing fixed or selectable angle of advancement or retardation.

A possible circuit arrangement to implement this hybrid arrangement is shown schematically in Figure 8.26(a) with the corresponding phasor diagram in Figure 8.26(b). The overall circuit arrangement includes the usual UPFC configuration with two coupling transformers, one connecting Converter 1 in parallel to the line and the other one Converter 2 in series with the line. This arrangement, in its simplest form, however, also includes an additional winding on the secondary (converter) side of each phase of the shunt-connected coupling transformer. The phase shifting could also be accomplished by a separate transformer. The transformer connection is such (e.g., delta-connected primary) that the voltage obtained at the phases a, b, and c of the secondary windings are in quadrature with the phase a, b, and c phase-to-neutral primary (line) voltages. Considering phase a [refer to Figure 8.26(a)], it is seen that this winding is connected in series with the secondary winding of the series-connected coupling transformer and the phase a output of Converter 2. As a result, the voltage injected in series with the line to control the flow of power through it is the vectorial sum of the fixed voltage provided by the shunt-connected coupling transformer and the controllable output voltage of Converter 2.

The operation of the hybrid power flow controller is summarized with reference to the phasor diagram shown in Figure 8.26(b). As explained above, the total injected voltage, used to control the power flow in the line is made up of two components: voltage component $\boldsymbol{V}_\sigma$, which is the fixed quadrature voltage provided by the shunt connected transformer to advance (or retard) the existing transmission angle by a fixed angle σ, and voltage component $\boldsymbol{V}_{pq}$, which is the controllable component provided by the UPFC. The magnitude of $\boldsymbol{V}_{pq}$ is variable in the range of $0 \leq V_{pq} \leq V_{pq\max}$ $(= V_\sigma)$ and its angle ρ is the range of $0 \leq \rho \leq 2\pi$ with respect to the fixed $\pi/2$ angle of $\boldsymbol{V}_\sigma$. The magnitude and angle of the controlled transmission voltage $\boldsymbol{V}_{seff}$ are obtained by vectorially adding the total injected voltage $\boldsymbol{V}_\sigma + \boldsymbol{V}_{pq}$ to the existing (sending-end) voltage $\boldsymbol{V}_s$. In the hybrid arrangement the circular operating region of the UPFC is centered around the end of the voltage phasor $\boldsymbol{V}_\sigma$ providing in effect a fixed angular shift for voltage phasor $\boldsymbol{V}_s$ and, with the stipulation of $V_{pq\max} = V_\sigma$, the total control region allows only unidirectional change (retardation or advancement) for the transmission angle.

The hybrid arrangement within its operating region maintains the flexibility of functional control characterizing the UPFC. For example, series reactive line compensation is accomplished in Figure 8.27(a) by choosing the angle ρ of voltage $\boldsymbol{V}_{pq}$ so that

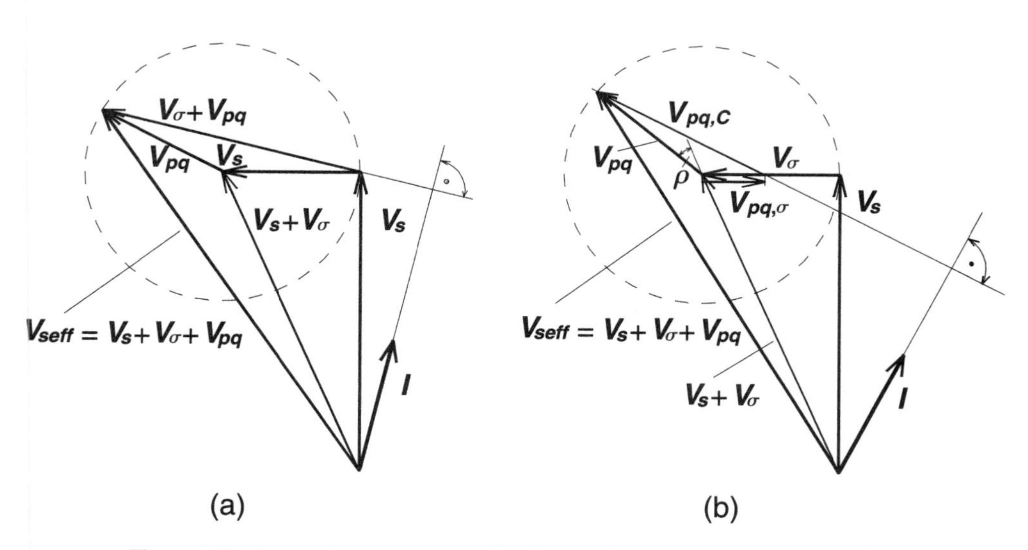

Figure 8.27 Phasor diagrams showing the UPFC with the phase shifter providing series capacitive compensation without (a) and with (b) angle control.

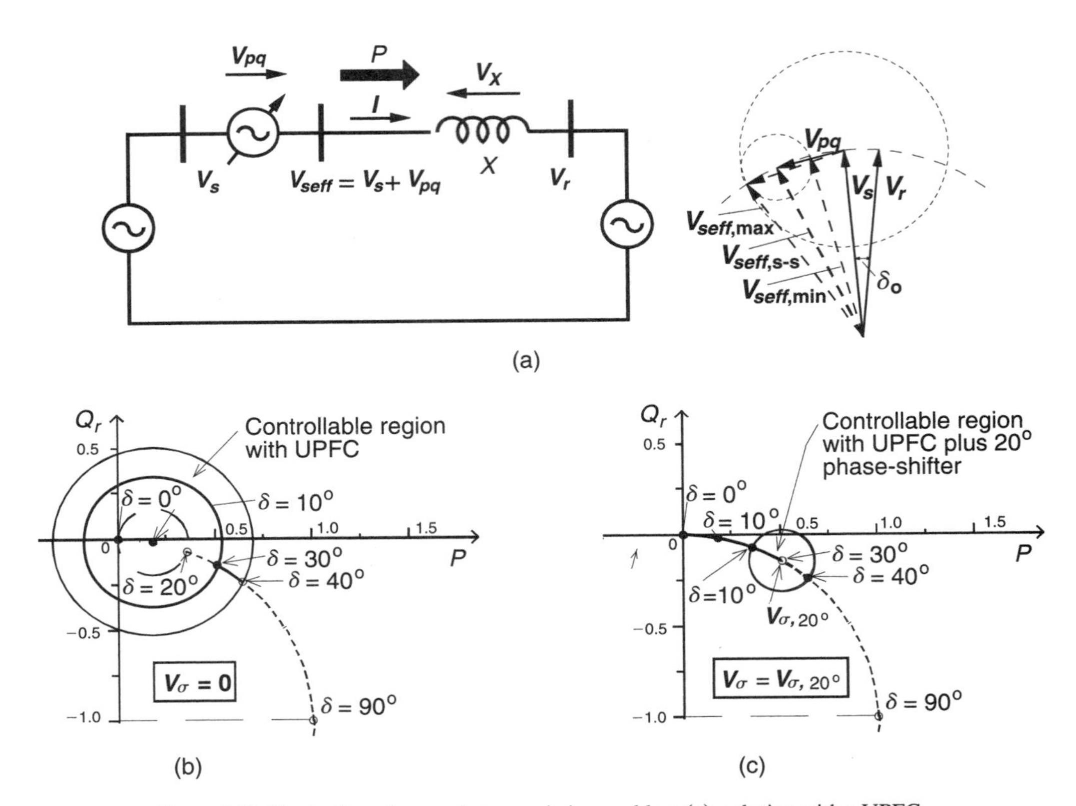

Figure 8.28 Illustration of example transmission problem (a), solution with a UPFC (b), and solution with a combined UPFC and a fixed phase shifter (c).

the total injected voltage ($V_\sigma + V_{pq}$), or a component thereof, is in quadrature with the line current. A general case for transmission angle control and series compensation for line impedance control is shown in Figure 8.27(b), where voltage component $V_{pq,\sigma}$ of V_{pq} is used to cancel part of the fixed quadrature voltage V_σ, and voltage component $V_{pq,C}$ is used to provide series reactive compensation for the existing reactive line impedance.

The benefits of the hybrid arrangement for certain applications can well be illustrated in the $\{Q_r, P\}$ plane. Consider the simple two-machine system, UPFC controlled intertie, shown in Figure 8.28(a). Assume that the attainable transmission angle is 10° and the required angle for the desired steady-state transmission is 30°. The control range estimated to handle load variations and dynamic disturbances is ±10°.

The above requirements can be met with a single UPFC of appropriate rating as illustrated by the relevant Q_r versus P plots in Figure 8.28(b). The circular arc drawn from the center identified by coordinates $Q_r = -1.0$ and $P = 0.0$ represents the Q_r versus P relationship with the uncompensated intertie line, which defines the transmitted power at 10° angle. The operating region of the UPFC is defined by a circle with a center at the 10° point of the Q_r versus P circular arc characterizing the uncompensated line. The radius of this operating region is determined by the distance between the 10° and 40° operating points of the uncompensated system defined by the stipulated operational requirements. As can be observed in Figure 8.28(b), the operating region thus obtained for the UPFC would facilitate very broad operating flexibility in controlling the transmitted real power from zero to maximum and beyond, while keeping the reactive power demand at the receiving end not only at zero but, if desired, varying it over a wide, lag to lead, range. Of course, this flexibility would come at a cost. The required *VA* rating of the UPFC would be large, about 0.5 p.u. of the maximum power transmitted.

The transmission requirements of the intertie can also be met with a more economical hybrid arrangement comprising a UPFC with a fixed 20 degree phase shift (provided by an integrated or separate transformer phase shifter) as illustrated by the relevant Q_r versus P plots in Figure 8.28(c). The phase shifter (assumed ideal for this example) increases the effective transmission angle from the prevailing 10 degrees to 30 degrees, which identifies the center of the UPFC's operating region. The radius of this circular operating region is determined by the relatively short distance between the 30 degree and 40 degree operating points on the Q_r versus P circular arc characterizing the uncompensated line. With the thus defined operating region, the latitude of transmission control is reduced, but is still significant: The real power can be controlled about ±35% around the steady-state value and the reactive power demand on the receiving end can be kept at zero over this range. However, the *VA* rating of the UPFC is only one third of that required in the previous case (i.e., about 0.167 p.u. vs. 0.5 p.u.).

8.3 THE INTERLINE POWER FLOW CONTROLLER (IPFC)

We have seen that the Unified Power Flow Controller is capable of independently controlling both the real and reactive power flow in the line. This capability of the UPFC is facilitated by its power circuit which is basically an ac-to-ac power converter, usually implemented by two back-to-back dc-to-dc converters with a common dc voltage link. The output of one converter is coupled in series, while the output of the other in shunt with the transmission line. With this arrangement, the UPFC can inject a fully controllable voltage (magnitude and angle) in series with the line and support the resulting generalized real and reactive compensation by supplying the real power required by the series converter through the shunt-connected converter from the ac bus.

The UPFC concept provides a powerful tool for the cost-effective utilization of individual transmission lines by facilitating the independent control of both the real and reactive power flow, and thus the maximization of real power transfer at minimum losses, in the line.

The Interline Power Flow Controller (IPFC), proposed by Gyugyi with Sen and Schauder in 1998, addresses the problem of compensating a number of transmission lines at a given substation. Conventionally, series capacitive compensation (fixed, thyristor-controlled or SSSC-based) is employed to increase the transmittable real power over a given line and also to balance the loading of a normally encountered multiline transmission system. However, independent of their means of implementation, series reactive compensators are unable to control the reactive power flow in, and thus the proper load balancing of, the lines. This problem becomes particularly evident in those cases where the ratio of reactive to resistive line impedance (X/R) is relatively low. Series reactive compensation reduces only the effective reactive impedance X and, thus, significantly decreases the effective X/R ratio and thereby increases the reactive power flow and losses in the line. The IPFC scheme, together with independently controllable reactive series compensation of each individual line, provides a capability to directly transfer real power between the compensated lines.

This capability makes it possible to: equalize both real and reactive power flow between the lines; reduce the burden of overloaded lines by real power transfer; compensate against resistive line voltage drops and the corresponding reactive power demand; and increase the effectiveness of the overall compensating system for dynamic disturbances. In other words, the IPFC can potentially provide a highly effective scheme for power transmission management at a multiline substation.

8.3.1 Basic Operating Principles and Characteristics

In its general form the Interline Power Flow Controller employs a number of dc-to-ac converters each providing series compensation for a different line. In other words, the IPFC comprises a number of Static Synchronous Series Compensators. However, within the general concept of the IPFC, the compensating converters are linked together at their dc terminals, as illustrated in Figure 8.29. With this scheme, in addition to providing series reactive compensation, any converter can be controlled to supply real power to the common dc link from its own transmission line. Thus, an overall surplus power can be made available from the under utilized lines which then can be used by other lines for real power compensation. In this way, some of the converters, compensating overloaded lines or lines with a heavy burden of reactive power flow, can be equipped with full two-dimensional, reactive and real power control capability, similar to that offered by the UPFC. Evidently, this arrangement mandates the rigorous maintenance of the overall power balance at the common dc terminal by appropriate control action, using the general principle that the underloaded lines are to provide help, in the form of appropriate real power transfer, for the overloaded lines.

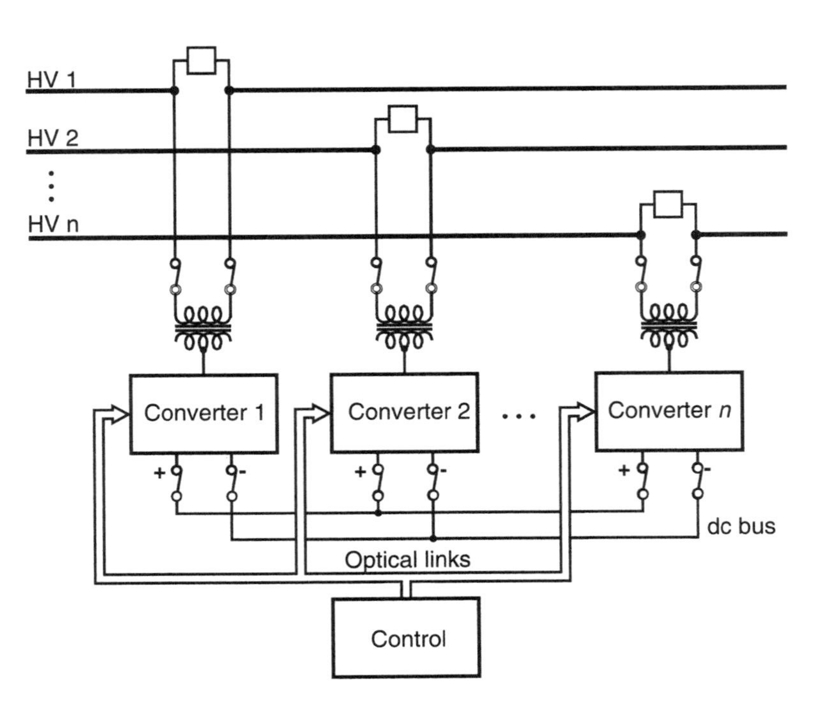

Figure 8.29 Interline Power Flow Controller comprising *n* converters.

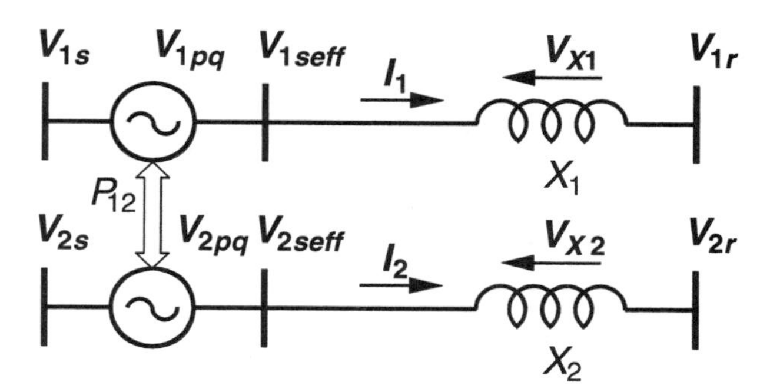

Figure 8.30 Basic two-converter Interline Power Flow Controller scheme.

Consider an elementary IPFC scheme consisting of two back-to-back dc-to-ac converters, each compensating a transmission line by series voltage injection. This arrangement is shown functionally in Figure 8.30, where two synchronous voltage sources, with phasors $\boldsymbol{V_{1pq}}$ and $\boldsymbol{V_{2pq}}$, in series with transmission Lines 1 and 2, represent the two back-to-back dc-to-ac converters. The common dc link is represented by a bidirectional link for real power exchange between the two voltage sources. Transmission Line 1, represented by reactance X_1, has a sending-end bus with voltage phasor $\boldsymbol{V_{1s}}$ and a receiving-end bus with voltage phasor $\boldsymbol{V_{1r}}$. The sending-end voltage phasor of Line 2, represented by reactance X_2, is $\boldsymbol{V_{2s}}$ and the receiving-end voltage phasor is $\boldsymbol{V_{2r}}$. For clarity, all the sending-end and receiving-end voltages are assumed to be constant with fixed amplitudes, $V_{1s} = V_{1r} = V_{2s} = V_{2r} = 1.0$ p.u., and with fixed angles resulting in identical transmission angles, $\delta_1 = \delta_2$ $(= 30°)$, for the two systems. The two line impedances, and the rating of the two compensating voltage sources, are also assumed to be identical, i.e., $V_{1pq\max} = V_{2pq\max}$ and $X_1 = X_2 = 0.5$ p.u., respectively. Although Systems 1 and 2 could be (and in practice are likely to be) different (i.e., different transmission line voltage, impedance, and angle), in order to make the relationships governing the operation of the IPFC perspicuous, the above stipulated identity of the two systems is maintained throughout this section. Note that the two lines are assumed to be independent and not in any phase relationship with each other.

In order to establish the transmission relationships between the two systems, System 1 is arbitrarily selected to be the prime system for which free controllability of both real and reactive line power flow is stipulated. The reason for this stipulation is to derive the constraints which the free controllability of System 1 imposes upon the power flow control of System 2.

A phasor diagram of System 1, defining the relationship between $\boldsymbol{V_{1s}}$, $\boldsymbol{V_{1r}}$, $\boldsymbol{V_{1X}}$ (the voltage phasor across X_1) and the inserted voltage phasor $\boldsymbol{V_{1pq}}$, with controllable magnitude $(0 \leq V_{1pq} \leq V_{1pq\max})$ and angle $(0 \leq \rho_{1pq} \leq 360°)$ is shown in Figure 8.31. As illustrated, this phasor diagram is identical to that characterizing the UPFC (see, e.g., Figure 8.3).

For example, the rotation of phasor $\boldsymbol{V_{1pq}}$ with angle ρ_{1pq} varies both the magnitude and angle of phasor $\boldsymbol{V_{1X}}$ in a cyclic manner and as a result, both the transmitted real power, P_{1r}, and the reactive power, Q_{1r}, also vary with ρ_{1pq} in a sinusoidal manner, as illustrated for the UPFC in Figure 8.7.

The variation of P_{1r} and Q_{1r} with rotating $\boldsymbol{V_{1pq\max}}$ can best be illustrated in the $\{Q_{1r};\ P_{1r}\}$ plane shown, together with the corresponding phasor diagram, in Figure 8.32. At the selected transmission angle, $\delta_1 = 30°$, the uncompensated System 1 $(V_{1pq} = 0)$ transmits $P_{r,30°} = 1.0$ p.u. real power to, and absorbs $Q_{r,30°} = 0.268$ p.u. reactive

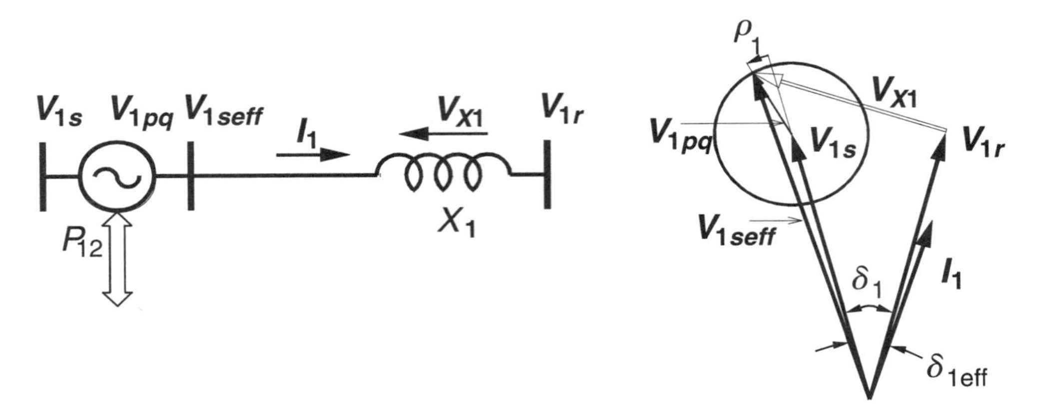

Figure 8.31 IPFC "prime" converter and corresponding phasor diagram.

power from, the receiving end. As shown for the UPFC, the rotation of $\boldsymbol{V}_{1pqmax}$ over 360 degrees produces a circular locus for P_{1r} and Q_{1r} with a radius of $[P_{1pq}^2 + Q_{1pq}^2]^{1/2}$ (= 0.5 p.u.) from the center defined by the coordinates $P_{1r,30°} = 1.0$ and $Q_{1r,30°} = 0.268$ in the $\{Q_{1r}; P_{1r}\}$ plane. This circular locus provides the boundary for the two-dimensional control range within which any corresponding Q_{1r} and P_{1r} values are achievable by the proper setting of magnitude V_{1pq} and angle ρ_{1pq}.

The compensation of the prime System 1 described above is identical to that characterizing the operation of the UPFC. However, in the case of the UPFC, the real power exchanged through the series voltage insertion is supplied via the shunt-connected converter from the sending-end bus. In the case of the simple IPFC considered here, this real power is obtained from the other line via the series-connected compensating converter of that line. In order to establish the possible compensation

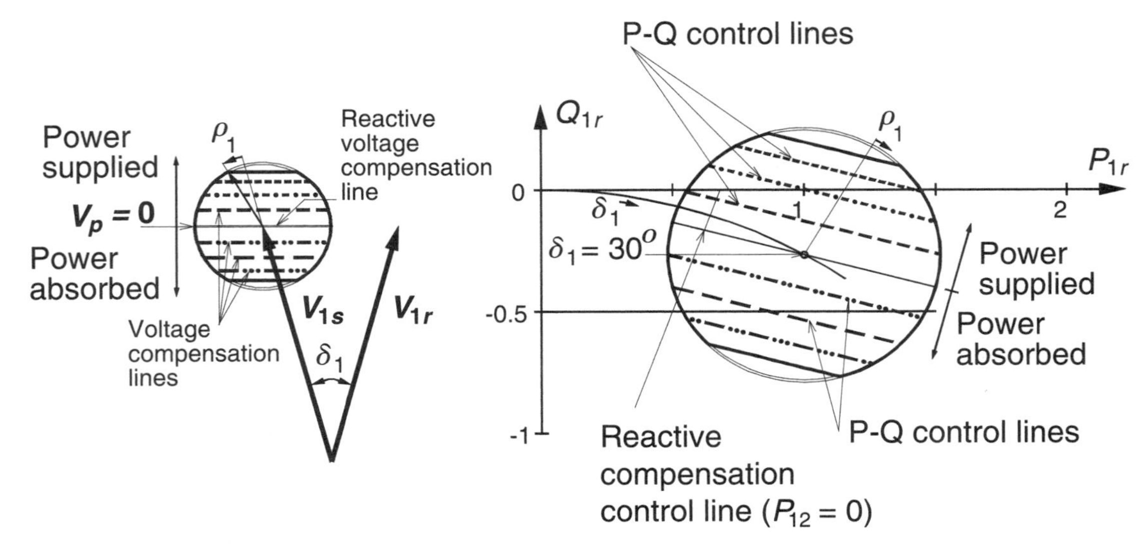

Figure 8.32 Variation of receiving-end real and reactive power as a function of the injected compensating voltage in Line 1.

range for Line 2, under the constraints imposed by the unrestricted compensation of Line 1, it is helpful to decompose the overall compensating power provided for Line 1 into reactive power Q_{1pq} and real power P_{1pq}. To this end, the injected voltage phasor $\mathbf{V}_{1pq}$ is decomposed into two components, one, $\mathbf{V}_{1q}$, in quadrature and the other, $\mathbf{V}_{1p}$, in phase with the line current. The products of these with the line current define Q_{1pq} and P_{1pq}. The component Q_{1pq}, generated internally by Converter 1, evidently provides series reactive compensation for Line 1. The component P_{1pq} provides real power compensation for Line 1, but this power must be supplied for Converter 1 by Converter 2 from Line 2.

It can be shown that if, within the circular operating region of the injected voltage phasor, the magnitude of $\mathbf{V}_{1pq}$ is controlled over the attainable range of angle ρ, so that its end stays on a straight line trajectory ("voltage compensation line"), parallel to the line connecting the ends of the sending-end and receiving-end phasors ("reactive voltage compensation line"), as illustrated at the left of Figure 8.32, then the component of $\mathbf{V}_{1pq}$ in phase with the prevailing line current will result in constant real power demand (that the converter must supply or absorb) independent of angle ρ. This means that within the operating region of $\mathbf{V}_{1pq}$, the reactive compensation is variable along a "voltage compensation line," independent of the real power compensation. The real power demand is, by definition, zero when the trajectory of $\mathbf{V}_{1pq}$ coincides with the "reactive voltage compensation line" (i.e., when $\mathbf{V}_{1p} = 0$ is maintained), which divides the circular operating region into two equal halves. An increasing amount of real power is to be supplied to System 1 as the compensation line is shifted higher and higher above the "reactive voltage compensation line" in the upper half of the control region. Conversely, increasing real power is to be absorbed from System 1 as the compensation line is shifted lower and lower below the "reactive voltage compensation line" in the lower half of the compensation region.

The "voltage compensation lines" prescribing the trajectory of the phasor $\mathbf{V}_{1pq}$ for constant real power demand in the phasor diagram of System 1, define a linear relationship between the receiving-end reactive and real power, Q_{1r} and P_{1r}, respectively, within the circular locus representing the boundary of the control region in the $\{Q_{1r}; P_{1r}\}$ plane, as shown at the right of Figure 8.32. Thus with $P_{12} = 0$ a "reactive compensation control line," which crosses the center of the boundary circle, defines the Q_{1r} versus P_{1r} relationship for purely reactive variable compensation. An infinite number of parallel Q_{1r} versus P_{1r} control lines of decreasing length, corresponding to the "voltage compensation lines" of the phasor diagram, can be established above and below the "reactive compensation control line," according to the real power exchanged. A number of such lines, corresponding to the "voltage compensation lines" of the related vector diagram, are drawn in the same manner (filled, dashed, dotted, etc.) in the Q_{1r} versus P_{1r} control region of the $\{Q_{1r}; \mathrm{P}_{1r}\}$ plane shown in the figure. Comparison of the left-hand and right-hand sides of Figure 8.32 indicates that the reactive power flow increases or decreases in proportion to the real power supplied to, or absorbed from, the line by series compensation.

With respect to the total IPFC scheme shown in Figure 8.30, the following conclusions can be drawn from the above discussion. The operating point of Converter 1 compensating the "prime" System 1 can always be considered a point of a particular "voltage compensation line." Consequently, a corresponding point on the related Q_{1r} versus P_{1r} control line in the $\{Q_{1r}; P_{1r}\}$ plane defines the resultant active and reactive power in the transmission line. In general, at a selected operating point, Converter 1 has to exchange both reactive and real power with Line 1. However, the converter can internally generate only the reactive power and thus it must be supplied with the real power it exchanges. The real power demand remains constant, while the internally generated reactive power changes, as the operating point of the converter is shifted along a selected "voltage compensation line." With the shifting operating point (i.e., with the resultant variable reactive compensation), the receiving-end real and reactive power moves along the relevant Q_{1r} versus P_{1r} control line in the $\{Q_{1r}; P_{1r}\}$ plane, changing primarily the transmitted real power P_{1r}. Moving the operating point from one "voltage compensating line" to another changes the real power demand of the converter and shifts the resulting receiving-end real and reactive power to a parallel Q_{1r} versus P_{1r} control line in the $\{Q_{1r}; P_{1r}\}$ plane, and thus primarily changing the reactive line power Q_{1r}.

It follows, therefore, that in order to satisfy the active power demand of Converter 1 operated along a selected "voltage compensation line," Converter 2 must be operated along a complementary "voltage compensation line" so as to precisely supply this demanded real power from Line 2 via the common dc link for Converter 1. In other words, the relationship $P_{2pq} = -P_{1pq}$ or $V_{2pq} \cos \psi_{2pq} = -V_{1pq} \cos \psi_{2pq}$ (where ψ_{2pq} and ψ_{1pq} are the angles between the injected voltages and the corresponding line currents, i.e., between $\boldsymbol{V}_{2pq}$ and $\boldsymbol{I}_2$, and between $\boldsymbol{V}_{1pq}$ and $\boldsymbol{I}_1$, respectively) must be continuously satisfied. (In practice, this condition could be satisfied by controlling Converter 2 to maintain the voltage of the common dc link in face of the variable real power demand of Converter 1.)

The operation of the two converter IPFC scheme is illustrated with the help of the complementary "voltage compensation" and Q_{1r} versus P_{1r} control lines in Figures 8.33(a) through (e). For this illustration, three particular operating points of Converter 1, marked 1A, 1B, and 1C, are located on an arbitrarily selected "voltage compensation line" of the prime System 1, shown at the left of Figures 8.33(a), (b), and (c). The corresponding Q_{1r} versus P_{1r} control line in the $\{Q_{1r}; P_{1r}\}$ plane, with the specific reactive and real power values obtained with the three operating points, are shown at the right of these figures. The phasor diagram and the corresponding $\{Q_{2r}; P_{2r}\}$ plane, with the complementary "voltage compensation" and Q_{2r} versus P_{2r} control lines, are shown for System 2 at the left and, respectively, at the right of Figures 8.33(d) and (e).

Operating point 1A [Figure 8.33(a)] is selected so that the injected voltage phasor $\boldsymbol{V}_{1pq}$ is perpendicular to the resultant voltage phasor, $\boldsymbol{V}_{1X}$, across the transmission line impedance X_1. As a result, the line current phasor $\boldsymbol{I}_1$ is in phase with $\boldsymbol{V}_{1pq}$ and thus Converter 1 provides strictly real power compensation to reduce the reactive line power, Q_{1r}, to zero. It is evident that Converter 2 can satisfy the real power demand of Converter 1 by operating at the complementary point 2A (which increases the reactive power flow in Line 2), as illustrated in Figure 8.33(d). (As is shown below, Converter 2 could be operated at another point of the same complementary "voltage compensation line" to provide reactive compensation to its own line in addition to supplying the required real power for System 1.)

Operating point 1B [Figure 8.33(b)] represents the maximum real power transfer attainable along the selected "voltage compensation line" of System 1. As seen, the real power P_{1r} is increased and the reactive power Q_{1r} is decreased by about a factor of two, relative to the values obtained with uncompensated ($\boldsymbol{V}_{1pq} = 0$) transmission. As compared to the case of Figure 8.33(a), the line current is significantly increased, reaching its maximum value, while the component of $\boldsymbol{V}_{1pq}$ in phase with $\boldsymbol{I}_1$, $\boldsymbol{V}_{1p}$, is proportionally decreased to its minimum value, yielding the same real power demand as that obtained at operating point 1A. Since there is no change in real power demand, the previously established operating point of Converter 2, shown in Figure 8.33(d), need not change.

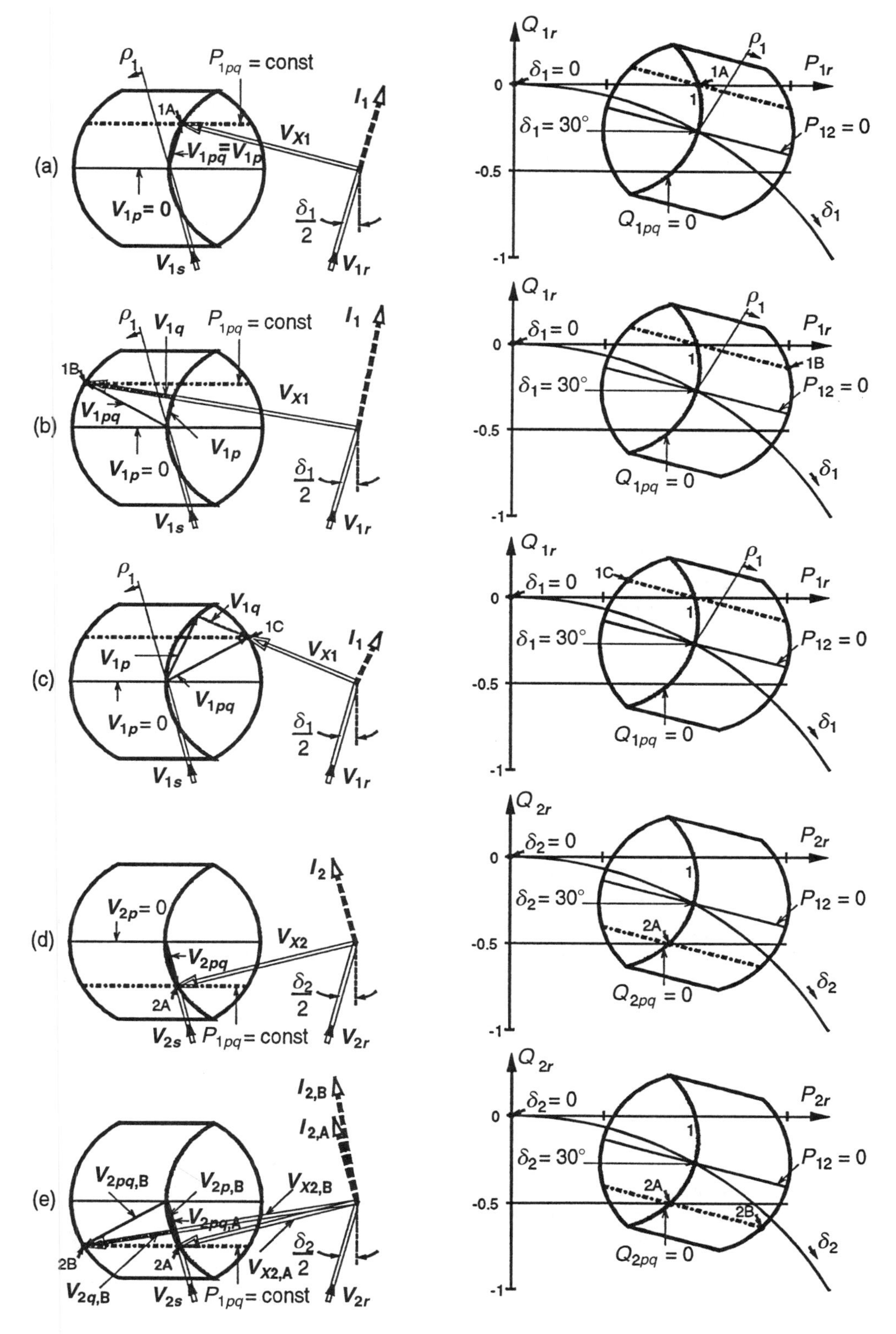

Figure 8.33 Illustration of the operation of a two-converter IPFC by coordinated phasor diagrams and $P - Q$ plots.

Operating point 1C [Figure 8.33(c)] represents the minimum real power transfer attainable with the "voltage compensation line" considered. In comparison with the uncompensated transmission, the real power P_{1r} is decreased from 1.0 p.u. to about 0.63 p.u. and the reactive power Q_{1r} is changed from −0.268 p.u. to about +0.1 p.u. At this boundary operating point, the line current is at its minimum, whereas the component $\boldsymbol{V}_{1p}$ in phase with $\boldsymbol{I}_1$ is the maximum obtainable with this "voltage compensation line." Again, the real power demand resulting from operating point 1C is the same as that obtained at 1A and 1B, requiring no change in the operating point of Converter 2.

From the established symmetry of the two systems, it follows that the operating point of Converter 2 can be freely changed along the complementary "voltage compensation line," as illustrated in Figure 8.33(e) without changing the real power flow between the two converters.

Inspection of the vector diagrams at the left-hand side of Figures 8.33(a) through (e) shows that the trajectory of phasor $\boldsymbol{V}_{1p}$ is always a circular arc. The center of this arc is in the middle between the ends of the sending-end and receiving-end voltage phasors. As can be observed in the figures, the intersection of the (actual or extended) transmission line voltage phasor $\boldsymbol{V}_X$ with the so-constructed arc precisely defines $\boldsymbol{V}_p$ as the component of $\boldsymbol{V}_{pq}$ that is in quadrature with $\boldsymbol{V}_X$ and, consequently, in phase with the line current $\boldsymbol{I}$. (Note that since the above observations are valid for both System 1 and System 2, subscripts 1 and 2 are omitted from the symbols used here.) It follows therefore that if the trajectory of the injected voltage phasor $\boldsymbol{V}_{pq}$ is made to coincide with the above defined arc (i.e., if the operating points of the compensating converter are restricted to this arc), then the converter provides only real power compensation and its reactive power exchange with the transmission line is zero ($Q_{pq} = 0$).

The operating points on the arc, prescribing zero reactive compensation ($Q_{pq} = 0$), translate into a similar circular arc defining the corresponding Q_r versus P_r relationship in the $\{Q_{1r}; P_{1r}\}$ plane, as shown in the right-hand side of Figures 8.33(a) through (e). To the right of this $[Q_r \text{ vs. } P_r]_{(Qpq=0)}$ arc in the overall Q_r versus P_r control area is the region of series capacitive compensation, which increases P_r. To the left of the arc, there is the region of series inductive compensation, which decreases P_r. From these figures the important observation can be made that the range of capacitive compensation is practically independent of the real power compensation provided. By contrast, the range of attainable inductive compensation decreases with the exchange of increasing real power (of either polarity) with the transmission line. The reason for this is that, with increasing inductive compensation, the line current decreases and the converter, having a limited voltage rating, is unable to provide the required total output VA with the available line current. The functional capability of the IPFC is not significantly affected by this in practical applications. This is because the required control range for reactive line power is usually appreciably smaller than that required for real power flow control and, therefore, the theoretical circular control region, in practice, can be limited by appropriate upper and lower boundary Q_r, versus P_r control lines, to a considerably greater degree than that illustrated in Figure 8.33. (This limitation also would often be necessary to keep the transmission voltage between specified limits.) Another consideration is that series compensators in practical applications are usually employed to increase, rather than to decrease, the transmitted power.

The relationships established for the two converter IPFC are summarized in Figure 8.34, where two phaser diagrams together with the corresponding Q_r versus P_r plots characterizing the two line power system are shown. The corresponding

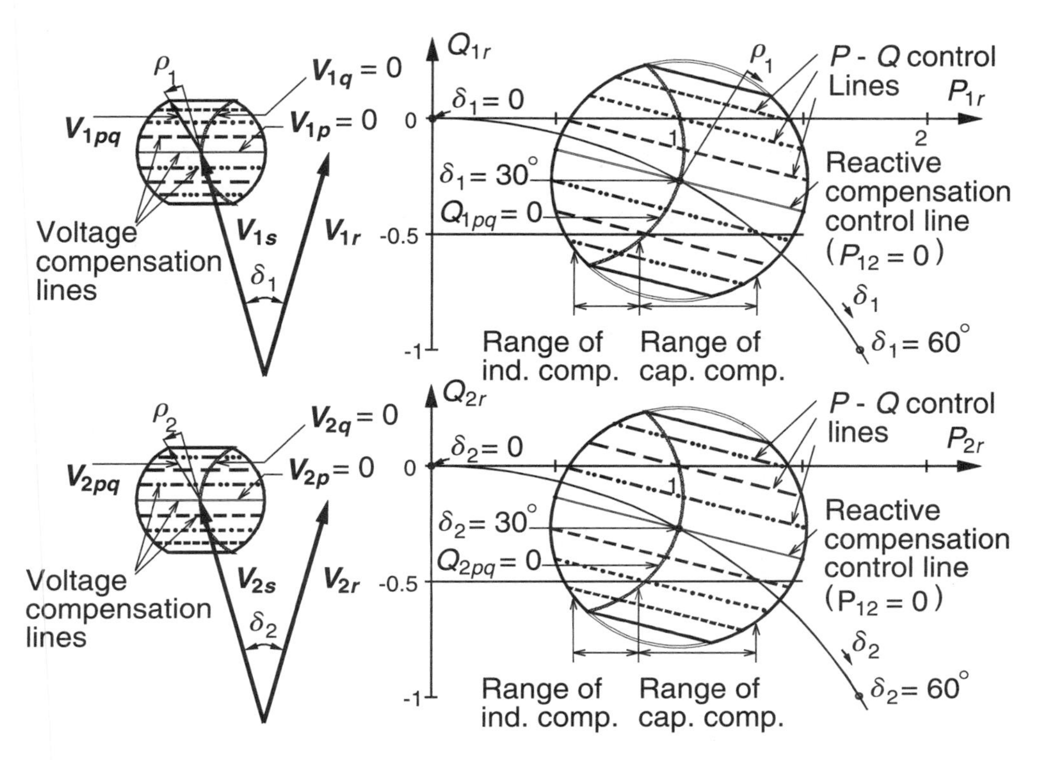

Figure 8.34 Illustration of the basic relationships governing the operation of a two-converter IPFC.

complementary "voltage compensation lines" and the related complementary Q_r versus P_r lines in the two control regions, including those of purely reactive and purely real compensation, are identified in the figure.

As illustrated in Figure 8.34, the complementary "voltage compensation lines" are located symmetrically above and below the two "reactive voltage compensation lines" ($V_{1p} = 0$ and $V_{2p} = 0$). Similarly, the corresponding complementary Q_r versus P_r control lines are located symmetrically above and below the purely reactive compensation line in the circular control region of the $\{Q_{1r};\ P_{1r}\}$ plane. As can be observed in the figure (where the related complementary lines for the two systems are drawn in a similar manner), the complementary lines of the two systems must be in the opposite (upper vs. lower) halves of the relevant compensation and control regions. The complementary lines clearly illustrate the meaning of the restriction $P_{2pq} = -P_{1pq}$, inherent for the simple two-converter IPFC scheme considered. That is, only in one of the two systems (i.e., in the prime system) is it possible to control both the real and reactive power flow. In the other, only the real power flow can be controlled within defined limits by available reactive compensation, while the prevailing reactive line power will be affected by the real power demand of the prime system.

In the illustration of the relationships characterizing the two converter IPFC, the transmission angle for both power systems is, for clarity, fixed at 30 degrees. The center of the circular locus defining the Q_r versus P_r control region is at the Q_{r0} and P_{r0} coordinates characterizing the uncompensated power system at a given transmission angle. As illustrated in Figure 8.35 for System 1, the variation of angle δ_1 moves Q_{1r0}

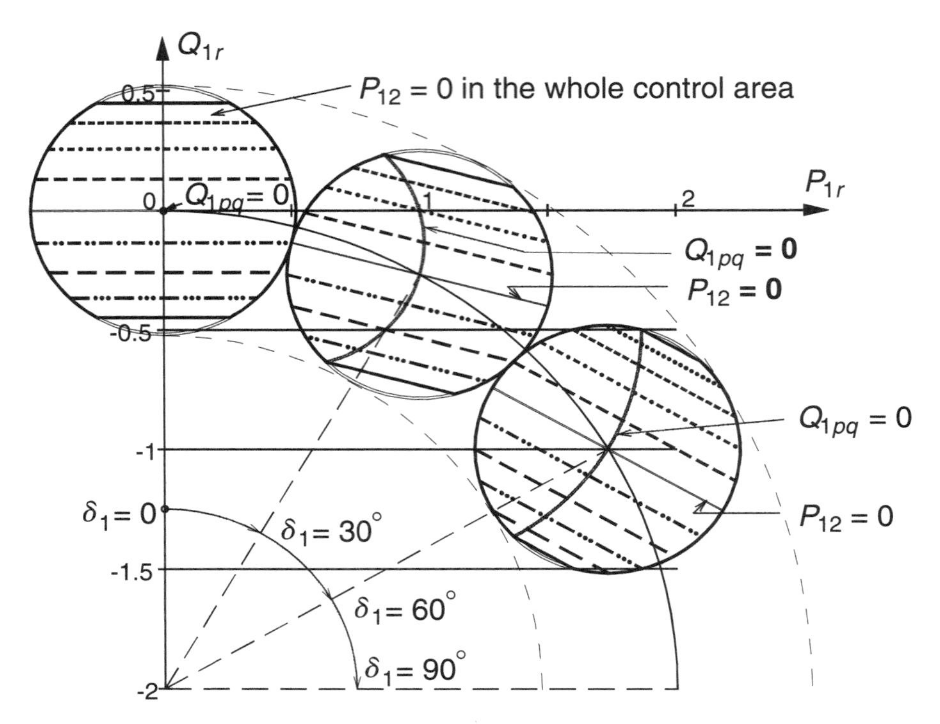

Figure 8.35 Relationship between transmission angle δ_1 and the compensation region in the $(P_{1r};\ jQ_{1r})$ plane for a two-converter IPFC.

and P_{1r0} on a large circular locus whose center is at $P_{1r} = 0$, and $Q_{1r} = -2.0$ in the $\{Q_{1r};\ P_{1r}\}$ plane. Consequently, the center of the boundary circle of the Q_{1r} versus P_{1r} control region in the $\{P_{1R};\ jQ_{1R}\}$ plane also moves along this circular arc with the varying δ_1, as illustrated for $\delta_1 = 0°$, 30° and 60°, respectively, in Figure 8.35.

With decreasing transmission angle the line current at a fixed compensated line impedance decreases. This means that the real power the series-connected converter (with a fixed voltage rating) can exchange with the uncompensated line decreases. For this reason, the range of real power compensation decreases with decreasing δ, unless simultaneous reactive compensation (to increase the line current and transmitted power) is carried out, as the four loci of the achievable Q_{r1} versus P_{r1} values, obtained with $Q_{1pq} = 0$ at $\delta_1 = 60°$, 37.5°, 15°, and 0°, illustrate in Figure 8.35.

It should be noted that the parallel "voltage compensation lines" and the corresponding Q_r versus P_r lines in the $\{Q_{1r};\ P_{1r}\}$ plane, introduced for the illustration of the principles governing the operation of the IPFC, do not imply a necessary framework for the IPFC control. Indeed, as indicated at the outset, the prime converter(s) could be controlled to provide any operating mode possible with the UPFC, including the closed-loop regulation of P_r and Q_r. This would take one degree of freedom (the control of reactive line power) from the other converter(s), but the other degree of freedom (the control of real power flow) would remain largely intact. The operation of secondary converter(s) could be visualized so that an appropriate closed-loop control (e.g., the one regulating the dc link voltage) would force the "real" component of the injected voltage phasor to move along the circular arc defining the $Q_{pq} = 0$ locus for the operating point of the converter(s). Another closed loop could regulate the real

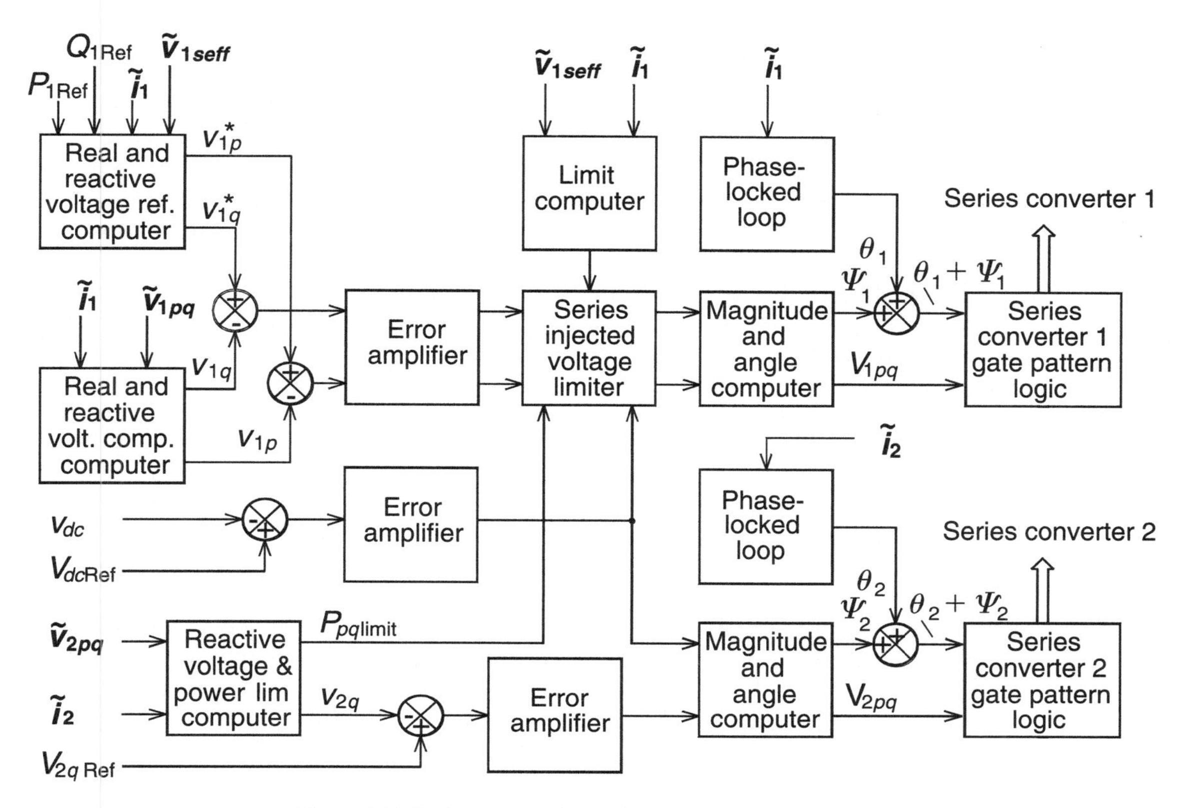

Figure 8.36 Basic control scheme for a two-converter IPFC.

power transfer in the secondary transmission line(s) by controlling the "quadrature" component of the injected voltage phasor along a "compensation line" intersecting the $Q_{pq} = 0$ arc at the point determined by the real power demand of the other converter(s).

8.3.2 Control Structure

The control structure of the IPFC is similar to that of the UPFC shown in Figure 8.16 with appropriate changes in the controlled variables and the necessary constraints imposed by the possible limitations of real power transfer. A possible IPFC control is shown in the form of a block diagram in Figure 8.36. For this structure the assumption of the previous section, stipulating System 1 with IPFC Converter 1 as the "prime" system requiring the independent control of both real and reactive power is, for clarity, retained. This stipulation makes the control of the two converters functionally somewhat different. However, the reader should note that in practice the two converter controls would be identical with control inputs putting either converter in the "prime" or "support" operating role.

As shown in Figure 8.36 the operation of Converter 1 is synchronized to line current $\tilde{i}_1$ and Converter 2 to line current $\tilde{i}_2$ by two independent phase-locked loops. This enables each converter to provide independent series reactive compensation and to keep operating under contingency conditions when the other line or converter is out of service.

The input to the "prime" control is either the desired real and reactive line power, P_1 and Q_1 (indicated in the figure by P_{1Ref} and Q_{1Ref}) or it could be the desired quadrature and real compensating voltage V_{1q} and V_{1p}, shown in the figure as internal references, V_{1q}^* and V_{1p}^*, derived from P_1 and Q_1. Voltage component V_{1q}^*, being in quadrature with the prevailing line current $\tilde{\boldsymbol{i}}_1$, represents series reactive compensation to control the transmitted real power, and component V_{1p}^*, being in phase with that, represents series real compensation to control the reactive power flow in the line.

The internally derived references, V_{1q}^* and V_{1p}^*, are compared to the actual voltage components, V_{1q} and V_{1p}, derived from the measured line current and injected voltage vectors $\tilde{\boldsymbol{i}}_1$ and $\tilde{\boldsymbol{v}}_{1pq}$. The thus obtained error signals after appropriate amplification and possible limitation provide the input for the computation of the magnitude, V_{1pq}, and angle, ψ_1 of the injected voltage vector $\tilde{\boldsymbol{v}}_{1pq}$.

The limitation preceding the computation of V_{1pq} and ψ_1 is an important function of the IPFC control to ensure system operation within predefined constraints. One set of constraints may be provided by the voltage and current limitations of Line 1. (This set of constraints would also apply to the series converter of the UPFC.) The second set of constraints are unique to the IPFC and related to the possible limitations of Line 2 to supply the real power demand resulting from the "prime" compensation of Line 1. As can be observed in Figure 8.36 these limitations may result from insufficient current in Line 2 to supply real power for the maintenance of the dc bus voltage (which is burdened by the real power demand of Line 1) or from limitations imposed for real power transfer by the allowed reactive power flow constraints on Line 2.

The control of Converter 2 is different from that of Converter 1 because it must support the operation of the "prime" Converter 1 by supplying the necessary real power from Line 2. This requirement means that, since the in-phase component of the injected compensating voltage is imposed on Line 2 by the real power demand of Line 1, the control of Converter 2 can influence only the transmitted real power in its own line by controlling the quadrature component, V_{2q}, of the injected voltage vector $\tilde{\boldsymbol{v}}_{2pq}$. Thus, the reference input to the control of Converter 2 is the desired quadrature compensating voltage, V_{2qRef}. (The reader should note that an equivalent input could be the desired compensating reactance. It would also be possible to provide a closed regulation loop for the desired real power P_2 that would yield V_{2q}^* as an internal reference.) Reference voltage V_{2q} is compared to voltage component V_{2q} derived from the measured injected voltage $\tilde{\boldsymbol{v}}_{2pq}$. From this, and from the amplified error representing the difference between the desired and actual dc bus voltage, the magnitude, V_{2pq}, and angle, ψ_2, of the injected voltage vector $\tilde{\boldsymbol{v}}_{2pq}$ are derived to generate the output voltage of Converter 2, as illustrated in Figure 8.36.

8.3.3 Computer Simulation

The operation of the IPFC has been simulated with the use of the *Electromagnetic Transients Program* (*EMTP*) simulation package. Referring to Figure 8.30, an elementary IPFC, composed of voltage-sourced Converters 1 and 2 with a common dc link, provides (real and reactive) series compensation for Lines 1 and 2. The two-bus lines are assumed identical: each has a sending-end voltage of 1.0 p.u., a receiving-end voltage of 1.0 p.u. with a 30° transmission angle, and a line reactance of 0.5 p.u. The converter in Line 1 is operated as the "prime" converter which can inject a voltage in series with the line at any angle with respect to the sending-end voltage. Converter 2, on the other hand, is operated as a "support" converter compelled to inject a

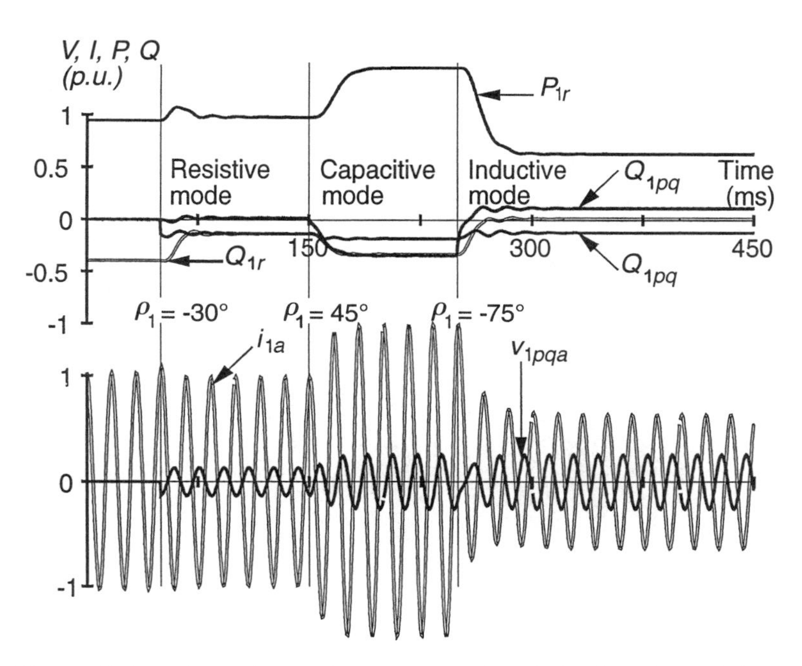

Figure 8.37 Waveforms showing the operation of the IPFC with the "prime" Converter 1 emulating real (resistive), capacitive, and inductive compensations of Line 1.

voltage in series with Line 2 so as to satisfy the real power demand (both supplied and absorbed) of Converter 1.

Consider the case, shown in Figure 8.33(a), where a voltage phasor $\boldsymbol{V}_{1pq}$ with a magnitude of 0.13 p.u. is injected at −30 degrees. In the corresponding simulation, Figure 8.37 in the 0 to 50 ms interval shows the uncompensated line current i_{1a} and the corresponding real power P_{1r} and reactive power Q_{1r} at the receiving end of Line 1. Figure 8.38 shows the (identical) uncompensated line current i_{2a} and the corresponding real power P_2 and reactive power Q_2 at the receiving-end bus of Line 2. At 50 ms, the voltage phasor $\boldsymbol{V}_{1pq}$, stipulated in Figure 8.33(a), is injected to provide only real power compensation. As a result, Converter 1 supplies P_{1pq} for Line 1. Converter 2 is operated so as to absorb real power P_{2pq} from Line 2. The sum of P_{1pq} and P_{2pq} for an ideal system must be zero. According to the ideal case illustrated in Figure 8.33(a), the real and reactive power, P_{1r} and Q_{1r}, should change from [1.0, −0.268] p.u. to [1.0, 0.0] p.u. According to the illustrations of Figure 8.33(d), P_{2r} and Q_{2r} should change from [1.0, −0.268] p.u. to [0.866, −0.5] p.u. The slight discrepancies in the simulation results are due to the finite resistance of the transmission line reactance and the leakage reactance of the coupling transformer of the converter.

At 150 ms, a voltage phasor $\boldsymbol{V}_{1pq}$ with a magnitude of 0.26 p.u. is injected at +45 degrees, as shown in Figure 8.33(b), to emulate a capacitive reactance in series with the transmission line. The line current, i_{1a}, is at its maximum value. The values of P_{1r} and Q_{1r} should, with an ideal lossless system, change to [1.5, −0.134] p.u.

At 250 ms, a voltage phasor $\boldsymbol{V}_{1pq}$ with a magnitude of 0.26 p.u. is injected at −75 degrees, as shown in Figure 8.33(c), to emulate an inductive reactance in series with the transmission line. The line current i_{1a} is at its minimum value. The values of P_{1r} and Q_{1r} should change to [0.634, 0.098] p.u. In all these cases, the injected voltage

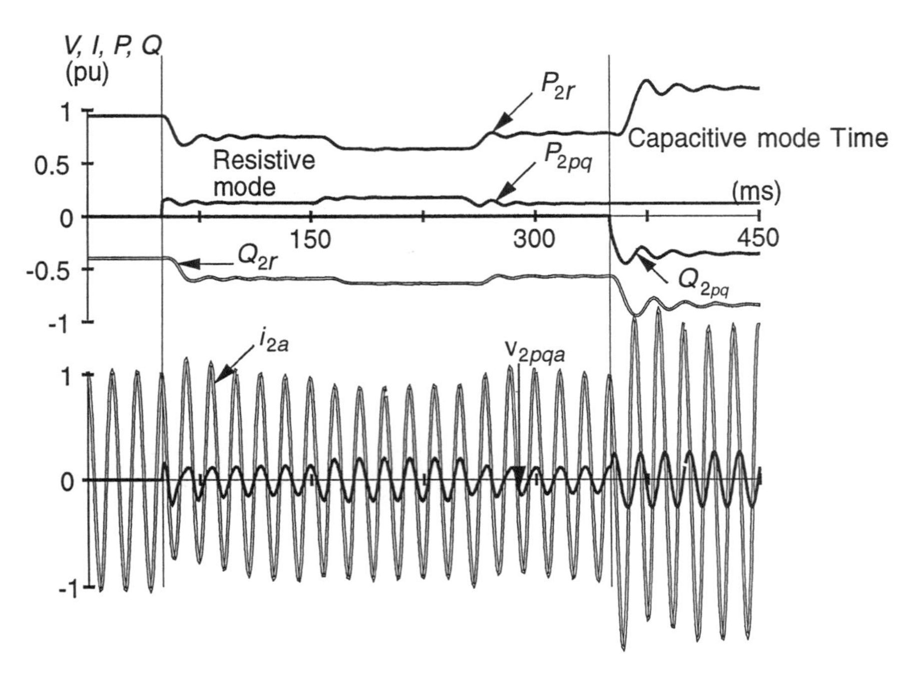

Figure 8.38 Waveforms showing the operation of Converter 2 providing the real power demand of Converter 1 and capacitive series compensation for Line 2.

phasor, $\boldsymbol{V}_{1pq}$, moves along the constant P_{1pq} line. Therefore, both P_{1pq} and P_{2pq} should maintain the same fixed value throughout this time. Also note that Converter 2 in these three illustrations is, by choice, operated with zero reactive compensation, exchanging no reactive power with Line 2 ($Q_{2pq} = 0$), and thus the values of P_{2r} and Q_{2r} should stay at [0.866, −0.5] p.u.

At 350 ms, the operation of Converter 2 is changed to inject a voltage phasor $\boldsymbol{V}_{2pq}$ with a magnitude of 0.26 p.u. so as to emulate a capacitive reactance, in addition to the resistance (i.e., the real power demand of Converter 1), in series with the transmission line. The real and reactive power, P_{2r} and Q_{2r}, should, in an ideal case, change to [1.366, −0.634] p.u., without affecting P_{1r} and Q_{1r}.

The real and reactive power plots in Figure 8.39 show the case in which the IPFC is controlling the real power flow in (prime) Line 1, in response to a given reference, while keeping the corresponding receiving-end reactive power practically at zero and maintaining a substantially constant real power transmission in Line 2. This is achieved by appropriately adjusting the real and reactive compensating voltage components in Line 2 to meet the real power demand of Line 1 and maintain constant real power flow in Line 2.

8.3.4 Practical and Application Considerations

The concept and basic operating principles of the IPFC are explained for clarity within the framework of two identical (and simple) systems in the previous section. In practical applications the IPFC would, in general, have to manage the power flow control of a complex, multiline system in which the length, voltage, and capacity of

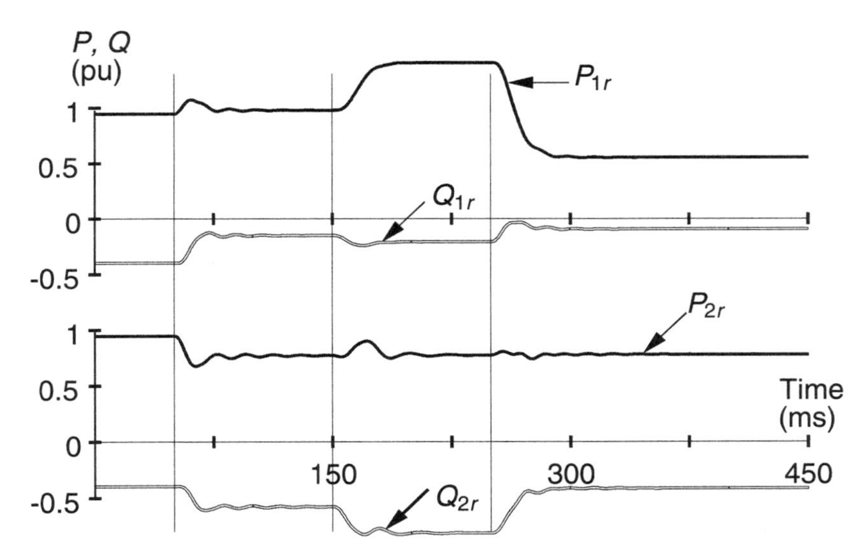

Figure 8.39 Real and reactive power in Lines 1 and 2 when the IPFC is controlling the real power in Line 1 at unity power factor and maintaining constant real power flow in Line 2.

the individual lines could widely differ. One of the attractive features of the IPFC is that it is inherently flexible to accommodate complex systems and diverse operating requirements. A few considerations relevant to practical applications are listed next.

(1) The IPFC is particularly advantageous when controlled series compensation or other series power flow control (e.g., phase shifting) is contemplated. This is because the IPFC simply combines the otherwise independent series compensators (SSSCs), without any significant hardware addition, and affords some of those a greatly enhanced functional capability. The increased functional capability can be moved from one line to another, as system conditions may dictate. In addition, the individual converters of the IPFC can be decoupled and operated as independent series reactive compensators, without any hardware change.

(2) Although converters with different dc voltage could be coupled via appropriate dc-to-dc converters ("choppers"), the arrangement would be expensive with relatively high operating losses. Therefore, it is desirable to establish a common dc operating voltage for all converter-based Controllers used at one location, which would facilitate their dc coupling and thereby an inexpensive extension of their functional capabilities. Reasonably defined common dc operating voltage should not impose significant restriction on the converter design, since at high output power multiple parallel poles are normally employed. Apart from the potential for dc coupling, common operating voltage would also be helpful for the standardization of the converter type equipment used at one location, as well as for the maintenance of spare parts inventory.

(3) The operating regions of the individual converters of the IPFC can differ significantly, depending on the voltage and power ratings of the individual

lines and on the amount of compensation desired. It is evident that a high-voltage/high-power line may supply the necessary real power for a low-voltage/low-power capacity line to optimize its power transmission, without significantly affecting its own transmission.

(4) The IPFC is an ideal solution to balance both the real and reactive power flow in multiline and meshed systems.

(5) The prime converters of the IPFC can be controlled to provide totally different operating functions, e.g., independent (P) and (Q) control, phase shifting (transmission angle regulation), transmission impedance control, etc. These functions can be selected according to prevailing system operating requirements.

8.4 GENERALIZED AND MULTIFUNCTIONAL FACTS CONTROLLERS

In the previous two sections of this chapter, two power flow controllers, the UPFC and IPFC, with the unique capability of providing both reactive and real series compensation are discussed. The UPFC can execute comprehensive compensation for a single line whereas the IPFC can provide comprehensive reactive and real compensation for selected lines of a multiline transmission system. However, there can be compensation requirements for particular multiline transmission systems which would not be compatible with the basic constraint of the IPFC, stipulating that the sum of real power exchanged with all the lines must be zero. This constraint could be particularly restrictive under an emergency contingency when those lines which were to support the "prime" lines would also be overloaded. This potential problem can be solved by combining the UPFC and IPFC concepts to realize a generalized Interline Power Flow Controller arrangement, in which a shunt-connected converter is added to the number of converters providing series compensation, as illustrated in Figure 8.40. With this scheme the net power difference at the dc terminal is supplied or absorbed by the shunt converter, and ultimately exchanged with the ac system at the shunt bus. This arrangement can be economically attractive because the shunt converter has to be rated only for the maximum real power difference anticipated for the whole system. Furthermore, it can also facilitate relatively inexpensive shunt reactive compensation, if this is needed at the particular substation bus, because the shunt converter would need to be rated only for the vectorial sum, i.e., the square-root of the squares, of real power it exchanges with the other converters and the reactive power it provides for shunt compensation.

It is also possible that the compensation and power flow control requirements at a given substation change due to system contingency, maintenance, and system operational changes. Under these conditions and scenarios, complete functional changes in compensation and control requirements may be possible. For example, in a weakened system or under heavy line loading, voltage support of a given bus may be more important than the power flow control of a given line. This type of problem, and the unpredictability of transmission requirements in the increasingly deregulated power industry, can be handled well by applying converter-based Controllers in an appropriate convertible configuration. The principle is illustrated in Figure 8.41 for a two converter, two-line arrangement, which could be expanded to any number of converters and lines. The inspection of this figure shows that with the appropriate closing and opening of the circuit switches connecting the converter outputs to the

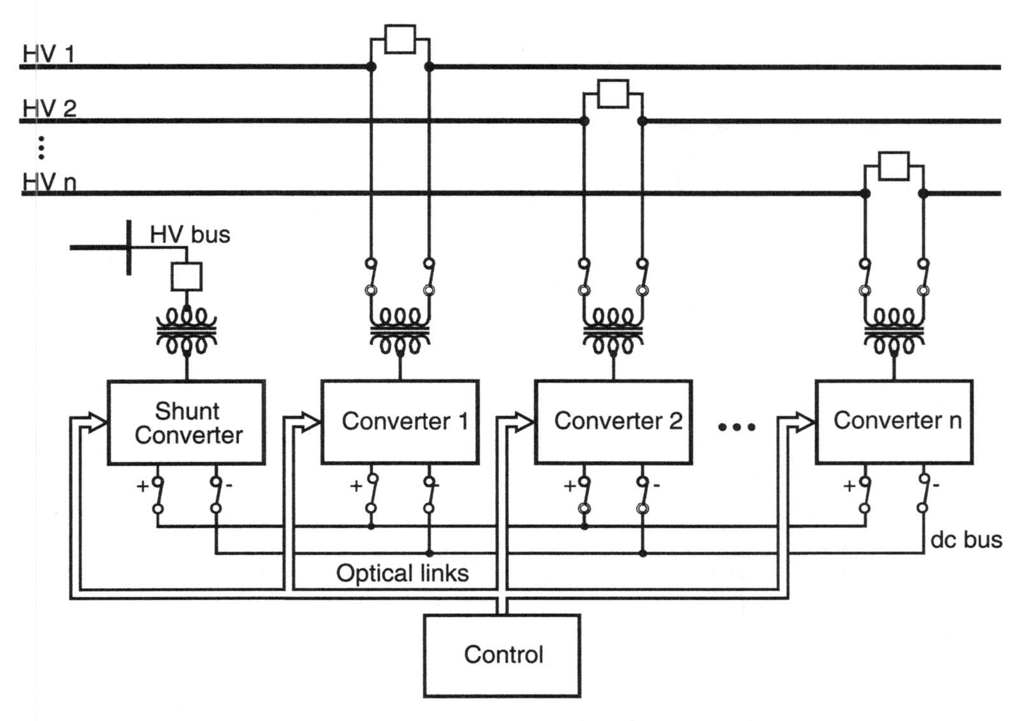

Figure 8.40 A Generalized Interline Power Flow Controller scheme for comprehensive power transmission control and management.

coupling transformers, the two converters, each with 1.0 p.u. *VA* rating, could be configured to function as:

- STATCOM (1.0 p.u. and 2.0 p.u. rating)
- SSSC (1.0 p.u. and 2.0 p.u. rating)
- STATCOM and SSSC (each 1.0 p.u. rating)
- UPFC (1.0 p.u. series and 1.0 p.u. shunt converter rating)
- IPFC (1.0 p.u. for each line)

Note that the per unit ratings shown are not the throughput ratings of the lines. Generally, the FACTS Controller ratings are smaller than the throughput ratings. They are more related to the per unit series inductance of the lines. The power electronic-based Controller ratings are defined by the product of the maximum voltage and maximum current the equipment handles, even if the maximum voltage and current do not occur simultaneously.

The importance of multifunctional and convertible FACTS Controllers for the optimal utilization of transmission systems cannot be overstated. At the beginning of this chapter, the functional use of two groups of FACTS Controllers, one employing thyristor-switched reactors, capacitors, and tap changers, and the other using voltage-sourced converters as synchronous voltage sources, were summarized and illustrated in Figures 8.1 and 8.2. These illustrations convey the traditional viewpoint of transmission control, that is, the transmitted real power is controlled by changing one of the three parameters, voltage, impedance, or angle, of the basic transmission equation. Thus,

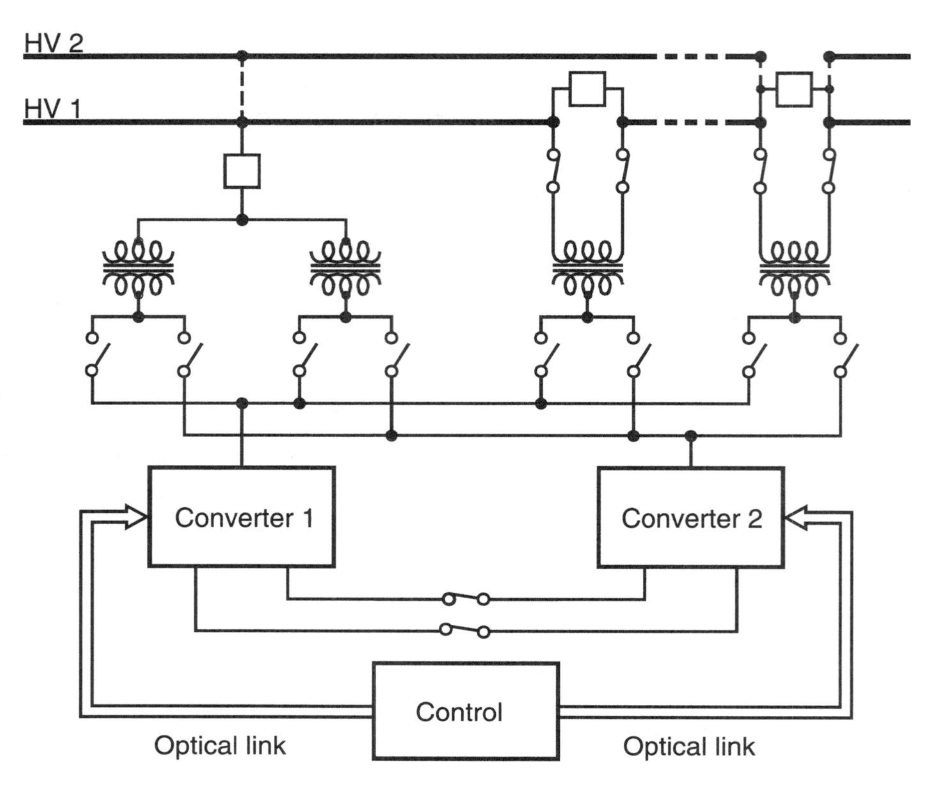

Figure 8.41 Illustration of the functional convertibility of a compensator scheme comprising two voltage-sourced converters.

usually one of the Controllers designed specifically for a single function voltage, impedance or angle control, is applied to solve the particular problem at hand. Whereas this approach may be economically savvy for the short term, it may not have the capability for broader optimization of the transmission assets, or the flexibility to handle changing transmission conditions, and ultimately it may result in a "stranded asset."

Figure 8.42 illustrates a contrasting approach of using standard voltage-sourced converter modules discussed above, which can function individually as conventional (voltage, impedance or angle) Controllers, but can also be converted from one functional use to another, and, more importantly, can be connected to a common dc link to provide comprehensive transmission control capabilities. As can be observed two voltage-sourced converter modules, used to control the power flow between bus 1 and bus 2, can be used individually as a STATCOM and an SSSC, or can be combined to function as a UPFC for the comprehensive control of both real and reactive power flow between bus 1 and bus 2. With the addition of the third converter module, the second line receives an independent series reactive compensator (SSSC). However, by connecting this converter to the common dc bus, a generalized IPFC is established which can control and optimize under the prevailing system condition the real and reactive power in both lines, from bus 1 to bus 2, and also from bus 1 to bus 3. This simple example shows the capability of the voltage-sourced converter-based approach to maintain full convertibility and individual functionality while also providing a powerful potential for an integrated transmission management system with comprehensive capacity of real and reactive power flow control and handling of dynamic disturbances.

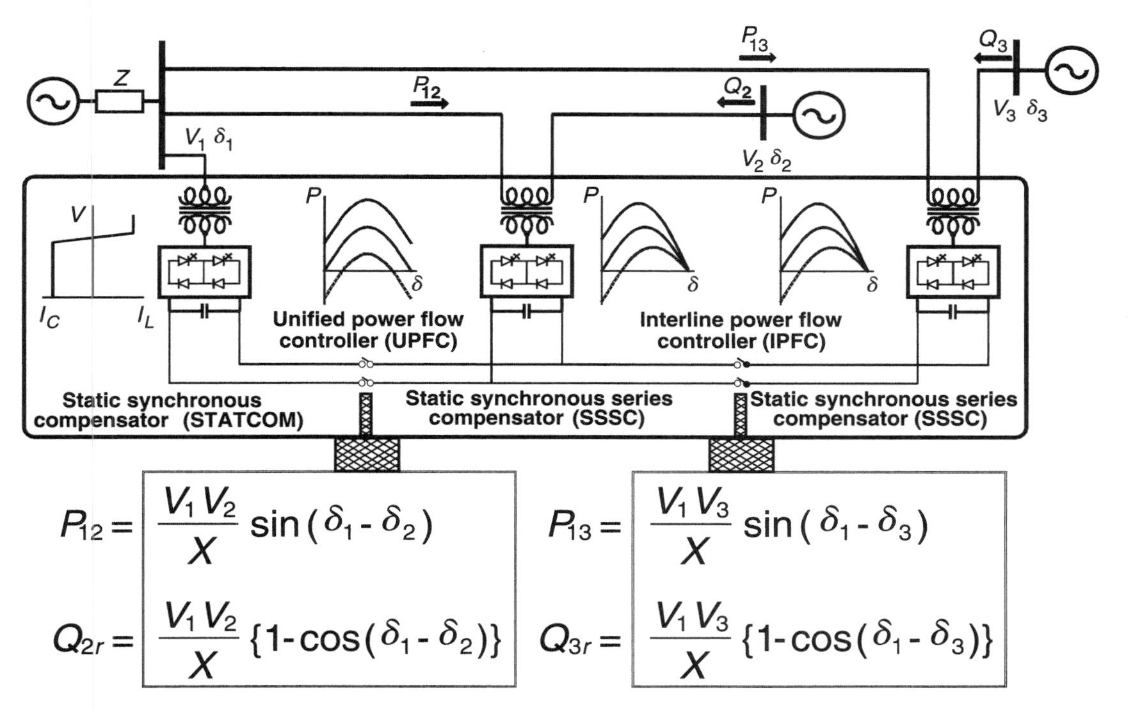

Figure 8.42 Illustration of a generalized IPFC concept for comprehensive real and reactive power flow control of a multiline transmission system.

In the multifunctional FACTS Controller arrangements discussed above, each Controller in a line can independently carry out limited compensation and control functions and thus the common dc connection does not represent a significant single point failure.

REFERENCES

Edris, A., et al., "Controlling the Flow of Real and Reactive Power," *IEEE Computer Applications in Power,* vol. 11, no. 1, January 1998.

Fardanesh, B., et al., "Convertible Static Compensator Application to the New York Transmission System," *CIGRE Paper 14-103,* 1998.

Gyugyi, L., "A Unified Power Flow Control Concept for Flexible AC Transmission Systems," *IEE Fifth International Conference on AC and DC Power Transmission,* London, Publication No. 345, pp. 19–26. Reprinted in *IEE PROCEEDINGS-C,* vol. 139, no. 4, July 1992.

Gyugyi, L., "Dynamic Compensation of AC Transmission Lines by Solid-State Synchronous Voltage Sources," *IEEE Trans. on Power Delivery,* vol. 9, no. 2, April 1994.

Gyugyi, L., et al., "The Interline Power Flow Controller Concept: A New Approach to Power Flow Management in Transmission Systems," *IEEE/PES Summer Meeting,* Paper No. PE-316-PWRD-0-07-1998, San Diego, July 1998.

Gyugyi, L., et al., "The Unified Power Flow Controller: A New Approach to Power Tranmission Control," *IEEE Trans. on Power Delivery,* vol. 10, no. 2, April 1995.

Lerch, E., et al., "Simulation and Performance Analysis of Unified Power Flow Controller," *CIGRE Paper No. 14-205,* 1994.

Mihalic, R., et al., "Improvement on Transient Stability Using Unified Power Flow Controller," *IEEE Trans. on Power Systems,* vol. 11, no. 1, January 1997.

Papie, I., et al., "Basic Control of Unified Power Flow Controller," *IEEE Trans. on Power Systems,* vol. 12, no. 4, November 1997.

Renz, B. A., et al., "World's First Unified Power Flow Controller on the AEP System," *CIGRE Paper No. 14-107,* 1998.

Schauder, C. D., et al., "Operation of the Unified Power Flow Controller (UPFC) Under Practical Constraints," *IEEE Trans. on Power Delivery,* vol. 13, no. 2, April 1998.

Schauder, C. D., and Mehta, H., "Vector Analysis and Control of Advanced Static Var Compensators," *IEE PROCEEDINGS-C,* vol. 140, no. 4, July 1993.

9

Special Purpose FACTS Controllers: NGH-SSR Damping Scheme and Thyristor-Controlled Braking Resistor

9.1 SUBSYNCHRONOUS RESONANCE

The question of subsynchronous resonance will arise in all FACTS applications, for brief or for in-depth consideration, for the basic reason that all high-speed, high-power Controllers have potential of enhancing or degrading subsynchronous phenomenon. This not only applies to the FACTS Controllers, but also HVDC, the Automatic Voltage Regulator (AVR), Power System Stabilizer, etc.

The subsynchronous problem is aggravated by series capacitor compensation and also high-speed reclosing of faulted lines with or without series capacitor compensation. It is therefore important for any power systems engineer to be familiar with this phenomenon, particularly those involved with the FACTS technology, because almost every FACTS Controller offers an opportunity for a SSR-neutral design and for value-added benefit in this area.

Electric power generation involves interaction between the electrical and mechanical energies coupled through the generators. It follows that any change in the electric power system results in a corresponding reaction/response from the mechanical systems and vice versa. Slow-changing load translates into slow-changing mechanical torque on the rotor shafts, which in turn is matched by slow-changing rotor angles to new steady-state angles between the rotors and the stators along with adjustment in the mechanical power input to the rotors through the turbines. Major disturbances such as faults and fault clearing, etc. result in high-transient torques on the mechanical system and corresponding transient twisting of the rotor shaft couplings between tandem turbines and generators.

A large turbine-generator unit acting as a whole mass, has its resonance frequency below about 5 Hz with other nearby large turbine-generators. A cluster of turbo-generators closely coupled in a region representing a large collective rotating mass, would have a resonance frequency in the range of 0.2–3 Hz. These frequencies represent the so-called power system stability swings, an issue of primary importance to the operation of a grid system.

Subsynchronous resonance has to do with the fact that internally, any turbine-generator mechanical system has inter-machine mechanical resonance frequencies between the masses along the shaft. These frequencies are below the main frequency, in the range of 10 Hz to 55 Hz for a 60 Hz system. The rotating shaft system of a large turbine-generator includes a number of large masses corresponding to several turbines, generators, and maybe even a coupled exciter, connected by rotor shafts which act like torsional springs.

For example, Figure 9.1(a) shows a representation of the high-pressure turbine generator unit of the Mohavi Power plant, the one damaged twice by the subsynchronous resonance, during 1970–71. This plant consists of two steam generating units. Each unit is made up of two turbine-generator units in cross-compound arrangement. One generator, 483 MVA high-pressure unit operates at 3600 rpm and the other generator 420 MVA low-pressure unit operates at 1800 rpm. The mechanical torsional frequencies are such that only the high-speed generators are subject to subsynchronous resonance problems. The low-pressure unit is hard to be excited by the SSR currents because it has relatively high mechanical damping at its own resonance frequencies and also provides damping at its companion high-pressure unit's resonance frequencies. Again by way of example, Figure 9.1(b) shows a diagram of mechanical gain versus frequency of the Mohavi high-pressure unit with four machines including two turbines, a generator and an exciter. Because the turbine-generator mechanical system comprises large masses connected by steel shafts, this rotating system has torsional resonance frequencies at which the adjacent masses tend to twist back and forth for any shock. Figure 9.1(b) shows three resonance modes with frequencies of approximately 26.7 Hz, 30.1 Hz, and 56.1 Hz, corresponding to the three couplings of four masses. Large masses coupled together have lower resonance frequency and the exciter being a little machine connected to the generator has the highest resonance frequency. The masses will experience relative motion when excited by a shock. The mechanical system has low but positive damping, which increases with load and without the impact of electrical system, these vibrations will die out slowly, with time constants in the range of ten or more seconds. The peak mechanical gains are limited by the amount of mechanical damping at each frequency and the lower the frequency the lower is the damping. It is worth noting that at low frequencies the three frequencies approach the response of a single inertia equal to the response of a single inertia of concern to the power system stability issues.

Problems arise because the unit is connected to the electrical system with which it interacts in a variety of ways.

A shock to a turbine-generator unit on its own and connected radially to a power system via a series compensated line, can cause the turbine-generator unit to oscillate at one or more of its modes of torsional oscillations. These oscillations can produce peak torques in the shaft system several times the normal torque corresponding to the rated power. Under some circumstances of a big shock such as a system fault followed by line tripping and high-speed reclose into a fault, shaft twisting may become excessive at one of the resonance frequencies and potentially damage the shaft, even without any series capacitor compensation.

Under some scenarios, such as an electrical system with high series capacitor compensation, the net damping at one of these frequencies can become negative and subsynchronous oscillation can build up from a very small disturbance. The excessive twisting of the shaft system may result in its loss of life, a cumulative phenomenon

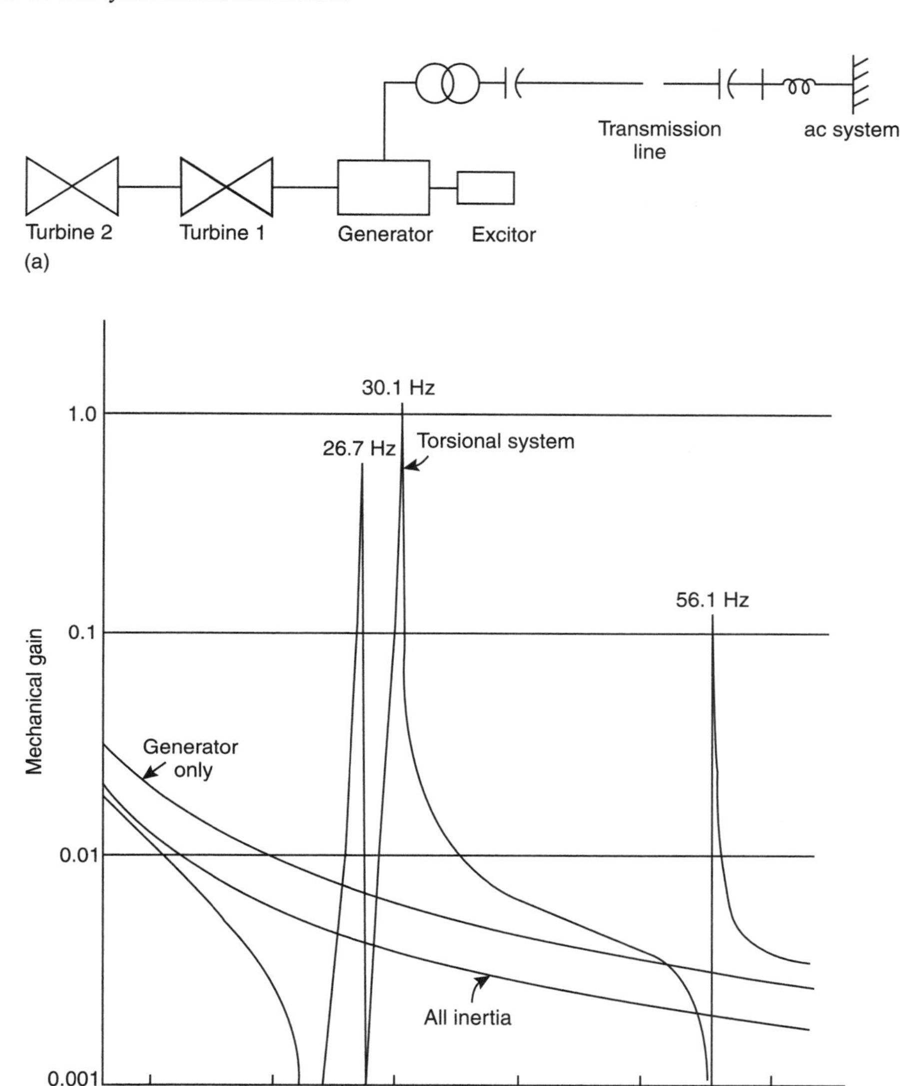

Figure 9.1 Representation of Mohavi Generator and Mohavi-Lugo line as an example of subsynchronous resonance: (a) Electrical-mechanical one-line diagram—Mohavi to Lugo; (b) Mechanical frequency response of Mohavi high-pressure unit.

of mechanical stresses and in a worst case the shaft can break. Such an event is usually catastrophic in the damage caused to the turbine-generator system and takes months to repair.

Figure 9.2 shows the modal electrical-mechanical block diagram for an SSR mode. The term B is the damping coefficient equal to the inverse of the mechanical gain shown in Figure 9.1(b). For example, at 26.8 Hz (mode 1), where the maximum gain is 1.23, the effective mechanical damping coefficient is 0.813 per unit torque per

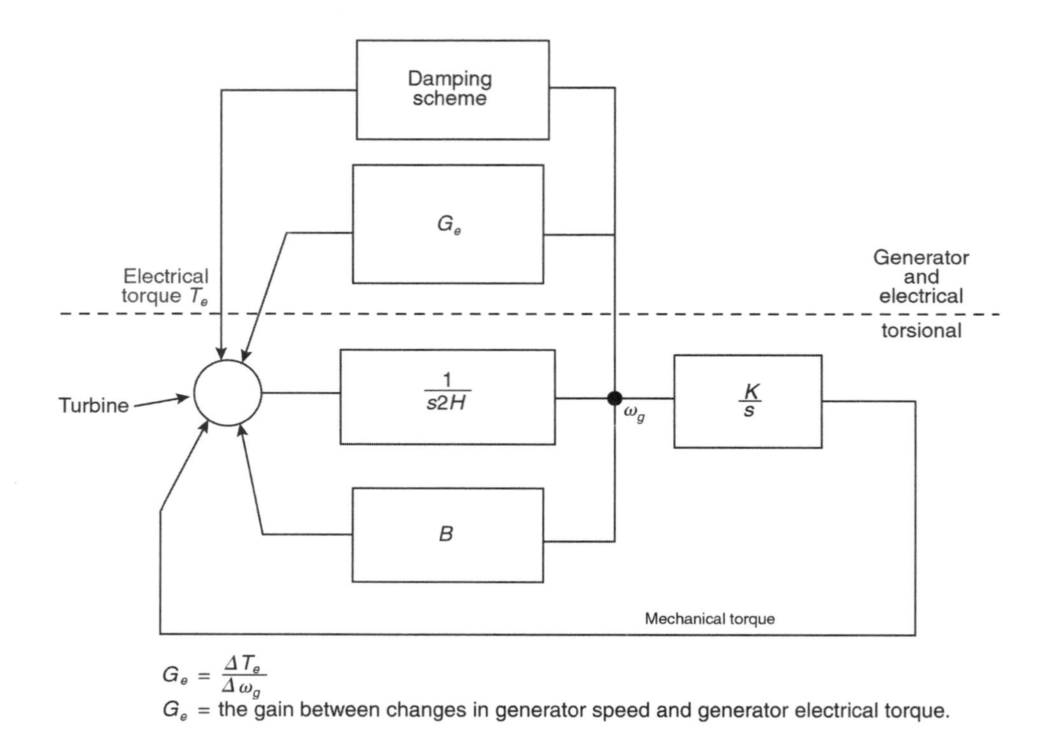

$G_e = \frac{\Delta T_e}{\Delta \omega_g}$

G_e = the gain between changes in generator speed and generator electrical torque.

Figure 9.2 Stability block diagram of a subsynchronous resonance mode.

unit speed. However, the total response of the system includes the impact of the electrical system as well and Figure 9.2 shows the closed loop block for the electrical system and any other damping scheme. Generally speaking, the electrical system without series capacitor compensation, being inductive, provides positive damping at the subsynchronous resonance frequencies. However, under some circumstances, the electrical system can have negative damping, which if negative enough to counter the positive damping of the mechanical system can result in a net destabilizing effect to create unstable conditions at a particular mode.

The electrical and mechanical systems are coupled through the generator at complementary frequencies. A 26.7 Hz mechanical frequency translates into 60 ± 26.7 Hz electrical frequency, 33.3 Hz being the subsynchronous electrical frequency and 80.6 Hz being the corresponding supersynchronous frequency. To explain the electrical interaction, if a change in the generator rotor occurs, say at frequency f_m, it produces a change in the generated voltage at the complementary frequencies of $f_e = f_o - f_m$ and $f_e = f_o + f_m$. These voltages in turn result in a current flow at the corresponding frequencies in accordance with the system impedance at these frequencies. Depending on the phase angles, these currents produce corresponding mechanical torques at the mechanical frequency f_m and thereby a further change in the rotor. This closed-loop response can be regenerative or degenerative depending on the phase of the dependent and independent variables. At the coincidence of the electrical resonance with the complement of a mechanical resonance frequency, the machine looks like a negative resistance as viewed from the electrical system and the electrical system looks like a negative damping as viewed from the mechanical system. Given typical power system impedance, the supersynchronous torque is highly damped whereas the subsynchronous torque may be negatively damped. If the electrical system's resonant frequency is close to the induced subsynchronous frequency then the net mechanical damping would be reduced, and the subsynchronous torque at that mechanical frequency will be amplified and will take a long time to decay. If the net damping becomes negative, as would be the case if the electrical system's resonant frequency is at or very close to the induced subsynchronous frequency, then the net mechanical damping would be negative and the shaft oscillation would build up with the slightest disturbance. There is a vast amount of literature explaining in-depth analyses of the relationships between the electrical and mechanical systems at subsynchronous frequencies and various other aspects of this phenomenon; excellent examples are a classic set of IEEE papers "Analysis and control of subsynchronous resonance," presented at the IEEE 1976 Tesla symposium by the Subsynchronous Resonance Task Force led by Richard Farmer and a recent book by Professor Padyar.

Series capacitor compensation along with the line inductance represents a resonant circuit. If X_L and X_C are the series inductance (including the line, transformers, generator subsynchronous reactance, and the ac system) and series capacitance values at fundamental frequency, respectively, then the resonance frequency f_s is given by

$$f_s = f_o \sqrt{(x_C/x_L)}$$

Series capacitor compensation is provided to partially offset the series inductance, and the ratio X_C/X_L may range from 0.05 to 0.75. Therefore the electrical resonance frequency f_s may range from 13 Hz to 50 Hz and corresponding mechanical frequency may range from 47 Hz to 10 Hz, for a 60 Hz system. High compensation means reaching lower and lower mechanical frequency.

As mentioned earlier, FACTS Controllers, installed for whatever reason, can alleviate the subsynchronous resonance problem, and as mentioned at the end of this section, there are a number of mitigating circumstances.

Another concern of importance for subsynchronous oscillations is the peak turbine generator shaft torques resulting from severe shocks from the electrical system even without the series capacitors. As mentioned before, electrical faults produce corresponding oscillating torques on the shaft system, which decay with a long time constant. This is quickly followed by fault clearing which induces another oscillating torque, which depending on the timing can coincide in phase and add to the initial torque. If the fault clearing is followed by high-speed closing with large angle across the breaker, it would produce yet another large oscillating torque, which may also coincide in phase with the existing oscillating torques. A much more severe condition would occur if the fault has not cleared and the breaker is reclosed into it. Since the peak torque is used as a basis for turbine-generator shaft design, high-speed reclosing is usually not practiced for the lines that leave the thermal power plants. Similar problems can occur during synchronizing and line switching. Whereas the synchronizing can be controlled, it is not normal to control the timing of the line switching and reclosing. Again FACTS Controllers can contribute in reducing and damping the transient torques induced by faults, fault clearing, synchronizing, and high-speed reclosing to safe levels.

The resonance phenomenon as it relates to series capacitor compensation may seem alarming; however, there are several mitigating circumstances.

1. As seen from Figure 9.1(b), the resonance points are rather sharp and the modes are significantly separated.

2. Series-compensated transmission lines have specific resonance frequencies which can be located between the modes.
3. Different plants connected to the system have different resonant frequencies and therefore tend to damp each other's oscillations. Thus the resonant conditions become severe only when a plant ends up in a radial mode through a long series compensated line. In the case of the Mohavi plant, the damage occurred when only the high-pressure unit was operating and another connection to Eldorado was opened by a fault. As a consequence the high-pressure unit was left operating in radial mode, via a 500 kV line with series capacitor of 60% of the line inductance. This series capacitor, and the total series inductance including that of the generator, transformer and the system, resulted in a electrical resonance frequency coincident with the compliment of the second subsynchronous mode of the high-pressure turbine.
4. The hydro turbine-generators have significantly larger damping than the fossil or nuclear plant turbine-generators.
5. Generally, this problem should not arise for small generating plants, which have quite a high, single-mechanical subsynchronous frequency.
6. Finally there are means available to provide significant damping which reduces the transient torques from all kind of shocks. Essentially, different FACTS Controllers discussed in previous chapters can all be provided with a value-added function of damping subsynchronous torques. Two other Controllers suitable for this purpose, the NGH SSR Damping Scheme and the Thyristor-Controlled Braking Resistors, are discussed in this chapter.

9.2 NGH-SSR DAMPING SCHEME

9.2.1 Basic Concept

Series capacitor compensation of medium and long ac transmission lines has been recognized as a powerful and cost-effective tool for optimum/economical use of transmission lines, and improving system stability and power flow through the intended routes. This technique is extensively used in Western United States, Canada, Brazil, and a number of other countries. However, technical issues of the protection of capacitors during line faults, subsynchronous oscillations, and consequent higher torque on the machine shafts have been a deterrent for many others. This section addresses one of the Controllers for active damping, referred to as the NGH-SSR scheme or just NGH Scheme, with NGH being the initials of its inventor, one of the authors of this book. The NGH Scheme is intended to:

1. Minimize subsynchronous electrical torque and hence mechanical torques and shaft twisting.
2. Contain build up of oscillations due to steady state subsynchronous resonance.
3. Suppress the dc offset of the series capacitor. Dc offset of a series capacitor occurs during faults, fault clearing, reclosing and other disturbances and this in turn feeds the subsynchronous electrical torques.
4. Protect series capacitors from over-voltages.
5. Reduce capacitor stresses, including overvoltages and the rate of discharge current and eliminate oscillations associated with capacitor discharge during bypass, thereby reducing capacitor cost.

The circuit diagram is shown in Figure 9.3, and the basic concept is described below with reference to the waveforms in Figures 9.3(b) and (c). In Figure 9.3(b), the 60 Hz voltage v_c is combined with a dc voltage and in Figure 9.3(c) the 60 Hz voltage is combined with a subsynchronous voltage v_{cc}. It is seen in both cases that some half-cycles are longer than the nominal 60 Hz half-cycle period of 8.33 ms. Similarly, any combination of the dc voltage, subsynchronous voltage, or voltage associated with any low-frequency stability-related oscillations, will result in some half-cycles being longer than the nominal half-cycle period. Conversely, if there were no dc or any other low-frequency component combined with the main frequency, then each half-cycle would be equal to the nominal half-cycle period (8.33 ms for 60 Hz). For the present discussions these distorted voltage waveforms represent the voltage across a series capacitor. The hypothesis behind the NGH scheme is that the unbalanced charge in the series capacitor interchanges with the system inductance to produce oscillations. If this unbalanced charge is eliminated from the series capacitor, the system would be effectively detuned to any frequency other than the main frequency.

The basic scheme shown for one phase in Figure 9.3(a) consists of an impedance in series with an ac thyristor switch connected across the capacitor. This impedance may be an inductance or a resistance or a combination of the two. Essentially the impedance should be as small as possible, with the intent to discharge the capacitor with the thyristor bypass switch. In practical terms the best impedance is a combination of a small resistor, the size of which is essentially limited by the peak transient current capability of the switch and a small inductor, the size of which is limited by the di/dt limit of the thyristors. With an inductor in series, some of the charge will be transferred to the next half-cycle, which should generally be helpful since the half-cycle following the longer half-cycle will most likely be shorter. For this discussion we can assume that the switch can discharge and remove the capacitor charge when ordered to do so.

Basically, the control of the thyristor switch is designed such that when zero crossing of the capacitor voltage is detected [t_o in Figure 9.3(c)], the succeeding half-cycle period is timed. As soon as the half-cycle exceeds the set time (e.g., 8.33 ms for 60 Hz), the corresponding thyristor is turned on to discharge the capacitor and bring about its current zero sooner than it would otherwise. The thyristor stops conducting when the thyristor current reaches zero soon after the capacitor voltage zero. At each capacitor voltage zero of each half-cycle, a new timing count starts with the intent to discharge the capacitor for the time that its half-cycle voltage exceeds the set period.

There are a number of possible variations and/or improvisations of this basic concept in order to meet other functions. If the problem faced is only the transient torque problem (no steady-state resonance), the set period can be slightly larger than 8.33 ms, say 8.5 ms, so that the thyristors will not conduct at all during steady state, 60 Hz, and small changes. Yet the damping scheme will be effective in removing capacitor charge during large changes from steady state which may lead to dc offset and significant subsynchronous components in the line current and capacitor voltage.

On the other hand, if there is a likelihood of steady-state resonance, the set period may be slightly less than 8.33 ms. The thyristors will now conduct during steady state at the tail end of each half-cycle of the capacitor voltage, and provide detuning effect against any gradual build up of oscillations holding it to a low level. There will of course be a continuous power loss but this will be very small and of little consequence to the thyristor rating or the cost of losses.

It should be noted that the operation of the NGH scheme is independent for each phase and does not require any signal from the power generator or specific detection of any particular subsynchronous frequency or presence of dc voltage bias

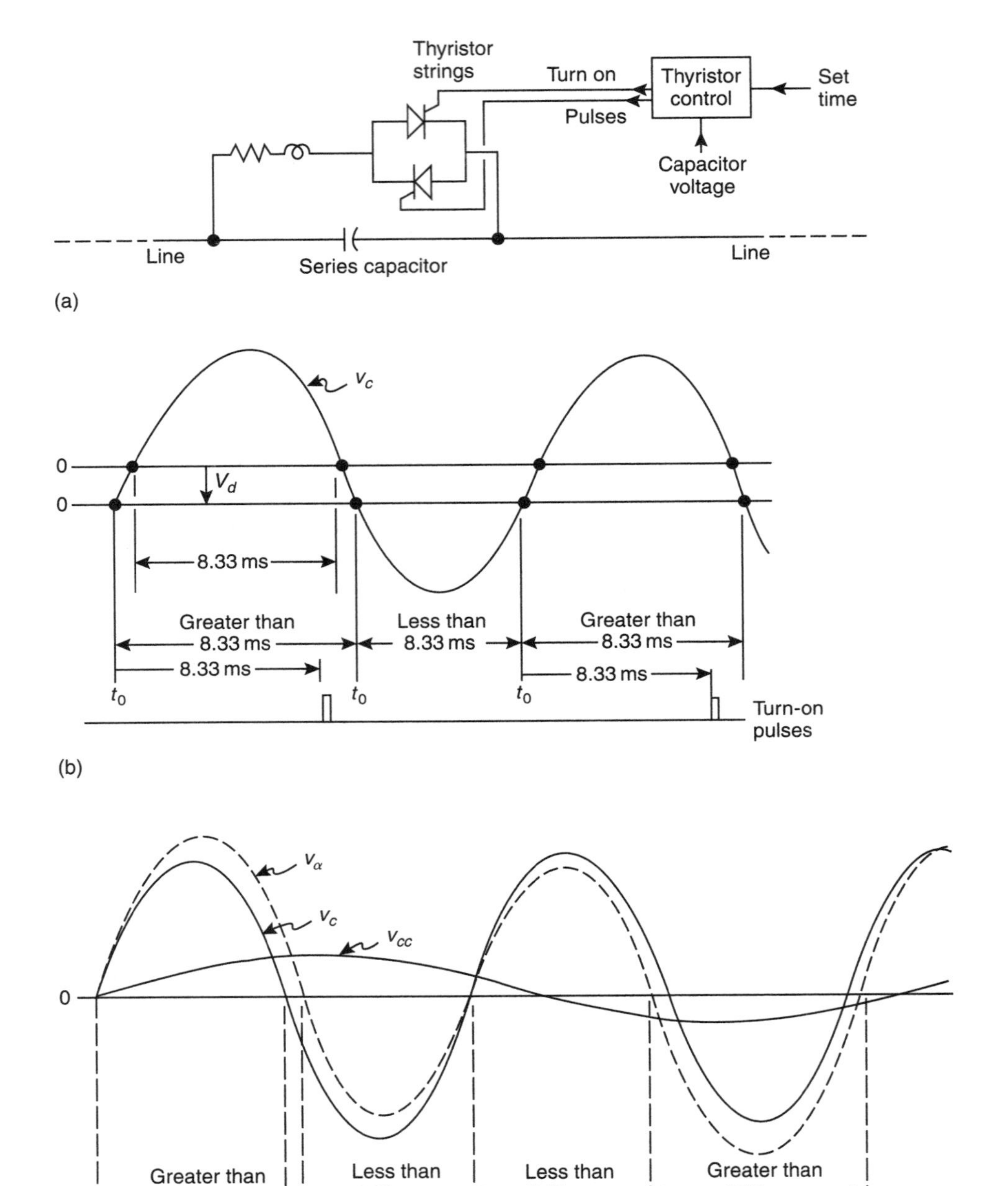

Figure 9.3 NGH-SSR Damping Scheme circuit diagram and principle of operation: (a) Basic circuit diagram; (b) 60 Hz combined with dc; (c) 60 Hz combined with subsynchronous.

in the capacitor. The control can be located at the capacitor platform level or at the ground level with the firing pulses transmitted via fiber-optic links.

A further strategy may be to change the set time depending on various circumstances, such as slow changes in the base frequency around nominal frequency. However, this generally should not be necessary, since timing for each half-cycle starts anew and basically the damping effect is not so sensitive to the precision in the set angle. Removal of the dc bias of the series capacitor is the most important function because this dc charge is what provides the energy for feeding the subsynchronous oscillations and also results in high over-voltages across the series capacitor. Additional control can be provided to more effectively remove the dc bias from the capacitor.

The thyristors can be assigned the additional function of protecting the capacitors and also themselves, by firing the valve, if the instantaneous forward voltage across the valve exceeds a set level. Firing of the valve on forward voltage also protects the thyristors against reverse voltage, since they are connected back to back. As mentioned earlier, the series resistor has to be dimensioned to limit the rate of current rise and peak capacitor discharge current.

Usually series capacitors are installed in modules and this damping scheme can be adapted exactly in accordance with the modular concept. The scheme is basically a passive scheme, i.e., there is no feedback as with other methods. It basically serves to drastically reduce the contribution or gain of the electrical system in the total electromechanical oscillatory system.

9.2.2 Design and Operation Aspects

Figure 9.4 shows the layout of the NGH damper installed at Lugo Substation across one of the series capacitor segments of the Mohavi-Lugo line. Figure 9.5 shows simulator waveforms with and without the NGH damping for the 500 kV Mohavi-Lugo line described in the previous section. The disturbance was caused by a three-phase, three-cycle fault on another line thus leaving the turbine-generator in radial operation with the Lugo line. The NGH scheme was applied on four out of eight segments of 8.7% compensation of the line inductance each. The thyristor switch can be reduced to half its size with the latest thyristors. The waveforms include series capacitor voltage for one of the segments, the line current, the generator electrical torque, and generator-exciter mechanical torque, one of the four mechanical torques. It is clearly seen that the subsynchronous components are rapidly damped out from the capacitor voltage and hence the line current and the electrical and mechanical torques are dramatically suppressed. Attention is drawn to the series capacitor voltage waveform during the first few cycles when the capacitor voltage is high. Not shown is the fact that by firing the switch earlier on a set over-voltage magnitude, the capacitor over-voltage would be effectively limited to a defined peak value.

The NGH Scheme later evolved into the Thyristor-Controlled Series Capacitor (TCSC) concept invented by Vithayathil and discussed in Chapter 6. The NGH scheme with essentially an inductance in series with the ac switch, is a TCSC rated for only a small current specifically tailored for the damping purposes. It is obvious that if a TCSC is required for control of the line current, then the TCSC can simultaneously take care of the subsynchronous phenomenon. In fact bypassing the TCSC capacitor with its thyristor-controlled inductor will turn the series combination into an inductor. Even when operating normally, the TCSC can be designed to have an inductive impedance at SSR frequencies, thus ruling out any chance of SSR. In general less than half of the series capacitors need a TCSC for damping purposes.

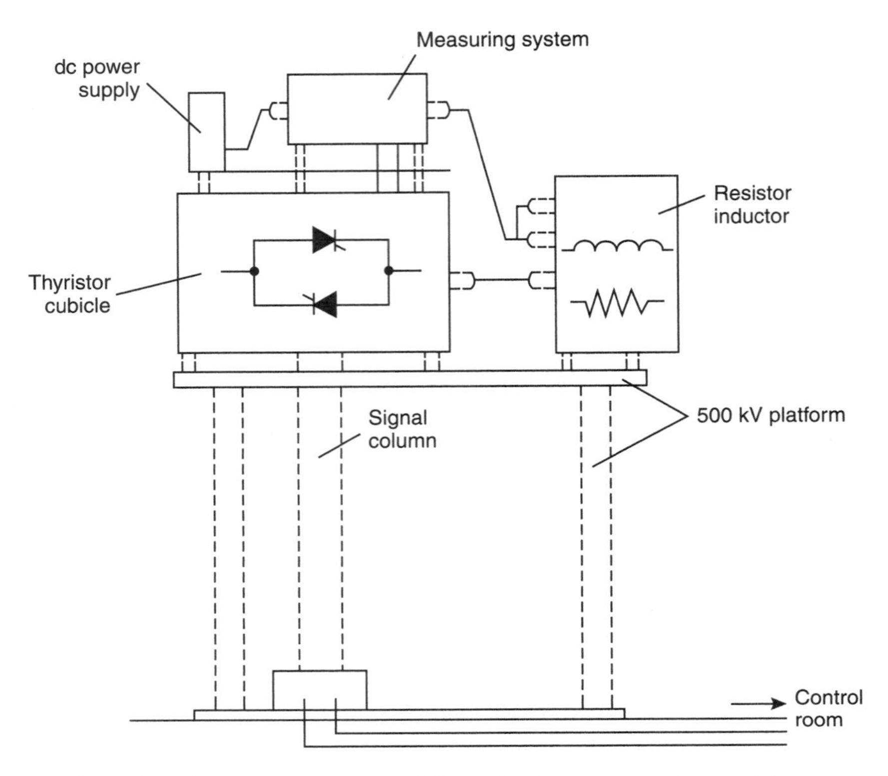

Figure 9.4 The main components of the NGH Damper prototype and its arrangement.

However, if the line current control, the primary function of TCSC, is not needed, then the NGH scheme represents a low cost alternative, largely because of the reduced duty and hence cost of the thyristor valves and inductors and much reduced stresses on the capacitors. In fact with an NGH damper, it is possible to upgrade the existing capacitor bank by 20% of its rating. The thyristors are not subjected to a high rate of recovery voltage because at the end of each current zero, the thyristor voltage rises from zero rather slowly as the capacitor charges in reverse polarity. No transient overshoots are experienced as for thyristors in converter applications. During turn-on, the rise in current can be shaped with appropriate values of inductor and resistor for a damped turn-on. Being generally a high-voltage application the switch would be made up of a string of series-connected thyristor switches. For example, a 10 Ω, 1000 A rms capacitor will have a peak voltage of 14.4 kV. If the maximum protection level of the series capacitor is set to 2 p.u., i.e., 28.4 kV, one would need a string of about four 10 kV or five 8 kV thyristor switches in series including one redundant thyristor switch.

9.3 THYRISTOR-CONTROLLED BRAKING RESISTOR (TCBR)

9.3.1 Basic Concept

The Thyristor-Controlled Braking Resistor (TCBR), Figure 9.6, is a shunt-connected thyristor-switched resistor (usually a linear resistor). Each leg of a three-phase

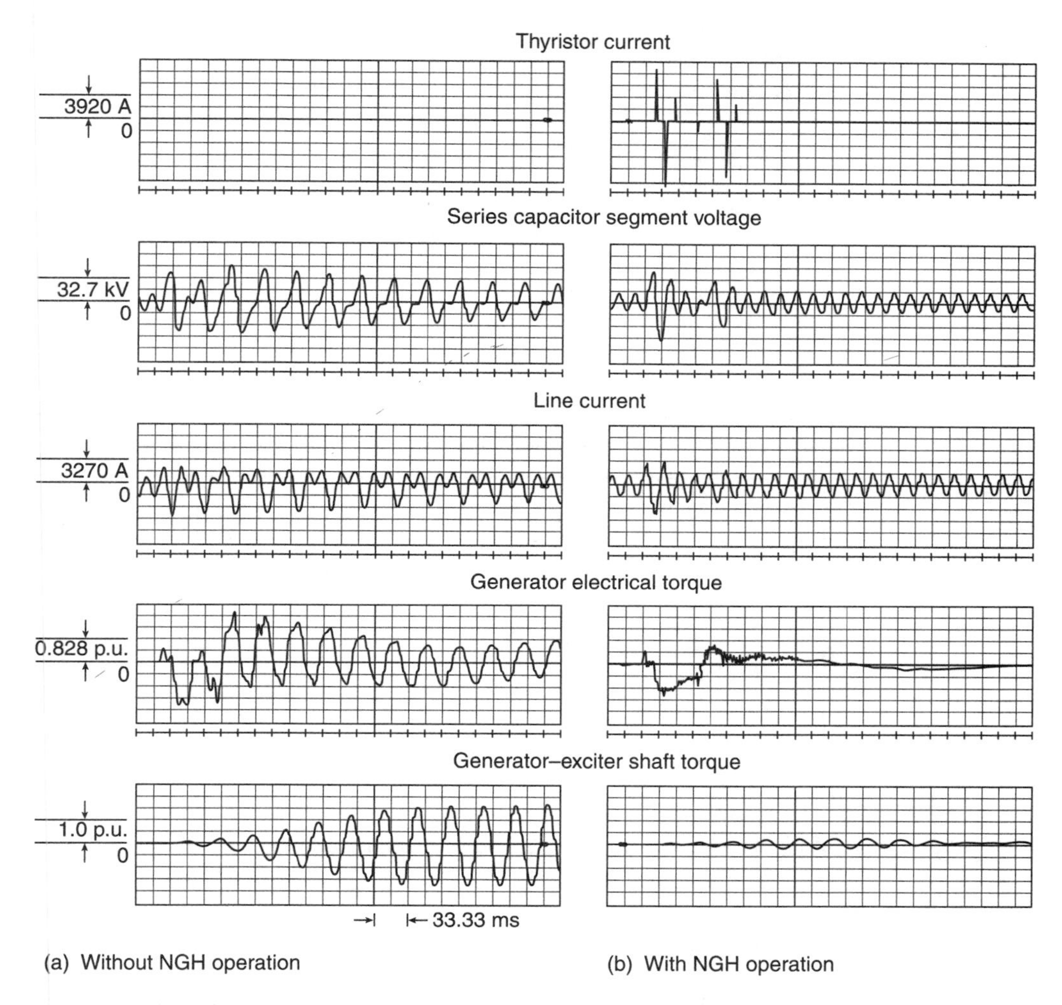

Figure 9.5 Subsynchronous torque damping with NGH-SSR Damping Scheme with 35% of the 70% series capacitors provided with the damping.

TCBR is controlled on or off, half-cycle by half-cycle to aid stabilization of power system transients and subsynchronous oscillations by reducing the net available energy for acceleration and hence speed deviation of the generating unit during a disturbance. TCBR is also referred to as Dynamic Brake.

TCBR can be utilized for variety of functions:

1. Prevent transient instability during the first power system swing cycle, by immediately taking away the power that would otherwise be used in accelerating the generator.
2. Enhance damping to prevent dynamic instability involving low frequency oscillations between interconnected ac systems.
3. Damp subsynchronous resonance (SSR) resulting from series capacitor compensation.
4. Reduce and rapidly damp subsynchronous shaft torques, thereby enabling

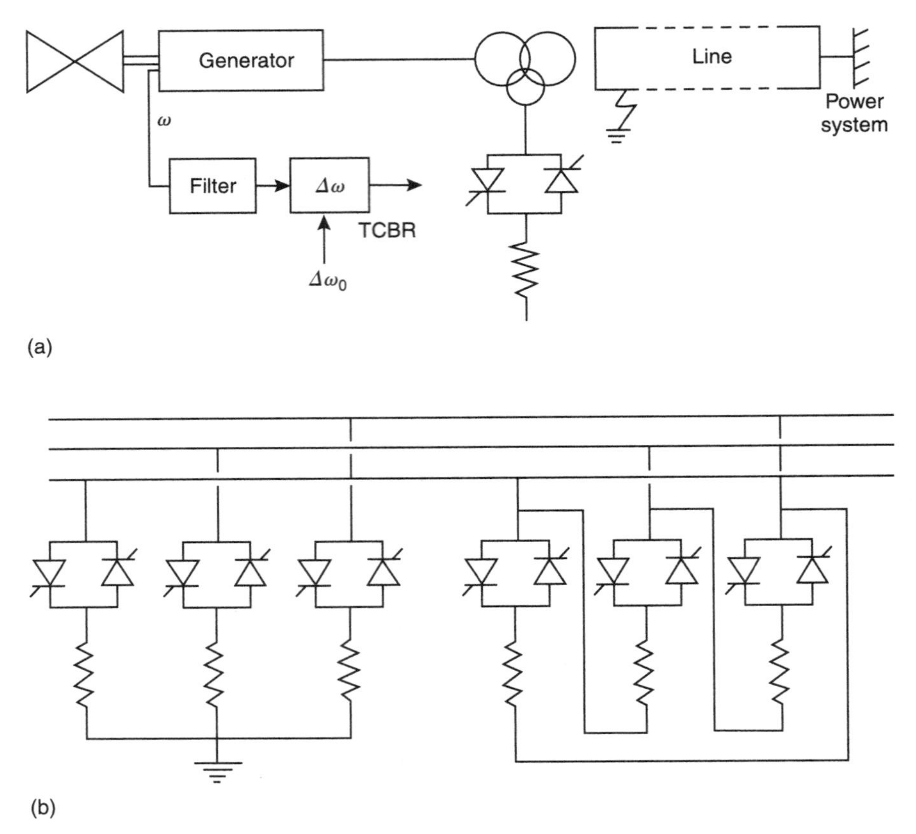

Figure 9.6 One-line diagram of Thyristor-Controlled Braking Resistor: (a) One line system diagram including TCBR; (b) Wye- and delta-connected TCBR.

safe high-speed reclosing of lines near a power plant. This is of significance with or without series capacitor compensation, although this problem is further aggravated by series capacitor compensation of lines leaving a power plant.

5. Facilitate synchronizing a turbine-generator. Out-of-phase synchronizing of a turbine-generator can produce shaft torques more severe than a bolted three-phase fault at the generator.

All these functions in turn enable maximizing the value of transmission and generation assets. Given appropriate rating and intelligent control it can be designed to simultaneously undertake multiple functions noted above. A TCBR can often be the lowest cost, and a simple, highly reliable FACTS Controller. The best location for a TCBR is near a generator that would need braking during transient instability conditions. Transformer tertiary at or near the generator bus would be desirable. The three phase legs may be connected in wye or delta, although delta connection would be more convenient since the ground path would not be needed. A large bank may be divided into two or more banks to achieve partial redundancy.

9.3.2 Design and Operation Aspects

Figure 9.7(a) shows a power angle diagram for a simple system of Figure 9.6(a), in order to first explain the role of TCBR in terms of equal area criteria for transient

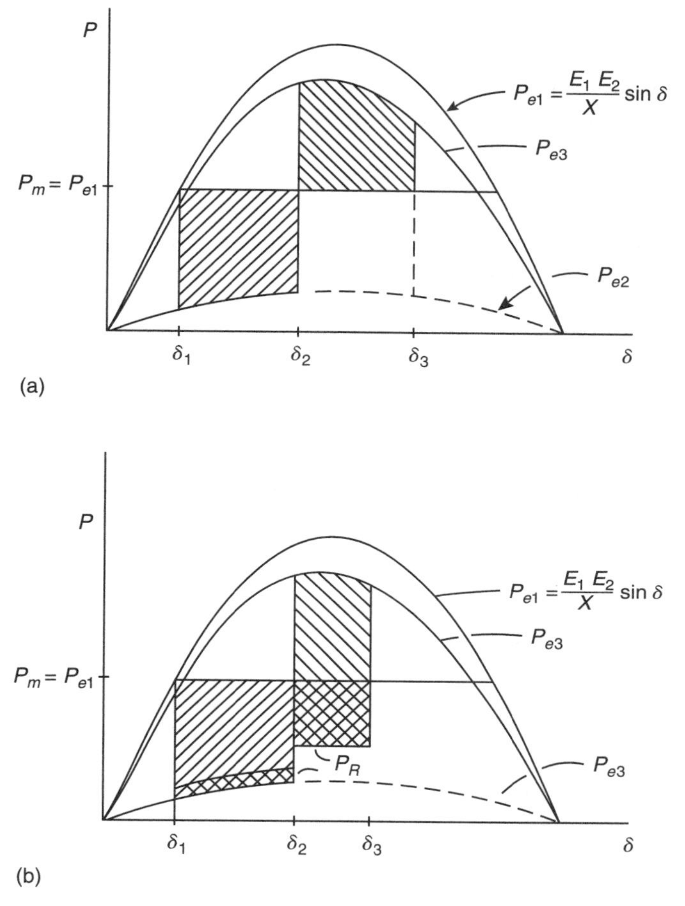

Figure 9.7 Power angle diagram for without and with TCBR: (a) without TCBR; (b) with TCBR.

stability (also discussed in Chapter 5 in relation to the application of STATCOM for improving transient stability). P_{e1} is electric power from the generator requiring a rotor-stator angle (power angle) δ_1. At steady state, the mechanical input power P_m is equal to the electrical power, neglecting losses. When a fault occurs on a line, the electrical output is greatly decreased during the fault and may even reach zero for a fault close to the generator. Assuming that the mechanical power remains constant during the first transient swing, the electric power during the fault drops to P_{e2}, and the excess mechanical power $P_m - P_{e2}$ (shaded area) begins to accelerate the turbine generator. As the machine speed increases, the rotor-stator angle also increases. The angle increases to δ_2 at the instant of fault clearing. Following the fault clearing, including removal of the faulted line, the electric power is restored through alternate transmission paths. The power angle follows a new lower curve of P_{e3}, determined by the new value of E_1E_2/X, which is lower because of increased impedance between the generator and the remote system.

Following the fault clearing, the electrical power exceeds the mechanical power because of the increase in the power angle. As a consequence of the excess electrical power, the generator starts to decelerate. When the post-fault shaded area of the excess electrical energy equals the pre-fault shaded area of excess mechanical energy, the machine angle reaches its transient maximum δ_3 and its speed equals the synchronous speed. The rotor then starts to turn back, accelerate and the angle decreases. If the excess electrical energy following the fault did not equal the excess mechanical energy, it would be too late and the angle would take off (continue to increase). The generator connection to the system would be severed by the protective means in order to save the generator and the system from high over-current. This also means that adequate margin has to be available for operation during the steady state operation, so that the generator does not fall out of step as a result of the most severe fault.

The TCBR can help to increase this margin, or conversely help in increasing the stability limit. Figure 9.7(b) shows the effect of the resistor in terms of the power angle curve. During the fault the resistor power, though small, helps to decrease the accelerating power and after the fault clearing it is much more effective in increasing the decelerating power. The maximum angle δ_3 in now smaller than the case without the TCBR. Extra margin available depends on the power rating of the TCBR and it is not hard to imagine that this may be the most cost-effective and simplest way to enhance transient stability, if it happens to be the limiting criteria for the generated power.

Figure 9.8 shows power swing waveforms in time domain for low-frequency system swing, including power angle, speed deviation, and resistor power. Speed deviation ($\Delta\omega = \omega - \omega_o$, in which ω is the actual generator speed and ω_o is the steady-state generator speed) is the basic control parameter, as shown in a simple one-line diagram of the TCBR. The resistor is switched on when the filtered speed deviation $\Delta\omega$ of a particular oscillation to be damped is positive, and the TCBR is switched off when the speed deviation is negative. In-phase current produces an electrical torque component in phase with generator speed oscillation. The electrical damping torque has the same phase as the frictional and speed damping torques, but is significantly

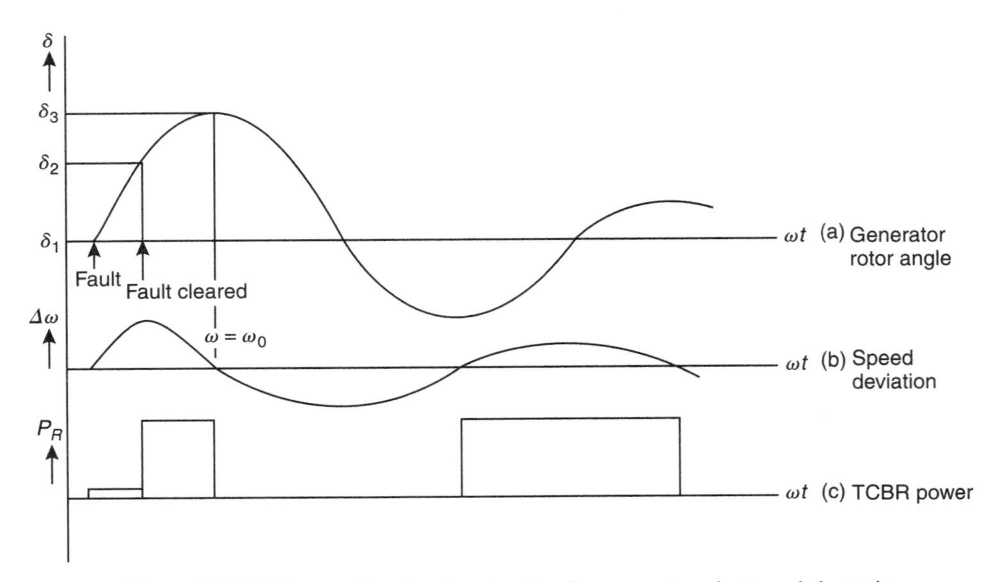

Figure 9.8 TCBR operation for damping low-frequency transients and dynamic instability: (a) generator rotor angle; (b) speed deviation; (c) TCBR power.

more powerful. During the fault the resistor power would be small or zero depending on the terminal voltage during the fault. Once the fault is removed the voltage will increase and so will the resistor power.

Since the resistor only consumes power and does not supply power, it helps to serve as a brake when the speed is higher than normal but cannot help to counter increase in speed when the speed is lower than normal. This is not a deficiency but rather is an operating feature of this Controller, a feature that makes it best suited for location at the generating end to provide rapid response during transient acceleration. Once the generator is drawn back with the help of the TCBR from the first transient acceleration, the TCBR will also effectively assist in rapid damping of the dynamic oscillations.

As discussed in Sections 9.1 and 9.2, the speed deviation during disturbances will also contain subsynchronous oscillations. Figure 9.9 shows a typical trace of undamped speed deviation versus time following a fault clearance near the high side of a generator transformer. The waveform for the generator speed deviation contains a significant component of approximately 2 Hz power system swing frequency, and mechanical torsional natural frequencies including the dominant frequency of 20 Hz and smaller components of 33 Hz and 43 Hz.

If the objective is to control power system swing oscillations, the speed deviation derived from the turbine-generator plant can be filtered through a low-pass filter to derive the low-frequency signal for controlling the low-frequency oscillations. This

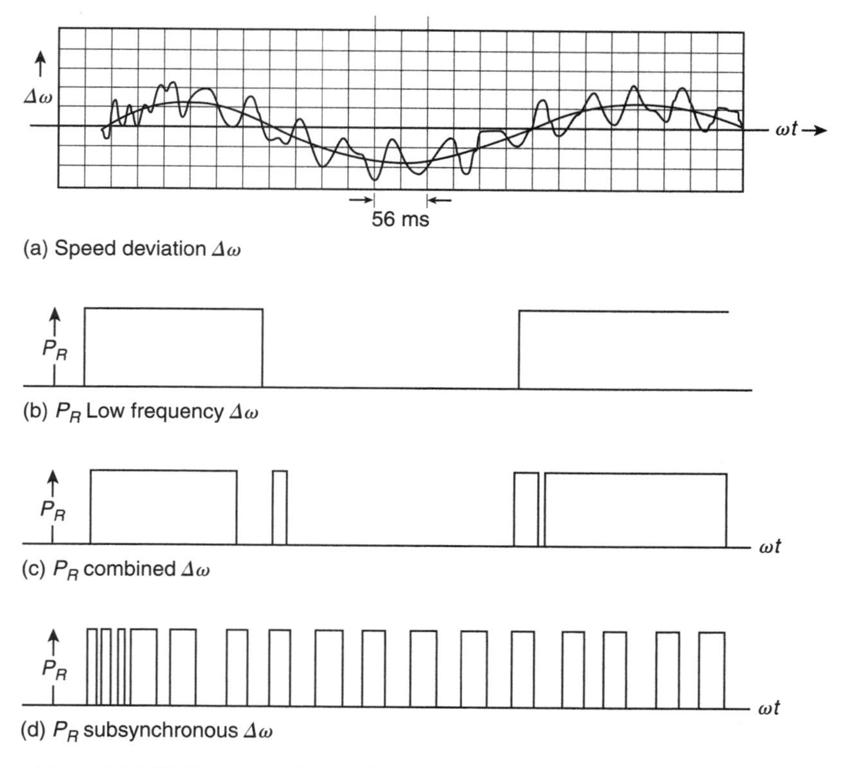

Figure 9.9 TCBR power pulse for different speed deviation input: (a) speed deviation $\Delta\omega$; (b) P_R with low-frequency $\Delta\omega$; (c) P_R with combined $\Delta\omega$; (d) P_R with subsynchronous $\Delta\omega$.

low-frequency component of speed deviation is also shown in the waveform in Figure 9.9(a) and the on–off sequence of the TCBR is shown by Figure 9.9(b). It is seen that these pulses are wide, corresponding to the low-frequency time period.

It should be mentioned that the dynamic brake using mechanical switches and one-shot on–off switching has been used in the past for control of the first transient swing cycle.

On the other hand, if the objective of a TCBR is to control the subsynchronous torques, whether to counter the impact of series capacitor compensation or to utilize high-speed reclosing without high-torsional torques on the shaft system, the speed deviation signal can be filtered through a high-pass filter designed to obtain signal for the entire subsynchronous speed deviation. For the composite subsynchronous speed deviation, the on–off sequence of the TCBR is shown by Figure 9.9(d). This strategy results in shorter TCBR load pulses corresponding to the subsynchronous oscillations. Figure 9.9(c) shows the load pulses corresponding to the unfiltered speed deviation input. It is evident that such a signal will essentially result in damping the low-frequency oscillation until the low-frequency swing is small enough for the unfiltered signal to frequently cross the zero line.

The preferred approach would be to first save the turbine-generator from the first transient swing and then switch to assisting damping of the subsynchronous oscillations. Considering that different power system events may result in one or the other frequency to be most threatening, the best approach to control of TCBR would be to provide individual filters to obtain speed deviation for each frequency and apply the signals selectively in accordance with their magnitude and priority.

As mentioned, the speed deviation waveform in Figure 9.9(a) is without any application of the TCBR. Application of TCBR load, as per Figures 9.9(b) and (d), will rapidly damp the corresponding oscillations.

The speed deviation signal is essential to the correct exploitation of the TCBR. This can be obtained from available measurement techniques of directly measuring the speed deviation of the turbine-generator shaft. However, it is quite feasible to derive such a signal from the generator output voltage or the voltage at the TCBR location.

Given that the function of TCBR is to damp oscillations, it is important to appreciate the nature of load switching by thyristor switches. Figure 9.10(a) shows the three resistor-leg voltages (phase-to-neutral voltages for wye-connected TCBR and phase-to-phase voltages for delta-connected TCBR). Being essentially three independent resistive loads, these waveforms may also be considered current waveforms. TCBR may be based on thyristors switched half-cycle by half-cycle without firing angle control. Given a three-phase TCBR, it involves six actions of power change per cycle. Suppose that the break is required to be on for time t_{B1}, which is somewhat shorter than 60 Hz half-cycle, and the thyristors are turned on for that period without firing angle control. It is seen that because a thyristor will turn off only when its current reaches zero, the TCBR conduction will last for about a half-cycle longer with decreasing power during that extended period beyond time t_{B1}. This is of no significance when the objective is to damp power system swings of low frequency, because the turn-on time would be relatively very long. However, if the required pulses are short, it would lead to less than optimum use of TCBR for damping higher subsynchronous frequency torques. The effect shown on the resistor power is somewhat exaggerated because for subsynchronous frequencies the half-cycles will be longer than the 60 Hz half-cycles. One approach to optimum use of TCBR would be to select the pulse width to be somewhat less than the half-cycle period of the individual speed deviation

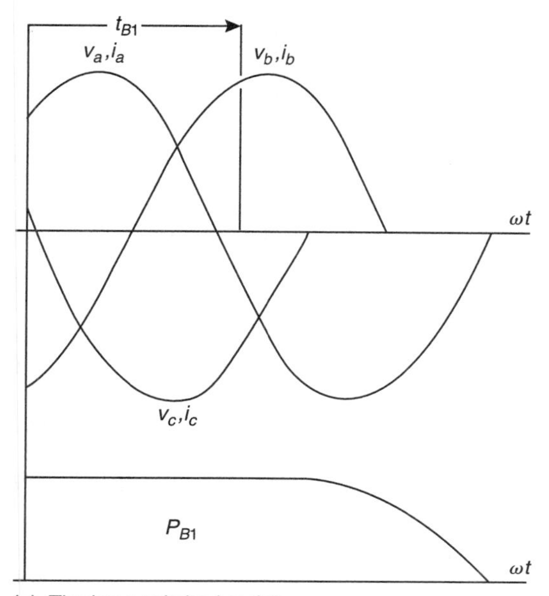

(a) Thyristor-switched resistor

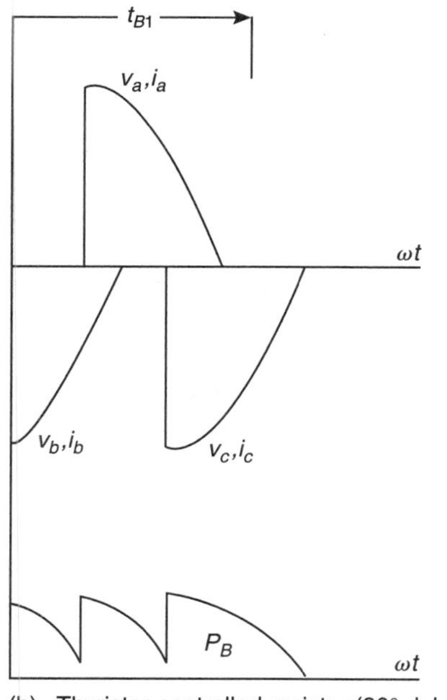

(b) Thyristor-controlled resistor (90° delay)

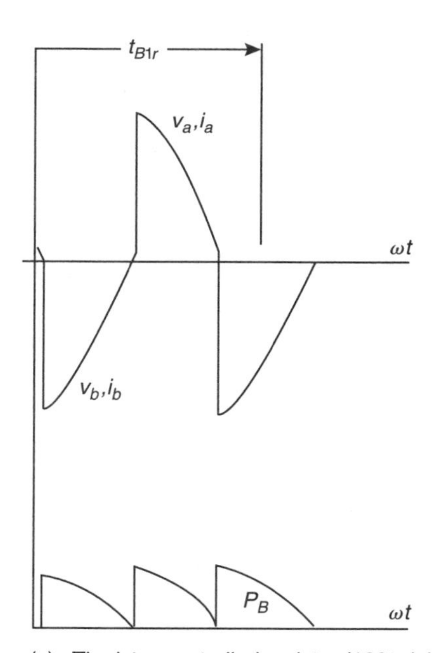

(c) Thyristor-controlled resistor (120° delay)

Figure 9.10 Effects of thyristor turn-on time and thyristor turn-off at current zero for TCBR: (a) Thyristor-switched resistor; (b) Thyristor-controlled resistor.

frequency to be damped. It should also be noted that the resistor load is extended or shortened in steps of 60 degrees of a 60 Hz cycle.

It is also important to understand the differences between thyristor-switched versus thyristor-controlled TCBR and appreciate any benefits of having firing angle control. Given that the thyristors are already there, additional cost of thyristor switches may result from higher switching losses, but given the short time duty this cost should not be significant. With turn-on control the load current can be controlled as shown by the waveforms for the 90 degree delay and 120 degree delay in Figure 9.10(b), for the same turn on time t_{B1} as for the thyristor-switched resistor. Turn-on angle control will be useful for reducing the damping and the dissipation in resistors when the oscillation magnitude is low. One can also contemplate a sustained low-level damping with turn-on angle say, in the range of 170°–180°, or even application of a criteria similar to that for the NGH damper, that if the line voltage duration exceeds a set time period the thyristor switch will turn on. These possibilities suggest a variety of options to exploit the availability of turn-on angle control.

Considering the transient and dynamic requirement of a typical system, a TCBR rating would generally have high MW/MWHrs ratio. For example, MW rating may be 100–200 MW for a 1000 MW power plant, and its energy duty requirement per event would be in the range of 100–200 MJ for a series-compensated line. If the duty required includes suppressing the first transient swing and low-frequency dynamics, the MW requirement may be in the range of 30–40% of the generator rating for several seconds. On the other hand, if there is no series capacitor compensation, and the objective is to only facilitate high-speed reclosing and synchronizing, then the MW requirement may be only 5% of the generator rating for 0.5–1.0 second. It is quite reasonable to consider small continuous duty as well, say a few percent of the TCBR rating to damp small signal low-frequency and subsynchronous oscillations.

It is also self-evident that a shunt Controller with a converter and energy storage, such as Superconducting Magnetic or Battery Systems, or even a dc capacitor-based storage, both the energy absorption and energy delivery can be utilized. Such Controllers would be more effective than the dynamic brake in damping but not in reducing the first transient swing. It is also necessary to recognize the difference between the dynamic brake and the storage systems, both of which actually subtract and/or add real energy, and the other Controllers, such as TCSC, STATCOM, UPFC, etc., which impact the transfer of energy between the two systems. For comparable MVA Controller ratings it would be appropriate to point out that from a stability point of view, adding and subtracting of real power can be several times more effective than transfer of power. Thus for countering transient and dynamic stability, particularly the former, and reduction of subsynchronous torque related to high-speed reclosing and synchronizing, TCBR may be the most cost-effective means.

REFERENCES

Bayer, W., Habur, K., Povh, D., Jacobson, D., Guedes, J., and Marshall, D. A., "Long Distance Transmission with Parallel AC DC Link from Cahora Bassa (Mozambique) to S. Africa and Zimbabwe," *CIGRE Session Group 14, Paper 14-306,* 1996.

Bhargava, B., "Effectiveness of Thyristor-Controlled Series Capacitors in Damping SSR on Mohave Generators and its Comparison with the NGH Device," *Proceedings, IEEE/Royal Institute of Technology Stockholm Power Tech: Power Electronics,* June 1995.

Bowler, C. E. J., Baker, D. H., and Moran, C. G., "FACTS and SSR—Focus on TCSC Application and Mitigation of SSR Problems," *Proceedings of Flexible AC Transmission Systems (FACTS) Conference,* Boston, MA, May 1992.

Farmer, R. G., et al., IEEE Task Force: Analysis and Control of Subsynchronous Resonance," *IEEE PES Winter Meeting and Tesla Symposium,* Publication No. 76 CH1066-0-PWR, 1976.

Hamouda, R. M., Iravani, M. R., and Hackam, R., "Torsional Oscillations of Series Capacitor in Compensated AC/DC Systems," *IEEE Trans. Power Systems,* vol. 4 no. 3, pp. 889–896, August 1989.

Hingorani, N. G., "A New Scheme for Subsynchronous Resonance Damping of Torsional Oscillations and Transient Torque—Part I," *IEEE PES Summer Meeting,* paper No. 80 SM 687-4, Minneapolis, MI, 1980.

Hingorani, N. G., Hedin, R. A., and Stump, K. P., "A New Scheme for Subsynchronous Resonance Damping of Torsional Oscillations and Transient Torque—Part II," *IEEE PES Summer Meeting,* paper No. 80 SM 688-2, Minneapolis, MI, 1980.

Hingorani, N. G., Hedin, R. A., Stump, K. P., and Bhargava, B., "Evaluation of the NGH Scheme Applied to Mohavi Generators," *IEEE PES Summer Meeting,* paper No. 81 TH0086-9 687-4, July 1981.

Hingorani, N. G., Hedin, R. A., Stump, K. P., Schwalb, A., and Mincer, "A New Scheme for Damping of Subsynchronous Resonance in Series Compensated AC Transmission Systems," *IEE International Conference on Thyristor and Variable Static Equipment for AC and DC Transmission Systems,* Proceeding No. 205, pp. 609–664, December 1981.

Hiyama, T., Mishiro, M., Kihara, H., and Ortmeyer, T. H., "Fuzzy Logic Switching of Thyristor-Controlled Braking Resistor Considering Coordination with SVC," *IEEE Transactions on Power Delivery,* vol. 10, no. 4, pp. 2020–2026, October 1995.

Hyman, E., "Thyristor Control for SSR Suppression: A Case Study," *EPRI Conference on Flexible AC Transmission System,* Baltimore, October 1994.

Iravani, M. R., Edris, A. A., "Eigen-analysis of Series Compensation Schemes Reducing the Potential of Subsynchronous Resonance," *IEEE Transactions on Power Systems,* vol. 10, no. 2, May 1995.

Iravani, M. R., and Mathur, R. M., "Application of Static Phase Shifters for Suppressing Shaft Torsional Oscillations," *CIGRE Group 37 Session Paper 37-01,* 1986.

Lombard, X., and Therond, P. G., "Series Compensation and Subsynchronous Resonance Details Analysis of the Phenomenon and of its Damping by TCSC," *Proceedings of IEE Sixth Annual Conference on AC and DC Transmission,* pp. 321–328, April–May 1996.

Nehrir, M. H., Donnelly, M. K., and Adapa, R., "Thyristor Base Damping of Turbine-Generator Shaft Torsional Oscillations Resulting form Power System Disturbances and Subsynchronous Resonance: A Review," *North American Power Symposium,* Auburn, AL, October 1990.

Othman, H. A., and Angquist, L., "Analytical Modeling of Thyristor-Controlled Series Capacitors for SSR Studies," *IEEE Transactions on Power Systems,* vol. 11, no. 1, February 1996.

Padiyar, K. R., *Analysis of Subsynchronous Resonance in Power Systems,* New York: Kluwer Academic Publishers, 1999.

Piwko, R. J., Wegner, C. A., Kinney, S. J., and Eden, J. D., "Subsynchronous Resonance Performance Tests of the Slatt Thyristor Controlled Series Capacitor," *IEEE Transaction on Power Delivery,* vol. 11, no. 2, April 1996.

Pourbeik, P., and Gibbard, M. J., "Damping and Synchronizing Torques Induced on Generators by FACTS Stabilizers in Multi-Machine Power Systems," *IEEE PES 1996 Winter Meeting, Paper 96 WM 253-5-PWRS,* Baltimore, January 1996.

Rajaraman, R., Dobson, I., Lasseter, R. H., and Shern, Y., "Computing the Damping of Subsynchronous Oscillations Due to a Thyristor-Controlled Series Capacitor," *IEEE Transactions on Power Delivery,* vol. 11, no. 2, April 1996.

Raschio, P., Mittelstadt, W. A., Haner, J. F., Spee, R., and Enslin, J. H. R., "Evaluation of Dynamically Controlled Brake for the Western Power System," *CIGRE 1995 Symposium on Power Electronics in Electric Power Systems,* Tokyo, May 1995.

Shelton, M. L., Winkleman, P. F., and Mittelstadt, W. A., "Bonneville Power Administration 1400MW Braking Resistor," *IEEE PAS-94,* 1975.

Takasaki, M., and Hayashi, T., "Subsynchronous Oscillation Analysis of Power System with Power Electronics Applied Facilities," *CIGRE 1995 Symposium on Power Electronics in Electric Power Systems,* Tokyo, May 1995.

Wang, L., "Simulations of Prefiring NGH Damping Scheme On Suppressing Torsional Oscillations Using EMTP," *IEEE PES 1996 Summer Meeting, Paper 96 SM 530-6-PWRS,* Denver, July–August 1996.

Wang, Y., Mittelstadt, W. A., and Maratukulam, D. J., "Variable Structure Braking-Resistor Control in a Multi-Machine Power System," *IEEE Transactions on Power Systems,* vol. 9, no. 3, 1994.

Wasynczuk, O., "Damping Shaft Tortional Oscillations Using a Dynamically Controlled Resistor Bank," *IEEE Transactions Power Apparatus and Systems,* vol. PS100, no. 7, pp. 3340–3349, July 1981.

10

Application Examples

This chapter is intended to describe four applications, in order to convey planning, functional specifications, design and operational aspects of FACTS projects. These projects described below include:

- WAPA's Kayenta Advanced Series Capacitor (ASC, manufacturer's name for TCSC)
- BPA's Slatt Thyristor Controlled Series Capacitor (TCSC)
- TVA's Sullivan Static Condenser (STATCON, acronym which was changed by IEEE to Static Compensator STATCOM)
- AEP's Inez Unified Power Flow Controller (UPFC)

There are also numerous SVC Installations; however, given the coverage that SVCs already have in papers and books, this Chapter focuses some of the new FACTS Technology.

10.1 WAPA'S KAYENTA ADVANCED SERIES CAPACITOR (ASC)

10.1.1 Introduction and Planning Aspects

Advanced Series Capacitor (ASC) is the name given by the manufacturer, Siemens, to their total series capacitor system including a TCSC and a conventional series capacitor. Accordingly, in this section, this installation will be referred to as Kayenta ASC or just ASC, for the total system and TCSC for the thyristor-controlled module. Dedicated in 1992, this first of its kind ASC was installed at the Kayenta 230 kV Substation in Western Area Power Administration in Northeast Arizona.

Like the Slatt-TCSC, Kayenta-ASC is part of the WSCC regional system, characterized by a long transmission line, a large number of power plants both hydro and thermal, and many series capacitor compensated lines. This ASC was needed to increase the reliable transmission capacity of a 230 kV line between Glen Canyon and Shiprock, as shown in Figure 10.1, and also to demonstrate this FACTS Controller as an acceptable planning option for future transmission capacity needs in the WAPA system. The Kayenta substation was selected for this unique project because of its location in the middle of 190 mile line. As can be seen from the one-line diagram, the surrounding system consists of many 500 kV and 230 kV lines connecting substations most of which are locations of large thermal power plants. There is a small local load of about 50 MW supplied from the Kayenta Substation.

The Glen Canyon-Shiprock 230 kV line was designed for an initial power transfer capability of 300 MW, but the line's effectiveness to carry scheduled power was diminished in the late 1960s due to the addition of parallel 345 kV and 500 kV paths. In 1977, a 230 kV phase-shifting transformer was installed at Glen Canyon Substation to reestablish effectiveness of the 300 MW path to Shiprock. With power transfers on the interconnected network approaching the transmission system's ability to reliably serve increasing loads, and with restrictions on building new lines, the economic benefits of adding series compensation became an attractive alternative to improve the line's power scheduling and transfer capability. An addition of 70% (110 Ω, 330 Mvar) of conventional series compensation was needed to increase the power scheduling capability by 100 MW, thus restoring its use to its full thermal rating. However, making part of this series compensation thyristor-controlled provided additional benefits, as described below.

Figure 10.2 shows the simplified one-line diagram of the series compensation at Kayenta Substation. It consists of two 55 Ω series capacitor banks each rated for 165 Mvar and 1000 amperes. One bank is operated in a conventional series compensation

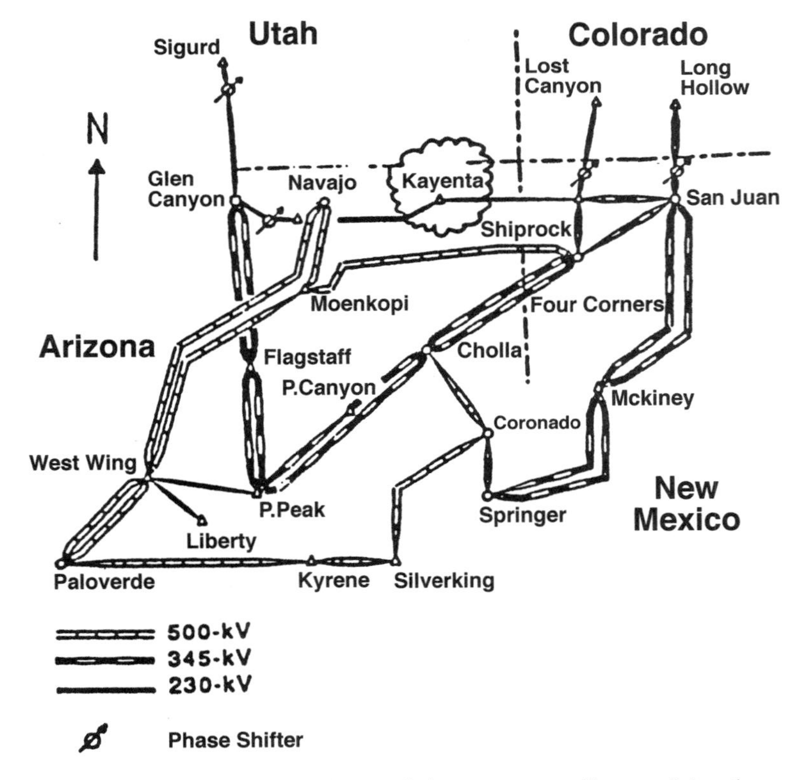

Figure 10.1 System diagram of transmission system near Kayenta Substation.

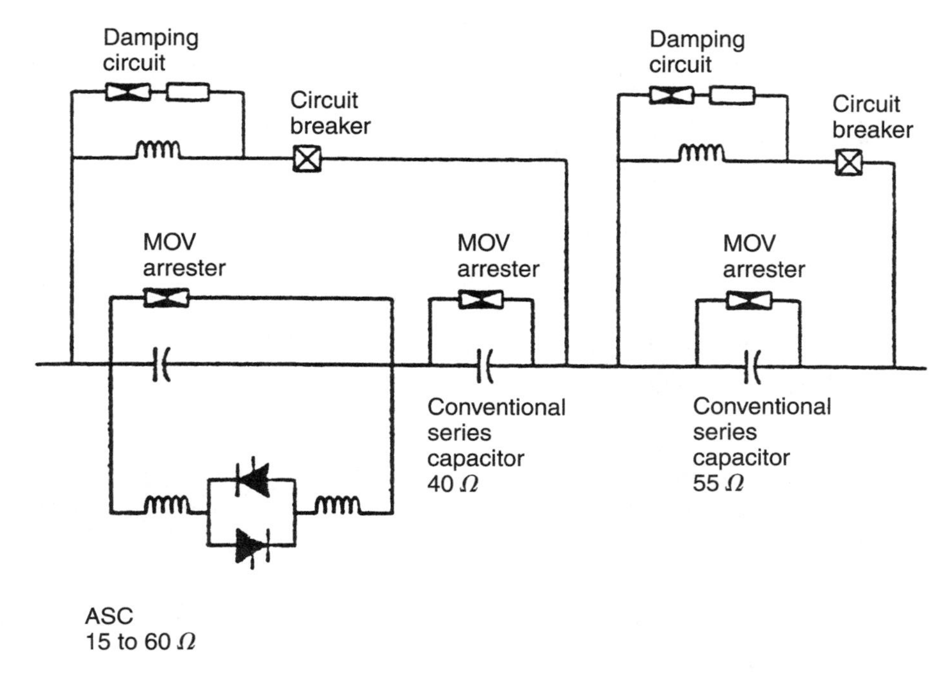

Figure 10.2 Single-line diagram of Kayenta ASC.

configuration with the second bank subdivided into a 40 Ω, 120 Mvar conventional segment and a 15 Ω, 45 Mvar TCSC. For completely conventional operation, the series compensation can be set to various levels by the fixed capacitors. These levels are established by 55 Ω, 40 Ω, and 15 Ω conventional series capacitor segments. By using operation of circuit breakers and thyristor switches, a combination of 0 Ω, 40 Ω, 55 Ω, and 110 Ω banks can be inserted into the transmission line based on operation needs. However, use of the 15 Ω segment as a TCSC provides the following advantages:

- Continuous control of series capacitor compensation, all the way up to 100%, as against discrete step control up to 70% compensation.
- Direct and dynamic control of power flow.
- Short-circuit current reduction by rapidly changing from a controlled capacitive to inductive impedance.
- SSR mitigation. The ASC takes on an inductive-resistive impedance characteristic at SSR frequencies, thus detuning and damping SSR oscillations.
- Improved protection and rapid reinsertion of the series capacitors during system faults.
- Reduces dc offset of the capacitor voltage within a few cycles.

Figure 10.3 shows a photograph of the ASC platforms, insulated from the ground by support insulators underneath. The 15 Ω TCSC portion occupies the front part of the entire ASC including the 40 Ω conventional capacitor located at the back. An inductor of approximately 3.0 Ω is divided in two inductors of approximately 1.5 Ω each, located on the two sides of the valve module.

Figure 10.3 ASC at WAPA Kayenta Substation. The photo shows one of the three platforms; each platform has a 15 ohms, 45/3 MVAR TCSC and 40 ohms, 120/3 MVAR conventional series capacitor. The thyristor switch is housed in the cabinet (courtesy Western Area Power Administration and Siemens).

10.1.2 Functional Specification

Overall TCSC parameters are:

■ Nominal capacitive reactance (9.5% series compensation with one module per phase)	15.0 Ω
■ Continuous effective capacitive reactance (with normal minimum thyristor control)	16.5 Ω
■ Inductive reactance (with thyristors fully conducting)	3.0 Ω
■ Maximum dynamic range of capacitive reactance (subject to 2.0 p.u. capacitor voltage limit)	60.0 Ω
■ Nominal three-phase mvar compensation	45 mvar
■ Rated current (1.0 p.u.)	1000 A rms
■ Continuous overload (1.1 p.u.)	1100 A rms
■ 30 minute overload current (1.33 p.u.)	1330 A rms
■ 1 minute overload current (1.66 p.u.)	1660 A rms
■ Operation with ambient temperature range	−40–+45°C

Figure 10.4 (a) shows the steady-state impedance characteristics of the TCSC in both the capacitive and inductive range. The inductive range can be obtained by varying the thyristor turn-on angle from 90 degrees up toward 180 degrees and the capacitive range is obtained by varying the angle from 180 degrees down toward 90 degrees. Dynamically the reactance in either inductive or capacitive range can be varied up to 60 Ω. However, the control for the variable inductive side is not installed and therefore the available inductive side operation is only at the fixed impedance of 3.0 Ω. Capacitive side operation is variable from −15 Ω to −60 Ω; however, this range is subject to the maximum capacitor voltage of 2.0 p.u. Therefore its continuous operation at −60 Ω will be permissible line current up to 250 A. At 250 A line current there would be 2000 A (2.0 p.u.) current in the capacitor inductor loop. As the impedance increases; the current-carrying capability becomes time dependent, as determined by the time-dependent over-voltage capability of the capacitors, as shown by Figure 10.4(b). Since the TCSC is a single module, it cannot cover the variable impedance range from inductive 3.0 Ω to capacitive 15 Ω. Thus, change in impedance from capacitive to inductive involves a jump from −15.0 Ω to +3.0 Ω, and vice versa; however, this impedance band is not needed for this line.

10.1.3 Design and Operational Aspects

10.1.3.1 Thyristor Switches. The Kayenta thyristor valves are of modular construction and housed in an outdoor weatherproof enclosure located on each phase of the series capacitor platforms, as shown in Figure 10.3. Thyristor valves are cooled by a mixture of 50% de-ionized water and 50% glycol, which is pumped through the fan-cooled heat exchangers at ground level. This system also provides heating for the platform analog-signal electronic processing equipment.

Each of the three thyristor valves has 11 bi-directional parallel-connected thyristor pairs forming individual levels. Of the 11 series-connected thyristor levels, one level is redundant, allowing the valves to provide normal service in the event of failure of any one thyristor level. The thyristors are 100 mm cell diameter, rated for 3.5 kA, 5.5 kV. The fault current at this site is only a few thousand amperes, too low to be a factor in the thyristor rating.

10.1.3.2 Control Protection and Monitoring. The thyristor arrays are linked to the ASC control system by fiber-optic interface, which is used to send low-energy firing signals and to monitor the status of individual thyristors.

The TCSC control and protection is digital microprocessor based. An open-loop control carries out the start-up sequence to ensure safe insertion of the TCSC. Monitoring of all basic system conditions, including line current, protection status, cooling system, and thyristor valve status is also carried out by an open loop control.

The control modes include Constant Impedance mode, Constant Current mode, Thyristor-Switched Reactor mode, and Wait mode. In the Constant Impedance mode, each phase is controlled independently to enable balancing of the phases but the maximum TCSC impedance unbalance is automatically limited to 15% (9 Ω). In the Constant Current mode, each phase has a closed loop current regulator, which has the ability to maintain constant current on the line. In the Thyristor-Switched Reactor mode the thyristors become fully conducting. Switchover from 3.0 Ω inductive to a controlled range of 15–30 Ω capacitive operation and vice versa can be used for power modulation to damp oscillations. While a study was carried out to simulate ASC in

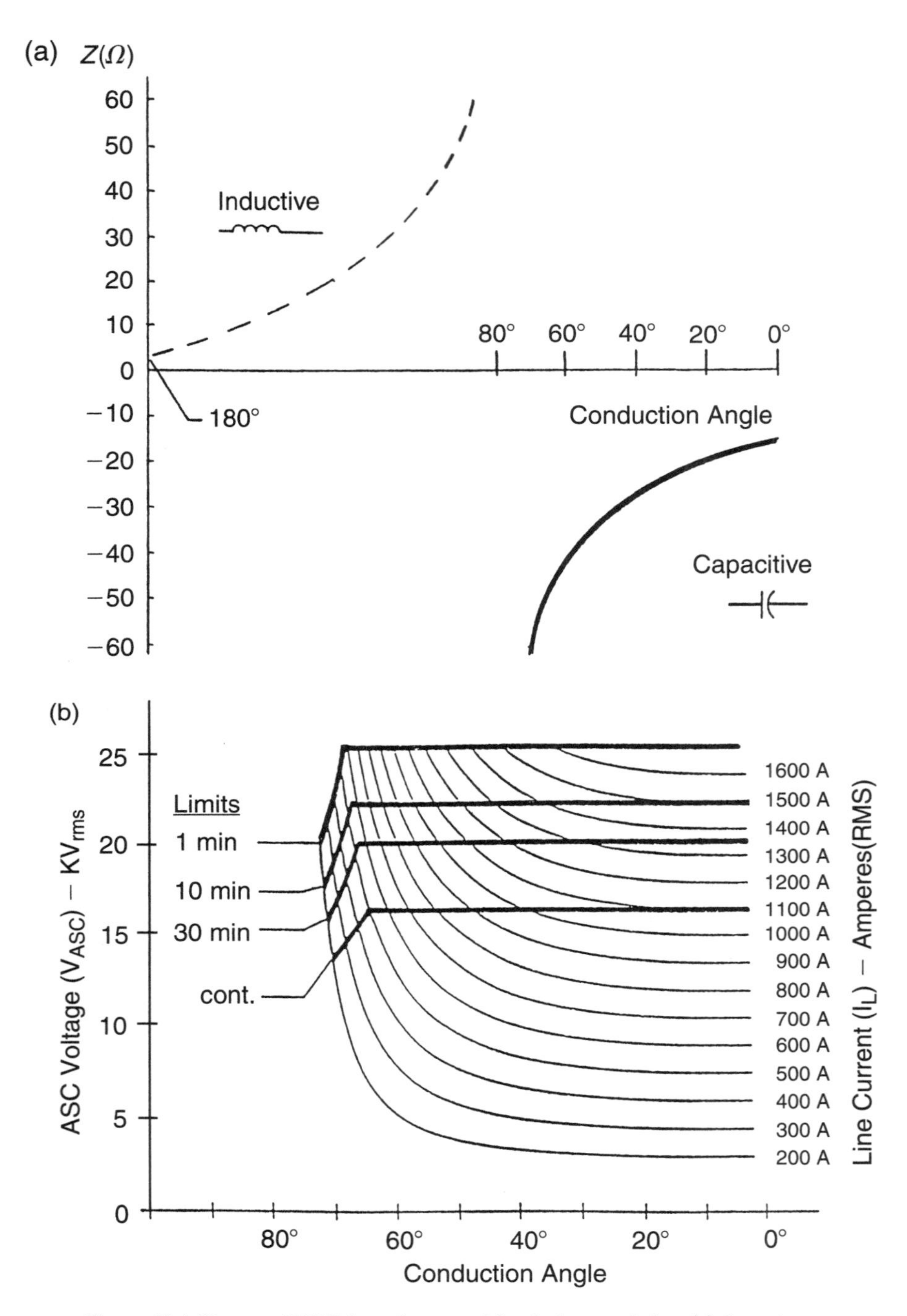

Figure 10.4 Kayenta TCSC impedance and load characteristics: (a) Impedance characteristic; (b) Time-dependent overload range.

the stability and transients program, stability control was not deemed necessary for this installation and was not installed. Performance of Kayenta ASC was also evaluated on a TNA simulator study for the SSR interaction. This study and the onsite field measurements at Kayenta and nearby Four Corners and San Juan generating stations confirmed that this TCSC appears inductive at subsynchronous frequencies. Since SSR with 70% conventional capacitor is not a problem for this line, and because the TCSC appears inductive at the SSR frequencies, no special damping modulation feature is incorporated. In the Wait mode the thyristor switches are turned off and the capacitor operates as a 15 Ω conventional capacitor.

The Thyristor Switched Reactor and the Wait modes may be entered automatically from the Constant Current and Constant Impedance modes if the system conditions are abnormal including faults and initial energization.

The Kayenta Substation is operated from WAPA control center in Montrose, Colorado and accordingly the ASC operation and control has been integrated into Montrose SCADA consoles. Tests showed that with an SCADA order to go from Wait mode to Constant Impedance, transition took place in four to five cycles for the impedance to reach a final steady-state value. An order to change from the Wait mode to a set Constant Current showed that the system required approximately one second to reach steady state. For an order to change from Thyristor-Switched Reactor mode to Constant Impedance mode, the current in the thyristors stops for about two cycles after which the current pulses start to rise and reach steady state in about 15 cycles.

The auxiliary power for the electronic circuits on the platform is derived from the secondaries of platform mounted current transformers. Therefore during low current operation this power source is inadequate and the platform control and protection is lost. The capacitor is not bypassed and remains in service as a conventional capacitor with the thyristors blocked during low current and transition during power reversal. It is recommended that for future projects a small auxiliary power be made available at the platform level.

The capacitor banks are protected by gapless Metal Oxide Varistors (MOVs) which limit the overvoltages.

The digital protection scheme as shown by Figure 10.5 includes:

- Line current supervision monitors the current to detect a bypass circuit breaker discrepancy.
- Capacitor overload protection monitors the accumulated overload and changes the operating point of the TCSC to reduce the voltage across the capacitor in accordance with the time-dependent overload capability.
- Capacitor unbalance protection to detect unbalance caused by blown internal fuses or shorted capacitors.
- MOV overload protection monitors the rise and the rate of rise of temperature. If the threshold level is exceeded, the capacitor and MOV are bypassed by the thyristor switch in order to reduce the voltage across the capacitor and the MOV. The conventional series capacitors on the other hand are bypassed by circuit breakers, once the overload capability limit is reached.
- MOV failure protection compares the currents through two split branches of each MOV.
- Thyristor-Controlled Reactor (TCR) overload protection includes independent overload protections for the valves and the reactors based on their thermal models. The heat produced by the thyristor switch is evaluated from the valve

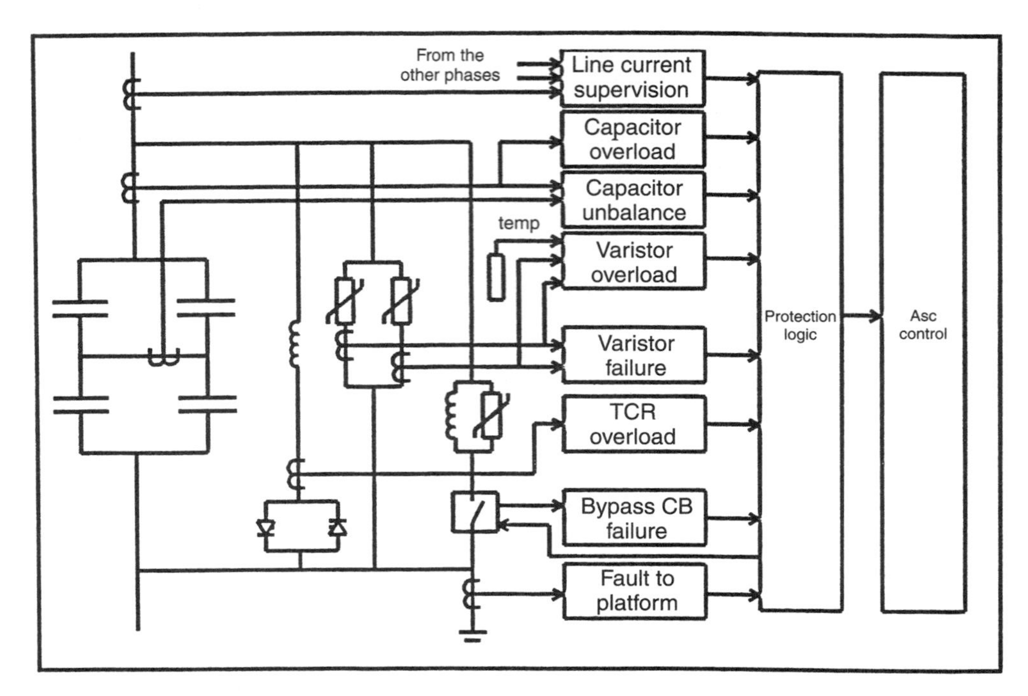

Figure 10.5 Kayenta TCSC protection.

current and the temperature of the cooling liquid and the air are used to calculate the heat dissipation.

- Bypass breaker failure protection initiates line breaker tripping if the bypass breaker fails after a closing command from the line protection.
- Fault-to-platform protection detects equipment-to-platform faults by monitoring the current in the bus connection between platform and one side of the transmission line.

10.1.3.3 Harmonics. It was confirmed that due to low-harmonic impedance of the internal loop of the TCSC compared to the system impedance, the harmonic currents essentially remain in the TCSC loop. Field measurements showed the individual harmonic current components propagating into the ac system to be less than 0.5%.

The inductor size was chosen to ensure that the series resonance frequency with the capacitor does not coincide with any characteristic harmonic. This resonance frequency is approximately 2.5×60 Hz. Figure 10.6 shows waveform of voltage across the fixed series capacitor and the TCSC for a line-to-ground fault on one side of the bank. It is seen that the voltage across the fixed series capacitor is essentially 60 Hz clamped by the MOV. However, the voltage across the TCSC is initially limited by the MOV, then reduced by the impedance control but has a significant 2.5×60 Hz harmonic. Had the resonance been at the third harmonic, the harmonic would have been greatly amplified.

10.1.4 Results of the Project

Results of the project confirmed that the ASC/TCSC:

- Provides continuous control of the transmission line current
- Provides dynamic control of power flow

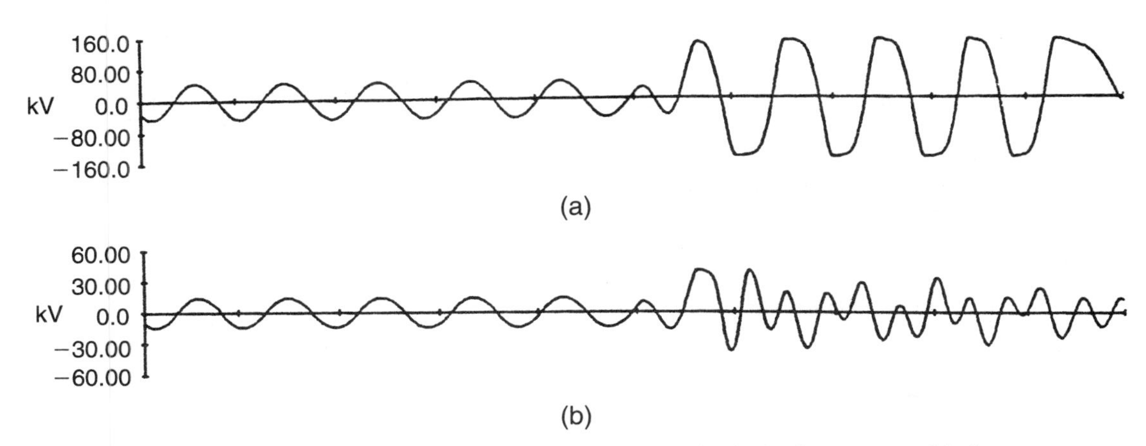

Figure 10.6 Voltage waveforms for Kayenta ASC for single-phase to ground fault: (a) Voltage across the fixed capacitor; (b) Voltage across the TCSC.

- Reduces fault current
- TCSC is inductive/resistive at the SSR frequencies and thus detunes and damps SSR oscillations
- Reduces and quickly eliminates the dc offset of the capacitor voltage
- Provides improved protection of the series capacitors and is an effective means of rapid reinsertion of the capacitors following bypass by the thyristors
- Is a reliable means of using existing transmission capacity while maintaining system security

REFERENCES

Agarwal, B. L., Hedin, R. A., Johnson, R. K., Montoya, A. H., and Vossler, V. A., "Advanced Series Compensation (ASC): Steady State, Transient Stability and Subsynchronous Resonance Studies," *Proceedings of Flexible AC Transmission Systems Conference,* Boston, MA, May 1992.

Christi, N., Hedin, R., Johnson, R., Krause, P., and Montoya, A., "Power System Studies and Modeling for the Kayenta 230 kV Substation Advanced Series Compensation," *AC and DC Power Transmission, IEE International Conference Publication Series 5,* pp. 33–37, September 1991.

Feldman, W., Juette, G., Montoya, A. H., Sadek, K., Schultz, A., Torgerson, D. R., and Vossler, B. A., "230 kV Advanced Series Compensation, Kayenta Substation (Arizona), Project Overview," *EPRI Conference on Flexible AC Transmission System,* Cincinnati, OH, November 1990.

Hedin, R. A., Henn, V., Weiss, S., Montoya, A. H., and Torgerson, D. R., "Advanced Series Compensation (ASC) Transient Network Analyzer Studies Compared With Digital Simulation Studies," *Proceedings of Flexible AC Transmission Systems Conference,* Boston, MA, May 1992.

Hedin, R. A., Weiss, S., Mah, D., and Cope, L., "Thyristor Controlled Series Compensation to Avoid SSR," *EPRI Conference on Flexible AC Transmission System,* Baltimore, October 1994.

Jalali, S. G., Hedin, R. A., Pereira, M., and Sadek, K., "A Stability Model for the Advanced Series Compensator," *IEEE Transactions on Power Delivery,* vol. 11, no. 2, April 1996.

Jalali, S. G., Hedin, R. A., Weiss, S., Johnson, R., and Vossler, B., "Applications of the TCSC for Inter-Area Mode Damping," *EPRI Conference on the Future of Power Delivery,* Washington DC, April 1996.

Jalali, S. G., and Lasseter, R. H., "Harmonic Instabilities in Advanced Series Compensators," *Proceedings of Flexible AC Transmission Systems Conference,* Boston, MA, May 1992.

Jalali, S. G., Lasseter, R. H., and Dobson, I., "Dynamic Response of a Thyristor-Controlled Switched Capacitor," *IEEE Transactions on Power Delivery,* vol. 9, no. 3, pp. 1609–1615, July 1994.

Juette, G., Luetzelberger, P., Schultz, A., McKenna, S. M., and Torgerson, D. R., "Advanced Series Compensation (ASC); Main Circuit and Related Components," *Proceedings of Flexible AC Transmission Systems Conference,* Boston, MA, May 1992.

Krause, P. E., Osburn, B., Torgerson, D. R., Renz, K. W., and Weiss, S., "Kayenta ASC System Performance Special Operating Characteristics," *EPRI Conference on Flexible AC Transmission System,* Baltimore, October 1994.

Pereira, M., Sadek, K., and Jalali, S. G., "Advanced Series Compensation (ASC) Model for Stability Programs," *CIGRE SC 14 International Colloquium on HVDC & FACTS,* Montreal, September 1995.

Torgerson, D., "230 kV Advanced Series Compensation Kayenta Substation (Arizona), Project Overview," *EPRI Proceedings, Conference on Flexible AC Transmission System,* Cincinnati, OH, November 1990.

Weiss, S., Cope, L., Dvorak, L., Krause, P. E., Mah, D., Torgerson, D. R., and Zevenbergen, G., "Kayenta Staged Fault Tests," *EPRI Conference on the Future of Power Delivery,* Washington DC, April 1996.

Weiss, S., Hedin, R. A., Jalali, S. G., Cope, L., Johnson, B., Mah, D., Torgerson, D. R., and Vossler, B., "Improving System Stability Using an Advanced Series Compensation Scheme to Damp Power Swings," *Proceedings of IEE Sixth Annual Conference on AC and DC Transmission,* pp. 311–314, April–May 1996.

10.2 BPA'S SLATT THYRISTOR-CONTROLLED SERIES CAPACITOR (TCSC)

10.2.1 Introduction and Planning Aspects

This TCSC, dedicated in September 1993, is installed at the Slatt 500 kV Substation in Bonneville Power Administration (BPA) system in Oregon. The project was sponsored by Electric Power Research Institute (EPRI), and designed, installed, and delivered by General Electric to BPA for system tests and operation and ownership. After extensive field tests and evaluation it was put into commercial operation in 1995. As shown in Figure 10.7, the TCSC is in series with the Slatt-Buckley 500kV line, such that it is essentially an integral part of the Pacific Northwest 500 kV AC transmission system. Pacific Northwest system is part of a very large regional system, the Western States Coordination Council (WSCC) grid system, which includes all of the Western United States and Western Canada. Connected to Slatt substation is Portland General Electric's Boardman coal-fired power plant located approximately 15 miles from the Slatt substation.

Being the first of its kind, the location for the installation was not primarily selected according to the maximum need on the BPA system, but rather to expose the TCSC to severe operating conditions and, also, to gain sufficient operating benefits and experience. The objectives of the project were to produce a commercial multimodule TCSC, to test it thoroughly under severe conditions, and to operate it continuously as an integral part of BPA transmission system. Apart from being the highest ac voltage (500 kV), in the WSCC system, this location has high short-circuit current (20.3 kA), and it also provides an opportunity for evaluating TCSC interaction with a thermal power plant. If

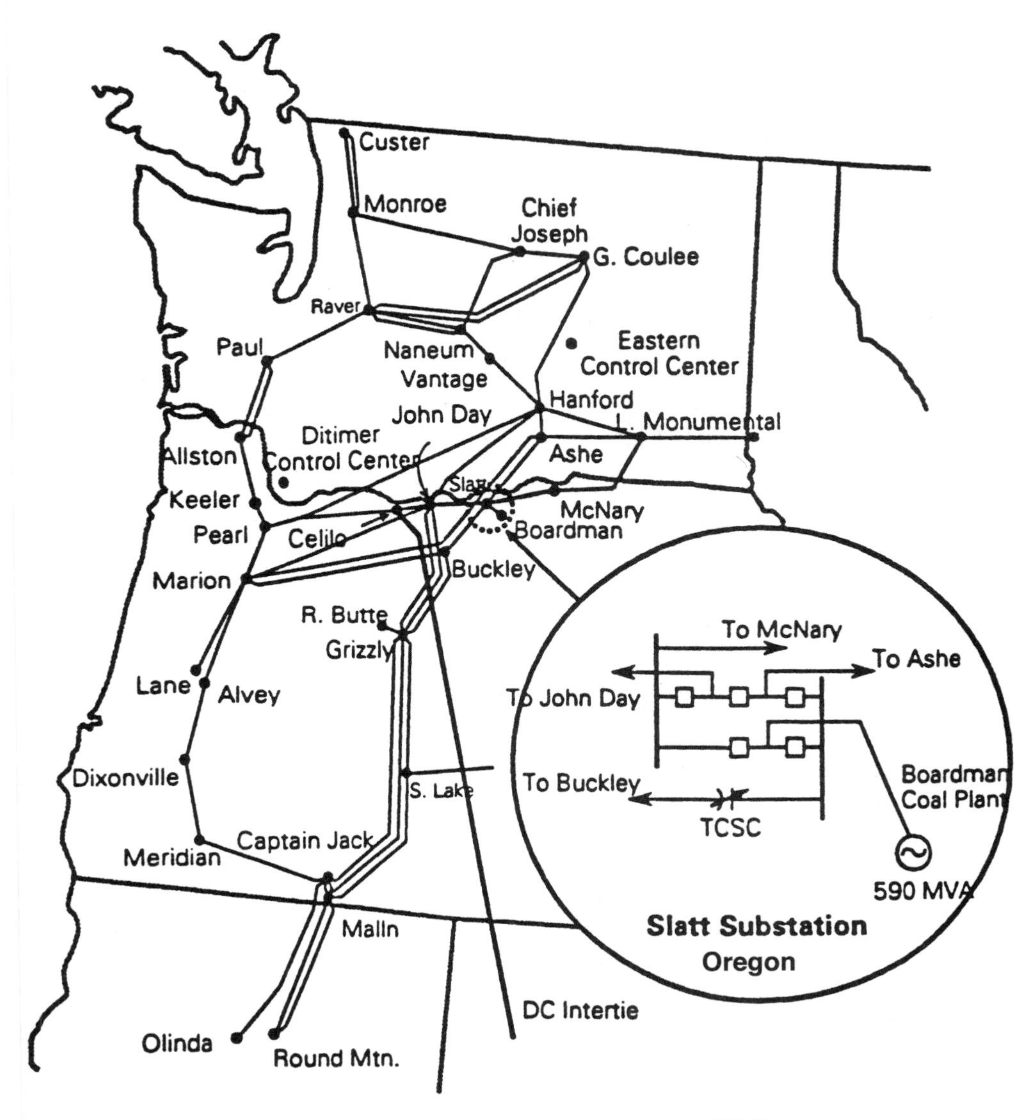

Figure 10.7 BPA's Slatt TCSC location in Pacific Northwest 500kV transmission system.

and when additional transmission capacity becomes necessary, there would be a proven alternative with adequate performance specifications to serve future transmission needs. Nevertheless this TCSC has provided benefits of transmitting more power from Northwest to Southwest and alleviated SSR concerns for the nearby Boardman thermal power plant. The key aspects of performance for this project include:

- Exacting technical and operating specifications
- Control of steady-state current sharing with respect to other parallel paths
- Reduce transient power system swing
- Damp dynamic and small signal power system oscillations

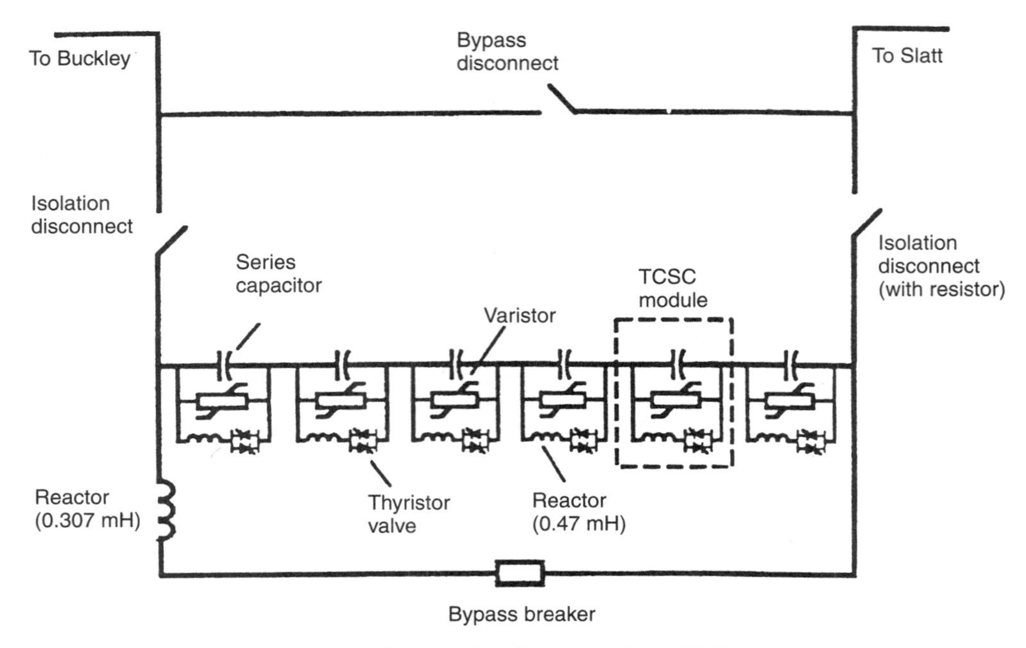

Figure 10.8 One-line diagram of Slatt TCSC.

- Damp SSR under a wide variety of conditions (including resonance) involving the nearby Boardman turbine generator
- Fault performance for a variety of fault conditions
- Possible interaction with nearby Celilo HVDC converter terminal
- Possible interaction with Keeler Static VAR Compensator

Figure 10.8 shows the one-line diagram of the Slatt TCSC. It consists of six identical modules connected in series, each consisting of a capacitor, an ac thyristor switch, reactor in series with the switch, a gapless metal oxide varistor, and module level control and protection. The entire TCSC is provided with a bypass breaker with a *di/dt* limiting inductor in series, for use in operational and protective functions with high speed mechanical bypass, and also disconnect switches for isolating the TCSC.

Figure 10.9 shows an aerial photograph of one platform of the installation. There are three such platforms corresponding to the three phases. On each platform the six thyristor switches are housed in three cabinets, each containing switches for two liquid-cooled modules. Directly behind the valve cabinets are the reactors, six for the thyristor switches and one for the bypass breaker. Behind the reactors are the varistors. At the other end of the platforms (opposite end to the thyristor cabinets) are the six capacitor racks. Each module has 70 capacitor units in parallel.

10.2.2 Functional Specifications

Overall TCSC parameters are:

- Nominal capacitive reactance (29% series compensation with all six modules) 8.0 Ω

Figure 10.9 Thyristor-Controlled Series Capacitor at BPA Slatt Substation.

■ Continuous effective capacitive reactance (with normal minimum thyristor control)	9.2 Ω
■ Net inductive reactance (all six modules bypassed)	1.2 Ω
■ Maximum dynamic range of capacitive reactance	24.0 Ω
■ Nominal three-phase mvar compensation	202 mvar
■ Rated current (1.0 p.u.)	2900 A rms
■ 30 minute overload current (1.5 p.u.)	4350 A rms
■ 10 second overload current (2.0 p.u.)	5800 A rms
■ 10 second three-phase mvar compensation	404 mvar
■ Maximum fault current through TCSC	20.3 kA rms
■ Maximum fault current in the thyristor valve	60 kA peak
■ Operation with ambient temperature range	−40–+45°C

Each module is independently controllable and can be operated with thyristors fully blocked, in which case its impedance is 1.33 Ω capacitive. Each can also be fully bypassed with its thyristor switch (thyristors conducting with zero delay angle), in which case its impedance is approximately 0.2 Ω inductive. In this case, the module impedance is a parallel combination of 0.18 Ω inductor and 1.33 Ω capacitor. For an input current of I_{ac}, a current of 1.15 I_{ac} flows through the inductor and −0.15 I_{ac} through the capacitor; i.e., a current of 0.15 I_{ac} circulates in the inductor-capacitor loop. Being inductive in this bypass mode, it can also serve as a current limiter, as required.

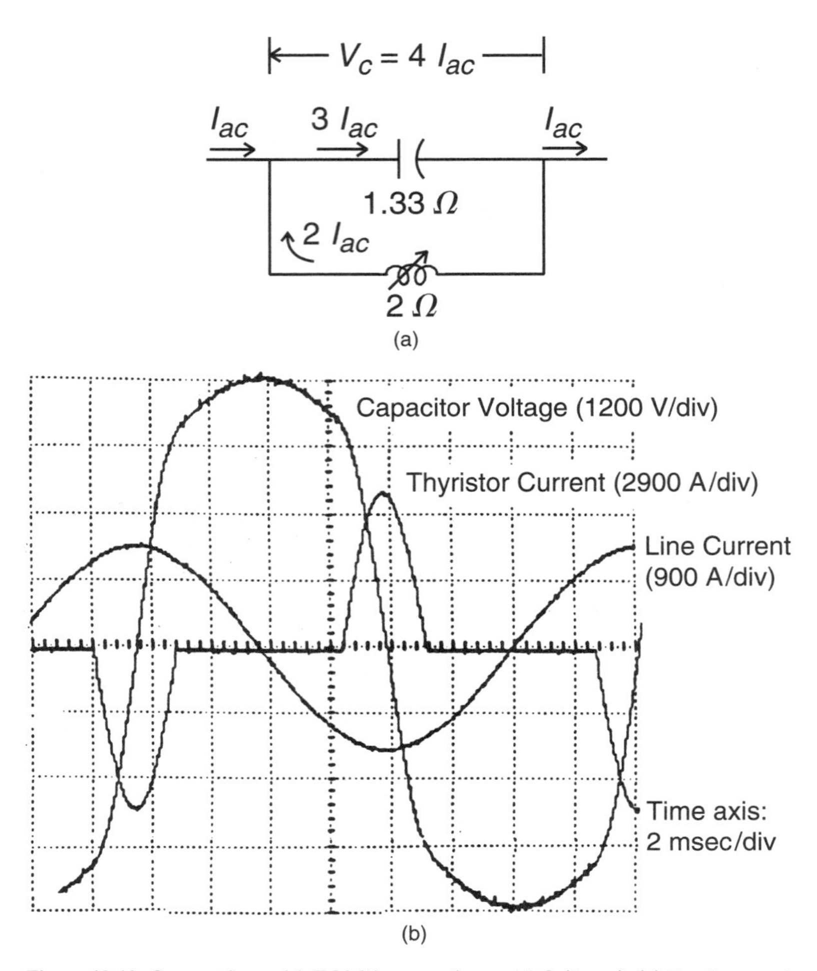

Figure 10.10 Current flow with TCSC in operation at 4.0 Ω (3 p.u.): (a) Fundamental current flow at 4.0 Ω impedance; (b) Module current and voltage waveforms at 4.0 Ω impedance.

Each module can also be operated with its thyristor switch turn-on angle controlled from 180 degrees (1.33 Ω capacitive) to lower than 180 degrees up to a range where its impedance is 4.0 Ω capacitive. In this case the module is an effective combination of a 2.0 Ω inductor in parallel with a 1.33 Ω capacitor at fundamental frequency. Impedance of the inductor with partial conduction has effectively decreased from infinity at 180 degrees to 2.0 Ω at some angle of advance from 180 degrees. For a fundamental input current of I_{ac}, a fundamental current of 2.0 I_{ac} flows in the inductor (and the thyristor switch) and 3.0 I_{ac} flows in the capacitor, i.e., 2.0 I_{ac} circulates in the inductor-capacitor loop, as shown by Figure 10.10(a).

At 4.0 Ω, the fundamental voltage across the capacitor would be $4I_{ac}$. Given the limitation of capacitor voltage capability, the continuous current rating at 4.0 Ω module-impedance is limited to a lower limit than the rated current of 2900 A. Figure 10.10(b) shows current and voltage waveforms of one module operating with impedance of 4.0 Ω (3.0 pu) at about 1000 A line current. The scales for these waveforms have been chosen such that, if the thyristor switch was off, the capacitor voltage waveform would be at the same peak level as the line current waveform.

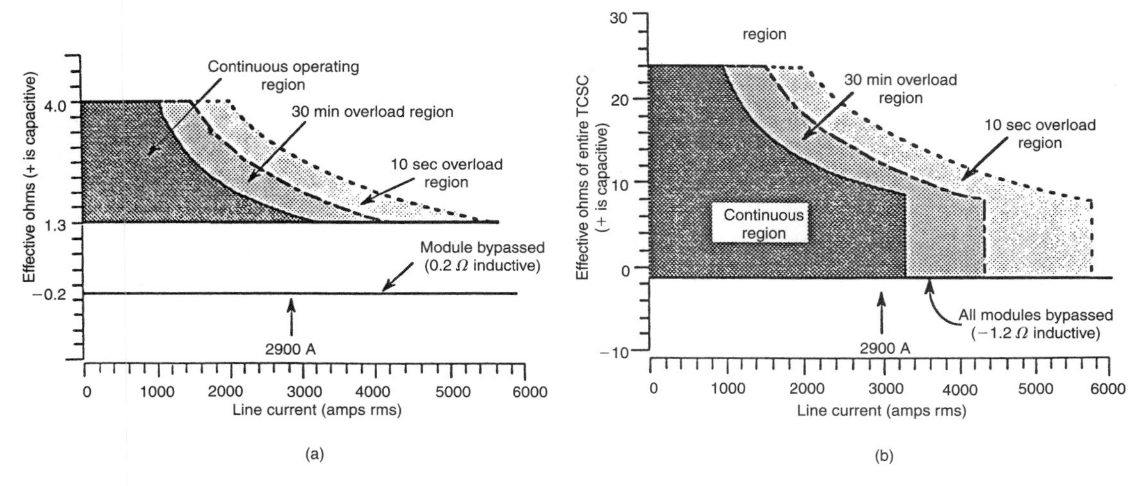

Figure 10.11 Capability curves for Slatt TCSC: (a) Capability curves for one TCSC module; (b) Capability curves for the entire six-module TCSC.

Figure 10.11(a) shows the impedance-current capability curve for one module. It shows that it can continuously operate at high ohmic value up to 4.0 Ω capacitive, but with declining current capability. Also shown are the areas of short time overload capabilities, which show that the TCSC's dynamic operating range is twice its continuous rating. It can operate at 16 Ω for 10 seconds and 12 Ω for 30 min. The temporary overload limit is determined by the IEEE standard 824, for series capacitors. A module can be operated at any point in the shaded areas and also on the line of −0.2 Ω inductive. An individual module cannot be operated between −0.2 Ω and 1.33 Ω.

However, since the individual modules can be operated independently at any point of their characteristic, including bypass with their thyristor switches, it is easy to realize that a combination of six modules would cover the entire range of operation from −1.2 Ω to 26 Ω, as shown in Figure 10.11(b). It is also evident that if all six modules were to be combined into one large unit, the operating characteristic would be the same as that for one module, Figure 10.11(a) with ohmic scale multiplied by 6. There would therefore be a prohibited impedance zone from −1.2 Ω to 8.0 Ω. The continuous ohmic control range would only be from 8.0 Ω to 26 Ω and along the bypass mode line of −1.2 Ω, as against the entire range of −1.2 Ω to 26.0 Ω. This is a clear advantage of the modular approach to the design of a TCSC, apart from the fact that the modular approach allows continued partial operation of the TCSC in the event of problems with individual modules.

It should be mentioned that the modules are not generally operated with thyristors blocked but rather at an angle somewhat less than 180 degrees. It is operated at 1.53 Ω instead of 1.33 Ω (9.2 Ω for all six modules) in steady-state minimum capacitive mode, to provide damping for SSR modes as per the NGH scheme discussed in Chapter 9.

This TCSC is not designed to operate as a thyristor-controlled series reactor.

10.2.3 Design and Operational Aspects

10.2.3.1 Thyristor Switches. Short-circuit current requirement of 60 kA peak for several cycles for the thyristors, rather than the load and overload requirement, was the determining factor for the thyristor design. These thyristors have 100 mm cell diameter and 3.3 kV blocking voltage capability. Because of the high short-circuit current requirement, thyristors have lower voltage rating than would be otherwise necessary. The thyristors are designed with a gate spread-out structure that ensures rapid spreading of the gate current and hence high *di/dt* capability and uniform turn-on with very high current level. Each thyristor switch has five thyristors in series including one redundant thyristor; thus for a two-way switching, each module has a total of 10 thyristors.

For some reason, the platform power supply for the thyristor control is derived from the platform current transformers. During low power on the ac line, there is not enough power to operate the TCSC and therefore it has 0.2 p.u. current as the minimum normal operating limit.

Valves being outdoors and located on 500 kV platforms, they are actively cooled by an ethylene glycol and water mix and the coolant is carried from the ground to the 500 kV platform through hollow porcelain insulators. There are two such insulators per platform; one carries the coolant up and the other down. The equipment for this primary cooling loop is located indoors at ground level. Heat from the coolant is transferred to the outside air via six dry heat exchangers located outdoors.

10.2.3.2 Control, Protection, and Monitoring. All control protection and monitoring is digital, located at ground level and housed in a control room. Signal transmission to and from the platform and the control room is via fiber optics. Figure 10.12 shows a one-line diagram of the control, protection, and monitoring system. The diagram is essentially self-explanatory. The Operator Interface System (OSI) and the Common Control and Protection (CCP) represent the master level functions, which supervise the six Module Control and Protection Systems (MCPs). Each Valve Interface System (VIS) sends control firing pulses to the thyristor switch units, and receives the status report back from the thyristor switch units using a fiber optic pair for each thyristor. The operation of six modules is automatically coordinated by CCP; i.e., each module receives its individual ohms-order from CCP.

The modular structure allows a module to be taken out of service, to carry out any repair and maintenance of a module's ground-based equipment with the remaining modules of the TCSC still in service.

Master level control functions include Ω or power order entry by local or remote operator, power flow regulator, power swing damping control, transient stability control, the module order distributor logic which computes the individual module orders based on higher level control functions, and the automatic TCSC connection and isolation sequences. Also provided is an automatic module rotation function, which ensures that all the modules see the same average duty.

The module level controls include generation of thyristor firing pulses, fast and stable firing angle control algorithm, and automatic overload limits.

Common level protections include the line over-current (timed and instantaneous), line under-current (timed), faults internal to each platform, platform to ground fault, cooling system failure, bypass breaker failure, station battery under-voltage, and control failure.

Module level protections include capacitor overvoltage (timed), capacitor failure, varistor over-temperature, varistor failure, module voltage imbalance, thyristor failure, thyristor gating failure, valve interface system failure, and control failure. Most module protections only cause the thyristor switch to bypass at module level. One thyristor failure in a string of five thyristors results only in an alarm so that the thyristor can later be replaced at an appropriate time, but the second failure causes a permanent

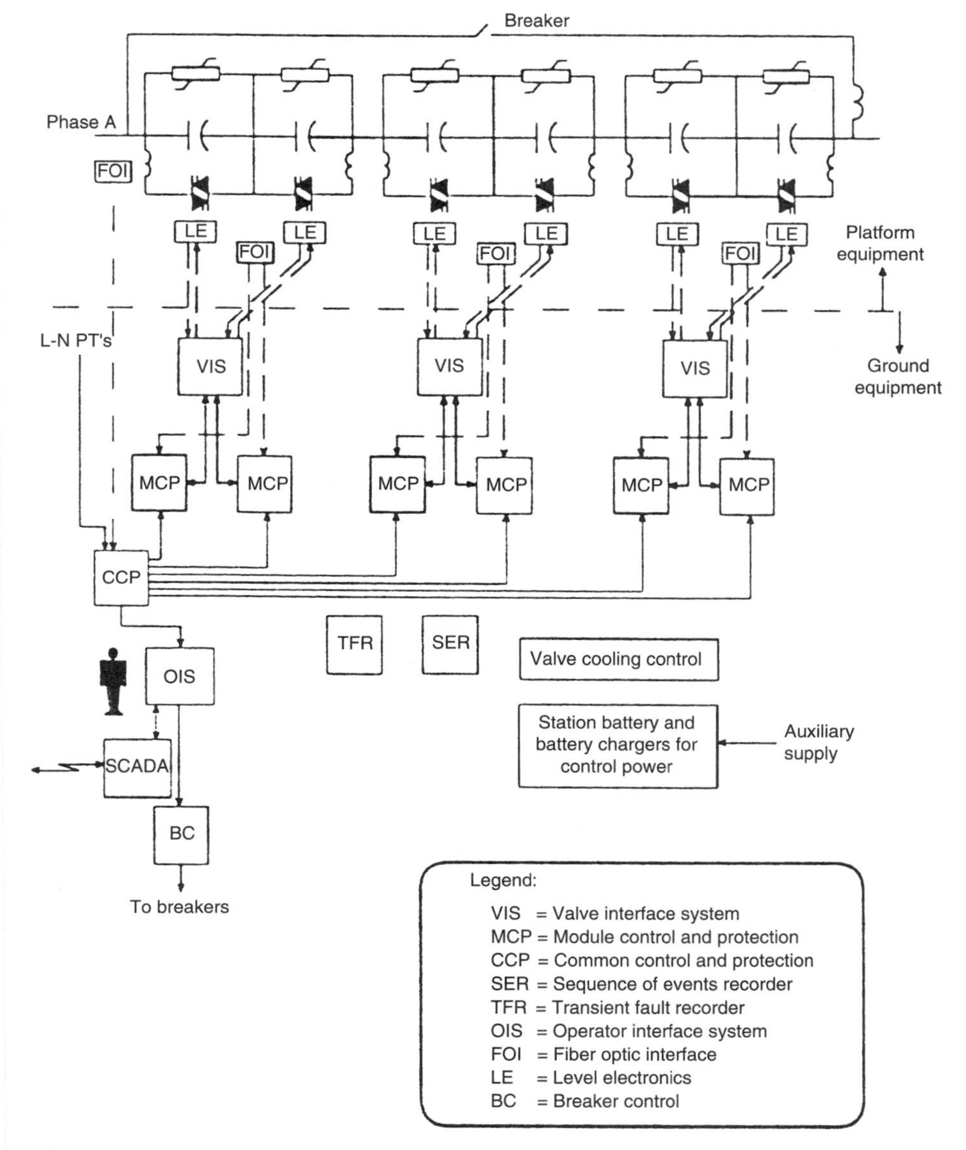

Figure 10.12 Major control and protection elements of Slatt TCSC.

bypass of the module. Thyristor gating-failure protection detects repetitive false firing and repetitive failure to fire and gives appropriate alarm.

10.2.3.3 Power System Stability. The Transient Stability Control, a fast open loop control, is provided, under which the TCSC immediately goes to full 24 Ω capacitive or any other set level, taking advantage of the short-term overload rating. It is also set to go to full-capacitive level when a pole trip occurs at the nearby Celilo HVDC Converter Station. TCSC then ramps back to 12 Ω after 9 sec. The objective is to maximize the power flow as fast as possible, to compensate for loss of electrical power flow through other lines and reduce acceleration of the machines. TCSC over-

TABLE 10.1 Maximum Individual Harmonic Current Generated by TCSC: All Six Modules in Operation

Harmonic Order	Loop Current A rms	Line Current A rms
3	1676	4.5
5	1074	2.4
7	450	1.2
9	176	0.2
11	174	0.6
13	127	0.8

load rating can be used for many other types of transient events and more complex controls to aid first swing stability.

Power Swing Damping control (PSDC) is also provided. Simulator tests showed this to be a very powerful damping means. This control has a dead band such that it stops further action at a low level of oscillations where the system damps oscillations itself with its inherent damping. The bandwidth of PSDC is greater than 5 Hz.

10.2.3.4 Harmonics. As mentioned before, in a thyristor control mode, current circulates in the loop consisting of the capacitor, the thyristor switch, and the inductor. Actually, the current in the thyristor valve is made up of partial-cycle waves, as seen in Figure 10.4(b), and consists of the fundamental and high level of harmonics as is typical of a Thyristor-Switched Reactor discussed in Chapter 5. There would be a natural concern that these harmonics may enter the system. Study results have shown, however, that as expected, this harmonic current remains within the loop and very little harmonic current enters the transmission line. Table 10.1 shows the maximum value of each harmonic that can be present in the loop and the corresponding level of harmonic entering the line for TCSC operation anywhere within its continuous rating. These harmonic levels are not simultaneous, but are maximum values at different operating points. Field measurements showed that the actual harmonics, adding up to a THD level of 1.4% in the line, were essentially from the Celilo converter station and more or less remained the same with the operation of TCSC up to 24 Ω.

10.2.3.5 SSR Performance. The worst case for the Boardman plant for its shaft torsional interaction with the electrical system is to be connected radially through the TCSC series capacitor and the Slatt Buckley line [Figures 10.13(a) and 10.7].

TNA measurements of the electrical damping (the transfer function from rotor speed to electrical torque) versus torsional frequency are shown in Figure 10.13(b). The Boardman plant has four torsional frequencies, 11.9 Hz, 23.3 Hz, 27.3 Hz, and 49.9 Hz; the last frequency corresponds to the coupling between the generator and exciter at the end of the shaft. The top trace is for the p.u. electrical damping with no series capacitor (series capacitor mechanically bypassed). The middle trace shows damping with the series capacitor in line but with the thyristor switches blocked. It shows a negative peak for electrical damping at which Boardman plant has its fourth torsional mode and is clearly unstable with this series compensation. It represents serious damaging contingency, if there were only five uncontrolled segments in service, and other transmission lines from Slatt were opened, thereby having the plant operate in a radial mode all the way to Buckley, as shown in Figure 10.7(a). Given the electrical system parameters, the other three torsional frequencies are too far away for possible

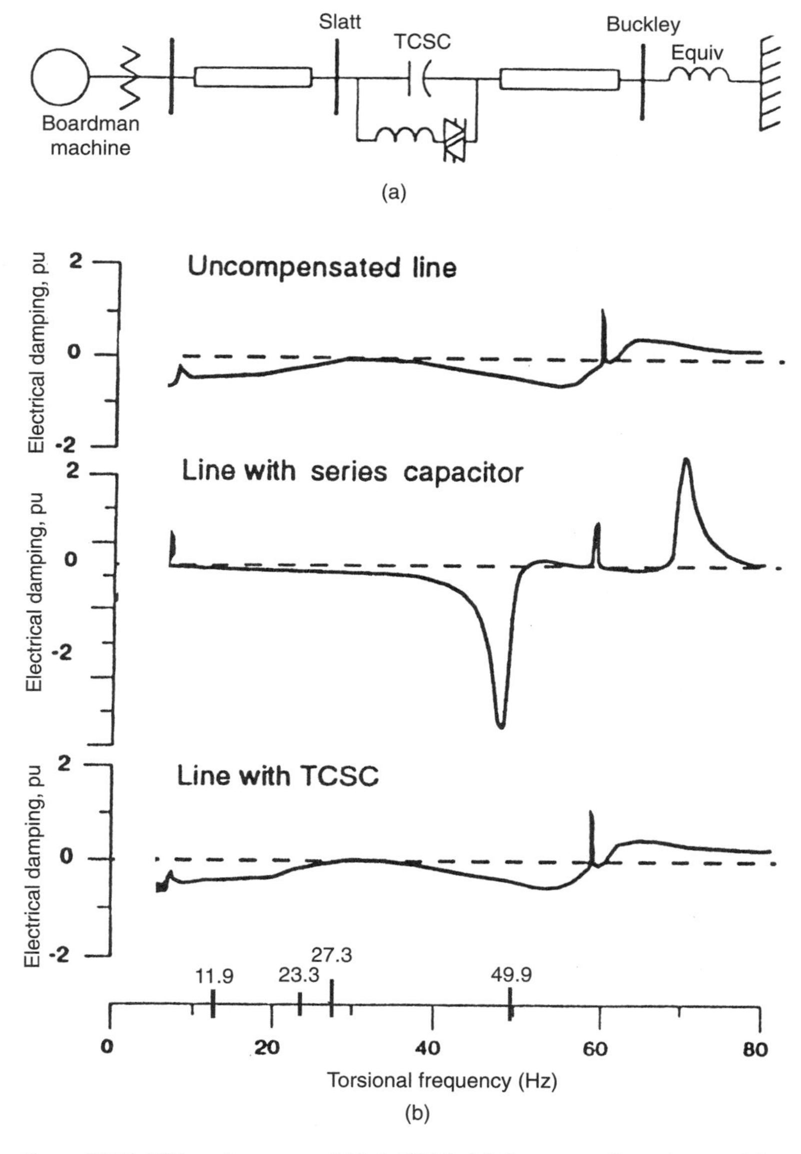

Figure 10.13 SSR performance of Slatt TCSC: (a) System configuration used for TCSC SSR performance tests; (b) Electrical damping plots comparing SSR performance—results of TNA tests.

interaction. The bottom trace in Figure 10.6(b) shows electrical damping with TCSC in operation with the same series compensation. It is seen that the electrical damping is essentially the same as the top trace without any series capacitor compensation.

Field tests were carried out by isolating the Boardman plant as per Figure 10.13(a). These tests showed that with the thyristor switch off, the decay time constant for deliberately excited oscillation at each modal frequency was approximately in the range of 2.5 to 4.0 sec, with 0, 1, 2, 3, 4, and 6 modules. However, with five modules and thyristor switches off, the decay time constant became infinity, i.e., the damping coefficient became zero. Thus with five series capacitor modules in series, mode 4, once excited by a disturbance would not decay. With thyristor switch turned on, the decay time constant remained in the range of 2.5 to 4.0, with any number of zero to six modules in service.

Tests were also carried out to find out what would happen if only one module had thyristor control in order to assess the possibility that only part of the series compensation may be TCSC. The finding was remarkable in that just one module, out of six, with thyristor control reduced the decay time constant to a safe level of a few seconds. This clearly indicates that the TCSC is not only SSR neutral but only a small part of the series compensation needs to have thyristor control, that is, if the purpose of compensation is to ensure security from subsynchronous resonance. If the value-added consideration requires control of current flow in order to impact transient, dynamic, or steady-state power flow, then more modules with thyristor control would be needed. As a matter of fact one of the reasons for selecting this site was the realization that the Boardman plant was subject to SSR at its mode 4.

10.2.4 Results of the Project

The results of this project can be summarized as:

- TCSC is SSR neutral.
- TCSC provides powerful damping against SSR in the presence of uncontrolled series capacitors, so that one can accommodate high percentage series compensation with only part of the series compensation provided with thyristor control.
- TCSC is an effective means of impedance and current control, and the TCSC has been used as such to transmit more power.
- TCSC is a powerful means of damping power system swings, at the transient, dynamic, and small signal basis.

REFERENCES

Hauer, J. F., Erickson, D. C., Wilkinson, T., Eden, J. D., Donnelly, M. K., Trudnowski, D. J., Piwko, R. J., and Bowler, C., "Test Results and Initial Operating Experience for the BPA 500-kV. Thyristor-Controlled Series Capacitor Unit at Slatt Substation. Part 2: Modulation, SSR, and Performance Modeling," *EPRI Conference on Flexible AC Transmission System,* Baltimore, October 1994.

Hauer, J. F., Mittelstadt, W. A., Piwko, R. J., Damsky, B. L., and Eden, J. D., "Modulation and SSR Tests Performed on the BPA 500-kV Thyristor-Controlled Series Capacitor Unit at Slatt Substation," *IEEE Transaction on Power Systems,* vol. 11, no. 2, May 1996.

Kinney, S. J., and Reynolds, M. A., "Slatt TCSC Operational Experience and Future Development," *EPRI Conference on the Future of Power Delivery,* Washington DC, April 1996.

Kinney, S. J., Reynolds, M. A., Hauer, J. F., Piwko, R. J., Damsky, B. L., and Eden, J. D., "Slatt Thyristor Controlled Series Capacitor System Test Results," *CIGRE 1995 Symposium on Power Electronics in Electric Power Systems,* Tokyo, May 1995.

Kinney, S. J., Mittelstadt, W. A., and Suhrbier, R. W., "Test Results and Initial Operating Experience for the BPA 500-kV Thyristor-Controlled Series Capacitor Unit at Slatt Substation. Part 1: Design, Operation, and Fault Test Results," *EPRI Conference on Flexible AC Transmission System,* Baltimore, October 1994.

Kosterev, D. N., Kolodziej, W. J., Mohler, R. R., and Mittelstadt, W. A., "Robust Transient Stability Control Using Thyristor-Controlled Series Compensation," *Proceedings IEEE Fourth Conference on Control Applications,* Albany, NY, pp. 215–220, September 1995.

Kosterev, D. N., Mittelstadt, W. A., Mohler, R. R., and Kolodziej, W. J. "An Application Study for Sizing and Rating Controlled and Conventional Series Compensation," *IEEE Transactions on Power Delivery,* vol. 11, no. 2, April 1996.

Larsen, E. V., Bowler, C., Damsky, B., and Nilsson, S., "Benefits of Thyristor Controlled Series Compensation," *CIGRE Paper 14/37/38-04,* 1992.

Larsen, E. V., Clark, K., Miske, Jr., S. A., and Urbanek, J., "Characteristics and Rating Considerations of Thyristor-Controlled Series Compensation," *IEEE Transactions on Power Delivery,* vol. 9, no. 2, April 1994.

Larsen, E. V., Clark, K., Wegner, C., Nyati, S., Hooker, J. K., Delmerico, R. W., and Baker, D. H. "Thyristor Controlled Series Compensation-Control Design and Dynamic Performance," *Proceedings of Flexible AC Transmission Systems Conference,* Boston, MA, May 1992.

McDonald, D. J., Urbanek, J., and Damsky, B. L., "Modeling and Testing of a Thyristor for Thyristor-Controlled Series Compensation (TCSC)," *IEEE Transactions on Power Delivery,* vol. 9, no. 1, pp. 352–359, January 1994.

Mittelstadt, W. A., Furumasu, B., Ferron, P., and Paserba, J., "Planning and Testing for Thyristor Controlled Series Capacitors," *Proceedings of Flexible AC Transmission Systems Conference,* Boston, MA, May 1992.

Nyati, S., Wegner, C. A., Delmerico, R. W., Piwko, R. J., Baker, D. H., and Edris, A. A., "Effectiveness of Thyristor Controlled Series Capacitor in Enhancing Power System Dynamics: An Analog Simulator Study," *IEEE Transactions on Power Delivery,* vol. 9, no. 2, pp. 1018–1027, April 1994.

Okamoto, H., Kurita, A., Clark, K., Larsen, E. V., and Miller, N. W., "Modeling and Performance of Multiple Multi-module TCSCs in ATP," *CIGRE Group 14 Session,* Paper 14-307, 1995.

Paserba, J. J., and Larsen, E. V. "A Stability Model for Thyristor Controlled Series tion (TCSC)," *Proceedings of Flexible AC Transmission Systems Conference,* Boston, MA, May, 1992.

Paserba, J. J., Miller, N. W., Larsen, E. V., and Piwko, R. J., "Thyristor Controlled Series Compensation Model for Power System Stability Analysis," *IEEE Transaction on Power Delivery,* vol. 10, no. 3, pp. 1471–1476, July 1995.

Piwko, R. J., Wegner, C. A., Furumasu, B. C., Damsky, B. L., and Eden, J. D., "The Slatt Thyristor-Controlled Series Capacitor Project—Design, Installation, Commissioning and tem Testing," *Proceedings of CIGRE 35th International Conference,* Paper 14-104, August September 1994.

Piwko, R. J., Wegner, C. A., Kinney, S. J., and Eden, J. D., "Subsynchronous Resonance Performance Tests of the Slatt Thyristor Controlled Series Capacitor," *IEEE Transaction on Power Delivery,* vol. 11, no. 2, April 1996.

Trudnowski, D. J., Donnelly, M. K., and Hauer, J. F., "Estimating Damping Effectiveness of BPA's Thyristor-Controlled Series Capacitor by Applying Time and Frequency Domain Methods to Measured Response," *IEEE Transactions on Power Systems,* vol. 11, no. 2, May 1996.

Urbanek, J., Piwko, R. J., Larsen, E. V., Damsky, B. L., Furumasu, B. C., and Mittelstadt, W., "Thyristor Controlled Series Compensator Prototype Installation at the Slatt 500 kV Substation," *IEEE Trans. Power Delivery,* vol. 8, no. 3, pp. 1460–1469, July 1993.

Urbanek, J., Piwko, R. W., McDonald, D., and Martinez, N., "Thyristor Controlled Series Compensation Equipment Design for the Slatt 500 kV Installation," *Proceedings of Flexible AC Transmission Systems Conference,* Boston, MA, May 1992.

Zhu, W., Spee, R., Mohler, R. R., Alexander, G. C., Mittelstadt, W. A., and Maratukulam, D. J., "An EMTP Study of SSR Mitigation Using the Thyristor-Controlled Series Capacitor," *IEEE Transactions on Power Delivery,* vol. 10, no. 3, July 1995.

10.3 TVA'S SULLIVAN STATIC SYNCHRONOUS COMPENSATOR (STATCOM)

10.3.1 Introduction and Planning Aspects

The first high-power STATCOM in the United States was commissioned in late 1995 at the Sullivan substation of the Tennessee Valley Authority (TVA) for transmission line compensation. The project was jointly sponsored by the Electric Power Research Institute and TVA, and designed and manufactured by the Westinghouse Electric Corporation.

The TVA is an agency of the federal government and also serves in the capacity of a large utility with an installed generation capacity of more than 30,000 MW and supplies electrical power over 16,000 miles (25,806 km) of transmission lines to other utilities and industrial customers in seven states. TVA is part of the Eastern grid of North America.

Northeast Tennessee is served by the Sullivan substation, near Johnson City, TN. Sullivan is supplied by the 500 kV bulk power network and by four 161 kV lines. The Sullivan 500 kV feed supplies about 55% of the winter peak load of 900 MW. Seven distributors and one large industrial customer are served from this substation. The Sullivan site was selected for the STATCOM installation for the following reasons:

1. The Sullivan substation is on the edge of the TVA service area and the ties into the system are, therefore, not as strong as substations inside the bulk power network. The 500 kV bus is tied to the 161 kV bus through a 1200 MVA transformer bank. The Sullivan 500 Kv bus is exposed to high voltages during light load conditions due to the inherent capacitance of the 500 kV transmission lines. The 161 kV bus is exposed to low voltages during peak load conditions. The Sullivan substation site is one of the few locations on the TVA network where the total symmetrical range of the reactive output of the STATCOM, from full inductive to full capacitive, could be utilized without applying fixed capacitor or reactor banks.
2. TVA has an interconnection with American Electric Power (AEP) at the Sullivan substation. AEP's investigation of power system oscillations indicated that some power oscillation may exist on their, and consequently, also on the TVA system. The Sullivan location thus offers the opportunity for TVA to utilize the ability of the STATCOM to damp these oscillations.

The TVA system at the Sullivan substation with the STATCOM is shown in Figure 10.14. When this project was built, STATCOM was called "STATCON," an abbreviation for "static condenser." The term STATCON was later changed to "static compensator" or STATCOM by IEEE and CIGRE. Also, the literature frequently uses the term "inverter" for a voltage-sourced converter instead of the more generally accepted term of "converter," used in this book. Therefore, the reader may frequently encounter these terms in the literature and illustrations. At this site the STATCOM is expected to provide two functions. One is the day-to-day voltage regulation, and the other is the contingency use during a major disturbance to the power system.

The day-to-day operation of the STATCOM is to regulate the 161 kV bus voltage

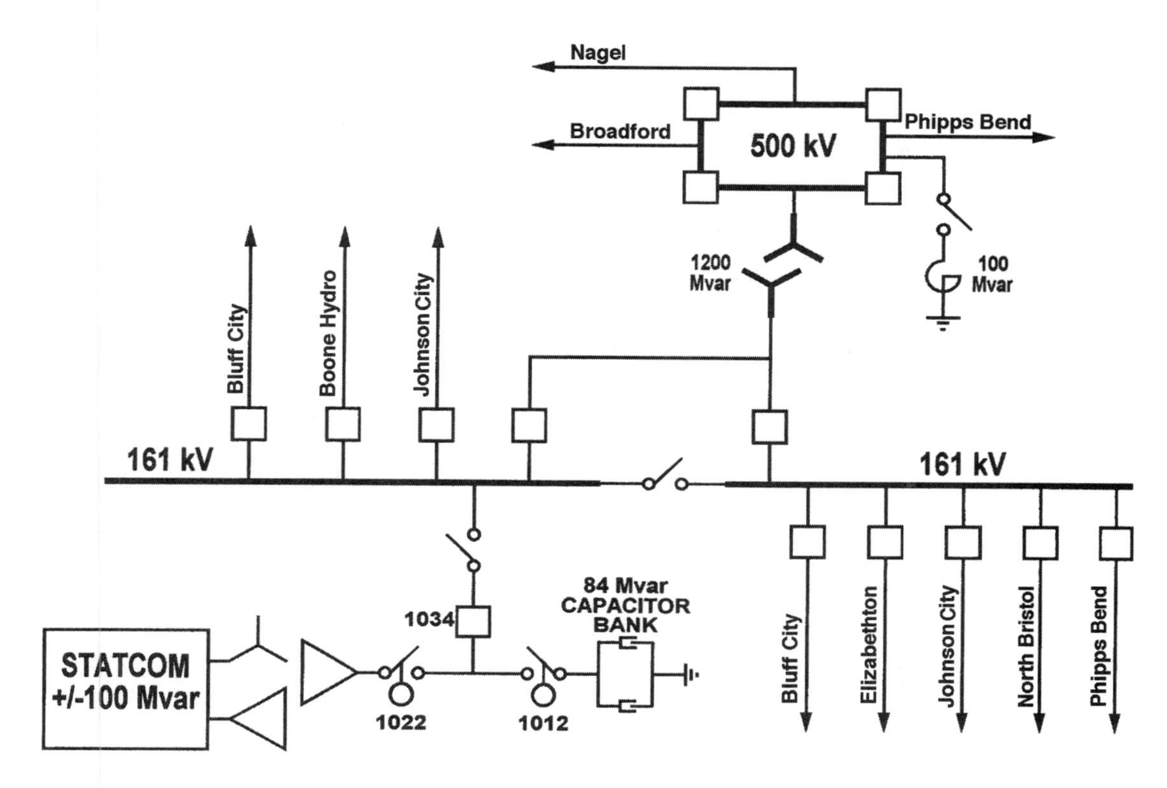

Figure 10.14 Simplified diagram of the TVA system at Sullivan sub-station, showing the STATCOM and adjacent 84 Mvar capacitor bank.

during the daily load buildup so that the tap changer on the transformer bank will be used less often. Most of the failures in the TVA's 500 kV transformer banks have occurred in the tap changer. By using the STATCOM to regulate the voltage, the expected life of the Sullivan transformer bank should be significantly extended. During off-peak periods, TVA experiences high voltage on the 500 kV bus at Sullivan. Originally, 100 Mvar of reactors were installed on one of the 500 kV buses to correct the voltage buildup. One of these reactors had failed and was not replaced, in anticipation that the STATCOM will provide this function.

TVA's standard capacitor bank is 84 Mvar. One switchable bank of capacitors will be installed at Sullivan along with the STATCOM. The switching of the capacitor bank will be controlled by the STATCOM. Thus the STATCOM with the capacitor bank will become an overall Static Var System (SVS) with an effective range of 100 Mvar inductive to 184 Mvar capacitive. This will allow rapid voltage control over a considerable range. If the Sullivan transformer bank is lost during winter peak conditions, the 161 kV bus voltage could drop 10 to 15%. As the load in the area grows over time this drop will become more pronounced and the voltage in this area could collapse. The STATCOM is expected to respond rapidly and maintain the voltage at a reasonable but low-voltage level until shunt capacitor banks at other substations in the general area switch on. Without the STATCOM, TVA would have had to initiate a project to either install a second Sullivan transformer bank or construct the fifth 161 kV line into the area. The STATCOM solution allowed TVA to avoid those large expenditures at least for several years.

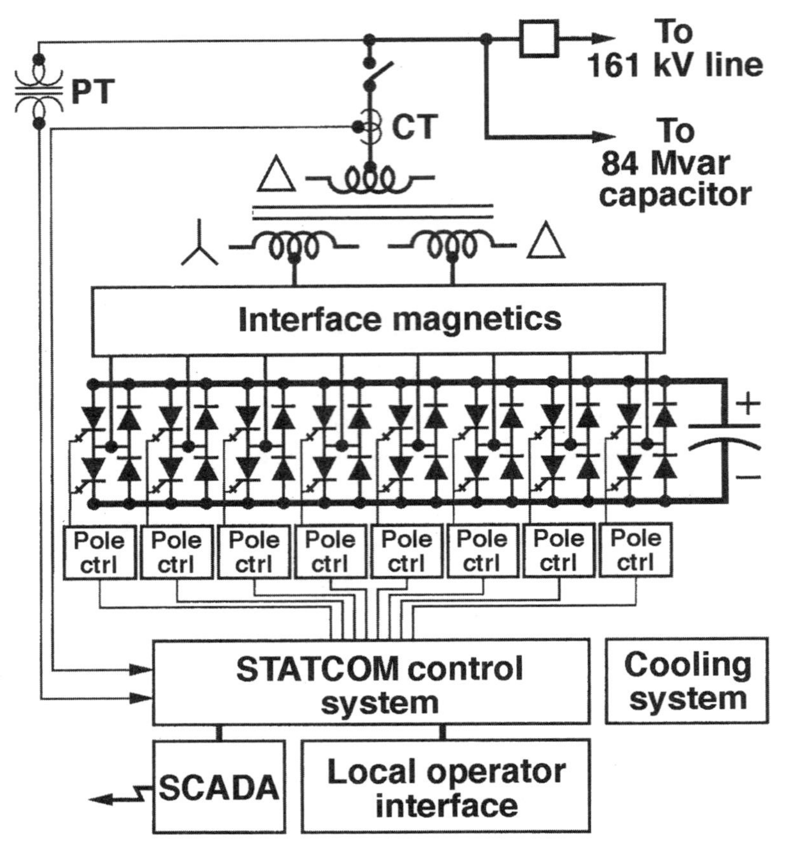

Figure 10.15 Single-line diagram showing the main elements of the TVA STATCOM installation.

10.3.2 STATCOM Design Summary

The TVA STACOM utilizes a harmonic neutralized converter composed of gate turn-off (GTO) thyristor valves, designed to meet utility harmonic requirements without filters and have sufficient redundancy in power semiconductors and other critical circuit elements to provide high operating availability. The main design parameters of the STATCOM converter are as follows:

Nominal capacity	±100 Mvar
Short term capacity	±120 Mvar
Transmission line voltage	161 kV
Converter nominal output voltage	5.1 kV
Number of converter pulses	48
Number of basic six-pulse, three-phase converters	8
Type of converter poles	two-level (two-valve)
Total number of converter poles	24
Number of series GTOs in each valve	5 (one redundant)

GTO rating	4.5 kV, 4kA (peak)
Total number of GTOs	240
Method of multipulse waveform generation	magnetic
Main coupling transformer	161 kV Δ to 5.1 kV Δ/Y
DC capacitor	65 kJ
Nominal dc voltage	6.6kV

Figure 10.15 is a single-line diagram showing the main elements of the STATCOM installation. The eight converter poles comprising 16 GTO valves, depicted symbolically, are associated with one of the three output phases the STATCOM generates. Each converter pole produces a square voltage waveform, progressively phase shifted from one pole to the next by an appropriately chosen angle. These eight square-wave pole voltages are combined by magnetic summing circuits into two voltage waveforms, displaced by 30 degrees. One of these waveforms feeds the wye and the other the delta secondary of the main coupling transformer. The final 48-pulse output voltage waveform is obtained at the transformer primary.

The STATCOM installation is considerably smaller in area than a conventional Static Var Compensator since the reactive power is generated by relatively compact power electronics rather than by capacitors and reactors. The complete STATCOM installation at the Sullivan substation is shown by an aerial photo in Figure 10.16, and

Figure 10.16 Aerial view of the STATCON site showing the building, outdoor components, and the adjacent capacitor bank (top left). (Courtesy of TVA.)

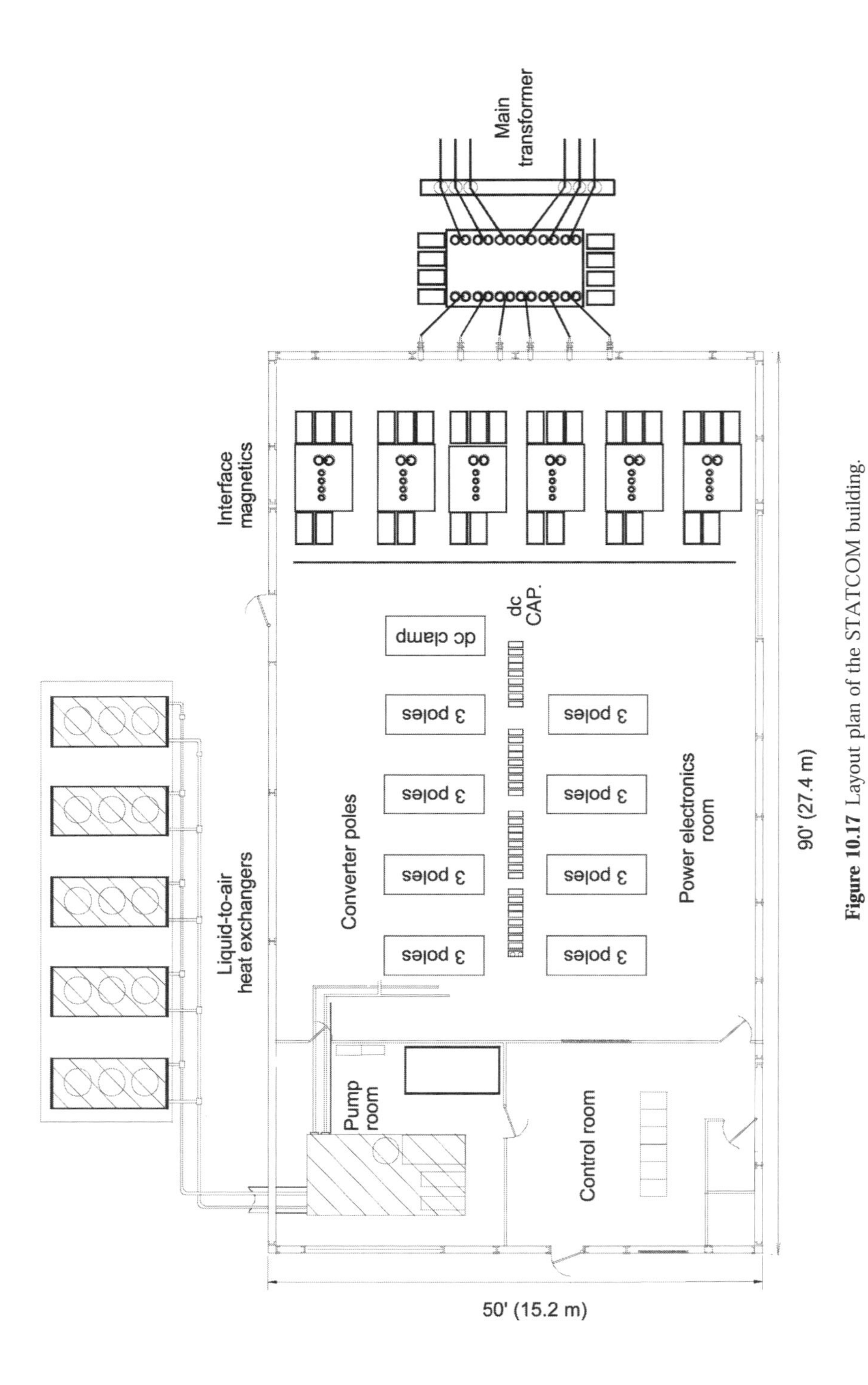

Figure 10.17 Layout plan of the STATCOM building.

the layout plan of the building is shown in Figure 10.17. The overall building size is 90′ × 50′ (27.4 m × 15.2 m), which contains all of the STATCOM converter and associated control and auxiliary equipment. The main transformer that couples the STATCOM to the 161 kV line is located outdoors. The building is a standard commercial design with metal walls and roof.

Figure 10.18 shows a view of the STATCOM valves. Each valve has associated electronic controls and gating circuits located adjacent to it and these communicate with the central control system through optical fiber links. The central control system receives feedback signals from potential transformers on the 161 kV bus and current transformers on the primary bushings of the main transformer. A sophisticated graphical display terminal with comprehensive diagnostic information is provided for local operator interface, but the STATCOM is normally controlled remotely via the SCADA. The power losses in the electronic valves generate heat that is removed by means of cooling fluid (deionized water and glycol mixture), circulated through the valves and then through liquid-to-air heat exchangers located outside the building.

The STATCOM control system, located in a separate room inside the main building, comprises two distinct subsystems, the *status processor* and the *real-time control*. The status processor is responsible for gathering, analyzing, and displaying status and diagnostic information from all of the STATCOM components, and also performs the sequencing tasks necessary for starting up the system, connecting it to the line, and shutting it down. It supports the local operator interface as well as the SCADA interface. The real-time control system is responsible for generating gating

Figure 10.18 View of the STATCOM converter hall showing the valve layout. (Courtesy of Siemens Power Transmission & Distribution, Orlando, FL.)

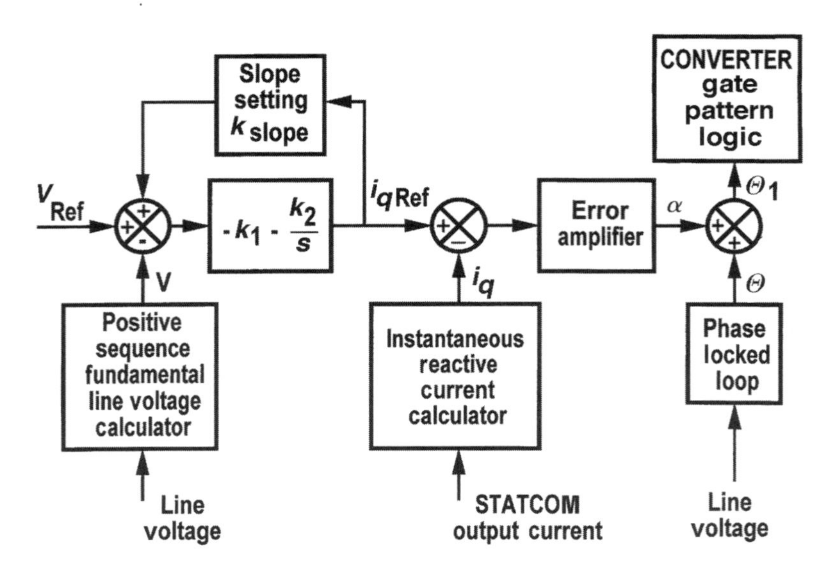

Figure 10.19 Functional block diagram of the STATCOM control system.

commands for the valves in such a way that the STATCOM maintains the correct level of reactive output current to regulate the 161 kV bus voltage to a set reference value. Figure 10.19 shows a block diagram of the algorithm employed in the real-time control system. An inner feedback loop is used to regulate the STATCOM instantaneous reactive current. Note that, as explained in Chapter 5, this control is achieved by varying the phase angle, α, of the converter output voltage relative to the transmission line voltage. This indirect voltage control technique makes it possible to maintain a constant maximum ratio between the converter output voltage and the dc-capacitor voltage. The reference value for the reactive current control loop is generated by an outer loop responsible for the system voltage control. This outer control loop is similar to that used in conventional static var compensators, and includes an adjustable slope setting that defines the voltage error at full STATCOM reactive output. There is an unavoidable delay in the feedback of the voltage regulating loop because of the time taken to compute the positive sequence fundamental bus voltage. Thus although very fast response can be achieved for the reactive current controller, the response time of the voltage regulator is typically about one cycle of the line voltage.

10.3.3 Steady-State Performance

The STATCOM generates a 48-pulse output voltage waveform by electromagnetically combining eight square wave outputs per phase. Figure 10.20 shows waveforms recorded on the wye secondary of the main transformer while the STATCOM was operating at full inductive output. The voltage waveform corresponds to the output of the converter. It closely approximates a sine wave, but the small residual steps can plainly be seen.

However, the current drawn from the 161 kV bus, as evident in Figure 10.20, is a very high quality sine wave, since the residual harmonics in the STATCOM output voltage are greatly attenuated by the leakage inductance of the transformer. Note

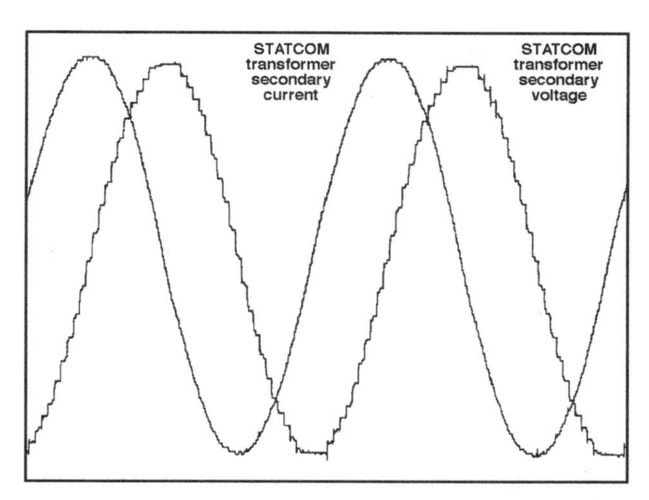

Figure 10.20 Voltage and current waveforms measured on the wye secondary of the STATCOM coupling transformer, operating at full inductive output (maximum var generation).

also that this high-quality current waveform is obtained without the use of any passive filter. Actually, spectral analysis of the line voltage carried out onsite showed that harmonics present on the line before the STATCOM is connected, are actually somewhat reduced when the STATCOM operates, due to the relatively low impedance of the STATCOM at harmonic frequencies.

Under steady-state conditions the STATCOM has operated as expected and provides a useful function in regulating the 161 kV bus voltage. In practice it is found that the STATCOM can typically regulate the voltage by 6 kV going between full inductive and full capacitive limits, since the system short-circuit impedance is about 5300 MVA with the 500 kV intertie present. (Since there are no other Static Var Compensators in the area, the STATCOM is operated with zero slope setting on the voltage control.) As expected, the STATCOM has taken over the voltage regulation at the Sullivan substation and, as a result, the on-load tap changing operations on the 500 kv transformer bank have decreased significantly. Figure 10.21 illustrates a typical example of daily load cycle encountered by the STATCOM (without the 84 Mvar capacitor bank).

10.3.4 Dynamic Performance

The dynamic performance of the STATCOM has been excellent in terms of response as well as settling times. This, as explained in Chapter 5, is partially due to the negligible internal transport lag and partially to the lack of resonant filter components at the output.

To measure the response of the instantaneous reactive current control, Figure 10.19, a test signal was injected into $i_{q\mathrm{Ref}}$ in place of the usual signal derived from the voltage controller. In effect this puts the STATCOM in a *var control mode* instead of the usual voltage control mode. The instantaneous reactive current reference was stepped from inductive to capacitive and vice versa. The resulting step responses are shown in Figures 10.22 and 10.23. These results show the ability of the STATCOM to make a 100 Mvar transition in a few milliseconds, with essentially no transient

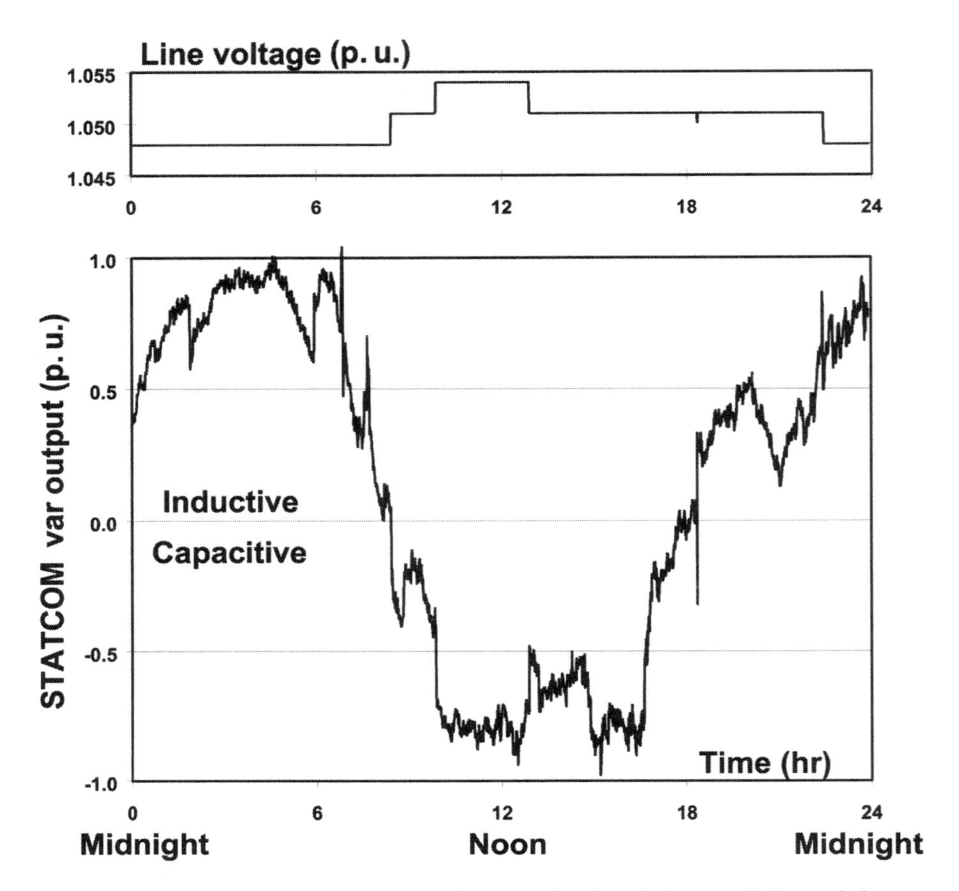

Figure 10.21 Log of TVA STATCOM operation for a 24-hour period recorded on July 4, 1996.

disturbance on the bus voltage (such as would be obtained with a capacitor switching or with the use additional passive filter networks). The rapid response of the STATCOM, as mentioned, is possible because of the very small transport delay in controlling the 48-pulse converter, which enables a very high bandwidth current control. Conventional thyristor-based static var compensators typically have much longer transport delays and are used in conjunction with passive filtering that can give rise to resonance problems.

The STATCOM was also extensively tested in normal operating mode of terminal voltage regulation. For this test the voltage reference signal V_{Ref} was stepped from 1.032 to 1.022 while operating in the normal voltage control mode, causing a 70 Mvar swing in the STATCOM output. The response is shown in Figure 10.24. This result shows stable operation with expected response time. It reflects the typical response of this type of voltage controller, taking about 25 ms (1.5 cycles). Clearly, from the previous results, the var output of the STATCOM can be changed extremely rapidly, but the voltage controller takes time to determine the desired var output based on line voltage measurement.

To determine the effect on the STATCOM of switching the associated 84 Mvar capacitor onto the line, the capacitor bank was switched onto the 161 kV bus with

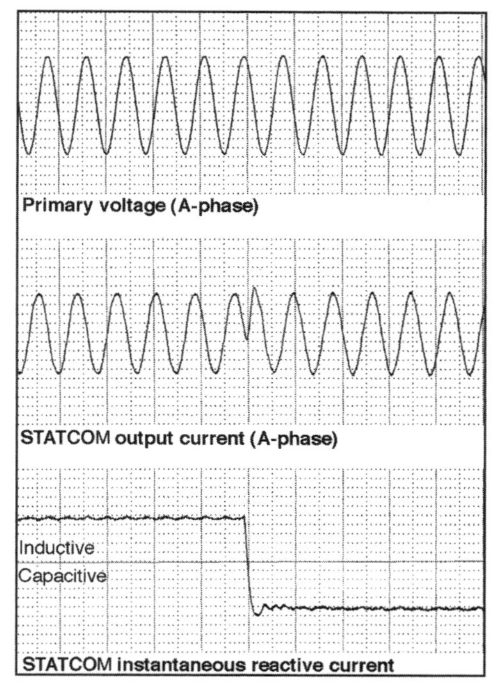

Figure 10.22 Response to a var reference step from 0.5 p.u. inductive to 0.5 p.u. capacitive, when STATCOM operated in var control mode.

the STATCOM operating capacitively in the normal voltage control mode. Figure 10.25 shows the recorded response as the STATCOM changes to the inductive region to maintain the bus voltage at the set value. Note the dc-bus voltage variation shown in Figure 10.25 and also a very small bus voltage transient observed when the capacitor is switched. Clearly, as seen from the effect on its dc bus, the STATCOM plays a part in this event, and serves to attenuate the bus disturbance when the local capacitor bank is switched. Figure 10.26 shows the reverse operation when the capacitor bank is switched out and the STATCOM returns to its original operating point.

The STATCOM performance was investigated for opening of the tie between the 161kV bus and the 500 kV bus at Sullivan (Figure 10.14). This relates to operation under the contingency condition where the 500 kV/161 kV transformer bank is lost. Without the 500 kV connection, the bus impedance seen by the STATCOM via the connections to grid system only through 161 kV is higher and consequently the gain of the voltage controller is increased. The STATCOM must maintain stable operation with the higher gain. The switching was accomplished by opening the tie breaker on the 161 kV side of the transformer bank. The STATCOM was energized prior to the switching, with its output set near zero by voltage reference adjustment. Figure 10.27 shows the measured response when the intertie is opened, and Figure 10.28 shows the reverse operation when the intertie is restored. The response in Figure 10.27 is faster (20 ms vs. 25 ms) because of the higher gain condition, but still stable. The STATCOM responds rapidly in both cases, moving to its capacitive limit in the first case, then back to its initial operating point in the second case.

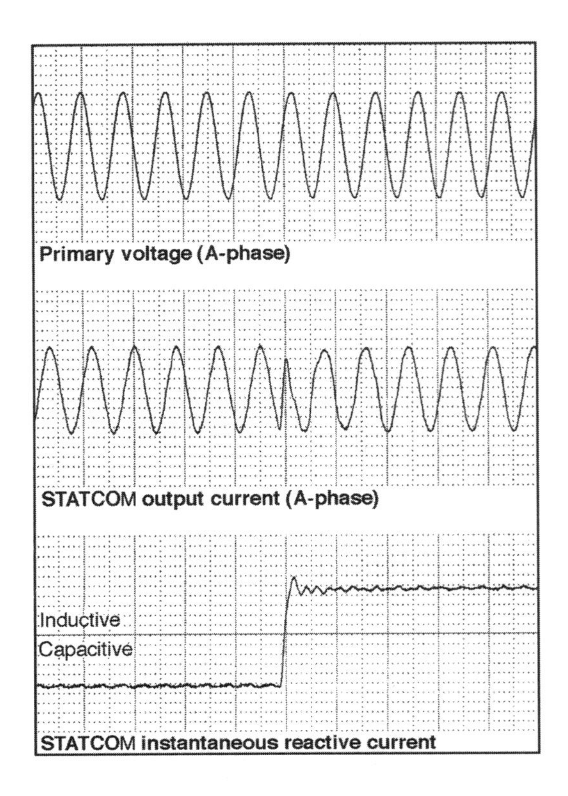

Figure 10.23 Response to a var reference step from 0.5 pu capacitive to 0.5 p.u. inductive, when STATCOM operated in var control mode.

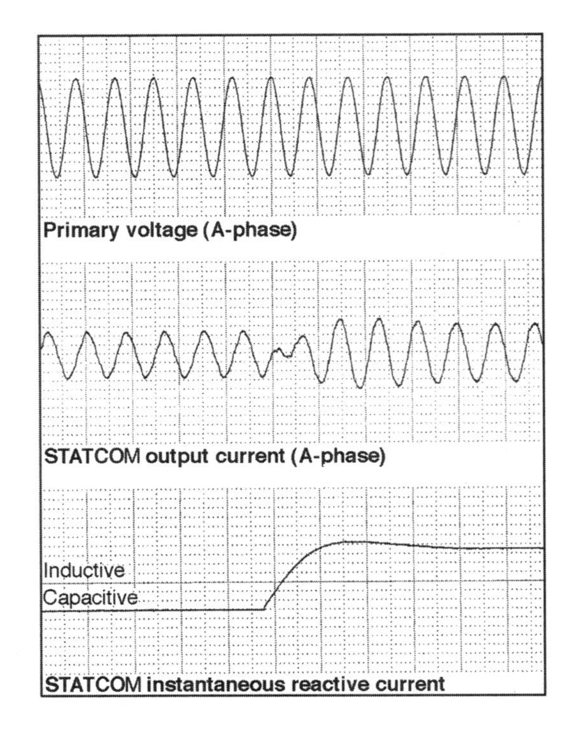

Figure 10.24 Measured response to a step in the voltage reference from 1.032 p.u. to 1.022 p.u., when STATCOM operated in voltage regulation mode. (STATCOM swings from 30 Mvar capacitive to 40 Mvar inductive.)

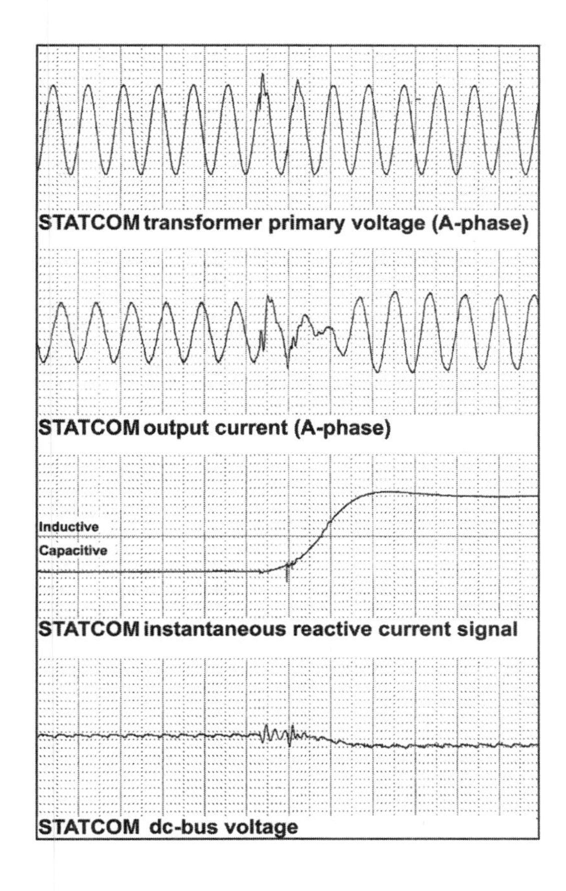

Figure 10.25 Measured STATCON response when switching the adjacent 84 Mvar capacitor bank to the line. STATCON swings from 42 Mvar capacitive to 55 Mvar inductive. Line voltage = 168.4 kV.

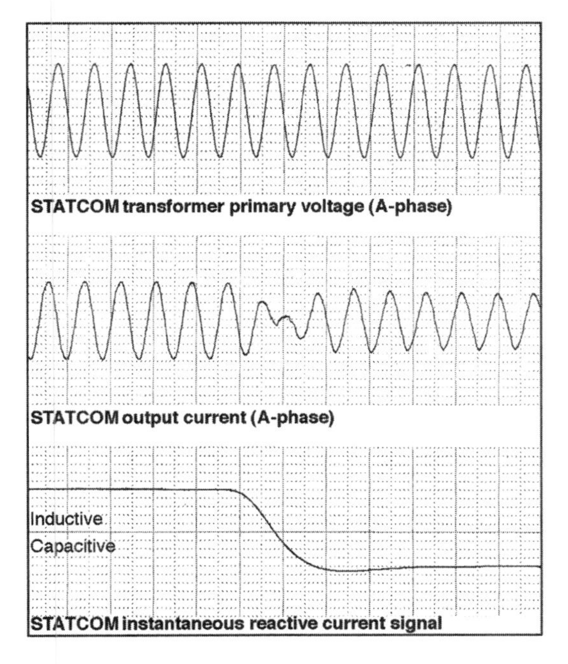

Figure 10.26 Measured STATCOM response when disconnecting the adjacent 84 Mvar capacitor bank from the line. STATCOM swings from 57.4 Mvar inductive to 40 Mvar capacitive. Line voltage = 164 kV.

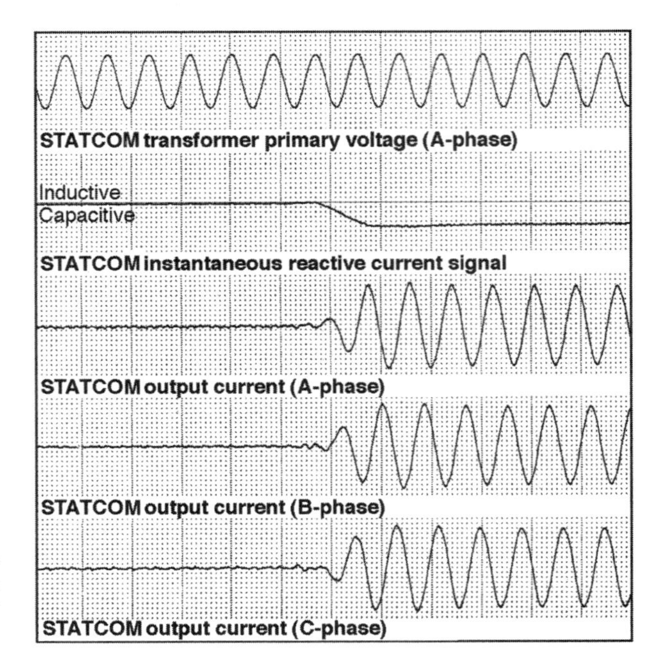

Figure 10.27 Measured response of the STATCOM to the opening of the 161 kV to 500 kV intertie at Sullivan, using the 161 kV circuit breaker.

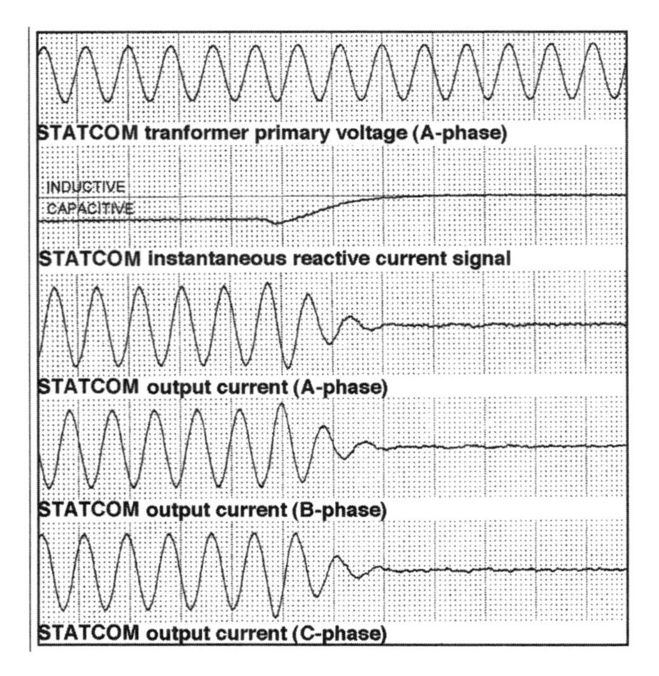

Figure 10.28 Measured response of the STATCOM to the re-closing of the 161 kV to 500 kV intertie at Sullivan, using the 161 kV circuit breaker.

10.3.5 Result of the Project

The STATCOM concept for transmission line compensation has been the subject of considerable interest. Studies have indicated potential advantages in performance, equipment size, installation labor, and total cost. The availability of large GTOs allowed the realization and practical evaluation of STATCOM of representative size in actual utility environment.

The TVA STATCOM has demonstrated the new GTO-based technology at a significant power level for the first time in the United States. In particular, the GTO devices have proved to be rugged and reliable components. This project has showed that the STATCOM is a versatile equipment, with outstanding dynamic capability, that will find increasing application in power transmission systems.

Other FACTS controllers using GTO-based technology have been conceived, and subsequently developed, offering unique potential benefits for power transmission systems. The performance tests at the Sullivan substation provided a valuable information base, and encouragement, for the design of this family of FACTS Controllers.

REFERENCES

Schauder, C. D., et al., "Development of a ± 100 MVAR Static Condenser for Voltage Control of Transmission Systems," *IEEE/PES Summer Meeting, Paper No., 94SM479-6 PWRD,* 1994.

Schauder, C. D., et al., "Operation of ±100 MVAR TVA STATCON," *IEEE, PES Winter Meeting, Paper No. PE-509-PWRD-0-01-1997.*

Schauder, C. D., et al., "TVA STATCON Project: Design, Installation and Commissioning," *CIGRE paper 14-106,* 1996.

10.4 AEP'S INEZ UNIFIED POWER FLOW CONTROLLER (UPFC)

10.4.1 Introduction and Planning Aspects

The first Unified Power Flow Controller (UPFC) in the world, with a total rating of ±320 MVA, was commissioned in mid-1998 at the Inez Station of the American Electric Power (AEP) in Kentucky for voltage support and power flow control. The project was jointly sponsored by the Electric Power Research Institute and AEP, and designed and manufactured by the Westinghouse Electric Corporation.

10.4.1.1 Background Information. American Electric Power is an investor-owned electric utility company that operates across seven midwestern states, from Michigan in the north and west, to Tennessee in the south, and Virginia in the east. These facilities, including the generating plants, 765 kV extra high voltage (EHV), 345 kV high voltage, and 138 kV transmission lines, distribution system, and associated stations, are all interconnected and operated as a major power network within the eastern U.S. power system. AEP provides electric service to approximately 1.7 million customers in an area of about 48,000 square miles (124,313 km^2).

AEP produces over 124 billion KWH a year, representing a peak system demand of over 26,000 MW. A major portion of the System's generation is located on the Ohio river and its tributaries. The areas south and east of these rivers depends on

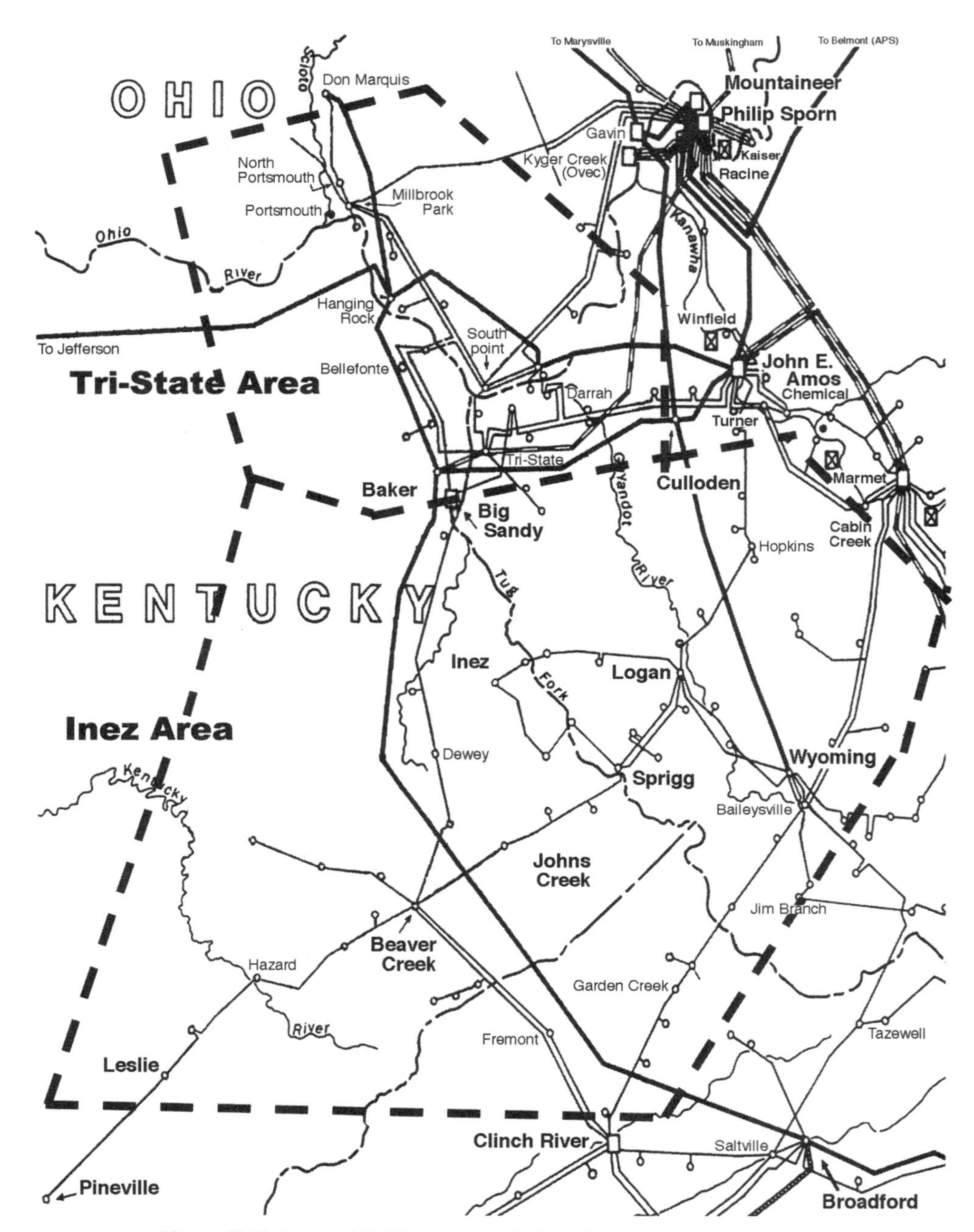

Figure 10.29 Inez and Tri-State areas of the AEP power system.

long transmission lines to meet the electric demand. The load area of concern regarding this project is one of these areas, the Inez Area identified in Figure 10.29.

The project involves the Inez Area and the Tri-State Area which is located just north of the Inez Area, as shown in Figure 10.29. These two areas are quite different in their physical and electrical characteristics. The reinforcement plan adopted to mitigate the problems strongly ties them as one area.

The Inez Area located in eastern Kentucky is rural in nature and has a population of about 670,000 which is spread over an area of about 6300 square miles (16,316 km^2).

Generating plants and EHV/138 kV substations are located only at the periphery of the area. The Inez Area's winter power demand of approximately 2000 MW is served by several long 138 kV transmission lines. System voltages and area reactive power requirements are supported by a $\pm$125 Mvar Static Var Compensator (SVC) and a large assortment of switched shunt capacitor banks located at several 138 kV and lower voltage subtransmission stations.

The Tri-State Area located adjacent to and north of the Inez Area has a population of about 480,000 people and is spread over an area of approximately 3500 square miles (9,064 square km). The Tri-State Area load is summer peaking with a requirement of about 1400 MW. It is served by a strong network of EHV lines, EHV/138 kV substations, and generating plants within and near its boundaries. The area transmission system supports power flows into the Inez Area via 138 kV inter-area line connections.

10.4.1.2 System Performance and Problems. The Inez Area depends on long 138 kV transmission lines to support its customers' demand. Even under normal system conditions, many 138 kV transmission lines carry power flows reaching 300 MVA. These power flow levels are well above the surge impedance loading for 138 kV lines and as such are indications of the system stress. Moreover, these high loadings as compared to the thermal capabilites of the lines left little margin for system contingencies. In addition, voltage levels in the Inez Area were generally as low as 95% of nominal. This is considered to be the lowest acceptable voltage level for reliable service. Heavy 138 kV line loadings and comparatively long line distances resulted in excessive voltage gradients across the transmission network, which was another indication of the stress on the system.

The Tri-State Area, on the other hand, is served by a strong network of EHV lines and an EHV/138 kV station. The load is concentrated in and around a comparatively smaller area. The 138 kV line exits are heavily loaded over relatively short distances. Voltage levels are maintained at all substations within a narrow bandwidth around 98% of nominal.

Several single contingency outages in the Inez Area could have resulted in depressed voltage and/or thermal overload conditions. As an example, a simulation of a 765 kV line outage, during winter peak load periods, indicated increases in system real and reactive power losses by about 85 MW and 700 MVars, respectively. The need to supply these increased system losses under a contingency condition would have further reduced the area voltages to levels which would certainly have resulted in wide area customer complaints. In addition, the underlying transmission system would be loaded to well above its winter emergency capabilities. A second contingency under these circumstances was intolerable.

Single-contingency outages in the Tri-State Area would not have caused any thermal overloads or depressed bus voltage conditions. However, double-contingency outages involving one or more EHV/138 kV transformers or 138 kV generating units would have resulted in thermal overloads of the EHV/138 kV transformer and/or 138 kV line exits.

10.4.1.3 System Reinforcement Plan. The analysis of the system performance indicated that additional voltage support and power supply facilities were required to resolve normal, single-contingency, and double-contingency system problems in the

Inez Area. In addition, the EHV/138 kV transformer and line exit capacities of the Tri-State Area needed to be increased.

Following extensive analysis, the conclusion was reached that constructing a high-capacity 138 kV line having thermal capabilities approaching that of a 345 kV line would provide an economical means of adding thermal capacity to the area. However, such a high-capacity line would not carry its share of line loading based on its high-capacity margin alone. Power flow on such a line would still be governed by its impedance and other ac transmission system parameters. Concurrent with thermal considerations, peak and off-peak voltage performances in the Inez Area also needed to be addressed. This dictated the need for a dynamic voltage support facility in the area.

Series capacitors together with SVCs and converter-based FACTS Controllers were then evaluated to enhance the power flow and provide sufficient voltage support. The evaluation showed that a combined converter-based FACTS Controller, such as the Unified Power Flow Controller (UPFC), which can provide both the voltage and line flow control capabilities, was a logical choice to be an integral part of the overall reinforcement.

The comprehensive reinforcement plan for the Inez and Tri-State areas include the following:

- A high-capacity 950 MVA, 138 kV line between the Big Sandy and the Inez Stations.
- A ±320 MVA UPFC at the Inez Station to fully utilize the high capacity of the new 138 kV line, provide dynamic voltage support, and control several mechanically-switched capacitors in the area.
- A 345/138 kV transformer at the Big Sandy Station in the Tri-State Area to provide for the power flow requirements of the new high capacity 138 kV line.
- Series reactors to constrain loadings on existing thermally-limited facilities.

The reinforcement plan was executed in two phases. The first phase of the plan included the installation of a 345/138 kV transformer bank at the Baker/Big Sandy Station and line power flow limiting series reactors at two stations. The last portion of this first phase involved the installation of the ±160 MVA shunt converter part of the UPFC at the Inez Station. The 345/138 kV transformers were designed to have sufficient capacity to provide the Tri-State Area needs and fulfill the future loading requirements of the high-capacity Big Sandy-Inez 138 kV line. The series reactors were installed to limit the loadings on critical, but low-capacity lines. During the first phase the shunt part of the UPFC functioned as a STATCOM and supported the reactive power and the dynamic voltage needs of the Inez Area. In addition, it provided signals to control the switching operations of several 138 kV shunt capacitor banks in the area.

The second phase of the plan included the construction of a high-capacity (950 MVA) 138 kV line between the Big Sandy and Inez substations, installation of the series part of the Inez UPFC and two 138 kV mechanically-switched shunt capacitor banks at the Inez Station. The series converter was specified to be identical to the shunt converter in order to increase the operating flexibility of UPFC.

The third and the last phase of the project included additional 138 kV line construction, switching configuration changes at several stations, and additional reactive power correction to supplement the existing system reactive power margins.

10.4.1.4 UPFC Operation Strategy. VOLTAGE CONTROL. The UPFC is required to regulate Inez substation 138 kV bus voltages and control six 138 kV shunt capacitor banks (a total of over 330 Mvar) located at the Inez and three other nearby stations. This way proper capacitor switching strategy is established to reduce daily and seasonal voltage fluctuations to within acceptable limits.

During system disturbances, mechanically-switched shunt capacitor banks and associated controls are generally slow to react. Over 60 shunt capacitor banks are connected on the 138 kV and lower voltage transmission system in the Inez Area. Under actual system contingency conditions, all of these banks may not switch on (hunting concerns) or some may over-correct the voltage and lock-out. To resolve this situation, the UPFC is required to maintain a predetermined reactive power margin to maximize the shunt converter's dynamic reactive power reserve for system contingency conditions. This ensures that the controllable reactive power range of the shunt converter (from −160 to +160 Mvar, i.e., a maximum control range of 320 Mvar) is available at all times to compensate for dynamic system disturbances. The shunt capacitor banks will be switched on and off to maintain the reserve UPFC and SVC margins during steady-state load fluctuations.

POWER FLOW CONTROL. With all transmission system facilities in service, the Big Sandy-Inez 138 kV line loading is to be maintained at a level which would minimize system losses. For the peak and near peak load conditions, load flow studies indicate that a 300 MW line loading would meet this objective. Actual flow on the line could be higher depending on other prevailing system conditions. The UPFC, therefore, may be required to slightly reduce the line loadings. Line reactive power flow and its direction will be monitored to help maintain the dynamic reactive power margin of the shunt inverter.

The series power flow control becomes important during contingency conditions. Major double contingency outages heavily load critical 138 kV facilities (especially the Big Sandy-Beaver Creek line). Controls of the UPFC have to monitor loadings on the three critical lower capacity lines. The control objective is to increase the Big Sandy-Inez 138 kV line loading to decrease loadings on the three lines. This control is to be activated as soon as any one of the three line loadings exceeds 90% of their respective emergency thermal ratings. The UPFC has to increase the Big Sandy-Inez line loading until the critical line loadings are reduced below the defined levels or the UPFC reaches its rating limit. Under severe contingency conditions, the UPFC-controlled Big Sandy-Inez line will be capable of transferring 950 MVA (4000 A).

10.4.2 Description of the UPFC

The Unified Power Flow Controller for the Inez Station was designed to meet the above defined system requirements, in particular, to provide fast reactive shunt compensation with a total control range of 320 Mvar (−160 Mvar to +160 Mvar) and control power flow in the 138 kV high-capacity transmission line, forcing the transmitted power, under contingency conditions, up to 950 MVA.

In order to increase the system reliability and provide flexibility for future system changes, the UPFC installation was required to allow self-sufficient operation of the shunt converter as an independent STATCOM and the series converter as an independent Static Synchronous Series Compensator (SSSC). It is also possible to couple both converters together to provide either shunt only or series only compensation over a doubled control range.

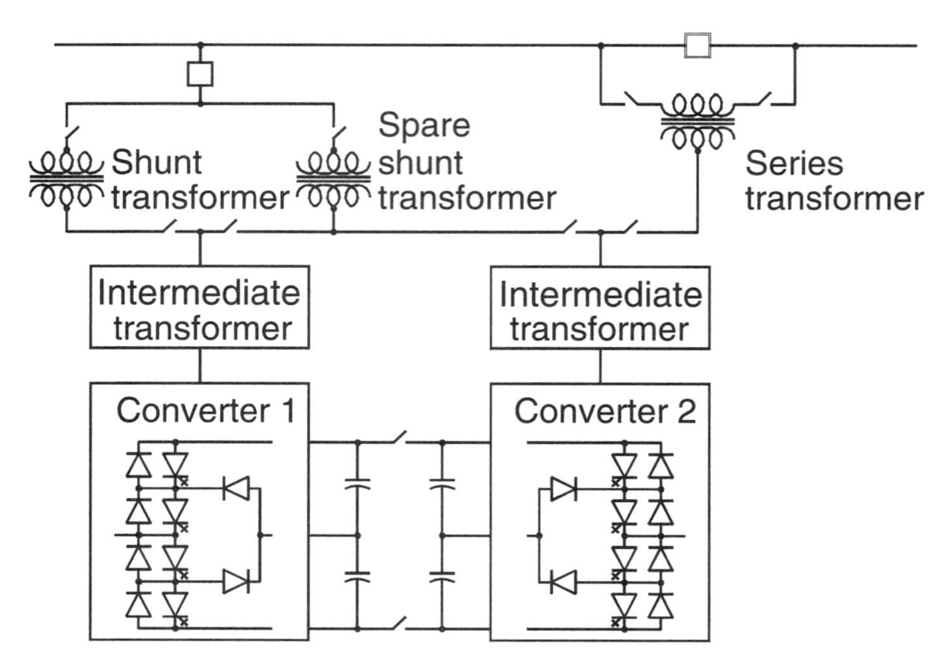

Figure 10.30 Simplified diagram of the UPFC installed at Inez.

10.4.2.1 Power Circuit Structure. The UPFC equipment comprises two identical GTO thyristor-based converters, each rated at ±160 MVA. Each converter includes multiple high-power GTO valve structures feeding an intermediate (low-voltage) transformer. The converter output is a three-phase voltage set of nearly sinusoidal (48-pulse) quality that is coupled to the transmission line by a conventional (three-winding to three-winding) main coupling transformer. The converter-side voltage of the main transformer is 37 kV phase-to-phase (for both shunt and series transformers). The shunt-connected transformer has a 138 kV delta-connected primary, and the series transformer has three separate primary windings each rated at 16 % of the phase voltage.

To maximize the versatility of the installation, two identical main shunt transformers and a single main series transformer have been provided, as illustrated in Figure 10.30. With this arrangement, a number of power circuit configurations are possible. Converter 1 can operate as a STATCOM with either one of the two main shunt transformers, while Converter 2 operates as an SSSC. Alternatively, Converter 2 can be connected to the spare main shunt transformer and can operate as an additional STATCOM. With the latter configuration a shunt reactive capability of ±320 MVA becomes available. The power circuit arrangement indicates the priority of shunt compensation at this location.

The converters are constructed from three-level poles, each composed of four valves. The basic circuit for a three-level type pole, together with a typical output voltage waveform, is illustrated symbolically in Figure 10.31. The three-level pole offers the additional flexibility of a step in the output voltage that can be controlled in duration, either to vary the fundamental output voltage or to assist in waveform construction. Each converter used 48 valves in 12 three-level poles with a nominal dc voltage of 24 kV (+12kV and −12kV with respect to the midpoint). The valves are composed of a number (eight in the outer and nine in the inner valves) of 4500 V,

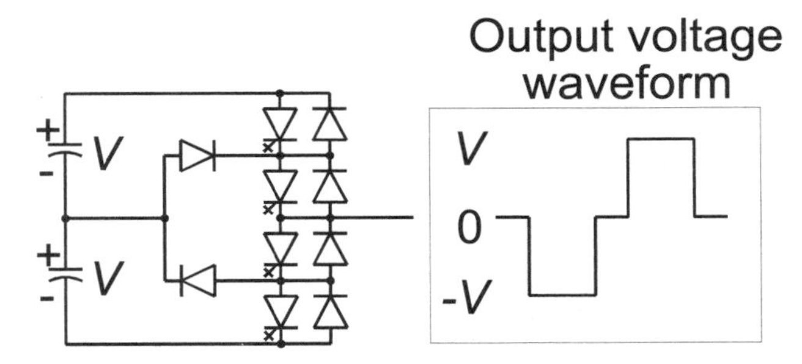

Figure 10.31 Basic three-level converter pole employed in the Inez UPFC and associated output voltage waveform.

4000 A GTOs, each with its associated antiparallel diode, and snubber components. They are operated in each pole at 60 Hz switching rate and the phase of the switching is strategically controlled from one pole to the next to facilitate harmonic elimination. The pole ac outputs are summed via the intermediate transformer at the secondary windings of the main coupling transformer. The voltage appearing across each winding is a nearly sinusoidal 48-pulse waveform, as illustrated in Figure 10.32. The total MVA rating of the intermediate transformer is approximately 50% of the main transformer rating. Figure 10.33 shows a sketch of the layout for the major UPFC components, Figure 10.34 a view of the actual converter valve hall, and Figure 10.35 shows an overall aerial view of the UPFC installation at the Inez Substation. The base area of the UPFC building, designed to accommodate additional auxiliary components for future system needs, is 200 × 100 ft or 61 × 30.5 m.

10.4.2.2 Control System. Both converters comprising the UPFC are controlled from a single central control system housed in three cabinets in the control room.

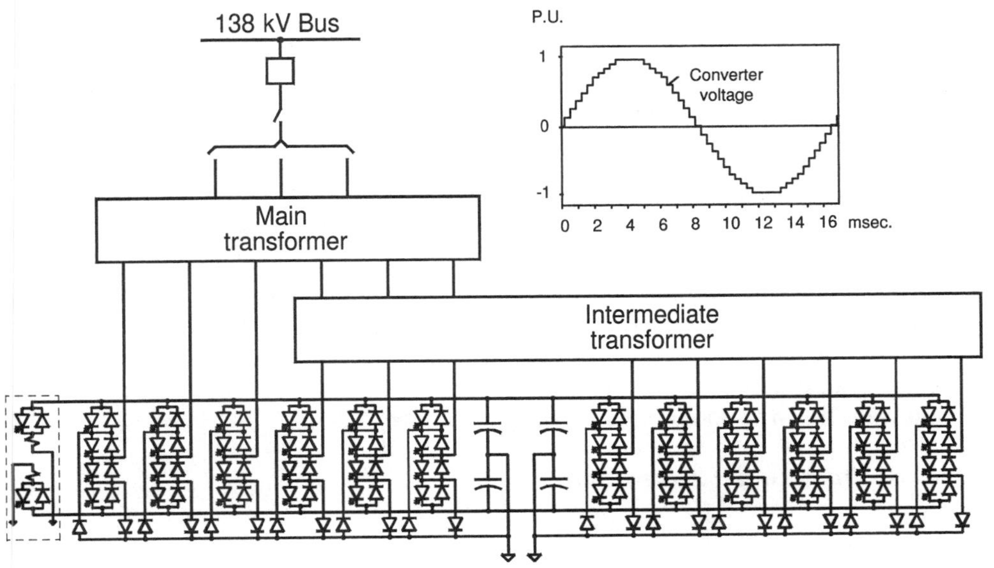

Figure 10.32 Simplified power circuit schematic of the 48-pulse converter used in the Inez UPFC and corresponding output voltage waveform.

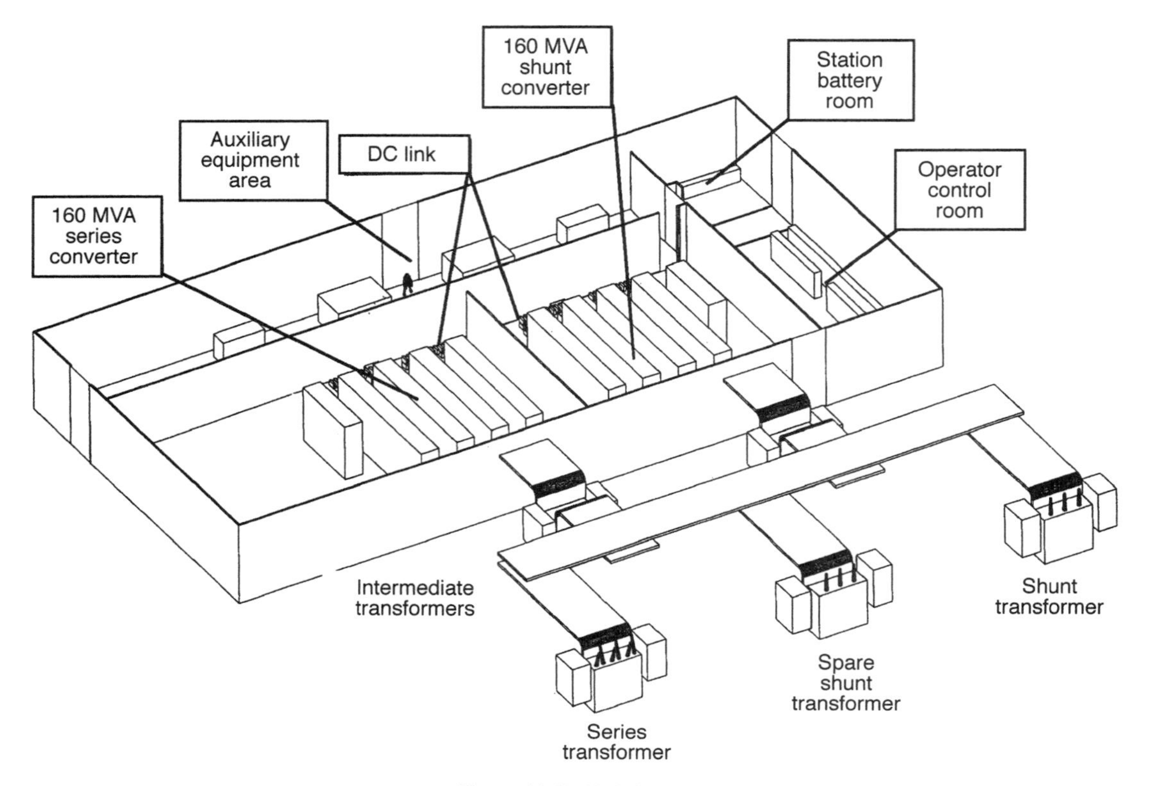

Figure 10.33 UPFC layout.

Two of the cabinets house the relay interface and signal conditioning, while a single cabinet contains the control electronics. The conceptual structure of the control system is shown in Figure 10.36.

The actual control algorithms that govern the instantaneous operation of the two converters are performed in the real-time control electronics which employs multiple digital signal processors. The real-time control communicates with the pole electronics mounted on each pole via the valve interface that is linked to the poles by fiber-optic cables. The status processor is connected to every part of the system, including the cooling system and all of the poles, by serial communications. During run time it continually monitors the operation of all subsystems, collecting and analyzing status information. It is responsible for all start-up and shutdown sequences and for the organizing and annunciation of all alarm conditions. The status processor is serially connected to a graphical display terminal which provides the local operator interface. A hierarchical arrangement of graphical display screens gives the operator access to all system settings and parameters, and provides extensive diagnostic information right down to the individual GTO modules.

10.4.3 Operating Performance

In the course of commissioning the UPFC, tests were performed to verify its predicted capability. For this purpose, the UPFC was directed to produce large controlled swings of real and reactive power on the Big Sandy line, and sizeable swings

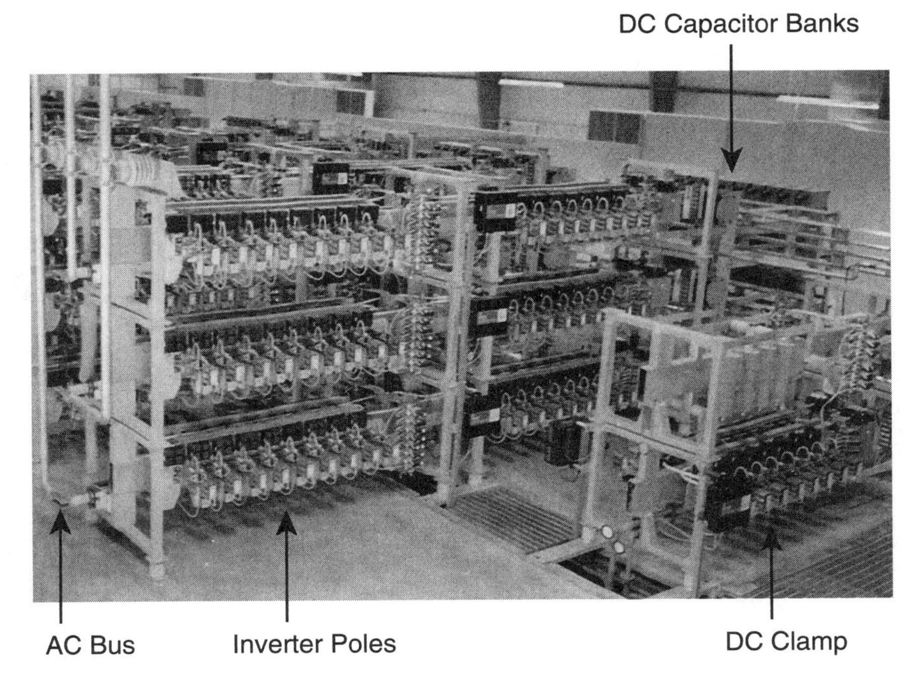

Figure 10.34 UPFC converter hall. (Courtesy of Siemens Power Transmission & Distribution, Orlando, FL.)

of voltage at the Inez station, while measurements were recorded. It is emphasized that these swings do not in any sense represent "normal" duty for the UPFC in this application. Arbitrary variations of this kind can be disruptive to the power system because they force real power flows to be redistributed in the network and affect voltage regulation at other stations. AEP system operators defined acceptable boundary limits for the tests and, in addition, slow ramping functions were applied to the control references for the UPFC automatic power flow controller.

Five representative cases have been selected for the purpose of this presentation. The first three cases show the UPFC independently controlling line P, line Q, and Inez bus voltage, respectively. The fourth case is for the UPFC maintaining unity power factor on the line, and the final case is a demonstration of the series converter operating as an SSSC. In all cases (except the SSSC), the UPFC is operating with the shunt converter in automatic voltage control mode and the series converter in automatic line power flow control mode. Each set of results is annotated using the sign convention as defined in Figure 10.37. Note in particular that P and Q for the line are measured at the line-side terminals of the series insertion transformer. This is the actual power at the end of the line and is defined as positive toward Big Sandy. The real and reactive power for each of the two converters is also shown. Note that the converters independently generate reactive power but that their real power is substantially equal and opposite. In each case a few stationary points (between major transitions) have been chosen and a phasor diagram drawn to represent the operating condition at that point. This is considered to be helpful because it is difficult to interpret each situation from the time plots alone. The interpretation is made even more difficult by the fact that the power network "adjusts" following each major transition, especially

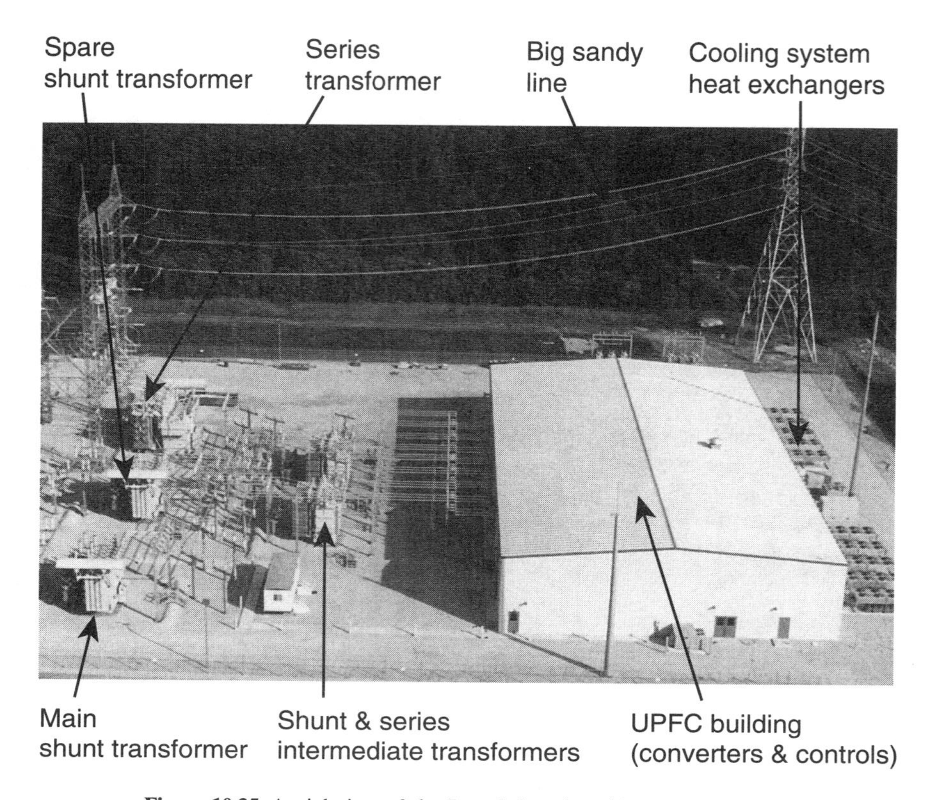

Figure 10.35 Aerial view of the Inez Substation. (Courtesy of AEP.)

with regard to the voltage phase angles at Inez and Big Sandy. The following five sections will take the form of a specific commentary for each case.

Case 1: UPFC Changing Real Power (P). Refer to Figure 10.38. This case starts with the UPFC idling near zero-injected voltage and the real power flow on the line near the "natural" level of 150 MW from Big Sandy. The shunt converter is regulating the Inez bus to 1.0 p.u. by generating about 60 Mvar capacitive, and about 36 Mvar are being delivered into the line. The objective for this case is to maintain the Inez bus voltage and the line Q unaltered while making big step changes in line P.

The UPFC is first commanded to raise the line power to 240 MW. It does this by injecting a voltage of about 0.16 p.u. roughly in quadrature (lagging) with the Inez bus voltage. To satisfy the required conditions, the shunt converter drops its capacitive output to about 20 Mvar, and the series converter delivers about 40 Mvar capacitive to the line. Real power exchange between the converters is about 8 MW.

The second transition commanded is a 170 MW drop in the line power to 70 MW. This is accomplished with little change in the magnitude of the injected voltage, but about 180° phase shift, so that the injected voltage is now still roughly in quadrature with the Inez bus voltage, but leading. The shunt converter produces about 85 Mvars capacitive, the series converter reverses its output to 10 Mvar inductive, and about 8 MW flows between the converters through the dc bus. The final transition returns the system to the initial operating point.

Throughout this case the Inez bus voltage is tightly regulated at 1.0 p.u. and Q

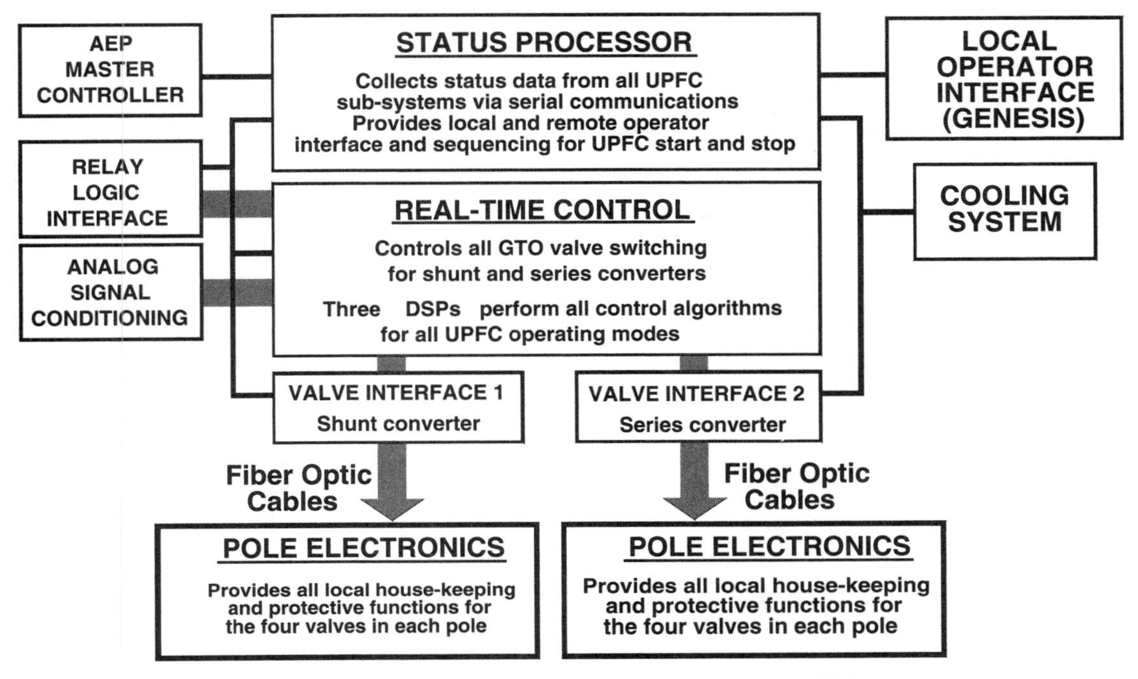

Figure 10.36 Conceptual control structure of the Inez UPFC.

on the line stays constant. Note, however, that the voltage, V_2 applied to the transmission line, is lowered by a few percent relative to V_1 to achieve the first swing and raised by a few percent for the second. It should also be noted that the large changes in real power arriving at Inez must, of course, be balanced by an equal and opposite total change in the power on the other lines leaving the station. For this to happen a change in the phase angle of the Inez bus voltage, V_1, is unavoidable. At the time when these tests were performed the natural power flow on the line was too high for the UPFC to demonstrate its unique ability to reverse power flow. On other occasions,

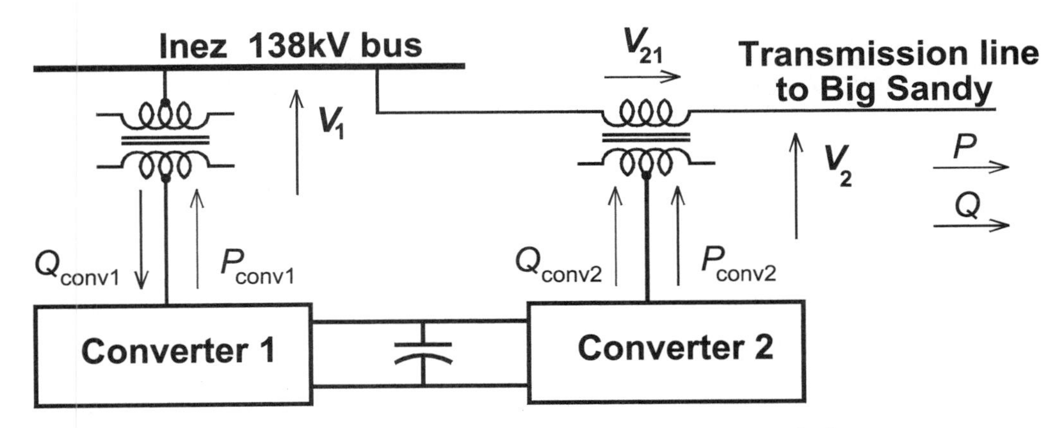

Figure 10.37 Definition of polarity conventions used in the test results for power and voltage measurements.

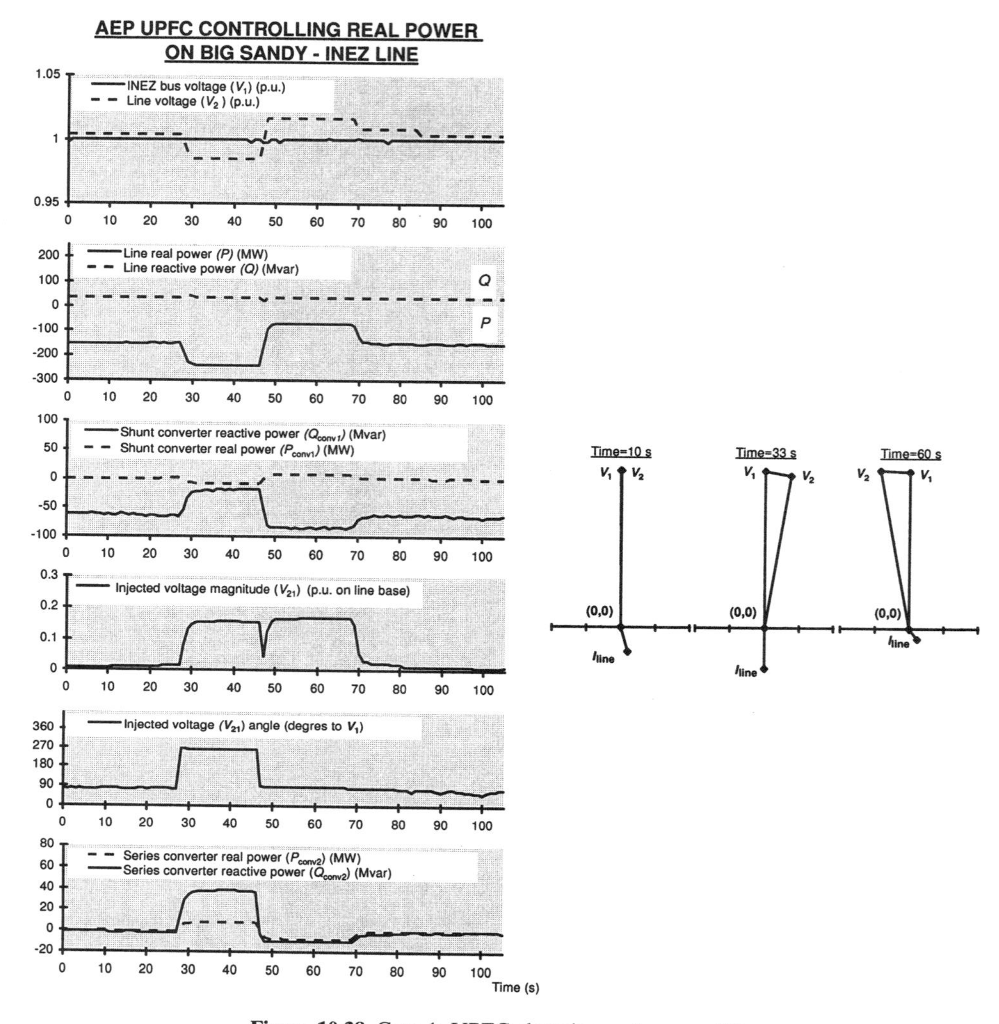

Figure 10.38 Case 1: UPFC changing real power (P).

however, when the line has been lightly loaded, this has been demonstrated successfully and the UPFC has driven real power back toward Big Sandy.

Case 2: UPFC Changing Reactive Power (Q). Refer to Figure 10.39. The initial conditions for this test are similar to Case 1 but the objective is to regulate the Inez voltage at 1.0 p.u. and keep the line real power, P, constant while causing large steps in the line reactive power, Q.

For the first swing, the UPFC reference for Q is changed from +30 Mvar to −30 Mvar. After the change, 30 Mvar is being received from the line, compared with the 30 Mvar delivered *to* the line initially. The UPFC forces the change by injecting about 0.05 p.u. voltage roughly in antiphase with $\mathbf{V}_1$. The line voltage, $\mathbf{V}_2$, is consequently reduced in magnitude by about 5%. For the second step, the Q reference is taken to +100 Mvar (i.e., 100 Mvar to the line). This time the injected voltage is in phase with V_1 so that V_2 is increased by about 5%. The final step reduces the Q

reference to zero. The line is now fed at unity power factor with V_2 reduced by about 2.5% relative to V_1.

It is interesting to note that the changes in Q at the line terminals are balanced almost entirely by equal and opposite changes in the reactive output of the shunt converter, which acts to maintain the Inez voltage. This brings to light a fascinating capability of the UPFC. In essence, it can "manufacture" inductive or capacitive Mvars using the shunt converter and "export" this reactive power into a particular transmission line (i.e., the one with the series insertion transformer), without changing the local bus voltage and without changing the reactive power on any of the other lines leaving the substation. It can therefore regulate the station bus voltages at both the sending and receiving ends of a transmission line, while still freely controlling the real power flow, P, on the line.

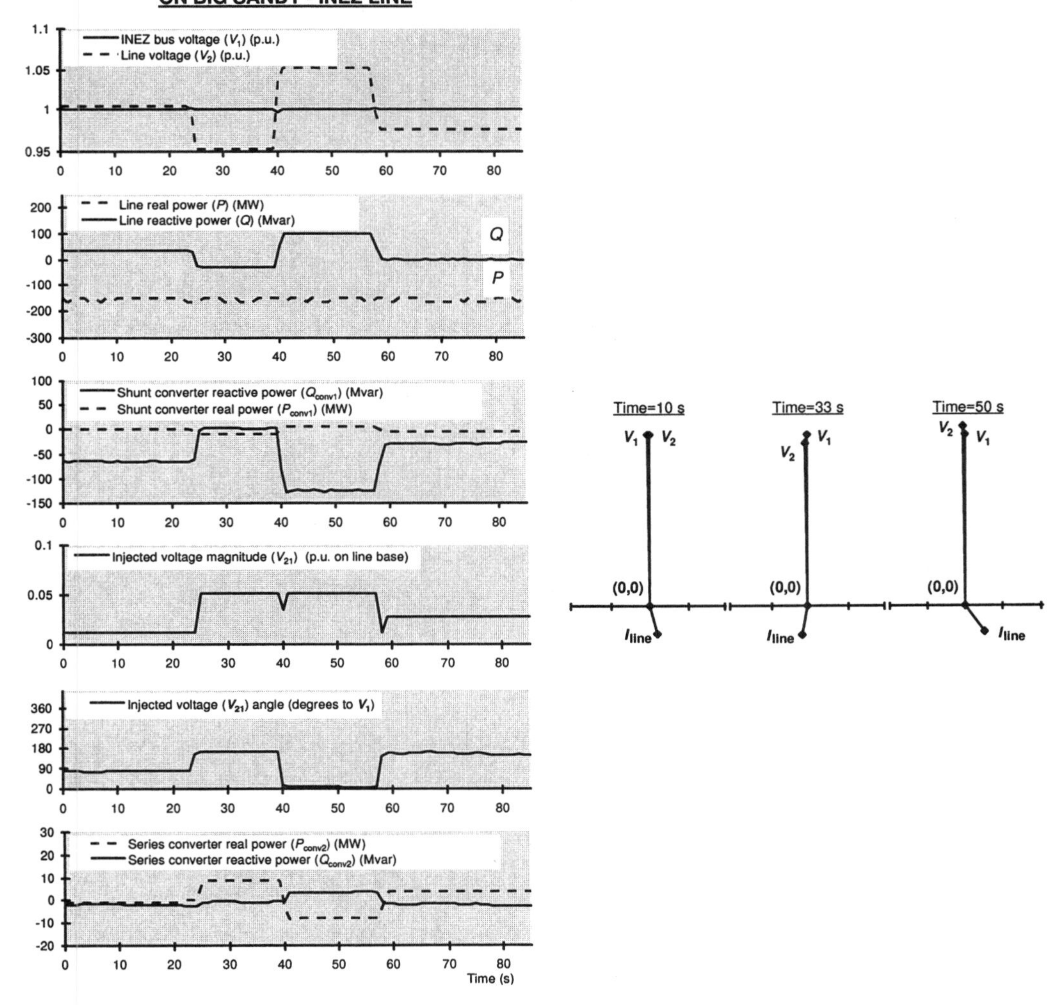

Figure 10.39 Case 2: UPFC changing reactive power (Q).

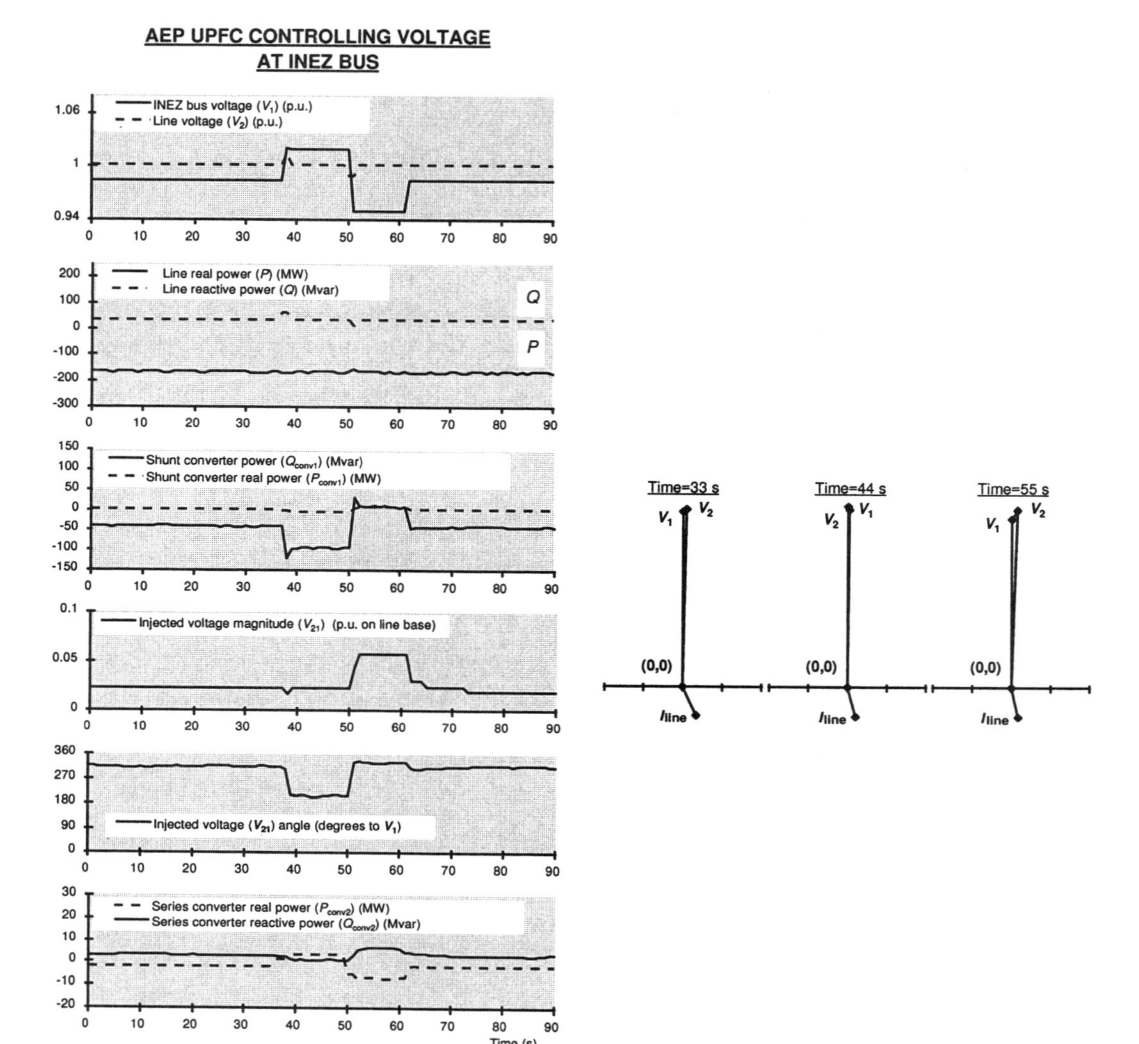

Figure 10.40 Case 3: UPFC changing Inez bus voltage.

Case 3: UPFC Changing Local Bus Voltage. Refer to Figure 10.40. The objective for this case is to produce a large voltage change at Inez on command, while maintaining an unaltered level of P and Q on the line. The voltage reference is stepped from an initial value of 0.985 p.u., to 1.02 p.u., to 0.95 p.u., and back to 0.985 p.u. The UPFC successfully holds the line P and Q constant (using very small changes in injected voltage), while the shunt converter goes from its initial output of 40 Mvar capacitive, to 100 Mvar capacitive, to 0 Mvar, and back to 40 Mvar capacitive.

In a sense, this case is the opposite of the previous one. In the previous case, the Inez bus and the other transmission lines were "insulated" from changes in the reactive loading of the Big Sandy line. In this case, the line is insulated from the large voltage swings at Inez, while the other lines leaving the station all experience changes in reactive loading.

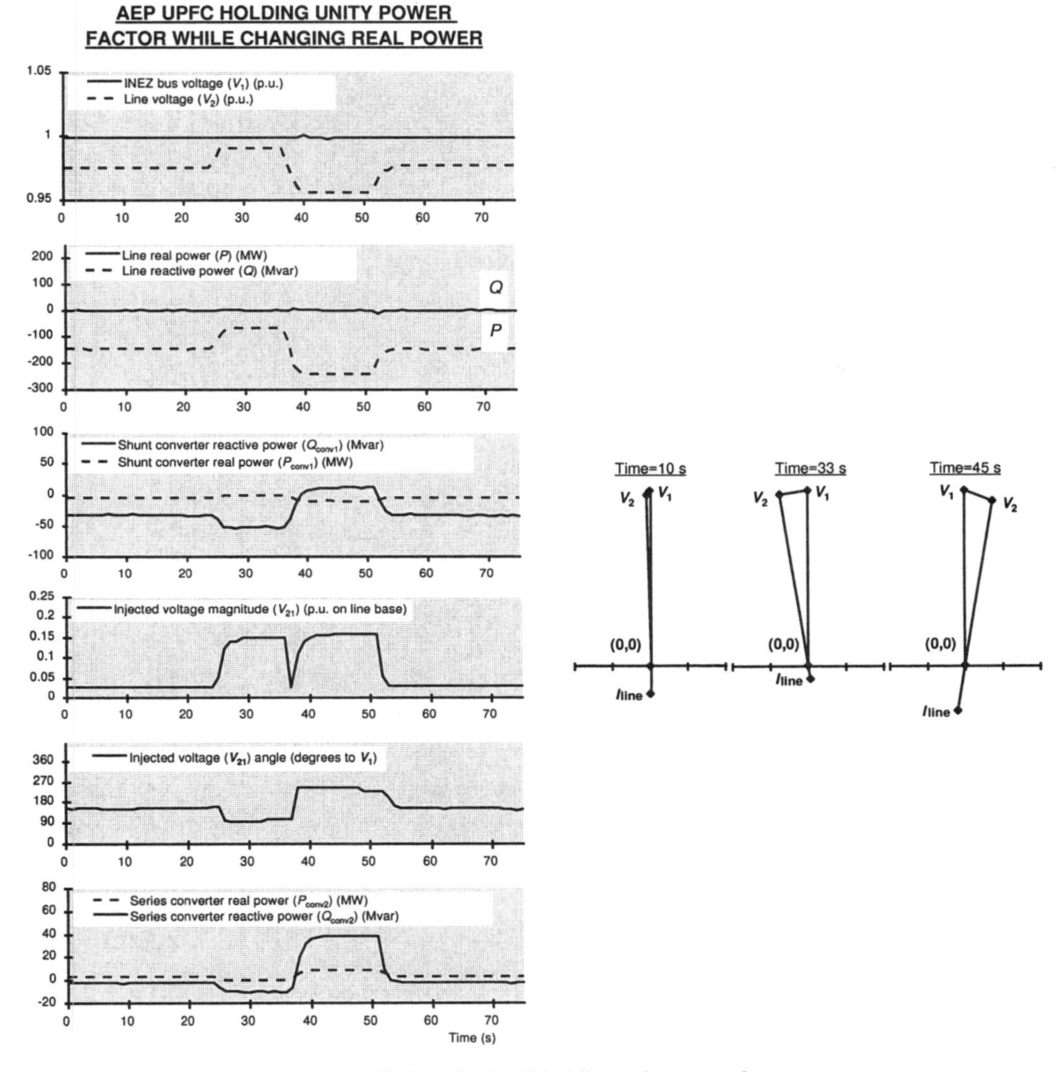

Figure 10.41 Case 4: UPFC holding unity power factor.

Case 4: UPFC Holding Unity Power Factor. Refer to Figure 10.41. The objective for this case is to maintain unity power factor looking into the transmission line, while producing large swings in real power, P, on the line, and also maintaining the Inez bus at 1.0 p.u. This case is really a combination of Cases 1 and 2. It is of particular interest because driving a line at unity power factor should, in principle, make it possible to deliver the largest amount of real power into the line for the lowest current. This should result in the most efficient use of the line from a thermal point of view. Naturally, the reactive power consumed by the line itself must now be supplied at the other end of the line, resulting in higher voltages at that end.

Case 5: Series Converter Operating in SSSC Mode. Refer to Figure 10.42. For this case, the shunt converter is disconnected from the dc terminals of the series

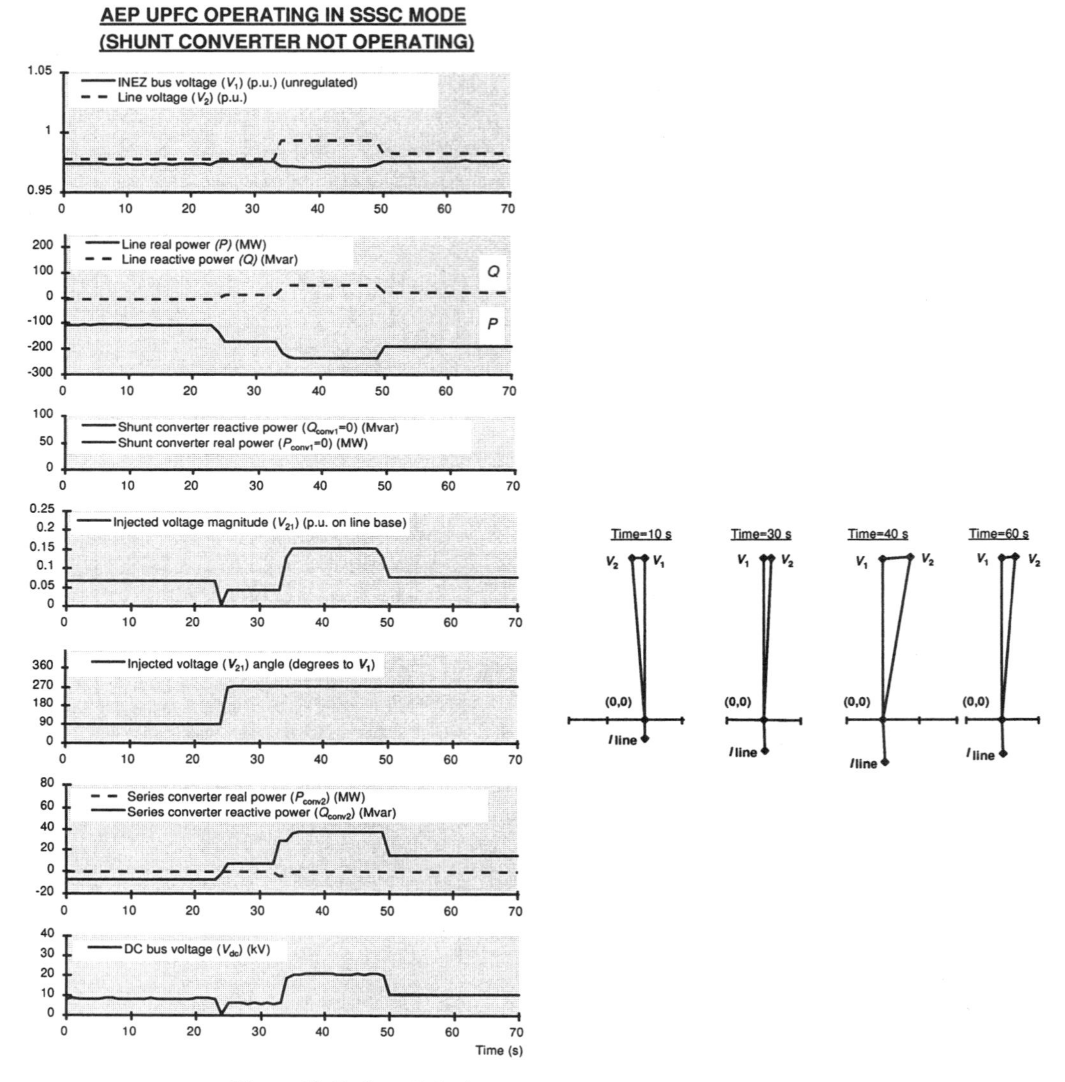

Figure 10.42 Case 5: Series converter operating in SSSC mode.

converter and is completely out of service. Consequently the Inez bus voltage is not regulated. The series converter injects voltage into the line essentially in quadrature with the prevailing line current. The injection angle deviates from true quadrature to draw real power from the line for converter losses and to charge and discharge the dc bus capacitor banks. By means of the quadrature voltage injection, the SSSC is able to raise or lower the line current, but cannot independently alter P and Q. In principle, the SSSC can reverse the direction of power flow on a line, but the control becomes difficult because the transition of the line current through zero, at which point real power cannot be drawn from the line. SSSC operation is an important subset of full UPFC operation, since it can be used when the shunt converter is not available. A single series-connected converter installation may be the most cost-effective solution for applications where a simpler form of power flow control is sufficient.

The objective of this case is simply to show the SSSC raising and lowering the power on the Big Sandy line. The SSSC is operated to control the magnitude and the polarity of the injected voltage (i.e., the line power is not automatically controlled). Voltage injections are selected to give a sequence of approximately 100 MW, 180 MW, 250 MW, and finally 200 MW on the line. Note the corresponding changes in Q. From the phasor diagrams it can be clearly seen how $\boldsymbol{I}_{\text{line}}$ maintains a constant phase relationship to V_1 and is always in quadrature with $\boldsymbol{V}_{21}$. The natural flow on the line is between 100 MW and 180 MW. Consequently the polarity of $\boldsymbol{V}_{21}$ is reversed from lagging $\boldsymbol{I}_{\text{line}}$ to leading in this transition. The polarity reversal is accomplished by taking the converter dc bus voltage to zero, then raising it again with 180 degree phase shift in the converter output voltage.

10.4.4 Results of the Project

Theoretical analyses and simulations predicted unique, hitherto unavailable functional characteristics and operating performance. The application of the concept in a major utility reinforcement plan demonstrated the recognition of the major practical advantages offered by this technology. This first of a kind AEP UPFC installation at Inez has fully proven that, as predicted, the UPFC:

- Has the unique capability to provide independent and concurrent control for the real and reactive line power flow, as well as for the regulation of the bus voltage.
- Has a flexible circuit structure to be reconfigured for independent shunt (STATCOM) and series (SSSC) compensation, as well as for only shunt or only series compensation at double rating.
- Has a rugged and reliable GTO-based converter structure that is capable of operating properly in utility environment.

The successful UPFC installation at Inez has also proven the concept of operating two or more voltage-sourced converters from a common dc bus, enabling the converters to exchange real power among them and thus being capable of providing two-dimensional compensation by injecting real as well as reactive power into the transmission line. This paved the way to the generalization of this concept to multiline transmission control and its practical implementation, as the Interline Power Flow Controller, in the Convertible Static Compensator project sponsored jointly by EPRI and the New York Power Authority (NYPA) with Siemens Power Transmission and Distribution, Orlando, Florida, for NYPA's Marcy substation.

REFERENCES

Maliszewski, R. M., and Rahman, M., "Planning and Operating Considerations for the World's First UPFC Installation," *CIGRE Report No. 37-96 (US) 13 (E),* Paris, 1996.

Mehraban, A. S., et al., "Application of the World's First UPFC on the AEP System," *EPRI Conference—The Future of Power Delivery,* Washington DC, April 9–11, 1996.

Rahman, M., et al., "UPFC Application on the AEP System: Planning Considerations," *IEEE Trans. on Power Systems,* vol. 12, no. 4, November 1997.

Renz, B. A., et al., "AEP Unified Power Flow Controller Performance," *IEEE/PES Winter Meeting, Paper No. PE-042-PWRD-0-12,* 1998.

Schauder, C. D., et al., "AEP UPFC Project: Installation, Commissioning, and Operation of the ±160 MVA STATCOM (Phase 1)," *IEEE/PES Winter Meeting,* Paper *No. PE-515-PWRD-0-12,* 1997.

Index

K

L

M

N

P

Q

R

S

About the Authors

Narain G. Hingorani received the B.S. degree in electrical engineering from Baroda University in India and the M.Sc., Ph.D., and Honorary Doctor of Science degree from the University of Manchester Institute of Science and Technology in England.

Dr. Hingorani is credited with originating the concepts of Flexible AC Transmission System (FACTS) and Custom Power, which are expected to revolutionize future ac power transmission and distribution systems, respectively. His many achievements were highlighted in an individual profile in the December 1990 issue of *IEEE Spectrum*.

In 1995 following a twenty-year career at EPRI in transmission and distribution areas, Dr. Hingorani retired from the position of vice president of Electrical Systems. He then became a consultant to help utilities plan and engineer studies and purchase power electronics technology, particularly HVDC, FACTS, and Custom Power. Prior to joining EPRI, Dr. Hingorani spent six years with the Bonneville Power Administration. His responsibilities included the commissioning of the Pacific DC Intertie and series capacitor compensation, and EHV ac transmission. Since 1972 he has helped many utilities in the purchase and commissioning of HVDC technologies, and helped industry in their business strategies and the U.S. Office of Naval Research on their R&D strategies in power electronics.

In 1988, Dr. Hingorani was elected to the National Academy of Engineering of the USA. From 1988 to 1996, he was chairman of CIGRE Study Committee 14. He is a Life Fellow of the Institute of Electrical and Electronic Engineers (IEEE), was a member of the Editorial Board of *IEEE Spectrum* from 1993 to 1995, a member of the Engineering Review Board for the Bonneville Power Administration from 1988 to 1994, and a member of the Editorial Board of *IEEE Proceedings* from 1991 to 1993. He is presently a member of the IEEE Foundation Board. In 1985 Dr. Hingorani was presented with the Uno Lamm Medal by the IEEE Power Engineering Society for outstanding contributions in High Voltage Direct Current Technology, and in 1995 received the prestigious IEEE Lamme Gold Medal for leadership and pioneering contributions to the transmission and distribution of electric power. He has authored over 150 papers and articles on HVDC and ac transmission, and coauthored a book on HVDC power transmission.

Laszlo Gyugyi received his basic technical education at the University of Technology, Budapest, Hungary. He further studied mathematics at the University of London, England, and electrical engineering at the University of Pittsburgh, where in 1967 he obtained the M.S.E.E. and at the University of Salford, England, where in 1970 he received the Ph.D.

In 1963 Dr. Gyugyi joined the Westinghouse Science & Technology Center, where he had a number of engineering and technical management positions of increasing responsibility. During the last 15 years, he has been responsible for the corporation's central research and development work in power electronics for utility, industrial, and defense applications. In 1995 he was appointed technical director for the corporation's power electronics activity, a position he presently holds at Siemens Power Transmission and Distribution.

In 1992 Dr. Gyugyi received the Westinghouse Order of Merit (the corporation's highest honor) for his creativity and achievements in power electronics, and for his personal efforts and leadership in establishing advanced power electronics-based technology for the compensation and control of utility transmission and distribution systems. In 1994 he received the William E. Newell Power Electronics Award, which the IEEE Power Electronics Society presents annually for outstanding achievement in power electronics. In 1999 Dr. Gyugyi received the first FACTS Award from the IEEE Power Engineering Society for his "outstanding contributions and leadership in advancing power electronics technology for ac transmission systems."

Dr. Gyugyi is a Fellow of the Institution of Electrical Engineers (IEE) and a member of various IEEE and CIGRE Working Groups in the area of power transmission system compensation and control. He also participates in the Professional Development Programs of the University of Wisconsin, giving lectures on the application of power electronics in utility and industrial systems. Dr. Gyugyi has carried out research and development in many areas of power electronics and control. He, in collaboration with B. R. Pelly, established the theoretical foundations of ac to ac switching converters in *Static Power Frequency Changers* (Wiley 1976; Energoatomizdat 1980). Subsequently, he has increasingly focused on the development of new power electronic technologies for power transmission and distribution, and pioneered the converter-based approach for FACTS. Dr. Gyugyi has contributed to engineering handbooks and authored more than 50 societal publications, including several invited and prize award papers. He holds 76 U.S. patents.